INDUSTRIAL MICROBIOLOGY

Current Progress and Novel Technologies

INDUSTRIAL MICROBIOLOGY

Current Progress and Novel Technologies

Prakash Kumar Sarangi, PhD
Akhilesh Kumar Singh, PhD
Rajesh Kumar Srivastava, PhD
Brijesh Pandey, PhD
Editors

AAP | APPLE ACADEMIC PRESS

First edition published 2027

Apple Academic Press Inc.
1265 Goldenrod Circle, NE,
Palm Bay, FL 32905 USA

760 Laurentian Drive, Unit 19,
Burlington, ON L7N 0A4, Canada

CRC Press
2385 NW Executive Center Drive,
Suite 320, Boca Raton FL 33431

4 Park Square, Milton Park,
Abingdon, Oxon, OX14 4RN, UK

Apple Academic Press exclusively co-publishes with CRC Press, an imprint of Taylor & Francis Group, LLC

Library and Archives Canada Cataloguing in Publication

CIP data on file with Canada Library and Archives

Library of Congress Cataloging-in-Publication Data

CIP data on file with US Library of Congress

ISBN: 978-1-77964-220-2 (hbk)
ISBN: 978-1-77964-221-9 (ebk)

DOI: 10.1201/9781779642219

About the Editors

Prakash Kumar Sarangi, PhD
Scientist (Microbiology), Central Agricultural University, Imphal, Manipur, India

Prakash Kumar Sarangi, PhD, is a Scientist with specialization in Microbiology at the Central Agricultural University in Imphal, India. Dr. Sarangi's research is focused on bioprocess engineering, biochemicals, biomaterials, and sustainable development. His expertise is in bioconversion of biomass into biofuels, biochemicals, and nutraceuticals, as well as environmental waste remediation. He has taken leading roles as the principal investigator in different R&D projects on biomass conversion and microbial biodegradation into value-added products. He has more than 15 years of teaching and research experience in biochemical engineering, microbial biotechnology, downstream processing, food microbiology, and molecular biology. He is serving as an editorial board member for many international journals, including *PLOS One, 3 Biotech,* and *Biofuels, Bioproducts and Biorefining*. He is also serving as guest editor of the journals *Biomass Conversion and Biorefinery, Chemical Engineering and Technology,* and *Biofuels, Bioproducts and Biorefining*. He has published more than 130 research articles in peer-reviewed journals and authored more than 75 book chapters. He has presented at many national and international conferences. He is associated with many scientific societies as a fellow member (Society for Applied Biotechnology) and life member (Biotech Research Society of India, Society for Biotechnologists of India, Association of Microbiologists of India, Indian Science Congress Association, Forum of Scientists, Engineers and Technologists, and International Association of Academicians and Researchers). He has published many books on the field of biomass conversion and value addition. Dr. Sarangi has a PhD degree in Microbial Biotechnology from the Ravenshaw University, India, and MTech degree from the Department of Agriculture and Food Engineering, Indian Institute of Technology Kharagpur, India.

Akhilesh Kumar Singh, PhD
Assistant Professor, Department of Biotechnology, Mahatma Gandhi Central University (MGCU), Motihari, Bihar, India

Dr. Akhilesh Kumar Singh is currently an Assistant Professor in the Department of Biotechnology, Mahatma Gandhi Central University (MGCU), Motihari, Bihar, India. Before joining MGCU, Dr. Singh worked as a faculty at the Amity Institute of Biotechnology, Amity University Uttar Pradesh, Lucknow Campus, Lucknow, India (September 2008 to October 2019). He has over 16 years of experience in teaching and multidisciplinary research, focusing on microbial physiology, industrial biotechnology, and nanobiotechnology. His research focuses on the production of polyhydroxyalkanoate bioplastics, metallic nanobiotechnology, and nitric oxide signaling. Dr. Singh holds a PhD in agricultural and food engineering and an MTech in agricultural biotechnology from the prestigious Indian Institute of Technology (IIT), Kharagpur, India, with a master's degree in biotechnology from another prestigious institute, i.e., IIT, Roorkee, India. He has qualified Graduate Aptitude Test in Engineering (GATE) in Life Science conducted by IIT Kanpur on behalf of the Ministry of Human Resource Development (MHRD), Government of India. He was awarded the prestigious Council of Scientific and Industrial Research (CSIR)-University Grants Commission (UGC) Junior Research Fellowship (CSIR-UGC JRF) by the Council of Scientific and Industrial Research, India. In addition, Dr. Singh has also qualified the Indian Council of Agriculture Research (ICAR)-National Eligibility Test (NET), i.e., ICAR-NET in Agricultural Biotechnology, conducted by ICAR, India. He has been credited with four patents, 35 research papers, 35 review articles in high-repute international journals, and 60 chapters in reputed books. He has also edited the book titled *Nanobiotechnology: Mitigation of Abiotic Stress in Plants*, published by Springer. Dr. Singh has been on the editorial board as well as the panels of reviewers of several reputed journals of Elsevier, Springer Nature, John Wiley & Sons, and so on. Dr. Singh has also supervised several PhD, MTech, MSc, and BTech students.

Rajesh Kumar Srivastava, PhD
Senior Assistant Professor,
Department of Biotechnology, School of Core Engineering, GITAM (Deemed to be University), Visakhapatnam, Andhra Pradesh, India

Rajesh Kumar Srivastava, PhD, is a Senior Assistant Professor in the Department of Biotechnology, School of Core Engineering, GITAM (Deemed to be University), Visakhapatnam, Andhra Pradesh, India. He focuses his research on applied microbiology, industrial biotechnology, and fermentative production synthesis. He has published 48 papers in international and national journals, 13 book chapters in international books, and five book chapters in national books, with several papers and book chapters currently in progress. He has also published two national-level books. He is a reviewer for the *Journal of Applied Microbiology, Archives of Microbiology*, *PLOS One*, and *Biomass Conversion and Biorefinery*. He received his PhD degree in fermentation technology from IIT Bombay, Mumbai, in 2009. Before this degree, he completed his MSc in Microbiology and MTech in Biochemical Engineering (from IIT BHU, Varanasi, UP). After earning his PhD degree, he had a postdoctoral fellowship from the University of Amsterdam, Netherlands.

Brijesh Pandey, PhD
Professor, Department of Biotech, School of Life Sciences, Mahatma Gandhi Central University, Motihari, Bihar, India

Brijesh Pandey, PhD, specializes in secondary metabolism and genetic engineering for regulatory manipulation of secondary metabolic pathways. He is a Professor in the Department of Biotechnology, School of Life Sciences at Mahatma Gandhi Central University, Motihari, Bihar, India. His research focuses on metabolic regulation and assessment of the therapeutic and industrial importance of secondary metabolites. He has more than three dozen international publications to his credit. He is a life member of the Indian

Science Congress and the Association, Society of Biological Chemists (I) and a founding member of the Florafauna Science Foundation, aimed at enabling entrepreneurial villages. He has more than 15 years of teaching and research experience at the university level. He received his PhD degree in Biochemistry from the University of Lucknow in India.

Contents

Contributors

Isra Aman Aziz
Amity Institute of Biotechnology, Amity University Uttar Pradesh, Lucknow Campus, Lucknow, Uttar Pradesh, India

Shreetama Bhattacharjee
Amity Institute of Biotechnology, Amity University Uttar Pradesh, Lucknow Campus, Lucknow, Uttar Pradesh, India

Nishat Fatima
Department of Zoology, Vermibiotechnology Laboratory, D.D.U. Gorakhpur University, Gorakhpur, Uttar Pradesh, India

Pawan Kumar
Department of Biotechnology, Mahatma Gandhi Central University, Motihari, Bihar, India

Vishwajit Kumar
Department of Microbiology, Kasturba Medical College, Manipal Academy of Higher Education, Manipal, India

Babli Kumari
Department of Biotechnology, School of Life Sciences, Mahatma Gandhi Central University, Motihari, Bihar, India

Piyush Behari Lal
Department of Microbiology, Kasturba medical College, Manipal Academy of Higher Education, Manipal, India

Manish Mishra
Department of Biomedical Sciences, Mercer University School of Medicine, Macon, USA

Priti Pal
Amity Institute of Biotechnology, Amity University Uttar Pradesh, Lucknow Campus, Lucknow, Uttar Pradesh, India

Brijesh Pandey
Department of Biotechnology, School of Life Sciences, Mahatma Gandhi Central University, Motihari, Bihar, India

Anand Prakash
Department of Biotechnology, School of Life Sciences, Mahatma Gandhi Central University, Motihari, Bihar, India

D. V. Surya Prakash
Department of Biotechnology, Meerut Institute of Engineering and Technology, Meerut, Uttar Pradesh, India

Om Prakash
Department of Biotechnology, School of Life Sciences, Mahatma Gandhi Central University, Motihari, Bihar, India

Ram Prasad
Department of Botany, School of Life Sciences, Mahatma Gandhi Central University, Motihari, Bihar, India

Saurabh Singh Rathore
Department of Life Sciences, Inter University Center for Teacher Education, Bananas Hindu University, Varanasi, Uttar Pradesh, India; Department of Biotechnology, School of Life Sciences, Mahatma Gandhi Central University, Motihari, Bihar, India

Prakash Kumar Sarangi
College of Agriculture, Central Agricultural University, Imphal, Manipur, India

Laxuman Sharma
Department of Horticulture, Sikkim University, Sikkim, India

Aditi Singh
Amity Institute of Biotechnology, Amity University Uttar Pradesh, Lucknow Campus, Lucknow, Uttar Pradesh, India

Akhilesh Kumar Singh
Department of Biotechnology, School of Life Sciences, Mahatma Gandhi Central University, Motihari, Bihar, India

Anjali Singh
Department of Zoology, Vermibiotechnology Laboratory, D.D.U. Gorakhpur University, Gorakhpur, Uttar Pradesh, India

Keshav Singh
Department of Zoology, Vermibiotechnology Laboratory, D.D.U. Gorakhpur University, Gorakhpur, Uttar Pradesh, India

Maharudra Pratap Singh
Department of Biotechnology, School of Life Sciences, Mahatma Gandhi Central University, Motihari, Bihar, India

Pankaj Kumar Singh
Department of Zoology, Vermibiotechnology Laboratory, D.D.U. Gorakhpur University, Gorakhpur, Uttar Pradesh, India

Yashvardhan Singh
Amity Institute of Biotechnology, Amity University Uttar Pradesh, Lucknow Campus, Lucknow, Uttar Pradesh, India

Sandeep Sirohi
Department of Botany, Hariom Saraswati P.G. College, Dhanauri, Haridwar (affiliated to Sri Dev Suman Uttarakhand University, Badshahithaul, Tehri Garhwal), Uttarakhand, India

Rajesh Kumar Srivastava
Senior Assistant Professor, Department of Biotechnology, School of Core Engineering, GITAM (Deemed to be University), Visakhapatnam, Andhra Pradesh, India

Mala Trivedi
Amity Institute of Biotechnology, Amity University Uttar Pradesh, Lucknow Campus, Lucknow, Uttar Pradesh, India

Nitika Vats
Department of Biotechnology, Meerut Institute of Engineering and Technology, Meerut, Uttar Pradesh, India

Ritu Vishnoi
Department of Botany, Hariom Saraswati P.G. College, Dhanauri, Haridwar (affiliated to Sri Dev Suman Uttarakhand University, Badshahithaul, Tehri Garhwal), Uttarakhand, India

Abbreviations

AAB	acetic acid bacteria
ABE	acetone–butanol–ethanol
ACP	acyl carrier protein
ALE	adaptive laboratory evolution
ATP	adenosine triphosphate
ADH	alcohol dehydrogenase
AD	anaerobic digester
ACE	angiotensin-converting enzyme
AS	aqueous suspension
ARTP	atmospheric and room-temperature plasma
BC	bacterial cellulose
BV	bacterial vaginosis
BAP	bioactive peptides
CS	capsule suspension
CCUS	carbon capture, utilization, and storage
CRISPR	clustered regularly interspaced short palindromic repeats
CCM	carbon concentration mechanism
CLA	conjugated linoleic acid
CNF	cooked non-fermented
DM	diabetes mellitus
DHA	docosahexaenoic acid
DPA	docosapentaenoic acid
EPA	eicosapentaenoic acid
EMP	Embden–Meyerhof–Parnas pathway
ELM	emulsion liquid membranes
EROI	energy return on investment
EPN	entomopathogenic nematode
US EPA	Environmental Protection Agency
ELISA	enzyme-linked immunosorbent assay
EPS	extracellular polymeric substance
FFDCA Act	Federal Food, Drug, and Cosmetic Act
FAME	fatty acid methyl ester
NHER	fixes the strand break
FBA	flux balance analysis
FDA	Food and Drug Administration
FFA	free fatty acids

GABA	gamma-aminobutyric acid
GRAS	generally recognized as safe
GMO	genetically modified organism
GDM	gestational diabetes mellitus
GEILM	green emulsion ionic liquid membrane
HMP	hexose monophosphates
HR	homologous recombination
HA	hydroxy alkanoic acids
IBD	inflammatory bowel disease
IPM	integrated pest management
IDF	International Diabetes Federation
IEA	International Energy Agency
KDC	ketoacid decarboxylase
LAB	lactic acid bacteria
LHC	light-harvesting antenna protein
LCL-PHA	long-chain-length polyhydroxyalkanoate
LDL	low-density lipoprotein
LDPE	low-density polyethylene
MGP	matrix-Gla protein
MODY	maturity-onset diabetes of the young
MCL	medium chain length
MEK	methyl ethyl ketone
MEC	microbial electrolysis cell
MFC	microbial fuel cell
MSP	minimum selling price
NADPH	nicotinamide adenine dinucleotide phosphate
NHR	non-homologous recombination
OD	oil dispersion
OTase	oligosaccharyltransferase
PQOR	plastoquinone oxidoreductase
PGA	polyglutamic acid
PHA	polyhydroxyalkanoate
PCR	polymerase chain reaction
PTM	post-translational modification
PSI	photosystem I
PNSB	purple non-sulphur bacteria
PFR	pyruvate ferredoxin oxidoreductase
RTK	receptor tyrosine kinase
RCP	recombinant protein
ROP	ring-opening polymerization
SCL	short chain length
SRP	signal recognition particle

SCP	single-cell protein
SNP	single-nucleotide polymorphism
SSF	solid-state fermentation
SMR	steam methane reforming
SmF	submerged fermentation
SC	suspension concentrate
SE	suspo-emulsion
SWP	sweet whey powder
TFF-MF	tangential flow microfiltration
TSIL	task-specific ionic liquid
TEA	technoeconomic assessment
TPC	total phenol content
TALE	transcription activator-like effector
TF	transcription factor
TR	transcriptional regulator
TGF	transforming growth factor
TCA	tricarboxylic acid
ULV	ultralow volume
UTI	urinary tract infection
VHP	very high polarity
WB	Western blotting
WG	wettable granules
WP	wettable powder
WPC	whey protein concentrate
WHO	World Health Organization
ZFN	zinc-finger nucleases

Preface

The role of microbial biomass cannot be avoided as far as industrial applications are concerned. Utilization of microbial biomass is the prime focus of this book. It is based on the industrial microbial processes for the production of value-added industrial compounds in food, fuel, and other related sectors. Various fermentation processes and systems are the key points of this book. Bioconversion technologies to recover energy and fuel from plant-based non-edible biomass and other organic wastes are also discussed. This book will serve as a platform to recognize the breakthrough researches in understanding the roles of microbial biomass towards biorefinery applications, available fermentation technologies, and advanced engineering for large-scale biomass processing utilizing microbial biomass. Algal biomass for value-added products is also discussed.

The chapters in this book will make a widespread impact on the current microbial research and development. This book seeks to provide information on recent advances in the field of biotechnology, catalysis, bioprocess engineering, and downstream processing for the environmentally friendly production of valuable biochemicals. This book supports providing updated information that will solve the gaps between microorganisms and biotechnology by use of traditional and modern process technologies. The application of microbial biotransformation tools is a promising alternative to obtain natural products.

The editors are grateful to all the authors for contributing their valuable chapters to this book. The editors are also grateful to Ashish Kumar and Sandra Jones Sickels from Apple Academic Press, Inc. for their kind support and coordination during the entire project.

—***Editors***

CHAPTER 1

Microbial Mediated Fermented Foods: Resents Trends Progress and Prospects

PRITI PAL,[1] PRAKASH KUMAR SARANGI,[2,*] BRIJESH PANDEY,[3] ANAND PRAKASH,[3] SAURABH SINGH RATHORE,[3,4] and AKHILESH KUMAR SINGH[3,*]

[1]*Amity Institute of Biotechnology, Amity University Uttar Pradesh, Lucknow Campus, Lucknow, Uttar Pradesh, India*

[2]*College of Agriculture, Central Agricultural University, Imphal, Manipur, India*

[3]*Department of Biotechnology, School of Life Sciences, Mahatma Gandhi Central University, Motihari, Bihar, India*

[4]*Department of Life Sciences, Inter University Centre for Teacher Education, Bananas Hindu University, Varanasi, Uttar Pradesh, India*

*Corresponding author

ABSTRACT

Fermented food is a growing industry that is experiencing rapid growth and innovation as a result of the need for more effective agricultural waste utilization and the development of fermentation technologies (microbial and enzyme based processing). As a result of the physical and biochemical changes brought about by fermentation, foods made with these microorganisms have a longer shelf life than those that haven't been fermented. Consumers may benefit from consuming fermented foods due to their increased levels of antioxidants, peptide production, organoleptic qualities, probiotic bacteria, and antimicrobial activity compared to simple foods. The levels of anti-nutrients and toxins are also reduced. Among the

Industrial Microbiology: Current Progress and Novel Technologies.
Prakash Kumar Sarangi, Akhilesh Kumar Singh, Rajesh Kumar Srivastava, and Brijesh Pandey (Eds.)

DOI: 10.1201/9781779642219-1

many health benefits of the world's fermented foods are the synthesis of nutrients and the reduction or elimination of the risks of cardiovascular disease, cancer, gastrointestinal problems, allergic reactions, and diabetes. In addition, the health benefits of eating fermented foods are becoming increasingly well-known. The purpose of this chapter is to provide a review of the literature on biochemical changes that occur during the fermentation of foods, as well as a discussion of the health benefits that consumers gain from consuming fermented foods and beverages.

1.1 INTRODUCTION

Fermentation is the process by which microorganisms like bacteria, yeasts, and mycelial fungi use their own enzymes to produce food (Bourdichon et al., 2012). There are many potential sources of microorganisms, including the substrate itself, a starter culture, the ingredients, the equipment, and the surrounding environment. Food fermentation is only possible under specific conditions, including thc presence of the right substrate, the right microorganism(s), and the right environmental conditions (like the right temperature, pH, and moisture content). During fermentation, energy is released from the partial oxidation of carbohydrates and related compounds without the presence of an external electron acceptor (Xiang et al., 2019). The rate of fermentation can be changed by a number of variables, including the sugar used, the amount of nutrients and oxygen available, the presence of competing microorganisms, and the length of time the process is allowed to proceed. Numerous raw materials, microorganisms, and production processes are used to create the estimated 1,000 distinct fermented foods and beverages that are produced worldwide. A large variety of fermented products can be made by using different raw materials or substrates for the fermentation process or by changing the parameters of the fermentation process. This diversity can be found in microorganisms, the primary biological factor in fermentation processes (Melini et al., 2019). Scientists have discovered that organisms from all three domains of life play important roles in the final product's aroma, texture, and flavor profile during fermentation. To give just one example, yeast strains from *Saccharomyces cerevisiae* are crucial for both bread making and beer making. Lactic acid bacteria (LAB) are essential for the fermentation of cheeses, yogurts, and fermented vegetables (Pessione, 2012). Recently, archaea's role in the production of high-salt fermented seafood has been uncovered (Samtiya et al., 2020). Biotechnology has been used for many different purposes since its inception, with one of the earliest being the fermentation of food. However, microorganisms are now also widely used in the production of medications, biofuels, and bioplastics, as well as in the purification of industrial oils and water. Probiotics, which are bacteria isolated from fermented food, have been shown to improve the health of humans, animals, and plants (Chadare et al., 2019). But to further enhance the efficacy of probiotic products, there is a continuing need for newly discovered strains and strain combinations that perform better or display novel traits.

Microorganisms such as bacteria (*Lactobacillus, Streptococcus, Enterococcus, Lactobacillus,* and *Bifidobacterium*), molds (such as *Aspergillus oryzae, Aspergillus sojae, Penicillium roqueforti,* and *Penicillium chrysogenum*), and yeasts. Since the 19th century, significant efforts have been made toward developing improved fermentation. Since consumers are becoming more aware of the connections between nutrition and disease, research into the health advantages of fermented foods has rapidly increased. In light of the recent worldwide food crisis, fermentation has emerged as a viable strategy for optimizing the use of bioresources (Tamang et al., 2020). It is possible for the microorganisms used to start the fermentation process and the probiotic microbes added to fermented meals to function as "microbial factories," creating nutrients and bioactive of interest while recycling byproducts. Milk fermented with bacteria containing galactosidase, for instance, can be used to make lactose-free or lactose-reduced products because lactose is broken down during the fermentation process. The enzymes found naturally in microorganisms can cause a variety of hydrolysis reactions, which can liberate nutrients and bioactive (like peptides and amino acids) from the raw materials with the desired molecular sizes and bio-availability (Gupta et al., 2015). In comparison to their unfermented counterparts, fermented foods have been found to provide several health benefits for humans, including improved digestibility, greater glucose tolerance (Zhang et al., 2017), prevention of pathogenic bacteria growth and bacterial toxin generation, reduced gastrointestinal disturbances, breakdown of plant toxins (e.g., cyanogenic glycosides), and decreased risks of numerous diseases and conditions, including cardiovascular disease, arthritis, and numerous other conditions (Penas, 2017). Due to their great microbial richness and variety, foods that have fermented on their own are a promising resource for the discovery of new, uncharacterized microbes. Additionally, fermentation systems are simple to establish, maintain, and replicate, for example, the reproducibility of spontaneously fermented carrot juices (Samtiya et al., 2020). There are microbes in them that are more suitable to being cultivated than, for instance, the strictly anaerobic gut bacteria, making for easier manipulation and in-depth investigation (Marco et al., 2017). There is a wide variety of fermentation processes, ingredients, and microbes that go into making these delectable foods. There are four basic forms of fermentation exploited in the production of new products: alcoholic fermentation, lactic acid fermentation, acetic acid fermentation, and alkaline fermentation. Most of the work in lactic acid fermentation is done by the LAB. Fermented foods are a broad category that includes products like kimchi, sauerkraut, gundruk, and fermented grains. Ethanol (also known as alcohol) is produced in part through the fermentation of sugars. Many foods and drinks rely heavily on yeasts for fermentation, including bread, wine, beer, vodka, whisky, brandy, and liqueurs. *Acetobacter* species are in charge of the fermentation process that results in acetic acid. *Acetobacter* requires oxygen (e.g., vinegar) to produce acetic acid. Condiments like soy sauce, fish sauce, and fermented seed paste all produce an alkaline byproduct during fermentation. Fermentation enhances the nutritional value of foods by increasing the concentration of proteins, vitamins, vital amino acids, and fatty acids, among other components. However, 80% of the world's population suffers from chronic hunger and malnutrition. Therefore, fermented food products can help solve issues with the nutritional stability of the global population (Wang et al., 2018). Because of differences in climate, culture, consumer tastes, raw material availability, and technological innovation, regionally specific fermented foods are the norm rather than the exception (García-Burgos et al., 2020). Food and nutrition insecurity can be combated by preparing a vast diversity of fermented food products, which are made from a variety of ingredients depending on the availability of those ingredients. Asia is renowned for its methods to store and stabilize food supplies despite seasonal changes brought on by the monsoons. Approximately 95% of the 150 million metric tons of paddy produced annually in Southeast Asian countries are consumed domestically. Since ancient times, people have been fermenting cereals and other plant products to make all sorts of foods. Many Asian countries consume various fermented food products, including rice wine, fermented cassava tubers, and many others (Tamang et al., 2016a). Functional qualities, such as probiotics, antibacterial properties, antioxidants, peptide synthesis, fibrinolytic activity, poly-glutamic acid, and so on, may play an important role in the selection of starter culture(s) to be used in the preparation of fermented foods. Microorganisms employed as starter cultures for industrial food fermentation are included in Table 1.1; the resulting products are sold as functional, health, therapeutic, and nutraceutical foods all over the world (Thapa and Tamang, 2015). The goal of this study is to provide a concise summary of the current state of knowledge concerning the beneficial effects of fermented foods and beverages on human health and the beneficial qualities of the microorganisms that are integral to the fermentation process.

TABLE 1.1 There Are Certain Useful Microorganisms That Are Widely Employed as Starters in the Industrial Food Fermentation Industry

Microorganisms Species	Application in Fermented Food
L. casei subsp. *casei*	Dairy starter; cheese ripening; green table olives
L. delbrueckii subsp. *bulgaricus*	Yogurt and other fermented milks, mozzarella
Acetobacter aceti subsp. *Aceti B. subtilis*	Vinegar Fermented soybeans, protease, glycolipids, riboflflavin-B2 (food additive)
B. licheniformis	Protease (food additive)
Corynebacterium ammoniagenes	Cheese ripening
E. faecium	Soybean, dairy, meat, vegetables
L. kimchii	*Kimchi*
L. oeni	Wine
L. pentosus	Meat fermentation and bio preservation of meat; green table olives; dairy, fruits, wine
Zymomonas mobilis subsp. *mobilis*	Beverages
S. cerevisiae	Beer, bread, invertase (food additive)
S. cerevisiae subsp. *boulard*	Used as probiotic culture
Aspergillus flavus	α-amylases (food additive)
A. niger	Beverages; industrial production of citric acid; amyloglucosidases, pectinase, cellulase, glucose oxidase, protease (food additives)
A. oryzae, A. sojae	Soy sauce, beverages; α-amylases, amyloglucosidase, lipase (food additives)
Rhizopus oligosporus	*Tempe* fermentation

1.2 FERMENTED FOODS AND THEIR MICROORGANISMS

Many fermented foods and drinks contain LAB (Tamang, 2010). *Alkalibacterium, Carnobacterium, Enterococcus, Lactobacillus, Lactococcus,*

Leuconostoc, Oenococcus, Pediococcus, Streptococcus, Tetragenococcus, Vagococcus, and *Weissella* are all examples of major LAB genera that have been isolated from fermented foods and beverages around the world. *Bacillus* can be found in Asian and African alkaline-fermented foods (Tamang, 2015). *Bacillus amyloliquefaciens, Bacillus coagulans, Bacillus firmus, Bacillus licheniformis, Bacillus megaterium, Bacillus pumilus, Bacillus subtilis, Bacillus subtilis variety natto,* and *Bacillus thuringiensis* have all been found in fermented foods made from legumes (Kubo et al., 2011) and strains (Oguntoyinbo et al., 2007). λ-Polyglutamic Acid (PGA) is a type of amino acid polymer found in many Asian fermented soybean foods that contributes to their stickiness (Nishito et al., 2010). Milk products, fermented sausages, meat, and fish products have all been linked to the presence of *Kocuria, Micrococcus,* and *Staphylococcus* (all of which belong to the Actinobacteria) (Coton et al., 2010). *Arthrobacter* and *Hafnia* are found in fermented meat products, while cheese is home to *Bifidobacterium, Brachybacterium, Brevibacterium,* and *Propionibacterium* species (Bourdichon et al., 2012). Many fermented foods around the world also contain pathogens like *Enterobacter cloacae, Klebsiella pneumoniae, K. pneumoniae* subsp. *ozaenae, Haloanaerobium, Halobacterium, Halococcus, Propionibacterium, Pseudomonas,* and so forth (Tamang, 2010). *Brettanomyces, Candida, Cryptococcus, Debaryomyces, Dekkera, Galactomyces, Geotrichum, Hansenula, Hanseniaspora, Hyphopichia, Issatchenkia, Kazachstania, Kluyveromyces, Metschnikowia, Pichia, Rhodotorula, Rhodosporidium, Saccharomyces, Saccharomycodes,* and *Saccharomyces* filamentous molds play a crucial role in the fermentation process by producing enzymes and degrading anti-nutritive factors in foods and alcoholic beverages (Aidoo and Nout, 2010). Many fermented foods, Asian non-food amylolytic starters, and alcoholic beverages are reported to contain the yeasts *Actinomucor, Amylomyces, Aspergillus, Monascus, Mucor, Neurospora, Parcilomyces, Penicillium, Rhizopus,* and *Ustilago* (Chen et al., 2014).

1.3 CATEGORIZATION OF FERMENTED FOODS

1.3.1 FERMENTED DAIRY PRODUCTS

Fermented dairy products are an example of a high-quality fermented food. Due to their beneficial properties, such as increased lactose tolerance, better nutrient absorption (including minerals), a stronger gut-associated immune response, lower cholesterol levels, and reduced inflammation, probiotics are becoming increasingly popular. The fact that probiotics are enriched with beneficial nutrients such as lactic acid, galactose, free amino acids, fatty acids, and vitamins (particularly B complex) also contributes to their meteoric surge in popularity. Worldwide, LAB is routinely used in the manufacture of fermented dairy products (Batish et al., 2004). For example, the making of yogurt, *Dahi*, *Srichand*, Bulgarian buttermilk, acidophilus milk, *kefir*, *koumiss*, and cheese. A good example of a cheese that includes health-promoting microbes like probiotics and gives advantages in nutritional content and health benefits over non-fermented milks (Hoque et al., 2010) is Iranian ultrafiltered Feta cheese, which contains *Lactobacillus casei* with NaCl partially replaced by KCl (Minervini et al., 2012). Cheese has a variety of useful minerals, including calcium for bone health and tooth decay (Kato et al., 2002), conjugated linoleic acid (CLA) with anticarcinogenic and antiatherogenic effects (Bhattacharya et al., 2006), and peptides with diverse biological activity (Zaheer et al., 2013). Health problems including type 2 diabetes and cardiovascular disease can be counteracted in part because yogurt is a nutrient-dense fermented meal (Sodini et al., 2004). This is due to the fermentation process, which imparts bacterial cultures as well as the numerous macro- and micronutrients and bioactive metabolites produced by the bacteria. *Kefir*, a self-carbonated, somewhat foamy alcoholic beverage derived from animal (like goat, cow, and camel) or plant (like soya, rice, and coconut) milk, is known for its unique flavor and nutritional benefits (Zaheer et al., 2013). Fermented milk modifies the quality and quantity of proteins, vitamins, carbohydrates, and some fats, while the composition of minerals remains unchanged (Figure 1.1).

1.3.2 FERMENTED STAPLE CROP PRODUCTS

Some populations, such as vegans, vegetarians, or people who are lactose intolerant, are pleased to have access to alternatives provided by the widespread category of fermented staple crop foods, both solid and liquid. Barley, maize, millet, oats, rice, rye, sorghum, and wheat are just some of the cereals that have been used as fermentation substrates. Fermented cereal foods are tastier, have fewer anti-nutritional effects, and increase mineral bioavailability compared to their unfermented counterparts (Katina et al., 2005). Probiotic microorganisms (such as *Lactobacilli and Bifidobacteria*) and raw materials containing prebiotic carbohydrates

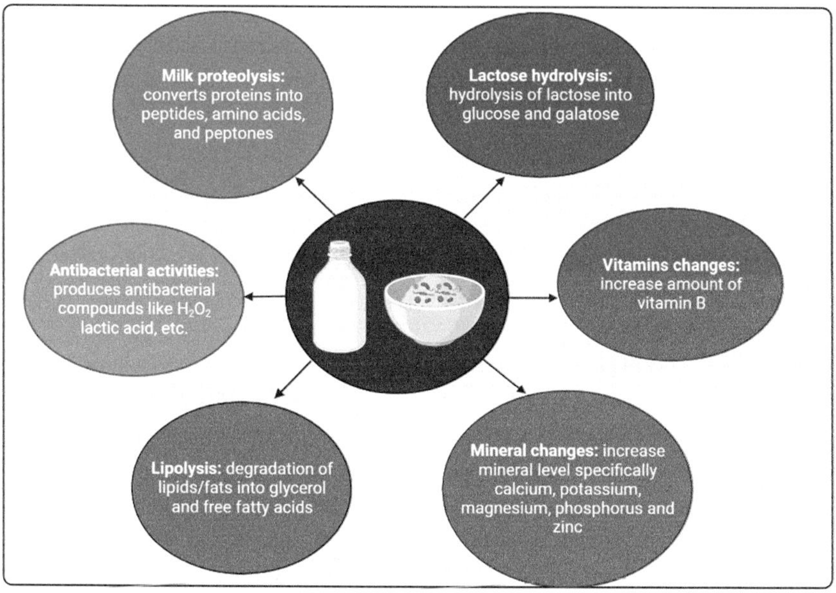

FIGURE 1.1 Modifications that take place in fermented dairy products.

(such as beta-glucan, arabinoxylan, galacto and fructo oligosaccharides, and resistant starch) can be used to boost the health benefits of fermented cereals. The preventive effects against type 2 diabetes, cardiovascular disease, Alzheimer's disease, cancer, and human B lymphomas have been attributed to the elevated quantities of sterols, unsaturated fatty acids, and phytochemicals (e.g., vitamins and phenolics) in fermented rice and rice bran (Kalaivani et al., 2010). Fermented cereal products like vinegar and beer are very common. Some examples of vinegars fermented from sugar and alcohol are Shanxi aged vinegar, Zhenjiang aromatic vinegar, and Kurosu vinegar (Howeida et al., 2015). These vinegars have a wide range of applications, including antioxidant, antibacterial, and anti-infectious properties, as well as effects on blood glucose, lipid metabolism, appetite, fatigue, and osteoporosis. In addition to the aforementioned *Kvass*, which is produced in Ukraine, Russia, and Belarus from fermented maize dough, other popular fermented cereal products include the Japanese sweet rice drink *Amazake*, the Nigerian fermented cereal porridge *Ogi*, the Mexican drink *Pozol*, and the Mexican drink *Tequila*, all of which are made from fermented maize dough (Omar and Ampe, 2000). Beer, despite its modest alcohol content, includes a variety of beneficial components and nutrients, such as nitrogen compounds, melanoidins, organic acids, vitamins, phenolic compounds, and mineral salts (Bamforth, 2002). Fermented sourdough goods, which are used as alternatives to gluten-free food, offer favorable textural, taste, and nutritional properties due to the proteolytic LAB and wheat flour endogenous enzymes used in the fermentation process (Rizzello et al., 2015).

1.3.3 FERMENTED SOY PRODUCTS

Due to their high protein content (40% on a dry basis), soybeans are a widely consumed and simple-to-grow crop. In 1999, the Food and Drug Administration (FDA) approved a health claim for soy, saying that "Diets low in saturated fat and cholesterol that include 25g of soy protein per day may reduce the risk of heart disease." As a result, the quality and variety of soybean products continued to improve. Fermented soy products have a distinct flavor and exceptional nutritional value due to the high levels of amino acids and fatty acids they contain (Steinkraus et al., 2004). Fermented soy foods like soy sauce, tamari, miso, tempeh, and natto are made by fermenting different parts of the soybean with yeast, bacteria, mold, or a combination of the three. Soybeans and cereal grains are soaked and cooked before being fermented with a mold to make tempeh, a traditional fermented food (usually of the genus *Rhizopus*). Enzymes in the mold catalyze hydrolysis, a process by which the soybean's components are broken down to produce distinct and desired food texture and flavor while lowering or eliminating antinutritional chemicals (Marco et al., 2017). Ganjang/soy sauce (brown liquid), Doenjang (brown paste formed by Meju), Gochujang (fermented red pepper paste, a spicy condiment), and *Cheonggukjang* (soybean paste fermented by *Bacillus* at 40–45 °C) are all fermented soy products used as seasonings because of their salty and umami flavors. In recent years, scientists have concentrated on enhancing the health benefits of fermented soy meals, such as vitamin K2 (menaquinone), which is necessary for bone and cardiovascular health and is exclusively present in fermented soy and animal products (Schurgers et al., 2007). There is evidence that consuming fermented soy products can lower the risk of developing diseases associated with age and hormones, including inflammation, infection, asthma, and cancer (Dong, 2011).

1.3.4 FERMENTED VEGETABLE PRODUCTS

Foods made from fermented vegetables have a positive effect on human health because they contain many substances that are good for people. Examples include dietary fiber, minerals, antioxidants, and vitamins. Fermented vegetables such as kimchi, sauerkraut, carrots, cauliflower, tomatoes, olives, green peas, and peppers are common in traditional food. Sauerkraut is made by salting and fermenting white cabbage (*Brassica oleracea var. capitata*) with lactic acid (Lu et al., 2003). Fermentation in brine (often 5%–8% NaCl) in open-faced tanks converts the cucumbers' saccharides into acids and other compounds throughout the pickling process (Fleming et al., 1978). Many varieties of kimchi are common in Korean cuisine, including *baechu kimchi* (cubed Chinese cabbage), *tongbaechu kimchi* (whole Chinese cabbage), *yeolmoo kimchi* (young oriental radish), *kakdugi* (cubed radish kimchi), *baek kimchi* (cubed radish kimchi with water), *dongchimi* (whole radish kimchi with water), and *nabak kimchi* (radish and Chinese cabbage cut into pieces). LAB contain significant amounts (107–109 CFU/g) and are used to *ferment baechu* cabbage and other vegetables to make these kimchi products. This means they are rich in prebiotic and probiotic microbes, dietary fiber, vitamins, and minerals. *Kimchi* has been proven to have anti-cancer, anti-inflammatory, antibacterial, anti-obesity, and anti-aging characteristics (Kun-Young et al., 2014).

1.3.5 FERMENTED FRUIT PRODUCTS

Fermented fruits may be produced from either whole fruits (such as apples, lemons, mangoes, palms, papayas, and pears) or fruit pulps and juices (such as bananas and grapes) via "spontaneous" fermentation by natural lactic bacterial surface microflora (such as *Lactobacillus* spp. and *Pediococcus* spp.). Yeasts and bacteria work together to ferment grapes, apples, and sometimes even pineapples to create wine, cider, and vinegar, three distinct alcoholic beverages derived from fruit. Fruit products are being fermented to create a new class of foods with improved health benefits and unique flavors. Certain red wines, for instance, are crafted to have higher concentrations of antioxidants, particularly polyphenols with known health benefits (Cocolin et al., 2000). Antioxidant power, including free radical scavenging ability, can be manufactured in a fermented papaya preparation (Imao et al., 2010). Apple cider vinegar is effective in lowering blood cholesterol and triglyceride levels and protecting LDL particles from oxidation (McGrath and Simone, 2015).

1.3.6 FERMENTED SEAFOODS

There are a variety of methods for producing fermented seafood, whether starting with a whole fish or a by-product. Fermenting fish with LAB such as *Lactobacilli*, *Pediococci,* and *Streptococci* is standard practice. *Nare-zushi* is a popular fermented sushi dish in Japan that is made with raw fish (such as mackerel, salmon, crucian carp, or sandfish) and rice (or another source of carbs) (Kuda et al., 2010). Fish sauces and seasonings are made through the long-term fermentation of fish mashes such as anchovies, small shrimp, icefish, and sand lance, as well as the by-products of tuna, sardine, and squid. Jeot-gal is a popular side dish or appetizer in Korean cuisine, as well as a key ingredient in *Kimchi*. Jeot-gal is made by fermenting raw fish that has been marinated with microorganisms that have proteolytic, lipolytic, and amylolytic activities at temperatures above 25°C. The antioxidant, angiotensin-II converting enzyme (ACE) inhibiting, gamma-aminobutyric acid (GABA) producing, cholesterol-reducing, and antitumor effects of jeot-gal, as well as the mineral (e.g., calcium, iron, phosphorus) and protein-based compound (e.g., amino nitrogen, soluble nitrogen, and volatile basic nitrogen's) content, can all be enhanced through proper processing (Kuda et al., 2010).

1.3.7 FERMENTED MEAT PRODUCTS

Fermentation has been used to preserve meat for thousands of years, so today there is a wide variety of fermented meat items available in grocery stores and restaurants all over the world. Fermented meats include sausage and ham, which use either naturally occurring microflora or commercial preparations of bacteria, yeasts, and molds (such as LAB, *Micrococci*, and *Staphylococci*). Meat fermentation is an acidification process that requires very little energy yet results in foods that are unique in their look, flavor, texture, and even health advantages (Ockerman et al., 2014). When making fermented meat, acids are crucial for keeping tabs on microbial activity and sugar consumption. Protein carbonylation and other protein-lipid and protein-protein interactions are examples of oxidation that occur during beef fermentation (Wójciak et al., 2012). Traditional fermented meat items include Italian salami, Spanish *salchichon* and *chorizo*, *Icelandic Slátur* (blood sausage), Irish black pudding (blood sausage) produced from pig's blood, beef sticks, and pepperoni. Long-fermented and dried sausage is likely to have a lower moisture level, a more concentrated flavor, a firmer consistency, and a higher nutrient content than commercially manufactured sausage.

1.3.8 ALCOHOLIC FERMENTED FOODS AND BEVERAGES

One of the earliest and most essential techniques for preserving food to be eaten and enjoyed is fermentation with alcohol. For the production of a wide range of alcoholic drinks, including beers, wines, and distilled liquors, yeast or sometimes yeast-like molds, such as *Amylomyces rouxii*, mold-like yeasts, such as *Endomycopsis*, and bacteria, such as *Zymomonas mobilis*, are used (Hossain and Kabir, 2016) (Figure 1.2). By fermenting sugars found in substrates like cereal grains, sugar cane juice, palm sap, fruit juices, diluted honey, or hydrolyzed starch, these microbes predominantly create ethanol and carbon dioxide. There are at least two different techniques used in Asia to ferment starch grains into alcohol. Starch can be converted into sugars by using amylase enzymes from molds like *Amylomyces rouxii*, and then the sugars can be fermented into ethanol using yeasts like *Endomycopsis fibuliger*. Substrates for alcoholic beverages are often rice, palm inflorescences like coconut and talipot, millet, and other grains. Wines made from coconuts and talipot palms are common all over Asia. In Vietnam, it is known as "*panam culloo*," in Indonesia as "*arak*," and in Malaysia, India, and Bangladesh as "*tuak*" or "*toddy*." The sap from coconut, palmyra, and toddy palm trees is fermented to produce a traditional alcoholic beverage called toddy. The unopened flower is slit at the tip to release the sap, which is then collected in a jar made of earthenware that is connected to the flower's stem. Sap can be taken either raw or after fermentation for up to 24 hours. Fresh sap, a dirty-brown sweet liquid with a sugar content of roughly 10%–18% w/w (Hossain and Kabir, 2016), can be fermented into an alcoholic beverage with an ethanol content of 9% (v/v). Palm wine is one of the most cost-effective ways for low-income communities to get vitamin B-12 and ascorbic acid because it contains 0.019%–0.028% (w/v) of both nutrients. Fermentation boosts B vitamin levels, particularly those of thiamine, riboflavin, and pyridoxine (Law et al., 2011). Total solids in fermented coconut toddy are 3.72 g/dL, with an alcohol content of 1.8%–7.9%, 0.29 g sucrose, and 3.0 g invert sugar. Ekanayake (2016) find 0.5 g/100 mL of nitrogen, phosphorus, potassium, calcium, and magnesium. The production of palm sap nectar requires the use of yeasts and LAB for the fermentation of ethanol, lactic acid, and acetic acid. Always present are yeasts like *Saccharomyces cerevisiae*; predominantly present are other LAB like *Lactobacillus plantarum* or *Lactobacillus mesenteroides*, or bacteria like *Zymomonas mobilis* or *Acetobacter* spp. (Hossain and Kabir, 2016). Coconut sap fermented alcohol is produced mostly by the yeast *Saccharomyces chevalieri* (Ekanayake, 2016).

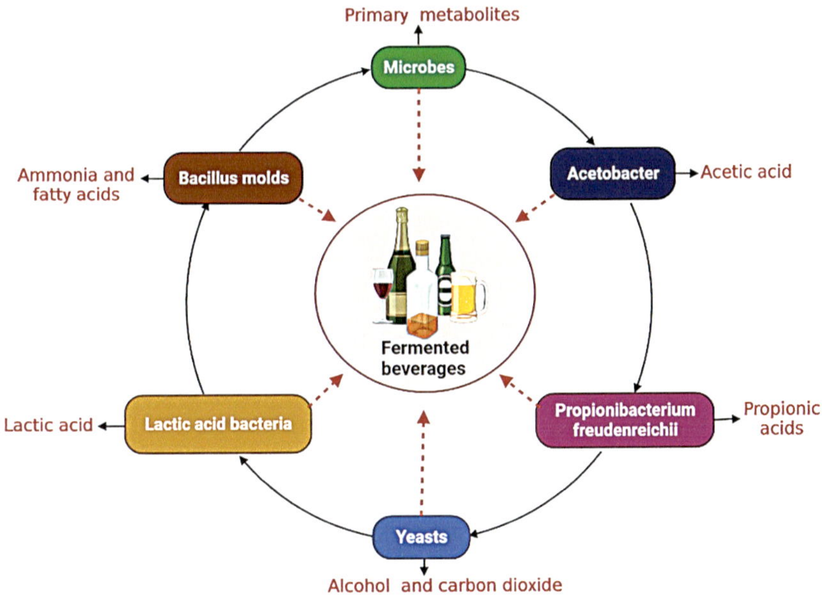

FIGURE 1.2 Involvement of microorganisms and their connection to the principal metabolites discovered during the fermentation of beverages.

1.3.9 OTHER FERMENTED PRODUCTS

There are many other fermented items available around the world besides the ones we listed in Table 1.2. Fermented tea has gained popularity because it may be infused with a variety of health benefits and enhanced

TABLE 1.2 Microbial Communities Associated with Ethnically Fermented Foods

Microbiota	Substrate	Traditional Fermented Foods	Sensory Property and Nature	Country	References
Lactobacillus plantarum	Soy milk	Tofu	Whey, soft, cheese like product, fried snacks, curry	Worldwide	Li et al. (2012)
Lactobacillus acidophilus	Cow/buffalo milk, starter culture	*Dahi*	Curd, savory	India, Nepal, Sri Lanka, Bangladesh, Pakistan	Balamurugan et al. (2014)
Lactobacillus sp.	Milk	*Koumiss*	Acid fermented milk, drink	Russia	Guo et al. (2015)
Lactobacillus plantarum	Radish, salt, water	Korean *kimchi*	Acidic, mild-sour, wet, soup, side-dish	Korea	Kwak et al. (2017)
Lactobacillus plantarum, *Lactobacillus fermentum*, and *Saccharomyces cerevisiae*	Raw cow milk	*Nunu*	Naturally fermented milk	Ghana	Behera et al. (2018)
Lactobacillus kefir, *Lactobacillus kefiranofaciens*, and *Lactobacillus kefirgranum*	Goat, sheep, Cow milk	*Kefir*	Alcoholic fermented milk, effervescent milk	Russia	Seo et al. (2018)
Lactobacillus helveticus R389	Animal milk	Swiss Cheese	Soft or hard, solid; side dish, salad	Worldwide	Ghosh et al. (2019)

flavors by using various strains of bacteria, yeasts, and fungi to modify the quantities of organic acids, vitamins, caffeine, and polyphenols in the tea (Jayabalan et al., 2007). Tea's ability to fight off pathogens like *Salmonella* and *Staphylococcus* and to help diagnose and treat gastrointestinal infections, diabetes, cardiovascular disease, and cancer is enhanced by the fermentation process and the processes associated with it (Fu et al., 2011). Fermented ginseng has been produced to enhance behavioral memory function; it has been shown to reverse memory impairment and reduce amyloid deposition in Alzheimer's disease in rats (Kim et al., 2016). Olives that have undergone fermentation are promoted as probiotic foods because they contain phenolic compounds and other bioactive substances that boost human performance and defense against disease (Peres et al., 2017).

1.4 NUTRITIONAL BENEFITS OF FERMENTED FOODS

1.4.1 FERMENTATION CAUSES A VARIETY OF CHEMICAL AND PHYSICAL CHANGES IN FOODS

Recently, bioreactors have been outfitted with cutting-edge technology like online sensors and *in situ* computer visualization and simulation applications to provide real-time measurements for accurate control of the production of fermented products; nevertheless, neither the fermentation process nor the subsequent handling and storage of fermented goods necessitate advanced equipment. Whether fermentation occurs naturally or is induced artificially, both the applied and naturally occurring microflora have a role in its inception and progression. During fermentation, a wide range of chemical, biological, and sensory changes take place, all of which ultimately affect the final product. The process of fermentation involves the enzymatic and chemical breakdown and subsequent change of food components in both edible and sometimes inedible raw materials (e.g., the removal of glycol residues). Therefore, the sensory quality, nutritional content, and health-promoting qualities of the fermented products are enhanced in a safe and effective manner. Fermentation produces small-molecule flavor compounds and texture-modifying substances like exopolysaccharides (EPSs), and fermentation produces lactic acid, which is responsible for the fermentation process and contributes to the development of distinctive flavors, aromas, and textures (Marco et al., 2017). Fermentation can increase the concentration of many micro- and macronutrients, including important amino acids, fatty acids, vitamins, and minerals. Fermentation can make foods more quickly digested and absorbed by the body, as well as decrease the amount of time spent preparing them. Fermentation makes nutrient enrichment easier to achieve. The process of fermentation allows for the production of foods with specific health advantages (either through the direct release of bioactive of interest or the conversion of parent chemicals into these bioactive) (e.g., probiotics, prebiotics, and non-nutrient bioactive like phenolic antioxidants). The growth of harmful microorganisms (including Gram-positive and Gram-negative bacteria) can be stifled by fermentation, and toxins (like mycotoxins) can be degraded, both of which contribute to increased food safety. Fermentation changes food in different ways depending on the microorganism and the raw materials used, as well as the surrounding environment and any additional processes that are associated with fermentation, such as the pre-roasting of buckwheat groats prior to solid-state fermentation with *Rhizopus oligosporus* (Pickens et al., 2011). Fermentation can change how foods behave in terms of temperature, viscosity, consistency, hardness, and stickiness. This, in turn, can change the chemical, biological, and sensory properties of the finished product.

The consumption, digestion, absorption, and utilization of additional food and feed components (such as enzyme inhibitors, flatulence causes, glucosinolates, lectins, tannins, polyphenols, phytic acid, and saponins present in mustard, rapeseed protein products, cereals, and legumes) might be hazardous to humans due to the presence of natural anti-nutritional factors (Tassou et al., 2003). Germinating, fermenting, enzymatic treatment, soaking, heating, canning, irradiating, selective extraction, fractioning, and isolating are just some of the mechanical, thermal, chemical, and biological techniques employed individually or in combination to lessen the presence of anti-nutritional components. By enhancing digestibility and nutritional value, enriching minerals and bioactive like amino acids and GABA, and removing undesired components like antinutrients, fermentation can improve food quality (compared to thermal, chemical, and mechanical procedures) (Mohd et al., 2012). Fermentation can be used to reduce or eliminate some undesirable intrinsic chemicals (such as protease, amylase, and other enzyme inhibitors) that may impair the rate and extent of key bioconversions (Singh and Benjakul, 2017). Fermentation and germination are two processes that, when used together, can have additive effects. If germinated rice were fermented with probiotic organisms, the resulting rice would be higher in natural fibers, GABA, and inositol hexaphosphate (Hung, 2005).

1.4.2 FERMENTATION AND PRODUCTION OF THE "BIOTICS"

Recent years have seen a rise in interest in a wide range of new "biotics," including probiotics, prebiotics, symbiotic, postbiotics, oncobiotics, para-probiotics, pharmabiotics, and psych biotics. High-viability and functional probiotics, prebiotics, and symbiotic are being made through fermentation. Increased attention has been paid to fermented foods and ingredients as of late due to their potential to foster a balanced microbiome in the digestive tract and improve human health.

In 2006, the term "probiotic" was officially coined to describe "live microorganisms that should be alive by the end of the shelf life of the product and when administered in adequate amounts, confer a measurable health benefit on the host at a defined dose" (Cousin et al., 2011). These days, LAB and bifidobacteria (Bifidobacterium) are the most common types of probiotics sold for human consumption. Up to 107 viable cells per gram or milliliter of probiotics should be preserved in food until it is consumed. For further probiotic advantages, fermented foods can have the probiotic potential of the naturally occurring microorganisms increased or additional probiotic cultures added. The latter group of bacteria may have an advantage in several respects due to their long history of adaptation to the dietary environment, including resistance to acidity and bile salts, colonocyte adhesion, antibiotic production, and lactase production (Bull et al., 2013). Probiotic fermented foods are safer because probiotic strains like *Lactobacillus* species produce antibacterial bacteriocins.

Probiotic fermented foods have health and nutritional benefits because of the enzymes (such as β-galactosidase, which aids in lactose breakdown and compensates for the decline in human lactase activity with age) and nutrients (bioactive) created during the fermentation process (e.g., enzyme cofactors, GABA, ACE peptide inhibitors, anti-hypertensive peptides, polyunsaturated fatty acids, isoflavone). Probiotic fermented foods have been proven to offer benefits for the treatment of irritable bowel syndrome, urinary tract infections (UTI), bladder cancer, allergies of the skin, stomach, and respiratory system (including food allergies in newborns), as well as the alleviation of symptoms of lactose digestion problems and the shortening of rotavirus diarrhea (West et al., 2010). Despite numerous reports detailing the positive effects of probiotics and fermented milk containing probiotics, the European Union has only approved a single health claim related to the lactose-digesting benefits of live yogurt cultures. The low number of effectiveness claims can be largely attributed to the diversity of the gut microbiota and effector strains regulating human health (Grüber et al., 2007). The probiotic effect, which occurs when live microorganisms, bacteria, or yeast are taken, and the biogenic effect, which occurs when microbial metabolites are consumed, are both thought to be beneficial to health (Stanton et al., 2005). It is standard procedure to combine probiotics and prebiotics in order to maximize the health advantages of the live bacterium and the prebiotic (Spacova et al., 2020). Synbiotic fermented foods are another type of fermented food used to manage microflora. Non-digestible prebiotics have a beneficial effect on the host by promoting the growth or activity of specific types of beneficial bacteria in the colon (Munekata et al., 2022). Prebiotics relate intestinal health to probiotics in this manner. Prebiotics, which can include fructo-oligosaccharides, lactulose, lactitol, inulin, glucan, and galacto-oligosaccharides, serve as a selective substrate for one or a limited number of potentially helpful commensal bacteria in the colon and are not absorbed in the upper gastrointestinal tract. Prebiotics are similar to probiotics in that they improve host health by changing the microbiota in the gastrointestinal system (via increasing the number of specific microbial strains or greatly altering the composition of the gut microflora). In addition, prebiotics may be combined with other bioactive such as organic acids (e.g., butyrate), active fatty acids (e.g., CLA), B-group vitamins, and bioactive peptides (BAP) (e.g., ACE inhibitory peptides) to promote the survival of probiotic strains and the creation of desirable bioactive substances, such as the increased production of ACE inhibitors with the addition of Aloe vera succulent plant powder (Basannavar et al., 2015).

1.5 ENRICHMENT AND MODIFICATIONS OF BIOLOGICAL SUBSTANCES IN FERMENTED FOODS

1.5.1 VITAMINS BIO-ENRICHMENT

Some processed foods have nutrients, primarily vitamins, added to them to boost public health. Vitamin C (ascorbic acid) can be added to fruit juices, and vitamin D (cholecalciferol) is added to milk (Figure 1.3). Due to its prohibitively high cost, this strengthening or enrichment method can only be implemented in the West. In light of this, it is recommended that the majority of nations employ this method of food fermentation in order to increase the foods' biological value (Chadare et al., 2019). Highly refined white rice contains very little of the B vitamin thiamine. Erysipelas is a disease caused by eating this rice (Wiley and Gupta, 2022). Babies who are breastfed by mothers deficient in thiamine (causing beriberi) can also experience sudden cardiac death at three months of age (Barennes et al., 2015). During the Ketan fermentation process, thiamine is synthesized by the microorganisms. The thiamine content of unprocessed rice is restored thanks to these microorganisms (Law et al., 2011). Therefore, this can be very useful for people who regularly consume rice. Soaked, dehulled, and partially cooked soy beans fermented with *Rhizopus oligosporus* or similar molds (Nout and Kiers, 2005) yield "*tempe*," a dish with deep roots in Indonesian cuisine. Cotyledons are knitted into slices, and then the mold is baked to create a firm cake. Proteins are slightly hydrolyzed, lipids are broken down into stachyose (a tetrasaccharides indigestible by humans) and fatty acids, riboflavin and niacin are increased by a factor of seven, and vitamin B-12, typically absent in vegetarian foods, is synthesized by a fermenting bacterium growing with the essential mold (Nout and Kiers, 2005). Many bowls of cereal and legume mixtures benefit from the tempe manufacturing process, which shortens cooking times and enhances digestibility and texture (Nout and Kiers, 2005). Inoculating Indian idli with *Klebsiella pneumoniae* (nonpathogenic strain) bacteria results in the production of vitamin B-12 (Ghosh and Chattopadhyay, 2011). Pulque, a traditional alcoholic drink from Mexico, is made by fermenting the juice of the cactus plant (Agave) (Escalante et al., 2016). Since pulque is rich in B vitamins like niacin, thiamine, pantothenic acid, riboflavin, p-aminobenzoic acid, biotin, and pyridoxine, it is a popular beverage among low-income Mexican families with children. Thin gruel-like alcoholic beverages, such as Kaffir beer, have a pleasant sour taste (Tamang et al., 2020). The Bantu people of South Africa traditionally make a brew called kafir, which ranges in alcohol content from 1 to 8%. Malted and un-malted kafir corn were used to brew the kafir beverage (*Sorghum caffrorum*). Millet or cornmeal can stand in for the kafir lime. Thiamin levels remain stable during fermentation, while riboflavin and niacin/nicotinic acid both increase in those who consume maize (Aka et al., 2016).

Approximately 10–12% of the sugar in palm sap is fermentable, making it a sweet, plump, milky white liquid (Amoa-Awua et al., 2007). In the tropics, it is a staple food. About 83 milligrams of ascorbic acid per liter can be found in this wine (Karamoko et al., 2016). The levels of thiamine, pyridoxine, and riboflavin are raised to between 25 and 150 micrograms per milliliter and 35 and 50 micrograms per milliliter, respectively, in fermented palm wine. Palm wine actually has a high vitamin B-12 content (190–280 mcg/mL), which is a bit of a surprise. Although tropical nutrition is extremely expensive, palm toddies serve an important role as the cheapest source of vitamin B (Anal, 2019).

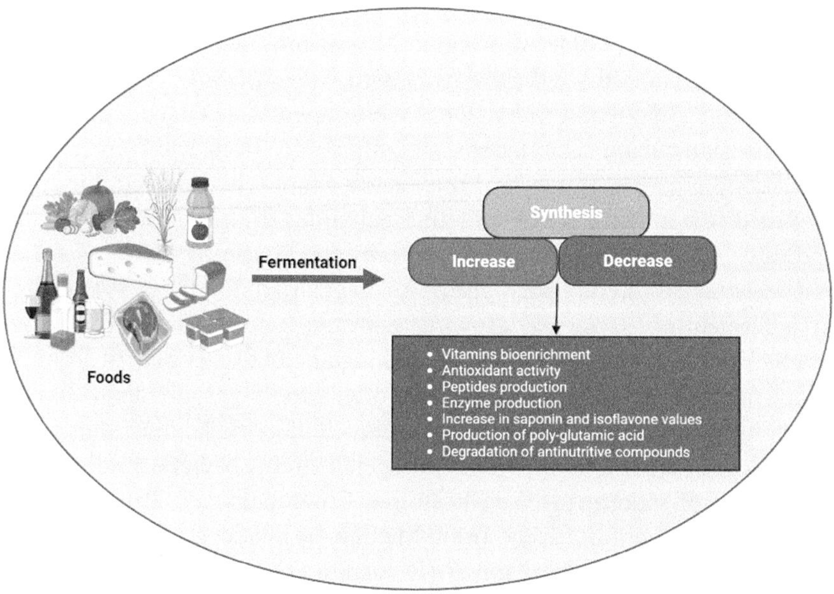

FIGURE 1.3 Fermented foods have greater nutrient density.

1.5.2 ANTIOXIDANT ACTIVITY

The reducing power test; 2,2′-azino-bis(3-ethylbenzothiazoline-6-sulfonic acid) (ABTS) assay; 1,1-diphenyl-2-picrylhydrazyl (DPPH) radical scavenging activity assay; and total phenol content (TPC) assessment are all examples of methods used to evaluate antioxidant activity in fermented foods (Abubakar et al., 2012). Natto (the antioxidant activity has also been found in yogurt and kimchi), *tempeh* (an Indonesian mold-fermented soybean food), *jang* and *chungkokjang* (Korean fermented soybean foods), *thuanao* (Thai fermented soybean foods), and *tungrymbai* and *bekang* (Indian fermented soybean foods). Abubakar et al. (2012) used the 1,1-diphenyl-2-picrylhydrazyl (DPPH) assay to determine the antioxidant activity of LAB during fermentation of skim milk, and they found free radical scavenging activity ranging from 14.7% (v/v) after fermentation up to 24 hours to 50.8% (v/v) after fermentation up to 72 hours. Antioxidant activity was found to be significantly ($p < 0.05$) influenced by both the type of LAB used and the length of time it was fermented. This study compared the increases in antioxidant value in fermented sausage produced using *L. curvatus* (SR6) and *L. paracasei* (SR10-1) (Zhang et al., 2017).

Due to the bacteria's efforts, the sausage's antioxidant activity was considerably enhanced, which improved its flavor and made it safer to eat. While *L. paracasei* SR10-1 (i.e., 285.67% ± 2.00% and 63.89% ± 0.93%, respectively) had a higher OH scavenging activity and a lower lipid peroxidation capacity, *L. curvatus* SR6 had a higher DPPH scavenging activity and reducing power (59.67% ± 6.68% and 47.31% ± 4.62%, respectively).

1.5.3 PEPTIDES PRODUCTION

Proteolytic microorganisms (mostly *Bacillus*) create BAP during food fermentation (De Mejia and Dia, 2010). Peptides have been shown to have a small effect on blood pressure (Phelan and Kerins, 2011), and they also serve an important purpose as antithrombotic (Hou et al., 2015) and immunomodulatory agents (Alcantara et al., 2016). Still falling short of expectations, BAP production is hindered by factors including the high cost of enzymes used in protein hydrolysis and a dearth of large-scale production facilities. On the other hand, microbial fermentation is a cheaper and more efficient way to make BAP. BAP have the same beneficial qualities as dairy products. There are four different forms of cell envelope protease: PrtB from *L. delbrueckii* subsp. *bulgaricus* (Vukotic et al., 2016); PrtP from *L. casei* and *L. paracasei* (Griffiths and Tellez, 2013); PrtR from *L. rhamnosus* and *L. plantarum* (Miyamoto et al., 2015); and PrtH from *L. helveticus* (Stefanovic et al., 2017). While most *Lactobacillus* species only have a single cell envelope proteinase (C.E.P.), *L. helveticus* has been found to have not one but four paralogs of PrtH, the distribution of which varies depending on strain (Singh et al., 2014). When it comes to the *Lactobacillus* genus, *L. helveticus* is the most proteolytic species, as it is the only one responsible for producing a wide range of BAP (Qian et al., 2011). This is due to the presence of several specialized C.E.P. paralogs. Angiotensin-converting enzyme (ACE) inhibitory action has been observed in fermented milk products such as koumiss, kefir, yogurt, cheese, fermented camel milk, and fermented fish products. BAP are made when soybeans are fermented, and Sanjukta and Rai (2016) found that they can help treat and prevent a number of metabolic disorders.

1.5.4 ENZYMES PRODUCTION THROUGH MICROORGANISMS

In the food fermentation process of Asian soya bean products, microorganisms, notably *Bacillus*, produce enzymes such as amylase, proteinase, mannanase, catalase, cellulose, and so forth (Chettri and Tamang, 2014). Amyloglucosidase, α-amylase, maltase, pectinase, invertase, cellulase, alkaline proteases, lipases, and β-galactosidase are enzymes used by mycelia fungi such as *Amylomyces, Actinomucor, Aspergillus, Mucor, Monascus, Rhizopus,* and *Neurospora.* An enzyme called taka-amylase A (T.A.A.) is produced by A. oryzae in koji and is used in a variety of different fields (Suganuma et al., 2007). In the Himalayas, alcohol is fermented from amylolytic starter cultures, which are used to create a stable, dry, cake-like product. These starting cultures comprise many yeast strains that synergistically increase amylase production (Tamang et al., 2007). *Saccharomycopsis capsularis, Saccharomyces fibuligera,* and *Pichia burtonii* are examples of such strains. Natto contains the *B. subtilis*-produced enzyme nattokinase, which has been shown to have fibrinolytic action (Kotb, 2012). Other bacteria isolated from fermented foods have also shown fibrinolytic activity (Zeng et al., 2013). These bacteria include *B. amyloliquefaciens, V. carniphilus, V. lutrae, P. acidilactici, E. faecalis, E. faecium,* and *E. gallinarum.* The *Virgibacillus halodenitrificans* SK1-3-7 strain, which was discovered in fermented fish sauce, exhibited fibrinolytic activity as well (Montriwong et al., 2015).

1.5.5 PRODUCTION OF POLY-GLUTAMIC ACID AND ENHANCED SAPONIN AND ISOFLAVONE CONTENT

There are four distinct chemical structures in soybeans that can be broken down into isoflavones like genistein, glycitein, and daidzein:

aglycones, malonylglucosides, and acetyl glucosides (Vitale et al., 2013). During fermentation, the isoflavone glucosides in several Asian fermented soybean products, such as Japanese *natto* and *miso*, Chinese *douchi* and *sufu*, Korean *doenjang* and *cheonggukjang*, Thai *thua nao*, and Indonesian *tempeh*, are converted into aglycones. Tempe's aglycone and isoflavone levels, especially Factor-II, were observed to increase during fermentation. There is evidence that the isoflavones in doenjang can boost LDL-C receptor activation and thereby protect against vascular diseases (Kwak et al., 2012). There are two types of soybeans saponins, D.D.M.P. (2, 3-dihydro-2, 5-dihydroxy-6-metyl-4H-pyran-4-one) and Group A (Paucar-Menacho et al., 2010), both of which are oleanane triterpenoid glycosides. Groups B and E saponins are derived from D.D.M.P. and have a number of health benefits, including protection against hypercholesterolemia and a decrease in the proliferation of colon cancer cells (Murata et al., 2006). When *B. subtilis* is used to ferment natto, the saponin content rises. Studies on Group B saponins have revealed that they have a high Kinema content, which is good for consumers (Omizu et al., 2011). Because ribosomal proteins are incapable of producing PGA, *Bacillus* sp. strains such as *B. licheniformis* and *B. subtilis* produce it in Asian fermented soybean products (Chettri et al., 2014). Because they are naturally biodegradable, water-soluble, and non-toxic to humans, the viscous components of Asian fermented soybean products are considered safe for human consumption (Chadare et al., 2019).

1.5.6 ANTI-NUTRITIVE COMPOUNDS DEGRADATION

Most food substrates contain anti-nutritional substances. In addition to being harmful to humans, these substances reduce the nutritional value of food, which in turn reduces the body's access to essential nutrients. Certain microorganisms found in fermented foods have been shown to degrade anti-nutritive substances, potentially making them safe for human consumption (Tamang et al., 2016b). Grating, dewatering, washing, peeling, fermenting, and roasting are all additional processing procedures that can be used to lower the cyanide level.

Gari and fufu, two fermented cassava foods popular in Africa, are particularly rich in phenolic compounds (Omolara, 2014). For traditional preparations of fufu and gari, the cyanogenic glycosides lotaustralin and linamarin found in bitter kinds of cassava tubers are removed using *Lactobacillus, Streptococcus,* and *Leuconostoc*. This leaves a product that is nontoxic and safe to eat (Bamidele et al., 2015). However, the *R. oligosporus* used in Tempe can eliminate gas by breaking down indigestible oligosaccharides like verbascose and stachyose into simpler sugars that the body can use (Sanchez, 2008). While *B. subtilis* degrades anti-nutritional compounds in kinema, phytic acid is reduced in idli and rabadi during fermentation (Tamang et al., 2016b).

1.6 APPLYING OMICS TECHNIQUES TO VARIOUS TYPES OF FERMENTED FOODS

The microbiomes and metabolomes of a wide variety of other fermented foods have been studied by scientists over the past decade. Foods such as bread, sausage, coffee, cocoa, wine, and beer (Zhu et al., 2018). Despite being produced in vastly diverse environments, the microbiomes of these fermented foods and beverages are strikingly comparable, even at the species level. This is true even for goods that are influenced by microbial succession.

The various Asian fermented foods have lately been subjected to shotgun metataxonomic and metagenomic techniques. Whole genome sequencing and functional analysis revealed that a number of *Lactobacillus plantarum* strains isolated from kimchi possessed promising probiotic capabilities (Beck et al., 2019). *Bacillus subtilis* VK161, a bacterium that produces an abnormally high amount of vitamin K2, was also fully sequenced after being isolated from natto (Parks et al., 2019). Another strain of *Bacillus subtilis*, BEST195, had its genome sequenced, and it, too, revealed natto-specific activities (Kamada et al., 2014). Shotgun metagenomics using nanopore technology was applied to kinema, a miso-like sticky fermented soybean snack popular in India, Nepal, and Bhutan. The results showed a *Bacillus* preponderance as well as many metabolic genes encoding for amino acid metabolism (Tamang et al., 2016b). Finally, metataxonomic and metagenomics have been applied to the study of Chinese soy sauce (Sulaiman et al., 2014) and Asian fermented vegetables in order to characterize the metabolic processes of the microbial communities that inhabit these foods (Peng et al., 2018).

1.7 THE BENEFITS OF FERMENTED FOODS TO HUMAN HEALTH

There are built-in mechanisms in ethnic foods that make them both nutritious and therapeutic, satisfying one's hunger while also curing various ailments (Thapa and Tamang, 2015). Okinawans in Japan have the longest life expectancy in the world (Willcox et al., 2004). This is due in large part to their traditional and cultural diets, which include foods like natto, miso, tofu, shoyu, fermented vegetables, cholesterol-free, low-fat, and high bioactive-compound foods, as well as other factors like vigorous physical activity, a healthy environment, happiness, and so on. It has been argued that the Korean condiment kimchi has positive effects on health (Han et al., 2015). The fact that kimchi can delay ageing is a perk (Kim et al., 2002). Nutritionally, natto is advantageous due to the presence of *B. subtilis*, which produces nattokinase, isoflavones, saponins, vitamin K, unsaturated fatty acids, probiotics, and immunomodulating properties (Nagai, 2015). Health-wise, kinesis can be beneficial as well (Tamang, 2015). The common Indian fermented milk product dahi was found to have anti-carcinogenic properties by Kumar et al. (2010). Kimchi's lactic acid has been linked to a reduction in fat storage and a reversal of the heart disease that often results from being overweight (Park et al., 2008). Evidence from clinical trials suggests that both kimchi and doenjang can help fight obesity (Park et al., 2012). Red wine's ability to slow down the ageing process can be attributed to the presence of melatonin, a hormone responsible for controlling the body's circadian rhythm (Walker, 2014). While many cultural groups hold traditional beliefs about the curative properties of certain foods, including fermented foods and beverages, more research is needed to confirm the health claims made about the vast majority of the world's naturally fermented foods and beverages (Jung et al., 2014). In Figure 1.4 find a few examples of how fermented foods can help your health.

1.7.1 SYNTHESIS OF NUTRIENT

Beneficial substances like vitamins, vital amino acids, and bioactive chemicals are more abundant in food substrates after fermentation (Thapa and Tamang, 2015). Tempe, the mold-fermented soybean food of Indonesia, contains vitamin B12 synthesized by nonpathogenic strains

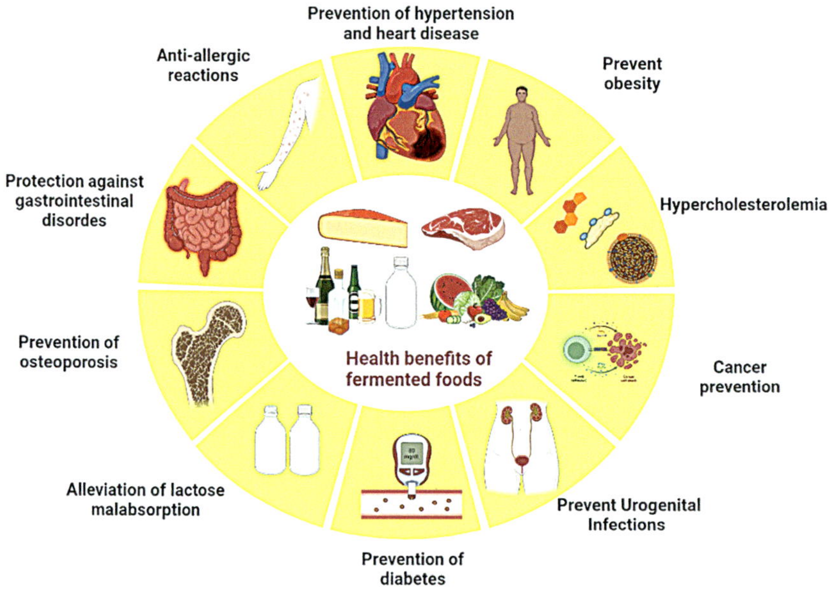

FIGURE 1.4 The positive effects of eating fermented foods on our health.

of *Klebsiella pneumoniae* and *Citrobacter freundii* (Astuti, 2015) and enhanced folic acid, niacin, riboflavin, nicotinamide, and pyridoxine by *Rhizopus oligosporus* (Keuth and Bisping, 1994). Idli, a dish made from fermented rice and legumes, is a staple in both India and Sri Lanka. Idlis have more thiamine, riboflavin, and methionine after fermentation, as discovered by Ghosh and Chattopadhyay (2011). The vitamins B complex and C, lysine, tryptophan, and iron levels of pulque, a Mexican alcoholic drink made from a cactus plant, are all known to increase during the fermentation process (Ramirez et al., 2004). Increased quantities of riboflavin and niacin are produced during the *Bacillus* fermentation process in several Asian fermented foods (Nagai, 2015). It has been discovered that *L. mesenteroides* and *L. sakei* are responsible for the riboflavin and folic acid synthesis in kimchi (Jung et al., 2013). Yeasts such as *Saccharomyces cerevisiae*, *Candida tropicalis*, *Aureobasidium* sp., and *Pichia manschuria* have been isolated from fermented cereal dishes like idli and jalebi in India and Pakistan, respectively, and are responsible for producing vitamin B12 (Syal and Vohra, 2013). Fermented soybean products are richer in free amino acids (Dajanta et al., 2011).

1.7.2 PREVENTION OF HYPERTENSION AND HEART DISEASE

Clinical experiments and animal studies have demonstrated the antihypertensive effects of a variety of fermented milk products (Sipola et al., 2002). Fermented milks or probiotic bacteria (Agerholm-Larsen et al., 2000) and fermented soy meals (Liu and Pan, 2010) have been linked to a lower risk of cardiovascular disease. Eating fermented whole grain meals has been shown to reduce serum LDL-cholesterol, hypertriacylglycerolemia, hypertension, coronary heart disease, insulin resistance, and hyperhomocysteinemia (Anderson, 2003). Consumption of fermented foods, including kefir, tempe (Hermosilla et al., 1993), and fermented soybean meals (Lee, 2005), is associated with lower cholesterol levels (Otes and Cagindi, 2003). The presence of two peptides, VPP and IPP, gives the Japanese fermented sour milk known as Calpis a hypotensive effect (Nakamura et al., 1996). Hypertension is reduced by *L. helvetius* in cultured milk (Shah, 2015). Mevinolin citrinin, which is present in *Monascus purpureus* (also known as angkak) in fermented red rice in China, inhibits the enzyme HMG-CoA reductase, which is essential for the production of cholesterol (Pattanagul et al., 2008). Studies have revealed that drinking Chinese fermented tea can lower one's chance of developing heart disease (Mo et al., 2008). There is evidence that the Asian fermented soybean foods natto (Nagai, 2015) and tempe (Astuti, 2015) have beneficial effects on blood pressure. A mold-fermented soybean snack popular in Korea called doenjang includes an isoflavone that has been found to reduce the risk of cardiovascular disease (Shin et al., 2015). Eating fermented whole grains may help prevent cardiovascular disease and type 2 diabetes (Anderson, 2003). In moderation, wine can be beneficial to your health (Walker, 2014). Polyphenols in red wine may be responsible for the synergistic effect of vitamins E and C in protecting cell membranes from oxidative damage (Fehér et al., 2005). Cardiovascular risk factors are improved in patients with hypertension and Type 2 diabetes who regularly consume Korean fermented soybean foods, including a reduced hypocholesterolemic impact (Jung et al., 2014). Hypertension can be treated with ACE inhibitory peptides isolated from dietary proteins (Jakubczyk et al., 2013). To prevent the spread of potentially fatal new cardiovascular disorders, thrombolytic therapy can make use of fibrinolytic enzyme-rich fermented foods (Singh et al., 2014).

1.7.3 PREVENTION FROM CANCER

Antimutagenic and anticarcinogenic properties have been found in some foods fermented using LAB (Lee et al., 2006). It has been proven that kefir can be effective in treating cancer (Wang et al., 2009). The fermented vegetable of Germany, sauerkraut, includes s-methylmethionine, which reduces the chance of stomach malignancies developing (Kris-Etherton et al., 2002). Live *L. acidophilus* cells in fermented milk products have been shown to decrease β-glucuronidase, azoreductase, and nitroreductase levels (all of which *catalyse* the conversion of procarcinogens to carcinogens), reduce exposure to procarcinogens, and boost the immunological response (Macouzet et al., 2009). Indian yogurt, or dahi, is rich in nutrients and has been shown to inhibit cancer growth (Mohania et al., 2013). Kimchi contains two bacteria, *W. cibaria* and *L. plantarum*, that have been related to a reduced risk of cancer (Kwak et al., 2014).

1.7.4 CONSUMPTION OF YOGURT CAN REDUCE BLADDER, COLON AND CERVICAL CANCER

Some gastrointestinal disorders, including cancer, have been observed (Verna and Lucak, 2010). Certain *Lactobacillus* strains help reduce the symptoms of inflammatory bowel illnesses such as Crohn's disease, paucities, and ulcerative colitis (Orel and Trop, 2014). *L. rhamnosus* GG has been demonstrated to be useful in treating acute diarrhea (Szajewska et al., 2007), while *L. helveticus*-fermented milk has been shown to enhance cognitive performance in otherwise healthy older persons (Chung et al., 2014). It has been demonstrated that consuming fermented milk products containing microorganisms can prevent diarrhea (Granier et al., 2013). This was found to be the case (Balamurugan et al., 2014). Korean kimchi is effective in treating inflammatory bowel disorders (Lim et al., 2011).

1.7.5 ANTI-ALLERGIC REACTIONS

Researchers found that *Lactobacillus kefiranofaciens* M1, which was isolated from kefir grains, had anti-allergenic characteristics (Hong et al., 2010). It has been shown that casein digestion during the ageing of fermented milk products aids in the elimination of allergenic reactivity, leading to greater tolerance (Alessandri et al., 2012). *Chongkokjang's* anti-allergic benefits include an increased skin thickness, a decreased ear thickness, a decreased auricular lymph node, and an infiltration of mast cells (Lee et al., 2014). *Lactobacillus* species isolated from kimchi have been shown to change the Th1/Th2 balance by producing high levels of IL-12 and IFN-, leading to a lower prevalence of atopic dermatitis and food allergies (Won et al., 2011). Due to its high concentration of Omega-3 polyunsaturated fatty acids, fermented fish oil can help mitigate allergy symptoms (Han et al., 2012).

1.7.6 PROTECTION FROM DIABETES AND OSTEOPOROSIS

Diets high in fiber may improve insulin sensitivity in those who are not diabetic and lower insulin requirements in those who are (Anderson, 2003). A probiotic Dahi diet is associated with a lower risk of developing diabetes by postponing the onset of glucose intolerance, hyperglycemia, hyperinsulinemia, oxidative stress, and dyslipidemia (Yadav et al., 2007). *Changkokjang*, if consumed on a daily basis, may improve insulin resistance and lead to better control of diabetes (Tolhurst et al., 2012). Natto's vitamin K2 promotes bone growth, which could make it useful for warding off osteoporosis in Japanese women of a certain age (Yanagisawa and Sumi, 2005). The protein in yogurt, along with the magnesium, calcium, phosphorus, and potassium it also contains, may work synergistically to promote bone health (Chandan and Kilara, 2013).

1.7.7 ALLEVIATION OF LACTOSE MALABSORPTION

Those with lactose malabsorption lack the enzyme ß-D-galactosidase, which is necessary for the entire digestion of lactose, the major carbohydrate in milk, into glucose and galactose (Shah, 2015). Yogurt is fermented using strains of *Lactobacillus delbrueckii* subsp. *bulgaricus* and *Streptococcus thermophilus* that generate high levels of ß-D-galactosidase, an enzyme that facilitates lactose malabsorption and reduces lactose intolerance symptoms (Shah et al., 2013). Studies have shown that eating fresh yogurt (with live yogurt cultures) aids in lactose digestion and absorption more so than eating a pasteurized version of the same food (Pedone et al., 2000). Kefir's high β-galactosidase levels aid in the reduction of lactose intolerance symptoms (Hertzler and Clancy, 2003).

1.7.8 INFLAMMATORY BOWEL DISEASES AND IRRITABLE BOWEL SYNDROME

Inflammatory bowel diseases (IBDs) have been associated with changes or disruptions in the normal intestinal microbiota, including *pouchitis,* Crohn's disease, and irritable bowel syndrome (Chmielewska and Szajewska, 2010). These disorders have been linked to the gut, and probiotics have been advocated as a treatment because they can either restore or alter the gut microbiota to make sure it is healthy to support the host (Zmora et al., 2019). Some research suggests that probiotics can help with gut inflammation by altering the gut microbiota; hence, they have been proposed as a possible therapy supplement (Wedajo, 2015). The composition of the intestinal microbiota is important in the etiology of IBD.

1.7.9 UROGENITAL INFECTIONS

More than one billion women worldwide are affected by UTIs, BV, and a variety of yeast infections (Waigankar and Patel, 2011). *Ureaplasma urealyticum, Mycoplasma hominis,* and *Gardnerella vaginalis* are some of the most often mentioned causes of BV (Hanson et al., 2016). According to research, the vaginal microbiome is ideal for the development and maintenance of over 50 distinct species, with the *Lactobacillus* group playing a crucial regulatory function. These LAB are commonly found in the human body, with *Lactobacillus vaginalis, Lactobacillus casei, Lactobacillus reuteri, Lactobacillus salivarius, Lactobacillus brevis, Lactobacillus rhamnosus,* and *Lactobacillus delbrueckii* subsp. *bulgaricus* being the most prevalent (Kerry et al., 2018). When there is a discordance in the vaginal microbial ecosystem, BV and other urogenital tract infections are more likely to occur. Probiotics are able to revitalize the body's microbiome by reintroducing a diverse population of LAB (Kerry et al., 2018).

1.8 CONCLUSIONS AND FUTURE PERSPECTIVES

Fermentation enhances the nutritional value and health advantages of the final product by inducing a number of chemical changes in the raw matrix's constituents. There was a greater concentration of bioactive compounds and improved antioxidant activity in fermented milks, cereals, fruit, vegetables, meat, and fish. There is evidence that eating fermented foods can improve health as well. Increasing the diversity of gut microbes through eating fermented foods and probiotics may have a moderating effect on brain and CNS function. Fermented foods contain a wide variety of components, some of which are beneficial to health, such as nutrients, bioactive, and enzyme-producing microorganisms, while others may be harmful. Because these medications can have both beneficial and detrimental impacts on the health of different people and groups, they require a case-by-case analysis of their efficacy and safety. In addition, the microbes in the fermented foods that people eat on a regular basis are viewed as "micro factories" that generate and enrich nutrients and bioactive with particular nutritional and health functions. Advances in molecular microbial ecology and characterization techniques, along with a comprehensive understanding of the microbial interactions that occur during food fermentation and the interplays between fermentation and other processing technologies, are expected to propel the fermented food industry to new heights in the coming years. To combat the growing public health crises, more research and development should be put into the manufacturing of wellness foods that are fermented and infused with beneficial microbes. Foods that promote health and wellness through fermentation can help us reach this goal, but they need to be environmentally friendly and tailored to different demographics and cultural preferences. Major obstacles include monitoring interactions between the fermentation process and various pre- and post-treatments and regulating metabolite products through individualized microbial fermentation (particularly those involving emerging processing technologies). All novel fermented products, such as those with high nutritional value and unique biological functionalities (fermentation-enabled wellness foods and functional fermented foods), must undergo a complete and rigorous safety assessment as novel foods before their health and nutritional benefits can be confirmed.

KEYWORDS

- **biochemical changes**
- **enzymes**
- **fermenting microorganisms**
- **food fermentation**
- **health benefits**
- **microorganisms**

REFERENCES

Abubakar, M. A. S., Zaiton, H., Imdakim, M. M. A., & Sharifah N. R. S. A. (2012). Antioxidant activity of lactic acid bacteria (LAB) fermented skim milk as determined by 1,1-diphenyl-2-picrylhydrazyl (DPPH) and ferrous chelating activity (FCA). *African Journal of Microbiology Research, 6*, 6358–6364.

Agerholm-Larsen, L., Bell, M. L., Grunwald, G. K., & Astrup, A. (2000). The effect of a probiotic milk product on plasma cholesterol: A meta-analysis of short-term intervention studies. *European Journal of Clinical Nutrition, 54*(11), 856–860.

Ahmed, Z., Wang, Y., Ahmad, A., Khan, S. T., Nisa, M., Ahmad, H., & Afreen, A. (2013). Kefir and health: A contemporary perspective. *Critical Reviews in Food Science and Nutrition, 53*(5), 422–434.

Aidoo, K. E., & Nout, M. J. R. (2010). Functional yeasts and molds in fermented foods and beverages. In J. P. Tamang & K. Kailasapathy (Eds.), *Fermented Foods and Beverages of the World* (pp. 127–148). CRC Press, Taylor and Francis Group.

Aka, S., Konan, G., Fokou, G., Dje, K. M., & Bonfoh, B. (2014). Review of African traditional cereal beverages. *American Journal of Research Communication, 2*, 103–153.

Alcántara, C., Bäuerl, C., Revilla-Guarinos, A., Pérez-Martínez, G., Monedero, V., & Zúñiga, M. (2016). Peptide and amino acid metabolism is controlled by an OmpR-family response regulator in *Lactobacillus casei*. *Molecular Microbiology, 100*(1), 25–41.

Alessandri, C., Sforza, S., Palazzo, P., Lambertini, F., Paolella, S., Zennaro, D., Rafaiani, C., Ferrara, R., Bernardi, M. L., Santoro, M., Zuzzi, S., Giangrieco, I., Dossena, A., & Mari, A. (2012). Tolerability of a fully maturated cheese in cow's milk allergic children: Biochemical, immunochemical, and clinical aspects. *PLOS ONE, 7*(7), e40945.

Amoa-Awua, W. K., Sampson, E., & Tano-Debrah, K. (2007). Growth of yeasts, lactic and acetic acid bacteria in palm wine during tapping and fermentation from felled oil palm (*Elaeis guineensis*) in Ghana. *Journal of Applied Microbiology, 102*(2), 599–606.

Anal, A. K. (2019). Quality ingredients and safety concerns for traditional fermented foods and beverages from Asia: A review. *Fermentation, 5*(1), 8.

Anderson, J. W. (2003). Whole grains protect against atherosclerotic cardiovascular disease. *The Proceedings of the Nutrition Society, 62*(1), 135–142.

Astuti, M. (2015). Health benefits of Tempe. In J. P. Tamang (Ed.), *Health Benefits of Fermented Foods* (pp. 371–394). CRC Press, Taylor and Francis Group.

Balamurugan, R., Chandragunasekaran, A. S., Chellappan, G., Rajaram, K., Ramamoorthi, G., & Ramakrishna, B. S. (2014). Probiotic potential of lactic acid bacteria presents in homemade curd in southern India. *The Indian Journal of Medical Research, 140*(3), 345–355.

Bamforth, C. W. (2002). Nutritional aspects of beer—A review. *Nutrition Research, 22*, 227–237.

Bamidele, O. P., Fasogbon, M. B., Oladiran, D. A., & Akande, E. O. (2015). Nutritional composition of fufu analog flour produced from cassava root (*Manihot esculenta*) and cocoyam (*Colocasia esculenta*) tuber. *Food Science & Nutrition, 3*(6), 597–603.

Barennes, H., Sengkhamyong, K., René, J. P., & Phimmasane, M. (2015). Beriberi (thiamine deficiency) and high infant mortality in northern Laos. *PLOS Neglected Tropical Diseases, 9*(3), e0003581.

Basannavar, S., Pothuraju, R., & Sharma, R. K. (2014). Effect of *Aloe vera* (*Aloe barbadensis* Miller) on survivability, extent of proteolysis and ACE inhibition of potential probiotic cultures in fermented milk. *Journal of the Science of Food and Agriculture, 94*(13), 2712–2717.

Batish, V. K., & Grover, S. (2004). Fermented milk products. In A. Pandey (Ed.), *Concise Encyclopedia of Bioresource Technology* (pp. 201–209).

Beck, B. R., Park, G. S., Lee, Y. H., Im, S., Jeong, D. Y., & Kang, J. (2019). Whole genome analysis of *Lactobacillus plantarum* strains isolated from kimchi and determination of probiotic properties to treat mucosal infections by *Candida albicans* and *Gardnerella vaginalis*. *Frontiers in Microbiology, 10*, 433.

Behera, S. S., Ray, R. C., & Zdolec, N. (2018). *Lactobacillus plantarum* with functional properties: An approach to increase safety and shelf-life of fermented foods. *BioMed Research International, 9361614*, 1–18.

Bhattacharya, A., Banu, J., Rahman, M., Causey, J., & Fernandes, G. (2006). Biological effects of conjugated linoleic acids in health and disease. *The Journal of Nutritional Biochemistry, 17*(12), 789–810.

Bourdichon, F., Casaregola, S., Farrokh, C., Frisvad, J. C., Gerds, M. L., Hammes, W. P., Harnett, J., Huys, G., Laulund, S., Ouwehand, A., Powell, I. B., Prajapati, J. B., Seto, Y., Ter Schure, E., Van Boven, A., Vankerckhoven, V., Zgoda, A., Tuijtelaars, S., & Hansen, E. B. (2012). Food fermentations: Microorganisms with technological beneficial use. *International Journal of Food Microbiology, 154*(3), 87–97.

Bull, M., Plummer, S., Marchesi, J., & Mahenthiralingam, E. (2013). The life history of *Lactobacillus acidophilus* as a probiotic: A tale of revisionary taxonomy, misidentification and commercial success. *FEMS Microbiology Letters, 349*(2), 77–87.

Chadare, F. J., Idohou, R., Nago, E., Affonfere, M., Agossadou, J., Fassinou, T. K., Kénou, C., Honfo, S., Azokpota, P., Linnemann, A. R., & Hounhouigan, D. J. (2019). Conventional and food-to-food fortification: An appraisal of past practices and lessons learned. *Food Science & Nutrition, 7*(9), 2781–2795.

Chandan, R. C., & Kilara, A. (2013). *Manufacturing Yogurt and Fermented Milks* (2nd edn.). John Wiley & Sons.

Chen, B., Wu, Q., & Xu, Y. (2014). Filamentous fungal diversity and community structure associated with the solid-state fermentation of Chinese Maotai-flavor liquor. *International Journal of Food Microbiology, 179*, 80–84.

Chettri, R., & Tamang, J. P. (2014). Functional properties of Tungrymbai and Bekang, naturally fermented soybean foods of North East India. *International Journal of Fermented Foods, 3*, 87–103.

Chmielewska, A., & Szajewska, H. (2010). Systematic review of randomized controlled trials: Probiotics for functional constipation. *World Journal of Gastroenterology, 16*(1), 69–75.

Chou, C. C., Lin, L. L., & Chung, K. T. (1999). Antimicrobial activity of tea as affected by the degree of fermentation and manufacturing season. *International Journal of Food Microbiology, 48*(2), 125–130.

Chung, Y. C., Jin, H. M., Cui, Y., Kim, D. S., Jung, J. M., Park, J. I., et al. (2014). Fermented milk of *Lactobacillus helveticus* IDCC3801 improves cognitive functioning during cognitive fatigue tests in healthy older adults. *Journal of Functional Foods, 10*, 465–474.

Cocolin, L., Bisson, L. F., & Mills, D. A. (2000). Direct profiling of the yeast dynamics in wine fermentations. *FEMS Microbiology Letters, 189*(1), 81–87.

Coton, E., Desmonts, M. H., Leroy, S., Coton, M., Jamet, E., Christieans, S., Donnio, P. Y., Lebert, I., & Talon, R. (2010). Biodiversity of coagulase-negative *Staphylococci* in French cheeses, dry fermented sausages, processing environments and clinical samples. *International Journal of Food Microbiology, 137*(2, 3), 221–229.

Dajanta, K., Apichartsrangkoon, A., Chukeatirote, E., Richard, A., & Frazier, R. A. (2011). Free-amino acid profiles of Thua nao, a Thai fermented soybean. *Food Chemistry, 125*, 342–347.

de Mejia, E. G., & Dia, V. P. (2010). The role of nutraceutical proteins and peptides in apoptosis, angiogenesis, and metastasis of cancer cells. *Cancer Metastasis Reviews, 29*(3), 511–528.

Ekanayake, S. (2016). Ethnic fermented foods and beverages of Sri Lanka. In J. P. Tamang (Ed.), *Ethnic Fermented Foods and Alcoholic Beverages of Asia* (pp. 139–150). Springer.

Escalante, A., López, S. D. R., Velázquez, G. J. E., Giles-Gómez, M., Bolívar, F., & López-Munguía, A. (2016). Pulque, a traditional Mexican alcoholic fermented beverage: Historical, microbiological, and technical aspects. *Frontiers in Microbiology, 7*, 1026.

Fabien, J., Cousin, D. G., Mater, B., Foligné, B., & Jan, G. (2011). Dairy propionibacteria as human probiotics: A review of recent evidence. *Dairy Science & Technology, 91*(1), 1–26.

Fehér, J., Lengyel, G., & Lugasi, A. (2005). A bor kultúrtörténete, a borterápia elméleti háttere [Cultural history of wine, the theoretical background of wine therapy]. *Orvosi Hetilap, 146*(52), 2635–2639.

Fleming, H. P., Thompson, R. L., Bell, T. A., & Hontz, L. H. (1978). Controlled fermentation of sliced cucumbers. *Journal of Food Science, 43*, 888–891.

Fu, D., Ryan, E. P., Huang, J., Liu, Z., Weir, T. L., Snook, R. L., & Ryan, T. P. (2011). Fermented *Camellia sinensis*, Fu Zhuan tea, regulates hyperlipidemia and transcription factors involved in lipid catabolism. *Food Research International, 44*, 2999–3005.

García-Burgos, M., Moreno-Fernández, J., Alférez, M. J., Díaz-Castro, J., & López-Aliaga, I. (2020). New perspectives in fermented dairy products and their health relevance. *Journal of Functional Foods, 72*, 104059.

Ghosh, D., & Chattopadhyay, P. (2011). Preparation of idli batter, its properties and nutritional improvement during fermentation. *Journal of Food Science and Technology, 48*(5), 610–615.

Ghosh, T., Beniwal, A., Semwal, A., & Navani, N. K. (2019). Mechanistic insights into probiotic properties of lactic acid bacteria associated with ethnic fermented dairy products. *Frontiers in Microbiology, 10*, 502.

Granier, A., Goulet, O., & Hoarau, C. (2013). Fermentation products: Immunological effects on human and animal models. *Pediatric Research, 74*(2), 238–244.

Griffiths, M. W., & Tellez, A. M. (2013). *Lactobacillus helveticus*: The proteolytic system. *Frontiers in Microbiology, 4*, 30.

Grüber, C., Wendt, M., Sulser, C., Lau, S., Kulig, M., Wahn, U., Werfel, T., & Niggemann, B. (2007). Randomized, placebo-controlled trial of *Lactobacillus rhamnosus* GG as treatment of atopic dermatitis in infancy. *Allergy, 62*(11), 1270–1276.

Guo, C., Zhang, S., Yuan, Y., Yue, T., & Li, J. (2015). Comparison of lactobacilli isolated from Chinese *Suan-Tsai* and *koumiss* for their probiotic and functional properties. *Journal of Functional Foods, 12*, 294–302.

Gupta, R. K., Gangoliya, S. S., & Singh, N. K. (2015). Reduction of phytic acid and enhancement of bioavailable micronutrients in food grains. *Journal of Food Science and Technology, 52*(2), 676–684.

Han, E. S., Kim, H. J., & Choi, H. K. (2015). Health benefits of kimchi. In J. P. Tamang (Ed.), *Health Benefits of Fermented Foods* (pp. 343–370). CRC Press, Taylor and Francis Group.

Han, S. C., Kang, G. J., Ko, Y. J., Kang, H. K., Moon, S. W., Ann, Y. S., & Yoo, E. S. (2012). Fermented fish oil suppresses T helper 1/2 cell response in a mouse model of atopic dermatitis via generation of CD4+CD25+Foxp3+ T cells. *BMC Immunology, 13*, 44.

Hanson, L., VandeVusse, L., Jermé, M., Abad, C. L., & Safdar, N. (2016). Probiotics for treatment and prevention of urogenital infections in women: A systematic review. *Journal of Midwifery & Women's Health, 61*(3), 339–355.

Hermosilla, J. A. G., Jha, H. C., Egge, H., & Mahmud, M. (1993). Isolation and characterization of hydroxymethylglutaryl coenzyme A reductase inhibitors from fermented soybean extracts. *Journal of Clinical Biochemistry and Nutrition, 15*(3), 163–174.

Hertzler, S. R., & Clancy, S. M. (2003). Kefir improves lactose digestion and tolerance in adults with lactose maldigestion. *Journal of the American Dietetic Association, 103*(5), 582–587.

Hong, W. S., Chen, Y. P., & Chen, M. J. (2010). The antiallergic effect of kefir *Lactobacilli*. *Journal of Food Science, 75*(8), H244–H253.

Hoque, M. Z., Akter, F., Hossain, K. M., Rahman, M. M., Billah, M. M., & Islam, K. M. (2010). Isolation, identification and analysis of probiotic properties of *Lactobacillus* spp. from selective regional yogurts. *World Journal of Dairy & Food Sciences, 5*, 39–46.

Hossain, M., & Kabir, Y. (2016). Ethnic fermented foods and beverages of Bangladesh. In J. P. Tamang (Ed.), *Ethnic Fermented Foods and Alcoholic Beverages of Asia* (pp. 73–89). Springer.

Hou, J. C., Liu, F., Ren, D. X., Han, W. W., & Du, Y. O. (2015). Effect of culturing conditions on the expression of key enzymes in the proteolytic system of *Lactobacillus bulgaricus*. *Journal of Zhejiang University Science B, 16*(4), 317–326.

Howeida, M. A., Idris, E. B., Almahdi, A. M., Sania, S. A., & Abdelwahhab, M. H. (2015). Antidiabetic and hypolipidaemic effects of *Citrus aurantifolin* leaves on hyperglycaemic and diabetic rats. *International Journal of Biological and Medical Research, 6*(3), 5095–5099.

Hung, C. (2005). Health food processing process using germinated rice to make health food containing natural eatable fibers, GABA, IP6, and probiotic: U.S. Patent 6,977,093[P].

Hwa, D. (2011). Utilization of soybean as food stuffs in Korea. In *Soybean and Nutrition*. https://doi.org/10.5772/21966 (accessed on 15th November 2025).

Imao, K., Wang, H., Komatsu, M., & Hiramatsu, M. (1998). Free radical scavenging activity of fermented papaya preparation and its effect on lipid peroxide level and superoxide dismutase activity in iron-induced epileptic foci of rats. *Biochemistry and Molecular Biology International, 45*(1), 11–23.

Jakubczyk, A., Karaś, M., Baraniak, B., & Pietrzak, M. (2013). The impact of fermentation and in vitro digestion on formation angiotensin converting enzyme (ACE) inhibitory peptides from pea proteins. *Food Chemistry, 141*(4), 3774–3780.

Jayabalan, R., Marimuthu, S., & Swaminathan, K. (2007). Changes in content of organic acids and tea polyphenols during kombucha tea fermentation. *Food Chemistry, 102*, 392–398.

Jung, J. Y., Lee, S. H., Jin, H. M., Hahn, Y., Madsen, E. L., & Jeon, C. O. (2013). Metatranscriptomic analysis of lactic acid bacterial gene expression during kimchi fermentation. *International Journal of Food Microbiology, 163*(2, 3), 171–179.

Jung, S. J., Park, S. H., Choi, E. K., Cha, Y. S., Cho, B. H., Kim, Y. G., Kim, M. G., Song, W. O., Park, T. S., Ko, J. K., So, B. O., & Chae, S. W. (2014). Beneficial effects of Korean traditional diets in hypertensive and type 2 diabetic patients. *Journal of Medicinal Food, 17*(1), 161–171.

Kalaivani, M., Sabitha, R., Kalaiselvan, V., & Rajasekaran, A. (2010). Health benefits and clinical impact of major nutrient, red yeast rice: A review. *Food and Bioprocess Technology, 3*, 333–339.

Kamada, M., Hase, S., Sato, K., Toyoda, A., Fujiyama, A., & Sakakibara, Y. (2014). Whole genome complete resequencing of *Bacillus subtilis natto* by combining long reads with high-quality short reads. *PLOS ONE, 9*(10), e109999.

Karamoko, D., Moroh, J. L. A., Bouatenin, K. M. J. P., & Dje, K. M. (2016). Biochemical and microbial properties of palm wine: Effect of tapping length and varietal differences. *Food and Nutrition Sciences, 7*(9), 763–771.

Katina, K., Arendt, E., Liukkonen, K., Autio, K., Flander, L., & Poutanen, K. (2005). Potential of sourdough for healthier cereal products. *Trends in Food Science & Technology, 16*(1–3), 104–112.

Kato, K., Takada, Y., Matsuyama, H., Kawasaki, Y., Aoe, S., Yano, H., & Toba, Y. (2002). Milk calcium taken with cheese increases bone mineral density and bone strength in growing rats. *Bioscience, Biotechnology, and Biochemistry, 66*(11), 2342–2346.

Kerry, R. G., Patra, J. K., Gouda, S., Park, Y., Shin, H. S., & Das, G. (2018). Benefaction of probiotics for human health: A review. *Journal of Food and Drug Analysis, 26*(3), 927–939.

Keuth, S., & Bisping, B. (1994). Vitamin B12 production by *Citrobacter freundii* or *Klebsiella pneumoniae* during tempeh fermentation and proof of enterotoxin absence by PCR. *Applied and Environmental Microbiology, 60*(5), 1495–1499.

Kim, B., Hong, V. M., Yang, J., Hyun, H., Im, J. J., Hwang, J., Yoon, S., & Kim, J. E. (2016). A review of fermented foods with beneficial effects on brain and cognitive function. *Preventive Nutrition and Food Science, 21*(4), 297–309.

Kim, J. H., Ryu, J. D., & Song, Y. O. (2002). The effect of kimchi intake on free radical production and the inhibition of oxidation in young adults and the elderly people. *Korean Journal of Community Nutrition, 7*(2), 257–265.

Kotb, E. (2012). *Fibrinolytic bacterial enzymes with thrombolytic activity*. Springer Briefs in Microbiology, 1–74.

Kris-Etherton, P. M., Hecker, K. D., Bonanome, A., Coval, S. M., Binkoski, A. E., Hilpert, K. F., Griel, A. E., & Etherton, T. D. (2002). Bioactive compounds in foods: Their role in the prevention of cardiovascular disease and cancer. *The American Journal of Medicine, 113*(Suppl 9B), 71S–88S.

Kubo, Y., Rooney, A. P., Tsukakoshi, Y., Nakagawa, R., Hasegawa, H., & Kimura, K. (2011). Phylogenetic analysis of *Bacillus subtilis* strains applicable to natto (fermented soybean) production. *Applied and Environmental Microbiology, 77*(18), 6463–6469.

Kuda, T., Kaneko, N., Yano, T., & Mori, M. (2010). Induction of superoxide anion radical scavenging capacity in Japanese white radish juice and milk by *Lactobacillus plantarum* isolated from *aji-narezushi* and *kaburazushi*. *Food Chemistry, 120*, 517–522.

Kumar, A., Singh, N. K., & Sinha, P. R. (2010). Inhibition of 1,2-dimethylhydrazine induced colon genotoxicity in rats by the administration of probiotic curd. *Molecular Biology Reports, 37*(3), 1373–1376.

Kwak, C. S., Park, S. C., & Song, K. Y. (2012). *Doenjang*, a fermented soybean paste, decreased visceral fat accumulation and adipocyte size in rats fed with high fat diet more effectively than nonfermented soybeans. *Journal of Medicinal Food, 15*(1), 1–9.

Kwak, M., Liu, R., & Kang, S. (2018). Antimicrobial activity of cyclic dipeptides produced by *Lactobacillus plantarum* LBP-K10 against multidrug-resistant bacteria, pathogenic fungi, and influenza A virus. *Food Control, 85*, 223–234.

Kwak, S. H., Cho, Y. M., Noh, G. M., & Om, A. S. (2014). Cancer preventive potential of kimchi lactic acid bacteria (*Weissella cibaria, Lactobacillus plantarum*). *Journal of Cancer Prevention, 19*(4), 253–258.

Law, S., Bakar, F. A., Hashim, D. M., & Hamid, A. A. (2011). Popular fermented foods and beverages in Southeast Asia. *International Food Research Journal, 18*(2), 475–484.

Lee, C. H. (2006). Creative fermentation technology for the future. *Journal of Food Science, 69*(2), 33–34.

Lee, Y. J., Kim, J. E., Kwak, M. H., Go, J., Kim, D. S., Son, H. J., & Hwang, D. Y. (2014). Quantitative evaluation of the therapeutic effect of fermented soybean products containing a high concentration of GABA on phthalic anhydride-induced atopic dermatitis in IL-4/Luc/CNS-1 Tg mice. *International Journal of Molecular Medicine, 33*(5), 1185–1194.

Li, S., Zhao, Y., Zhang, L., Zhang, X., Huang, L., Li, D., Niu, C., Yang, Z., & Wang, Q. (2012). Antioxidant activity of *Lactobacillus plantarum* strains isolated from traditional Chinese fermented foods. *Food Chemistry, 135*(3), 1914–1919.

Lim, J., Seo, B. J., Kim, J. E., Chae, C. S., Im, S. H., Hahn, Y. S., et al. (2011). Characteristics of immunomodulation by a *Lactobacillus sakei* proBio65 isolated from kimchi. *Korean Journal of Microbiology and Biotechnology, 39*(3), 313–316.

Liu, C. F., & Pan, T. M. (2010). In vitro effects of lactic acid bacteria on cancer cell viability and antioxidant activity. *Journal of Food and Drug Analysis, 18*(2), 77–86.

Lu, Z., Breidt, F., Plengvidhya, V., & Fleming, H. P. (2003). Bacteriophage ecology in commercial sauerkraut fermentations. *Applied and Environmental Microbiology, 69*(6), 3192–3202.

Macouzet, M., Lee, B. H., & Robert, N. (2009). Production of conjugated linoleic acid by probiotic *Lactobacillus acidophilus* La-5. *Journal of Applied Microbiology, 106*(6), 1886–1891.

Marco, M. L., Heeney, D., Binda, S., Cifelli, C. J., Cotter, P. D., Foligné, B., Gänzle, M., Kort, R., Pasin, G., Pihlanto, A., Smid, E. J., & Hutkins, R. (2017). Health benefits of fermented foods: Microbiota and beyond. *Current Opinion in Biotechnology, 44*, 94–102.

McGrath, M., & Simone, K. (2015). *Apple cider vinegar for health and beauty: Recipes for weight loss, clear skin, superior health, and much more – the natural way, a hand book.*

Melini, F., Melini, V., Luziatelli, F., Ficca, A. G., & Ruzzi, M. (2019). Health-promoting components in fermented foods: An up-to-date systematic review. *Nutrients, 11*(5), 1189.

Minervini, F., Siragusa, S., Faccia, M., Dal Bello, F., Gobbetti, M., & De Angelis, M. (2012). Manufacture of Fior di Latte cheese by incorporation of probiotic lactobacilli. *Journal of Dairy Science, 95*(2), 508–520.

Miyamoto, M., Ueno, H. M., Watanabe, M., Tatsuma, Y., Seto, Y., Miyamoto, T., & Nakajima, H. (2015). Distinctive proteolytic activity of cell envelope proteinase of *Lactobacillus helveticus* isolated from *airag*, a traditional Mongolian fermented mare's milk. *International Journal of Food Microbiology, 197*, 65–71.

Mo, H., Zhu, Y., & Chen, Z. (2008). Microbial fermented tea – A potential source of natural food preservatives. *Trends in Food Science & Technology, 19*(3), 124–130.

Mohania, D., Kansal, V. K., Sagwal, R., & Shah, D. (2013). Anticarcinogenic effect of probiotic *Dahi* and piroxicam on DMH-induced colorectal carcinogenesis in Wister rats. *American Journal of Cancer Therapeutics and Pharmacology, 1*, 8–24.

Mohd Ali, N., Mohd Yusof, H., Long, K., Yeap, S. K., Ho, W. Y., Beh, B. K., Koh, S. P., Abdullah, M. P., & Alitheen, N. B. (2013). Antioxidant and hepatoprotective effect of aqueous extract of germinated and fermented mung bean on ethanol-mediated liver damage. *BioMed Research International, 2013*, 693613.

Montriwong, A., Rodtong, S., & Yongsawatdigul, J. (2015). Detergent-stable salt-activated proteinases from *Virgibacillus halodenitrificans* SK1-3-7 isolated from fish sauce fermentation. *Applied Biochemistry and Biotechnology, 176*(2), 505–517.

Munekata, P. E. S., Pateiro, M., Tomasevic, I., Domínguez, R., da Silva, B. A. C., Santos, E. M., & Lorenzo, J. M. (2022). Functional fermented meat products with probiotics – A review. *Journal of Applied Microbiology, 133*(1), 91–103.

Murata, M., Hodai, T., Yamamoto, H., Matsumori, N., & Oishi, T. (2006). Membrane interaction of soyasaponins in association with their antioxidation effect – Analysis of biomembrane interaction. *Soy Protein Research, 9*, 82–86.

Nagai, T. (2015). Health benefits of *natto*. In J. P. Tamang (Ed.), *Health Benefits of Fermented Foods* (pp. 433–453). CRC Press, Taylor and Francis Group.

Nakamura, Y., Masuda, O., & Takano, T. (1996). Decrease of tissue angiotensin I-converting enzyme activity upon feeding sour milk in spontaneously hypertensive rats. *Bioscience, Biotechnology, and Biochemistry, 60*(3), 488, 489.

Nishito, Y., Osana, Y., Hachiya, T., Popendorf, K., Toyoda, A., Fujiyama, A., Itaya, M., & Sakakibara, Y. (2010). Whole genome assembly of a *natto* production strain *Bacillus subtilis natto* from very short read data. *BMC Genomics, 11*, 243.

Nout, M. J., & Kiers, J. L. (2005). Tempe fermentation, innovation and functionality: Update into the third millennium. *Journal of Applied Microbiology, 98*(4), 789–805.

Ockerman, H. W., & Basu, L. (2014). Production and consumption of fermented meat products. In H. W. Ockerman & L. Basu (Eds.), *Handbook of Fermented Meat and Poultry*.

Oguntoyinbo, F. A., Sanni, A. I., Franz, C. M., & Holzapfel, W. H. (2007). *In vitro* fermentation studies for selection and evaluation of *Bacillus* strains as starter cultures for the production of *okpehe*, a traditional African fermented condiment. *International Journal of Food Microbiology, 113*(2), 208–218.

Omar, N., & Ampe, F. (2000). Microbial community dynamics during production of the Mexican fermented maize dough *pozol*. *Applied and Environmental Microbiology, 66*(9), 3664–3673.

Omizu, Y., Tsukamoto, C., Chettri, R., & Tamang, J. P. (2011). Determination of saponin contents in raw soybean and fermented soybean foods of India. *Journal of Scientific and Industrial Research, 70*, 533–530.

Omolara, B. O. (2014). Cyanide content of commercial *gari* from different areas of Ekiti State, Nigeria. *Journal of Natural Sciences Research, 2*, 58–60.

Oppermann-Sanio, F. B., & Steinbüchel, A. (2002). Occurrence, functions and biosynthesis of polyamides in microorganisms and biotechnological production. *Die Naturwissenschaften, 89*(1), 11–22.

Orel, R., & Kamhi, T. T. (2014). Intestinal microbiota, probiotics and prebiotics in inflammatory bowel disease. *World Journal of Gastroenterology, 20*(33), 11505–11524.

Otes, S., & Cagindi, O. (2003). Kefir: A probiotic dairy – Composition, nutritional and therapeutic aspects. *Pakistan Journal of Nutrition, 2*, 54–59.

Park, J. A., Tirupathi, P. P. B., Yu, J. J., Oh, S. H., Daily, J. W. 3rd., & Cha, Y. S. (2012). Anti-obesity effect of kimchi fermented with *Weissella koreensis* OK1-6 as starter in high-fat diet-induced obese C57BL/6J mice. *Journal of Applied Microbiology, 113*(6), 1507–1516.

Park, J. E., Moon, Y. J., & Cha, Y. S. (2008). Effect of functional materials producing microbial strains isolated from kimchi on anti-obesity and inflammatory cytokines in 3T3-L1 preadipocytes. *The FASEB Journal, 23*(S1), 111.

Park, K. Y., Jeong, J. K., Lee, Y. E., & Daily, J. W. 3rd., (2014). Health benefits of kimchi (Korean fermented vegetables) as a probiotic food. *Journal of Medicinal Food, 17*(1), 6–20.

Parks, D., Chung, I. H., Lee, I. K., Kim, E. J., Niraula, S., & Chang, W. S. (2019). Genome sequence of *Bacillus subtilis natto* VK161, a novel strain that produces vitamin K2. *Microbiology Resource Announcements, 8*(35), e00444-19.

Pattanagul, P., Pinthong, R., Phianmongkhol, A., & Tharatha, S. (2008). Mevinolin, citrinin and pigments of *adlay angkak* fermented by *Monascus* sp. *International Journal of Food Microbiology, 126*(1, 2), 20–23.

Paucar-Menacho, L. M., Amaya-Farfan, J., Berhow, M. A., Mandarino, J. M., Mejia, E. G., & Chang, Y. K. (2010). A high-protein soybean cultivar contains lower isoflavones and saponins but higher minerals and bioactive peptides than a low-protein cultivar. *Food Chemistry, 120*, 15–21.

Pedone, C. A., Arnaud, C. C., Postaire, E. R., Bouley, C. F., & Reinert, P. (2000). Multicentric study of the effect of milk fermented by *Lactobacillus casei* on the incidence of diarrhea. *International Journal of Clinical Practice, 54*(9), 568–571.

Penas, E., Martinez-Villaluenga, C., & Frias, J. (2017). Sauerkraut: Production, composition, and health benefits. In *Fermented Foods in Health and Disease Prevention* (pp. 557–576). Academic Press.

Peng, Q., Jiang, S., Chen, J., Ma, C., Huo, D., Shao, Y., & Zhang, J. (2018). Unique microbial diversity and metabolic pathway features of fermented vegetables from Hainan, China. *Frontiers in Microbiology, 9*, 399.

Peres, C., Peres, C., & Malcata, F. X. (2017). Role of natural fermented olives in health and disease. In *Fermented Foods in Health and Disease Prevention* (pp. 517–542).

Pessione, E. (2012). Lactic acid bacteria contribution to gut microbiota complexity: Lights and shadows. *Frontiers in Cellular and Infection Microbiology, 2*, 86.

Phelan, M., & Kerins, D. (2011). The potential role of milk-derived peptides in cardiovascular disease. *Food & Function, 2*(3, 4), 153–167.

Pickens, L. B., Tang, Y., & Chooi, Y. H. (2011). Metabolic engineering for the production of natural products. *Annual Review of Chemical and Biomolecular Engineering, 2*, 211–236.

Qian, B., Xing, M., Cui, L., Deng, Y., Xu, Y., Huang, M., & Zhang, S. (2011). Antioxidant, antihypertensive, and immunomodulatory activities of peptide fractions from fermented skim milk with *Lactobacillus delbrueckii* ssp. *bulgaricus* LB340. *The Journal of Dairy Research, 78*(1), 72–79.

Ramirez, J. F., Sanchez-Marroquin, A., Alvarez, M. M., & Valyasebi, R. (2004). Industrialization of Mexican *pulque*. In K. Steinkraus (Ed.), *Industrialization of Indigenous Fermented Foods* (2nd ed., pp. 547–586). Marcel Dekker.

Rizzello, C. G., Cavoski, I., Turk, J., Ercolini, D., Nionelli, L., Pontonio, E., De Angelis, M., De Filippis, F., Gobbetti, M., & Di Cagno, R. (2015). Organic cultivation of *Triticum turgidum* subsp. *durum* is reflected in the flour–sourdough fermentation–bread axis. *Applied and Environmental Microbiology, 81*(9), 3192–3204.

Samtiya, M., Aluko, R. E., & Dhewa, T. (2020). Plant food anti-nutritional factors and their reduction strategies: An overview. *Food Production, Processing and Nutrition, 2*, 1–14.

Sanchez, P. C. (2008). *Philippine Fermented Foods: Principles and Technology*. UP Press.

Sanjukta, S., & Rai, A. K. (2016). Production of bioactive peptides during soybean fermentation and their potential health benefits. *Trends in Food Science and Technology, 50*, 1–10.

Schurgers, L. J., Teunissen, K. J., Hamulyák, K., Knapen, M. H., Vik, H., & Vermeer, C. (2007). Vitamin K-containing dietary supplements: Comparison of synthetic vitamin K1 and *natto*-derived menaquinone-7. *Blood, 109*(8), 3279–3283.

Seo, M. K., Park, E. J., Ko, S. Y., Choi, E. W., & Kim, S. (2018). Therapeutic effects of kefir grain *Lactobacillus*-derived extracellular vesicles in mice with 2,4,6-trinitrobenzene sulfonic acid-induced inflammatory bowel disease. *Journal of Dairy Science, 101*(10), 8662–8671.

Shah, N. P. (2015). Functional properties of fermented milks. In J. P. Tamang (Ed.), *Health Benefits of Fermented Foods* (pp. 261–274). CRC Press, Taylor and Francis Group.

Shah, N. P., Da Cruz, A. G., & Faria, J. D. A. F. (2013). *Probiotics and probiotic foods: Technology, stability and benefits to human health*. Nova Science Publishers.

Shin, D. H., & Jeong, D. (2015). Korean traditional fermented soybean products: *Jang*. *Journal of Ethnic Foods, 5*(1), 2–7.

Singh, A., & Benjakul, S. (2017). Serine protease inhibitors from squid ovary: Extraction and its effect on proteolysis and gel properties of surimi. *Journal of Food Science and Technology, 54*(1), 267–275.

Singh, T. A., Devi, K. R., Ahmed, G., & Jeyaram, K. (2014). Microbial and endogenous origin of fibrinolytic activity in traditional fermented foods of Northeast India. *Food Research International, 55*, 356–362.

Sipola, M., Finckenberg, P., Korpela, R., Vapaatalo, H., & Nurminen, M. L. (2002). Effect of long-term intake of milk products on blood pressure in hypertensive rats. *The Journal of Dairy Research, 69*(1), 103–111.

Sodini, I., Remeuf, F., Haddad, S., & Corrieu, G. (2004). The relative effect of milk base, starter, and process on yogurt texture: A review. *Critical Reviews in Food Science and Nutrition, 44*(2), 113–137.

Spacova, I., Dodiya, H. B., Happel, A. U., Strain, C., Vandenheuvel, D., Wang, X., & Reid, G. (2020). Future of probiotics and prebiotics and the implications for early career researchers. *Frontiers in Microbiology, 11*, 1400.

Stanton, C., Ross, R. P., Fitzgerald, G. F., & Van Sinderen, D. (2005). Fermented functional foods based on probiotics and their biogenic metabolites. *Current Opinion in Biotechnology, 16*(2), 198–203.

Stefanovic, E., Fitzgerald, G., & McAuliffe, O. (2017). Advances in the genomics and metabolomics of dairy *Lactobacilli*: A review. *Food Microbiology, 61*, 33–49.

Steinkraus, K. H. (2004). Origin and history of food fermentations. In Y. H. Hui, et al. (Eds.), *Handbook of Food and Beverage Fermentation Technology* (pp. 1–7). CRC Press.

Suganuma, T., Fujita, K., & Kitahara, K. (2007). Some distinguishable properties between acid-stable and neutral types of alpha-amylases from acid-producing *koji*. *Journal of Bioscience and Bioengineering, 104*(5), 353–362.

Sulaiman, J., Gan, H. M., Yin, W. F., & Chan, K. G. (2014). Microbial succession and the functional potential during the fermentation of Chinese soy sauce brine. *Frontiers in Microbiology, 5*, 556.

Syal, P., & Vohra, A. (2013). Probiotic potential of yeasts isolated from traditional Indian fermented foods. *International Journal of Microbiology, 5*(2), 390–398.

Szajewska, H., Skórka, A., Ruszczyński, M., & Gieruszczak-Białek, D. (2007). Meta-analysis: *Lactobacillus GG* for treating acute diarrhea in children. *Alimentary Pharmacology & Therapeutics, 25*(8), 871–881.

Tamang, J. P. (2010). Diversity of fermented foods. In J. P. Tamang & K. Kailasapathy (Eds.), *Fermented Foods and Beverages of the World* (pp. 41–84). CRC Press, Taylor & Francis Group.

Tamang, J. P. (2015). Naturally fermented ethnic soybean foods of India. *Journal of Ethnic Foods, 2*(1), 8–17.

Tamang, J. P., Cotter, P. D., Endo, A., Han, N. S., Kort, R., Liu, S. Q., Mayo, B., Westerik, N., & Hutkins, R. (2020). Fermented foods in a global age: East meets West. *Comprehensive Reviews in Food Science and Food Safety, 19*(1), 184–217.

Tamang, J. P., Dewan, S., Tamang, B., Rai, A., Schillinger, U., & Holzapfel, W. H. (2007). Lactic acid bacteria in *Hamei* and *Marcha* of Northeast India. *Indian Journal of Microbiology, 47*(2), 119–125.

Tamang, J. P., Shin, D. H., Jung, S. J., & Chae, S. W. (2016b). Functional properties of microorganisms in fermented foods. *Frontiers in Microbiology, 7*, 578.

Tamang, J. P., Watanabe, K., & Holzapfel, W. H. (2016a). Review: Diversity of microorganisms in global fermented foods and beverages. *Frontiers in Microbiology, 7*, 377.

Tassou, C. C., Panagou, E. Z., & Katsaboxakis, K. Z. (2003). Microbiological and physicochemical changes of naturally black olives fermented at different temperatures and NaCl levels in the brines. *Food Microbiology, 19*, 605–615.

Thapa, N., & Tamang, J. P. (2015). Functionality and therapeutic values of fermented foods. In J. P. Tamang (Ed.), *Health Benefits of Fermented Foods* (pp. 111–168). CRC Press, Taylor and Francis Group.

Tolhurst, G., Heffron, H., Lam, Y. S., Parker, H. E., Habib, A. M., Diakogiannaki, E., Cameron, J., Grosse, J., Reimann, F., & Gribble, F. M. (2012). Short-chain fatty acids stimulate glucagon-like peptide-1 secretion via the G-protein-coupled receptor FFAR2. *Diabetes, 61*(2), 364–371.

Verna, E. C., & Lucak, S. (2010). Use of probiotics in gastrointestinal disorders: What to recommend? *Therapeutic Advances in Gastroenterology, 3*(5), 307–319.

Vitale, D. C., Piazza, C., Melilli, B., Drago, F., & Salomone, S. (2013). Isoflavones: Estrogenic activity, biological effect and bioavailability. *European Journal of Drug Metabolism and Pharmacokinetics, 38*(1), 15–25.

Vukotić, G., Strahinić, I., Begović, J., Lukić, J., Kojić, M., & Fira, D. (2016). Survey on proteolytic activity and diversity of proteinase genes in mesophilic lactobacilli. *Microbiology, 85*, 33–41.

Waigankar, S. S., & Patel, V. (2011). Role of probiotics in urogenital healthcare. *Journal of Mid-Life Health, 2*(1), 5–10.

Walker, G. M. (2014). Microbiology of winemaking. In C. Batt & M. A. Tortorello (Eds.), *Encyclopedia of Food Microbiology* (2nd ed., pp. 787–792). Elsevier Ltd.

Wang, X., Du, H., Zhang, Y., & Xu, Y. (2018). Environmental microbiota drives microbial succession and metabolic profiles during Chinese liquor fermentation. *Applied and Environmental Microbiology, 84*(4), e02369-17.

Wang, Y., Xu, N., Xi, A., Ahmed, Z., Zhang, B., & Bai, X. (2009). Effects of *Lactobacillus plantarum* MA2 isolated from Tibet kefir on lipid metabolism and intestinal microflora of rats fed on high-cholesterol diet. *Applied Microbiology and Biotechnology, 84*(2), 341–347.

Wedajo, B. (2015). Lactic acid bacteria: Benefits, selection criteria and probiotic potential in fermented food. *Journal of Probiotics & Health, 3*, 1–9.

West, C. E., Hammarström, M. L., & Hernell, O. (2009). Probiotics during weaning reduce the incidence of eczema. *Pediatric Allergy and Immunology, 20*(5), 430–437.

Wiley, K. D., & Gupta, M. (2022). *Vitamin B1 Thiamine Deficiency*. In StatPearls. StatPearls Publishing.

Willcox, B. J., Willcox, D. C., & Suzuki, M. (2004). *The Okinawa Diet Plan*. Three Rivers Press.

Wójciak, K. M., & Dolatowski, Z. J. (2012). Oxidative stability of fermented meat products. *Acta Scientiarum Polonorum. Technologia Alimentaria, 11*(2), 99–109.

Won, T. J., Kim, B., Song, D. S., Lim, Y. T., Oh, E. S., Lee, D. I., Park, E. S., Min, H., Park, S. Y., & Hwang, K. W. (2011). Modulation of Th1/Th2 balance by *Lactobacillus* strains isolated from kimchi via stimulation of macrophage cell line J774A.1 *in vitro*. *Journal of Food Science, 76*(2), H55–H61.

Xiang, H., Sun-Waterhouse, D., Waterhouse, G. I., Cui, C., & Ruan, Z. (2019). Fermentation-enabled wellness foods: A fresh perspective. *Food Science and Human Wellness, 8*, 203–243.

Yadav, H., Jain, S., & Sinha, P. R. (2007). Antidiabetic effect of probiotic *Dahi* containing *Lactobacillus acidophilus* and *Lactobacillus casei* in high fructose fed rats. *Nutrition, 23*(1), 62–68.

Yanagisawa, Y., & Sumi, H. (2005). *Natto* bacillus contains a large amount of water-soluble vitamin K (menaquinone-7). *Journal of Food Biochemistry, 29*(3), 267–277.

Zeng, W., Li, W., Shu, L., Yi, J., Chen, G., & Liang, Z. (2013). Non-sterilized fermentative co-production of poly(γ-glutamic acid) and fibrinolytic enzyme by a thermophilic *Bacillus subtilis* GXA-28. *Bioresource Technology, 142*, 697–700.

Zhang, Y., Hu, P., Lou, L., Zhan, J., Fan, M., Li, D., & Liao, Q. (2017). Antioxidant activities of lactic acid bacteria for quality improvement of fermented sausage. *Journal of Food Science, 82*(12), 2960–2967.

Zhu, Y., Zhang, F., Zhang, C., Yang, L., Fan, G., Xu, Y., Sun, B., & Li, X. (2018). Dynamic microbial succession of Shanxi aged vinegar and its correlation with flavor metabolites during different stages of acetic acid fermentation. *Scientific Reports, 8*(1), 8612.

Zmora, N., Suez, J., & Elinav, E. (2019). You are what you eat: Diet, health and the gut microbiota. *Nature Reviews Gastroenterology & Hepatology, 16*(1), 35–56.

CHAPTER 2

Microalgal Biohydrogen an Insight into Recent Advances Glitches and Solutions

PRITI PAL,[1] RAM PRASAD,[2] BRIJESH PANDEY,[3] ANAND PRAKASH,[3] SAURABH SINGH RATHORE,[3,6] PRAKASH KUMAR SARANGI,[4,*] AKHILESH KUMAR SINGH,[3,*] and RAJESH K. SRIVASTAVA[5]

[1]*Amity Institute of Biotechnology, Amity University Uttar Pradesh, Lucknow Campus, Lucknow, Uttar Pradesh, India*

[2]*Department of Botany, School of Life Sciences, Mahatma Gandhi Central University, Motihari, Bihar, India*

3Department of Biotechnology, School of Life Sciences, Mahatma Gandhi Central University, Motihari, Bihar, India

[4]*College of Agriculture, Central Agricultural University, Imphal, Manipur, India*

[5]*Department of Biotechnology, School of Core Engineering, GITAM (Deemed to be University), Visakhapatnam, Andhra Pradesh, India*

[6]*Department of Life Sciences, Inter University Centre for Teacher Education, Bananas Hindu University, Varanasi, Uttar Pradesh, India*

*Corresponding author

ABSTRACT

Microalgae are a renewable resource that can be used to manufacture biohydrogen (bio-H_2). As a result, this contributes to both the energy problem and global warming. The discovery of various valuable metabolites as byproducts of the bio-H_2 generation process has received a lot of attention in recent years. The use of microalgae for bio-H_2 production is exceptional because

Industrial Microbiology: Current Progress and Novel Technologies.
Prakash Kumar Sarangi, Akhilesh Kumar Singh, Rajesh Kumar Srivastava, and Brijesh Pandey (Eds.)

DOI: 10.1201/9781779642219-2

of their carbon-mitigating effects and their use of solar energy. bio-H_2 production can be achieved through microbial methods. In this chapter, the authors discussed the enzymes and influencing parameters necessary for bio-H_2 production, as well as the state-of-the-art techniques currently employed to produce bio-H_2 from microalgal feedstock. Commercial certification of processes based on microalgal bio-H_2 production is still lacking. There are significant obstacles to implementing bio-H_2 production on a commercial scale, including low bio-H_2 yield and high production costs. This chapter proposes a method for commercializing bio-H_2 production using microalgae that involves an integrated production process. Further, we discuss the importance of process integration to boost efficiency, as well as the application of cutting-edge strategies to increase bio-H_2 production and economic viability, and the demand for further study.

2.1 INTRODUCTION

The International Encrgy Agency (IEA) estimates that annual production of green hydrogen (H_2) must reach 306 million metric tons by 2050 if the world is to achieve net-zero CO_2 emissions by that year (IEA, 2020). To create environmentally friendly H_2, electrolysis of water is used, which consumes a lot of power. The electricity used must come from carbon-neutral sources like solar or wind for the process to be considered carbon-neutral. As a result, this method is only responsible for producing 4% of the world's H_2 supply. This course of action is ecologically justifiable; however, it is more expensive and complex than alternative methods of obtaining H_2, such as steam methane reforming (SMR), which is used to produce the vast majority of H_2 used in industry today. Roughly 95% of all H_2 is currently produced as either blue (with CO_2 capture) or grey (without CO_2 capture) hydrogen using SMR technology. H2 production, according to the IEA, needs to increase by 66% from 2020 to 2030 and by 23% from 2030 to 2050 (IEA, 2020). The rise of renewable energy sources holds the promise of displacing some of the world's reliance on fossil fuels and ushering in a more sustainable energy system, as shown in Figure 2.1. One promising new technology is the production of biohydrogen (bio-H_2) from microalgae. Several renewable energy systems have zeroed in on microalgae as a primary source of fuel in recent years (Limongi et al., 2021). Microalgae are good for making bio-H_2 because they produce a lot of biomass, have a lot of lipids, clean up flue gases (CO_2 and NOx) well, can use different kinds of water, and do not take up any arable land (Ubando et al., 2021). The potential of biological hydrogen as a carbon-neutral fuel with long-term environmental benefits has attracted a lot of attention (El-Dalatony et al., 2020a). Given the right physiological conditions, like high light intensity and an anoxic atmosphere, the metabolic activities of green microalgae can produce bio-H_2. To produce microalgal biomass, microalgae can use carbon dioxide (CO_2). For example, 1 kilogram of dry cell biomass requires about 1.83 kilograms of CO_2. Additionally, green microalgae were first identified as the biological source of H2 by Gaffron (1939). Other green microalgae strains, such as *Chlamydomonas Reinhardtii*, have shown evidence of H_2 production capacity in recent years (Fan et al., 2016). Recent research has shown that biological hydrogen (H_2) can be used as a carbon-neutral and environmentally sustainable fuel (El-Dalatony et al., 2020a). Distribution, capture, storage, and transformation methods, for example, have an impact on industrial processes for the generation of bio-H_2. So, this chapter looked at how the processes of photosynthesis, Photofermentation, dark fermentation, and bio photosynthesis affect microalgae's ability to make bio-H_2. Microalgal dark fermentation is a better way to make bio-H_2 than Photofermentation, bio photolysis, and microbial electrolysis, which all use sunlight. This chapter also gave a brief summary of green microalgae and cyanobacteria as possible algal strains for making bio-H_2. Also, a thorough review was done to show enzymes that make bio-H_2 and are involved in biosynthesis and enzyme localization in microalgae. The activity of [Fe-Fe]-hydrogenases is roughly 10–100 times higher than that of [Ni-Fe]-hydrogenases and 1000 times higher than that of nitrogenases, making them the most powerful bio-H_2 producing enzymes (Mona et al., 2020). This study previously obscured the fact that bio-H_2 producing enzymes are notoriously inefficient at converting energy and are extremely sensitive to oxygen. Compared to phosphorus, nitrogen, and magnesium deficiency, sulfur deprivation was found to be the most viable pathway for bio-H_2 generation. This chapter also gives a short summary of the latest developments in making bio-H_2 from microalgae, such as genetic engineering, microalgae-bacteria consortiums, electro-bio-hydrogenation, and the use of nanomaterials to improve enzyme stability and hydrolytic efficiency. There is also a strong focus on the opportunities and obstacles inherent in the current state of bio-H_2 production from microalgae.

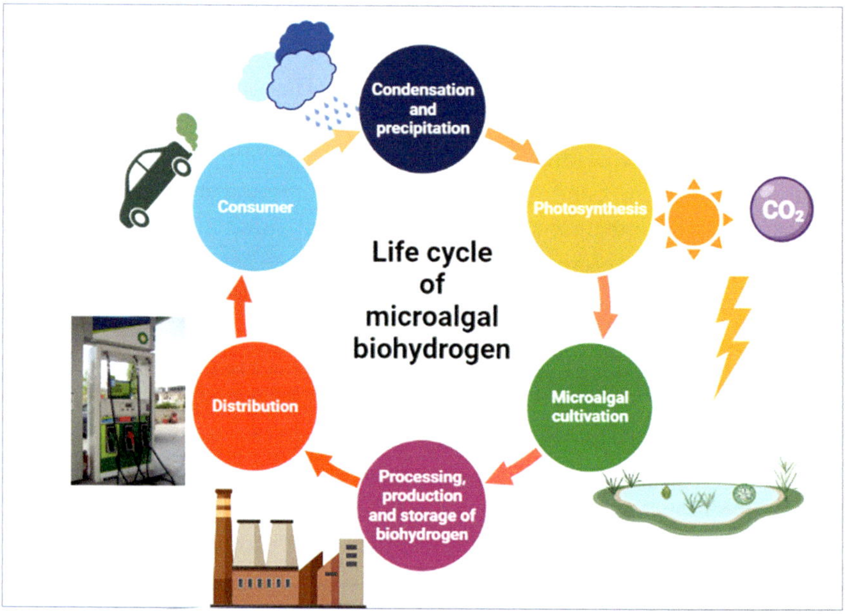

FIGURE 2.1 Life cycle of microalgal biohydrogen.

2.2 MICROALGAE AS A SUITABLE SOURCES OF BIO-H_2 PRODUCING BIOCATALYSTS

Microalgae are photosynthetic prokaryotic or eukaryotic microorganisms that, like higher plants, use the energy from light to convert carbon dioxide from the air into organic biomass. Photosynthesis yields carbohydrates, which are used extensively in cellular metabolism as a source of energy. During the night, stored carbohydrates are used as a fuel source, and the primary energy-producing metabolic process is respiration. Carbohydrates are critical for cellular defense because they are a key structural component of cell walls. Carbohydrates serve as anti-oxidants (such as ascorbate) and as solutes for osmotic balance within cells (e.g., mannitol). Due to their ability to store large amounts of carbohydrates for use in times of environmental stress, microalgal carbohydrates can be used as a fermentation feedstock. Multiple types of microalgae, such as *Chlorella, Scenedesmus, Monoraphidium, Platymonas, Anabaena, Tetraspora, Coelastrella, Micractinium,* and *Chlamydomonas,* have the potential to produce bio-H_2 (El-Dalatony et al., 2020a). Potentially useful green microalgae include *Botryococcus, Chlorella, Chlamydomonas,*

Chlorococcum, and *Scenedesmus.* In addition, *Synechocystis, Anabaena,* and *Synechococcus* are beneficial cyanobacteria. The bacterium *Nostoc* is an example of a potentially useful hydrogenase-containing organism (Deviram et al., 2020). In the sections that follow, we will look at how likely it is that green microalgae and cyanobacteria could be used to make bio-H_2. Renewable fuels from green microalgae hold great promise (Wang et al., 2017a, b). Since the discovery of green microalgae producing H_2 in the 1930s (Jim'enez-Llanos et al., 2020), however, only about 70 species from more than 30 genera have been studied. The species of *Chlorella,* such as *C. vulgaris var. vulgaris, C. pyrenoidosa, C. sorokiniana, C. fusca, C. lewinii, C. homosphaera, C. protothecoides, C. salina, Parachlorella kessleri,* and *C. vacuolate,* have the highest bio-H_2 production potential among microalgae shown in Table.2.1 (Lin and Shei, 2008). *Scenedesmus obliquus* and *Chlorella fusca* are also capable of producing bio-H_2, but *C. reinhardtii* serves as a model microalga for studying bio-H_2 metabolism in photosynthetic eukaryotes. Under strict photoautotrophic conditions, Tsygankov et al. (2006) found that *C. reinhardtii* could keep making bio-H_2 (56.4 mL/L culture) even when there was no acetate or other organic substrate in the medium. It has been noted that a high bio-H_2 yield rate could be detected in the light for brief periods of time (Khetkorn et al., 2017). Within 2.5 s of exposing dark-adapted *C. vulgaris* cells to a light density of about 0.03 W/m^2, the H_2 production rate reaches its maximum value, and the kinetic model is linear for at least 1 minute (Khetkorn et al. 2017). Cyanobacteria, which are also called blue-green algae like cyanophytes or cyanophycean, are also interested in making bio-H_2 and doing photosynthesis with oxygen (Khetkorn et al., 2017). Specifically, cyanobacteria are heterocyst-containing filamentous nitrogen-fixing organisms (Mona et al., 2020). Traditional nitrogen-fixing genera include *Anabaena, Calothrix, Nostoc*, and *Oscillatoria.* Heterocyst's provide the anaerobic environment required for nitrogen fixation, where bio-H_2 production is a byproduct (Equation 1). *Anabaena sp*., discovered in the late 19th century, was the first *cyanobacterium* shown to be capable of bio-H_2 production. Malek and Motamedian (2020) say that the genus *Anabena* has gotten a lot of attention for making bio-H_2 when fixing nitrogen, but its metabolic models have not yet been used to learn more about how bio-H_2 is made. *A. variabilis* ATCC 29413 is still a popular candidate for producing bio-H_2 products due to its high bio-H_2 yield among microalgae and cyanobacteria (Tiwari and Pandey, 2012). Even though it has been shown that increasing the number of heterocyst's in cyanobacteria when they do not have enough nitrogen increases bio-H_2 production, there is no direct link between the number of heterocyst's and the amount of bio-H_2 produced (Mona et al., 2011). Some cyanobacteria, such as *Synechocystis, Synechococcus,* and *Gloeobacter,* that do not fix nitrogen produce bio-H_2 as well. These bacteria have two types of hydrogenases, each of which has a unique set of roles and characteristics (Mona et al., 2020).

TABLE 2.1 Analysis of the Relative Efficiency of Different Microalgal Strains in Producing Bio-H_2

Microalgal Species	Pathway of (Bio-H_2) Production	Growth Condition	Yield of (Bio-H_2)	References
C. reinhardtii Stm6	Photofermentation	anaerobic conditions	540 mL/L	Kruse et al. (2005)
Chlorella salina Mt	Indirect biophotolysis	Sulfur-deprived conditions	0.5 mL/L/h	Chader et al. (2009)
Immobilized Chlorella vulgaris NIER-10003	Photofermentation	Anaerobic and absence of sulfur	238 mL/L/h	Song et al. (2011)
C. reinhardtii L159I-N230Y	Direct biophotolysis	anaerobic conditions	11.1 mL/L/h	Scoma et al. (2012)
Chlorella vulgaris strains YSL01	Direct biophotolysis	aerobic conditions	1.9 mL/L	Hwang et al. (2014)
Chlorella sorokiniana KU204	Direct biophotolysis	N-limitation and S-deprivation	89.64 mL/L	Pongpadung et al. (2015)
Chlorella pyrenoidosa	Direct biophotolysis	sulfur-deprivation	93.86 mL/L	Liu et al. (2016)
Chlorella sp. IOAC707S	Direct biophotolysis	photoheterotrophic conditions and (TAP)	38.00 mL/L	Batyrova et al. (2017)
C. reinhardtii CC-503	Direct and indirect biophotolysis	co-culture at sulfur deficiency conditions	255 mmol/ mgChl	He et al. (2018)
C. reinhardtii	direct and indirect biophotolysis	Sulfur deprivation and HSM + O_2 absorbent	2.58 mL/L/h	Nagy et al. (2018)
Chlorella vulgaris MACC360	Dark fermentation	tris-acetate-phosphate (TAP) co-culture	56.0 mL/L/d	Shetty et al. (2019)
Immobilized *Scenedesmus obliquus*	Direct and indirect biophotolysis	anaerobic condition sulfur deprivation	204.8 mL/L/d	Ruiz-Marin et al. (2020)
C. reinhardtii	Photofermentation	sulfur (S) and chloride ion (Cl^-) deprivation	0.38 mmol/L/ min	Hwang et al. (2021)
Chlorella vulgaris BEIJ (G-120)	Direct and indirect biophotolysis	HM + glucose	5.0 mL/L/h.	Touloupakis et al. (2021)

2.3 ENZYMES INVOLVED IN BIO-H_2 PRODUCTION FROM MICROALGAE

All methods of bio-H_2 production rely on enzymes native to specific microorganisms that convert sugar and oxygen into hydrogen (Mathews and Wang, 2009). Major enzymes involved in bio-H_2 production from different microalgae are hydrogenases and nitrogenases in Figure 2.2 (Bhatia et al., 2020). The Hup uptake hydrogenase is another enzyme involved in bio-H_2 metabolism, but instead of producing bio-H_2, it absorbs it (Wang et al., 2020b). Hydrogenase and nitrogenase are the only two enzymes covered in this review, both of which are synthesized and localized in microalgae. "Challenges of bio-H_2 Production from Microalgae: A Review" provides background on other hydrogen-producing enzymes (Show et al., 2019).

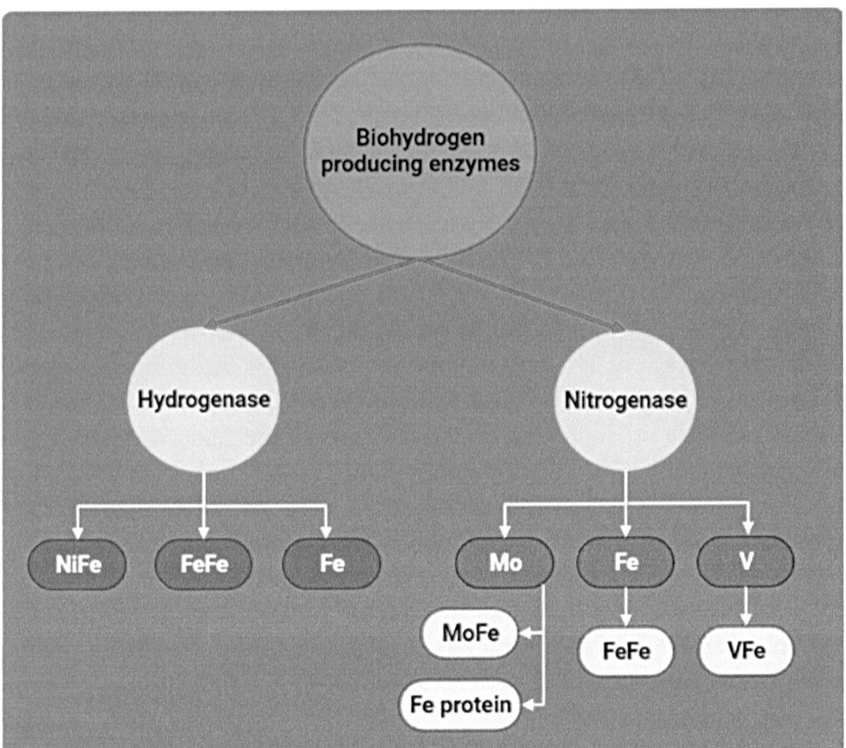

FIGURE 2.2 Enzymes involved in bio-H_2 production.

2.3.1 HYDROGENASE

Hydrogenase catalyzes proton reduction to H_2: $2H^+ + 2e^- \leftrightarrow H_2$, acting as the ultimate electron acceptor in the absence of oxygen. Hydrogenase genes have so far been discovered in *S. obliquus, C. moewusii, C. reinhardtii,* and *C. fusca*, all of which belong to the Chlorococcales and Volvocales (Chlorophyceae) (Mona et al., 2020). Only cyanobacteria and a few genera of green microalgae (such as *Chlamydomonas, Chlorella,* and *Scenedesmus*) among oxygenic phototrophic microorganisms have functional hydrogenases (Petrova et al., 2020). Different species may have varying degrees of enzyme maturity, activity, and structural variation. Hydrogenases are a type of metalloenzyme that catalyzes the reversible oxidation of electrons and protons by using iron (Fe) and sulfur (S) clusters at their active sites (Nagarajan et al., 2020b). This is the only one that has the ubiquitous iron-sulfur clusters [2Fe-2S], [3Fe-4S], and [4Fe-4S]. Hydrogenase's redox partners and bio-H_2 activation sites are connected by these classical metal sites (Vignais et al., 2001). Hydrogenase is an enzyme that can be divided into subclasses based on the metals in the active site. These subclasses are called [Fe-Fe] hydrogenase, [Ni-Fe] hydrogenase, and metal-free hydrogenase (Gadhe et al., 2015). Since [Fe-Fe] hydrogenases catalyze reversible bio-H_2 activation at very high rates, they are attractive targets for bioengineering attempts to integrate microbial bio-H_2 generation with aerobic photosynthesis due to their unique prosthetic group. Proton reduction in [Fe-Fe] hydrogenase has been found to occur at the H cluster, a complicated bridging FeS cluster (Swanson et al., 2015). Through protein cysteine thiolates, the H cluster connects to the organometal 2Fe subclusters in the same way that other conventional [4Fe-4S] subclusters do. The 2Fe subcluster is coordinated by non-protein ligands comprising CO, CN, and dithiomethylamine (Swanson et al., 2015). Ni-Fe bimetallic active sites, Mg^{2+} ions, and iron-sulfur clusters are all reported to be present in the centers of [Ni-Fe]-hydrogenases, which were first discovered by Azwar et al. (2014). The cyanobacterial [Ni-Fe]-hydrogenase has five subunits. Sulfur, iron, nickel, and cobalt make up the large one found in the catalytic subunit of the pentameric hydrogenase HoxH. Hydrogenase's tiny subunit, HoxY, is made up of a cluster of iron sulphide atoms (4Fe-4S), which is responsible for transferring electrons to the larger catalytic subunit. The remaining three components of the complex—HoxF diaphorase,

HoxU, and HoxE—serve as electron channels, allowing nicotinamide adenine dinucleotide phosphate (NADP) to donate an electron to the hydrogenase active site. It is true that [Ni-Fe]-hydrogenases are superior at bio-H_2 oxidation, but [Fe-Fe]-hydrogenases are more effective bio-H_2 producers (Lu and Koo, 2019). PSI's reduced ferredoxin is employed to help transport electrons from the degraded carbohydrate to the [Fe-Fe] active hydrogenase's site (Posewitz et al., 2004). Bio-H_2 is generated by [Ni-Fe]-hydrogenases, which can also serve as uptake hydrogenases by transferring electrons from bio-H_2 to decrease (NADP) (Mathews and Wang, 2009). The microalgal [Fe-Fe] hydrogenase is principally responsible for light fermentation, which is exemplified by a substantial rise in bio-H_2 release after illumination. In the same way that many [Ni-Fe]-hydrogenases are inhibited by O_2, they can be reactivated once O_2 is no longer present (Yuan and Koo, 2019). However, [Fe-Fe]-hydrogenases can be permanently inhibited by even trace amounts of O_2 (Kubas et al., 2017). According to previous research (Horner et al., 2002), the activity of [Fe-Fe]-hydrogenases is roughly 10–100 times greater than that of [Ni-Fe]-hydrogenases and 1000 times greater than that of nitrogenases.

2.3.2 NITROGENASES

Nitrogenase exploits MgATP (2ATP/e-) and low-energy electrons from reduction flavodoxin or ferredoxin to decrease a variety of substrates (Hallenbeck and Benemann, 2002). This is especially effective when N_2 gas is present. Blue-green algae (cyanobacteria) are the only microalgae that contain these enzymes because they contain heterocyst's, a specialized cell type found only in blue-green algae. Hydrogen is created by these enzymes via an irreversible mechanism that hydrolyzes at least four adenosine triphosphates (ATP) for every molecule of bio-H_2 created (Benemann, 1997). Nitrogenase, as stated by Hallenbeck and Benemann (2002), is capable of converting protons into bio-H_2 even in the absence of other substrates. This lays the groundwork for nitrogenase-based bio-H_2 production. The transfer of one electron to MoFe-protein and the associated dissociations occur during the hydrolysis of MgATP complexes containing Fe-protein in combination with MoFe-protein and 2ATP at each cycle (Hallenbeck and Benemann, 2002). Biosynthesis needs a lot of energy and ATP for catalysis, so a slow turnover rate is a problem. In general, nitrogenase is a relatively inefficient metabolic pathway for bio-H_2 production. In order to prevent O_2, a byproduct of photosynthesis, from deactivating the enzyme, nitrogenase-containing organisms have developed the following spatial strategies: (i) Reduced oxygen permeability in the heterocyst's of heterocystous cyanobacteria separates oxygenic photosynthesis from nitrogenase activity in algal cells; (ii) In heterocyst-free cyanobacteria, O_2-evolution and nitrogenase activity are two different things that only happen during the light phase and the dark phase, respectively (Ghirardi et al., 2000).

2.3.3 ENZYME PRODUCTION AND SUBCELLULAR LOCALIZATION IN MICROALGAE

Green unicellular microorganisms have hydrolase-producing genes in their nuclei (Florin et al., 2001). However, the chloroplast stroma is where mature proteins are stored and perform their functions (Melis and Happe, 2001). Genes like *C. reinhardtii's HYDA1* and *HYDA*2 encode two enzymes with identical structures but distinct roles in bio-H_2 production (Petrova et al., 2020). The proteins encoded by *HYDA1* and *HYDA*2 are both 49 kDa in size; *HYDA1* is made up of 497 amino acids, and *HYDA*2 encodes a protein of 505 amino acids that is 73% identical to *HYDA1* (Wang et al., 2020b). RNA silencing experiments demonstrated that *HYDA1* catalyzes *in vitro* hydrogenase activity; however, there is no indication that *HYDA*2 plays a substantial role in microalgal bio-H_2 generation (Godman et al., 2010). *HYDA1* was found to be the primary hydrogenase, responsible for 75% of bio-H_2 segregation, and *HYDA*2 was found to be an accessory hydrogenase, responsible for about 25% of bio-H_2 segregation, in a separate study by Meuser et al. (2012). An earlier study reported that [Fe-Fe] hydrogenase is extremely reactive to molecular oxygen and needs hypoxic conditions to keep performing at optimal levels (Meuser et al., 2012). Oxygen molecules inhibit hydrogenase activity in microalgae by preventing the *HYD* genes from being expressed, preventing the enzyme from maturing, and blocking catalytic sites (Forestier et al., 2003). Furthermore, the *hoxEFUYH* gene, which codes for Hox-hydrogenase, has been shown to produce NAD^+-reductase (Razu et al., 2019).

There are two protein complexes involved in this system: the diaphorase complex (*HoxE, HoxF*, and *HoxU*) and the hydrogenase complex (*HoxY* and *HoxH*). The bidirectional Hox-hydrogenase, which catalyzes both the uptake and release of molecular bio-H_2 is found in virtually all cyanobacteria (Razu et al., 2019). Dinitrogenase reductase (an iron protein) and dinitrogenase (a polyprotein) work together to convert nitrate to nitrite (MoFe protein). In contrast, three structural *nif* genes are responsible for encoding the two nitrogenase proteins (Mathews and Wang, 2009). Both *nifD* and *nifK* produce the MoFe protein, and *nifH* produces the Fe protein (Tamagnini et al., 2002). Dinitrogenase is a α2β2-heterotetramer. Both the *nifD* gene and the *nifK* gene code for the alpha subunit. The resulting proteins typically range in size from about 220–240 kDa. It is the catalyst for the formation of ammonia (NH_3) by breaking down N_2 bonds (Razu et al., 2019). Besides a [4Fe-4S] cluster, the heterotetrametric has two ATP binding sites. The *nifH* component of dinitrogenase reductase, a homodimer with a molecular weight of about 60–70 kDa, transfers electrons from flavodoxins (as external electron donors) to dinitrogenase (Wang et al., 2020b).

2.4 METABOLIC PATHWAYS OF BIO-H_2 PRODUCTION FROM MIROALGAE

bio-H_2 production is expected to become a serious competitor on the energy market as an alternative to a number of current energy sources (Foong et al., 2020). Microalgae have the potential to provide a wide range of bioenergy sources, with bio-H_2 being one of the most exciting (Ge et al., 2020) and is frequently produced using microalgae biomass. Microalgae can be used to produce bio-H_2 in two different ways: (i) by the microalgae themselves, producing bio-H_2 directly; and (ii) by microorganisms that ferment the microalgal biomass. Rates of bio-H_2 production range from 0.015 to 1.084 mmol/L/h for the first process and from 0.35 to 10.26 mmol/L/h for the second, depending on the feedstocks, growth conditions, and microorganisms employed (Oncel, 2015). However, by combining them, we can increase the rate at which bio-H_2 is produced. Figure 2.3 depicts the metabolic pathways involved in bio-H_2 production from microalgae. Specifically, the following sections elaborate on four primary bio-H_2 production pathways utilized by microalgae.

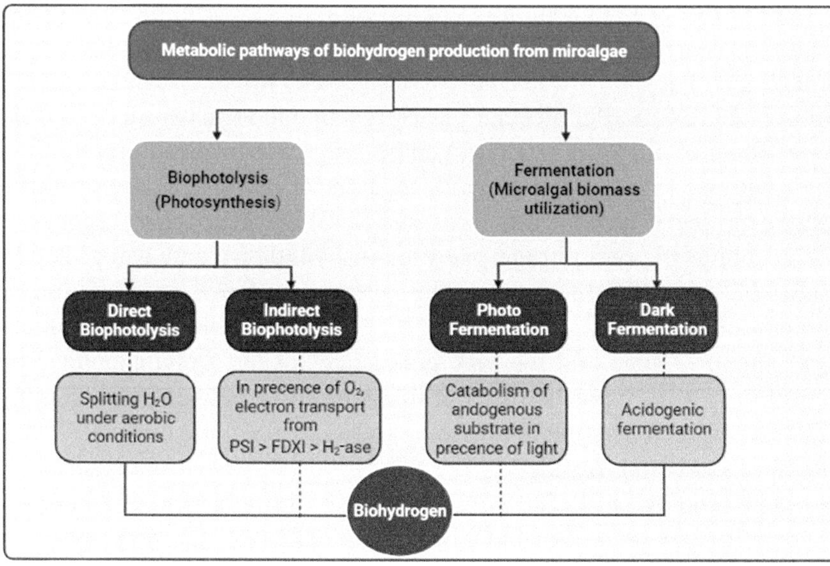

FIGURE 2.3 Metabolic pathways of bio-H_2 production by microalgal biomass.

2.4.1 FERMENTATION PROCESS FOR BIO-H_2 PRODUCTION FROM MICROALGAL BIOMASS

Producing bio-H_2 through fermentation is beneficial because it reduces our reliance on fossil fuels and keeps our air clean. As a result, this hydrogen-production technique shows great promise as a renewable energy option. Dark fermentation and Photofermentation are two distinct methods of the fermentation process.

2.4.1.1 PHOTOFERMENTATION

Fermentation techniques that use sunlight for photosynthesis are collectively known as "Photofermentation." This route converts light into energy instead of sugar. Photofermentation has advantages over alternative processes. The following equation describes photofermentation' s hydrogen production (Baeyens et al., 2020):

$$16ATP + N_2 + 16H_2O + 10H^+ + 8e^- \rightarrow 16ADP + 2NH_4^+ + 16pi + bio\text{-}H_2 \quad (1)$$

This process is powered not by water but by electrons that microalgae get from heterotrophic fermentation and the catabolic reaction of the endogenous substrate. Photocatalysis is the process by which electrons are extracted from the byproducts of organic substrates. The extraction of electrons also involves the use of light energy and oxidative carbon metabolism. In this case, e is synthesized via catabolism from endogenous substrates. Plastoquinone (PQ), a pool of electrons in both Photosystem I (PSI) and Photosystem II (PSII), is where these electrons are processed. The transfer of electrons to the PQ pool in the chloroplasts of different vascular plants is catalyzed by NADPH-plastoquinone oxidoreductase (PQOR) (Dalena et al., 2017). During this procedure, PSI absorbs light energy and releases electrons. By moving electrons to the appropriate [Fe]-H_2ase and ferredoxin (FDX) levels, the redox potential of this electron can be increased (Mosebach et al., 2017).

The high yield of bio-H_2 is made possible by the induction of ([Fe]-H_2ase) in the absence of oxygen. Incubating microalgal culture in the presence of 3-(3,4-dichlorophenyl)-1,1-dimethylurea (DCMU) is helpful for bio-H_2 production (Mu et al., 2020). In order to produce bio-H_2, photofermentation is preferred over the conventional dark fermentation. Zhang et al. (2020) compared Photofermentation, dark fermentation, and dark-Photofermentation to find out which method produced bio-H_2 more efficiently. Since Photofermentation resulted in the highest hydrogen content (maximum 58.90%) and the highest energy conversion efficiency (10.12%), it was deemed the most effective method. Additionally, photo-fermentation was superior to the other two pathways in terms of bio-H_2 yield. The bio-H_2 yield was highest 141.42 ml(g TS)$^{-1}$ during photofermentation and lowest 36.08 ml (g TS)$^{-1}$ during dark fermentation. As a result, the photofermentation method is a viable option for making bio-H_2. While it does produce bio-H_2, its efficiency in converting sunlight is low when compared to biophotolysis and electrochemical processes.

2.4.1.2 DARK FERMENTATION

Fermentation in the dark requires a complete lack of light. Using micro-organisms or microalgae, it undergoes a process known as heterotrophic fermentation. The majority of microalgae used to create bio-H_2 are green algae, which are particularly well-suited to the task because of their high carbohydrate content (Rajesh et al., 2020). Bio-H_2 production rates can be increased with the help of substrates like glucose, sucrose, starch, and cellulose. When completely converted, 1 mol of glucose yields 12 mol of hydrogen molecules.

$$C_6H_{12}O_6 + 6H_2O = 12H_2 + 6CO_2 \quad (2)$$

Through a process called "dark fermentation," 4 mol of hydrogen is produced. Organic acids are produced as waste in this reaction (Mona et al., 2020):

$$C_6H_{12}O_6 + 6H_2O = 2CH_3COOH + 4H_2 + 2CO_2 \quad (3)$$

For example, complex carbohydrates can be broken down into reducing sugars through hydrolysis in the dark fermentation process (Kucharska et al., 2019). Enzymes initiate an acidification process in which these elementary compounds participate. These enzymes are produced by bacteria that participate in fermentation. These acid byproducts then undergo a process called acctogenesis, where they are converted into bio-H_2 and acetate. In order to produce bio-H_2 along the same pathway, it is necessary to halt methanogenesis and collect bio-H_2 from earlier steps (Osman et al., 2020).

Dark fermentation uses two different pathways to generate bio-H_2. The H_2ase pathway catalyzes two routes: the breakdown of lactic acid and the reoxidation of NADH (El-Dalatony et al., 2020). The oxidation of pyruvate is catalyzed by the enzyme pyruvate ferredoxin oxidoreductase (PFR). So, it is important to find a way to steady the oxidation-reduction process. For example, you can use the conversion of NADH into compounds like butanol and ethanol to achieve this (Wang et al., 2020a). The rate and quantity of bio-H_2 produced via the dark fermentation process are superior to those of alternative methods. The problem lies in the relatively low concentration of hydrogen, which is only about 40–60%. As a result, the fuel cells cannot use the gas produced during fermentation unless it is first purified (Mona et al., 2020).

Song et al. (2020) used *enzymolysis* and acid pretreatment to release reducing sugars from *A. philoxeroides*, which improved the efficiency of the dark fermentation process. Under the best pretreatment conditions, the sugar-reducing yield was determined to be 0.35 g/g *A. philoxeroides*; after *enzymolysis*, this value was raised to 0.56 g/g. Therefore, the bio-H_2 yield from the pretreated *A. philoxeroides* was raised by 59.9%. An improvement in bio-H_2 output of 42.8%, from 59.2 to 89.8 ml/gVS, can be attributed to the optimization of the bacterial suspension, which boosted the maximum rate of bio-H_2 production from 1.42 to 4.64 ml/gVS/h. However, the by-products of *A. philoxeroides* pretreatment and their impact on bio-H_2 generation need to be studied.

2.4.2 PHOTOBIOLOGICAL PROCESS FOR BIO-H_2 PRODUCTION FROM MICROALGAL BIOMASS

Bio photosynthesis is the process by which photosystem II converts solar energy into oxygen, energy, and a reducing agent from water. This compound is employed in the reduction of protons by nitrogenase or hydrogenase enzymes, resulting in the production of hydrogen. The electrons obtained from water splitting are transported via the PSI and PSII electron transport chains. When this occurs, ATP is created while ferredoxin is reduced. These are involved in several metabolic reactions that lead to the formation of hydrogen (Razu et al., 2019). Indirect biophotolysis and direct biophotolysis are the two main types of this process. Indirect biophotolysis uses electrons generated by catabolizing endogenous substrates, while direct biophotolysis makes use of electrons generated by photosystem II's water-splitting reaction.

2.4.2.1 DIRECT BIOPHOTOLYSIS

Water is directly biophotolysis in the PSII, where solar energy is used to split water molecules. After the excess e- has been transferred to photosystem I, hydrogenase contributes to hydrogen production without contributing to greenhouse gas emissions. In addition to releasing carbon dioxide, oxygen is also produced in this way (Jiménez-Llanos et al., 2020). Following is the reaction that occurs during this procedure (Kumar et al., 2019):

$$2H_2O + \text{light energy} \rightarrow 2H_2 + O_2 \quad (4)$$

Electrons are created when water molecules are oxidized by the PSII in the direct biophotolysis pathway. The PSII converts solar photons into electrical energy for the PSI. The electron transport chain is the path that these charged electrons take. Electron transport is facilitated by ferredoxin

(Fd) and protein selenocysteine (PSI). Plastoquinol is synthesized via electron transfer from NADP+, which is then reduced to nicotinamide NADPH. When water is oxidized, it loses an electron, forming oxygen and hydrogen ions. In photosynthesis, these water-derived electrons are used by photosystem II. The enzyme ATP synthase can only produce ATP in the presence of a proton gradient. The electrons in the chloroplasts of microalgae are accepted by these protons at their terminals. Because of this, both hydrogen and oxygen gases are created during this process (Show et al., 2019).

Green microalgae are unique among microorganisms in that they are able to produce biohydrogen by splitting water through the process of direct biophotolysis, even in the absence of oxygen. Due to their ability to absorb sunlight and split the water molecule, green microalgae and cyanobacteria are classified as "oxygenic photosynthetic microorganisms" (Yilmaz et al., 2016). When exposed to light, water-PQOR (PSII) transfers its energy to PSI, which in turn transfers it to Fd (Eroglu and Melis, 2016); Fd-oxidoreductase generates reduced Fd, which then donates an electron to [FeeFe]-hydrogenase, reducing the proton (H^+) to a hydrogen molecule (H_2), resulting in O_2 production at PSI.

$$\text{Fd reduction in PSI: } 2H^+ + 2FD_{(\text{reduced})} \leftrightarrow 4H_2 + 2FD_{(\text{oxidized})} \tag{5}$$

$$\text{bio-}H_2 \text{ production in the presence of sunlight: } 2H_2O + \text{light energy} \rightarrow 2H_2 + O_2 \tag{6}$$

Direct biophotolysis is an environmentally friendly process. Water, sunlight, and carbon dioxide are its principal reactants, and all three are abundant and cheap. Furthermore, it produces H_2 and O_2 by transforming solar energy into chemical energy (Jiménez-Llanos et al., 2020). However, there are still a number of obstacles that must be addressed before this pathway can be considered a viable option for bio-H_2 production. The accumulation of O_2 is one of the major drawbacks of this pathway because it inhibits bio-H_2 production. Low photochemical efficiency, the need for intense light, and oxygen usage risks are additional disadvantages (Kumar et al., 2019). Microalgae must be cultivated on a massive scale if they are to absorb enough solar energy ($\Delta G = +237$ kJ mol^{-1}) to power the world. The ratio of respiration to photosynthetic capacity as well as the processing and concentration of cell biomass are also obstacles (Jiménez-Llanos et al., 2020).

2.4.2.2 INDIRECT BIOPHOTOLYSIS

Microalgae can perform indirect biophotolysis in the absence of oxygen. In these conditions, a wide range of microalgae can produce bio-H_2 through fermentation or respiration. The indirect approach is not permanent because when the light cycle comes back, photosynthesis kicks in and H_2ase activity drops (El-Dalatony et al., 2020). During the first stage of indirect biophotolysis, oxygen is generated and carbon dioxide is fixed into chemical energy in the form of carbohydrates and lipids. The second stage, which is essentially the same as the first, involves the use of a small, sealed photobioreactor and periodic division of the CO_2 synthesis process that takes place when no light is present (Razu et al., 2019).

Oxygen depletion halts the first electron transport chain in the anaerobic phase, and the electrons liberated during the catabolism of the endogenous substrate are then utilized for a non-photochemical reduction. After the electron is transferred, bio-H_2 is produced by [FeeFe]-hydrogenases (Lam et al., 2019). In the initial tests of this biological photolysis method for producing hydrogen, a non-heterocystous microalga, such as the *cyanobacterium Plectonemaboryanum*, was used. The algae were subjected to repeated "aerobic light-driven CO_2 fixation and O_2" cycles (Razu et al., 2019). The following are some responses to this procedure:

$$12H_2O + 6CO_2 + \text{light energy} \rightarrow C_6H_{12}O_6 + 6O_2 \quad (7)$$

$$C_6H_{12}O_6 + 12H_2O + \text{light energy} \rightarrow 12H_2 + 6CO_2 \quad (8)$$

The microalgae use their glycogen stores to create bio-H_2 (Acar and Dincer, 2018a). By temporarily decoupling oxygen and hydrogen evolution reactions mediated by CO_2 fixation and evolution, this strategy also helps address the oxygen sensitivity problem. During the aerobic phase, microalgae grow by transforming the energy from sunlight and carbon dioxide into lipids, carbohydrates, and other forms of biomolecular energy (Lam et al., 2019). Lack of O_2 prevents electrons from being transported along the chain. An anaerobic environment is produced when O_2 is not present in a medium. To convert the electrons generated in this pathway into bio-H_2, the oxygen-sensitive hydrogenase enzyme must be activated.

For the purpose of creating a method for the simultaneous generation of bio-H_2 and lipids, a number of microalgae (oleaginous) were screened using crude glycerol as a cheap source of exogenous carbon (Sengmee et al., 2017). All microalgae developed a high lipid content of over 20% when grown on crude glycerol, but only *Chlorella* sp. could produce a significant amount of bio-H_2 in anaerobic conditions. Bio-H_2 production peaked at 11.65 ± 0.65 ml/L in the 1L bioreactor and 10.31 ± 0.05 ml/L in the serum bottle under optimal conditions, and the lipid content of the recovered microalgal biomass was greater than 40%. Research into bio-H_2 generation coupled with the production of lipid by oleaginous microalgae is necessary to assure the long-term viability of biofuel production employing microalgae.

2.5 FACTORS INFLUENCING ALGAL BIO-H2 PRODUCTION

The production of bio-H_2 at low cost and in a sustainable manner requires many optimization and multi-parameter stages (Wirth et al., 2018). Bio-H_2 production in microalgae requires the induction of very specific metabolic pathways and metabolites (Aslam et al., 2018). Researchers are intrigued by the prospect of using microalgal species that can be grown on a laboratory scale to create bio-H_2 (Yang et al., 2019).

2.5.1 *MICROALGAL CELL DENSITY*

Growing conditions have a significant impact on cell density. Cell density is influenced by many parameters during the heterotrophic, autotrophic, or mixotrophic fermentation of microalgae. These include the availability of inorganic or organic substances, the concentration of CO_2, and the amount of light that penetrates the cells (Kumar et al., 2018). When there are not enough cells in a culture, the oxygen in the solution goes to waste because of respiration. Due to self-shading, anaerobiosis transition may occur rapidly at high cell densities, leading to the accumulation of a negligible amount of starch and, in turn, a detrimental effect on the bio-H_2 evolution rate within the vessel. When the concentration of microalgae was high to begin with, the rate at which they consumed oxygen increased, and PSII was completely inhibited, leading to a faster activation of H_2ase activity and increased bio-H_2 production for a longer time with a lower rate of anaerobic to aerobic transition (Li et al., 2014). However, maintaining anaerobic conditions in culture can be challenging when microalgal concentrations are low to begin with. This is due to the possibility that dissolved O_2 will build up during the process and not be consumed quickly enough by cell respiration (El-Dalatony et al., 2020). Also, the density of microalgal cells is the most important thing in this process of making bio-H_2.

2.5.2 TYPES OF VITAL NUTRIENTS

During the culturing process, a vitamin solution is typically added, which contains nutrients like biotin, cyanocobalamin, and thiamine. Ammonia and nitrate salts, two forms of nitrogen, are typically employed. Apart from photoautotrophic conditions, a carbon source may be required for microalgal growth. Microalgal cells are profoundly impacted by the accessibility of any of these substrates or nutrients, whereas microalgal cells are particularly vulnerable to stress when these nutrients are lacking. It is generally accepted that sulfur deficiency is the primary trigger for stress responses in biophotolysis for long-term bio-H_2 production (Antal et al., 2011). Reduced photosynthetic activity and carbon fixation may be the result of a lack of sulfur, while increased lipid and starch catabolism may be the result of this mechanism.

2.5.3 LIGHT DENSITY

Bio-H_2 production and biophotolysis are very similar to photosynthesis in their dependence on light intensity. Many scientists believe that the discovery of bio-H_2 production was made when sudden light was introduced to a dark microalgal culture, causing the algae to release bio-H_2. Most commonly, high incident light is used to promote PSI, or H_2O splitting, in *C. reinhardtii* cultures. Since PSII causes photoinhibition and PSI is responsible for H_2O splitting, the electrons released from the former reaction can be directed toward the generation of bio-H_2, increasing the rate at which the latter reacts (Martín del Campo et al., 2014).

2.5.4 SUBSTRATE AVAILABILITY

A carbon source (CO_2) is thought to be necessary for microalgal growth in dark heterotrophic fermentation. Microalgae can develop via a variety of different growth mechanisms, including heterotrophic, photoheterotrophic, mixotrophic, and photoautotrophic growth. *Chlorella* uses both exogenous and endogenous substrates in its heterotrophic metabolism to generate bio-H_2 (Bolatkhan et al., 2019). The concentration of carbon sources shows how economically viable the process is and can increase bio-H_2 production. They are an important part of external substrates because of this (Azwar et al., 2014). Starch, the most important endogenous substrate, is synthesized during the light stage of the process and consumed to provide electrons to boost bio-H_2 production (Lee et al., 2020). As well as acting as a carbonate replacement and buffer, phosphorus is essential for the production of ATP, the body's primary energy molecule. Bio-H_2 production is impacted by a number of factors, including the type of microalgae used, the microalgal cell concentration, and the type of carbon sources used (malic acid, acetic acid, and acetate). The initial glucose concentration is very important for bio-H_2 production. It is found that a glucose concentration of 5 g L^{-1} is optimal for effective bio-H_2 production (Rashid et al., 2012). The study found that the delay in bio-H_2 evolution could be reduced by 10 hours after 30 mM glucose was added (Song et al., 2020).

2.6 INNOVATION APPROACHES FOR THE PRODUCTION OF MICROALGAL BIO-H_2

The potential of algae for carbon capture, conversion, and utilization has made it an attractive feedstock for the production of next-generation biofuels (Nagarajan et al., 2020a). Improved pretreatment processes to boost the bioavailability of simple sugars from complex organic substrates, process factor optimization to boost the bio-H_2 yield, and the application of synthetic biological and genetic engineering techniques have all been implemented so far to advance bio-H_2 production. Placing algal cells in a nutrient-depleted medium (low in sulfur, nitrogen, phosphorus, magnesium, etc.) is a common technique, and the underlying mechanism has been the subject of extensive research as of late (Fan et al., 2016). The following sections of this review focus on four of these cutting-edge methods in turn.

2.6.1 GENETIC ENGINEERING TO IMPROVE BIO-H_2 PRODUCTION

As mentioned before, the key challenge to commercializing the microalgal bio-H_2 generation platforms is the sensitivity of enzymes to O_2. *Chlorella* sp. *HYDA* DT strain (CsHydAc.DT, encoded by CshydAc.DT) was genetically modified to alter amino acid residues A105I, G113I, V265W, or

V273I near the gas tunnel to prevent O_2 from entering the enzyme active site and boost bio-H_2 production by as much as a factor of 30-fold times (Yang et al., 2019). Hankamer et al. (2007) reported that knock-in and knock-out gene manipulation of bio-H_2 -producing microorganisms may play a crucial role in resolving the issue of low bio-H_2 production. Another study found that genetically modified *Anabaena* (with the *hupL* gene knocked out) produced four- to seven-times as much bio-H_2 as wild strains (Masukawa et al., 2002). Since PsbO is involved in the O_2 evolution of the PSII system, inhibiting *PsbO* gene expression results in the suppression of O_2 discharge and the induction of *HYDA*. Lin et al. (2013) discovered that after *PsbO* gene knockout, mutant biological bio-H_2 yield increased ninefold. Also, increasing hydrogenase's tolerance to oxygen radicals (O_2) by altering its structure is proposed (Wang et al., 2021). In order to prevent O_2 from entering the gas channel and protect the active center of hydrogenase, scientists have recently substituted certain amino acids (Khosravitabar, 2019). Increased bio-H_2 production in microalgae can also be achieved through the overexpression of particular genes. Codon optimization genes in *lba* and *hemA* were expressed in *C. reinhardtii*, leading to a 6.8-fold increase in *hemH-lba* protein expression and a 20% increase in direct photolysis bio-H_2 generation. Increased bio-H_2 production can be achieved through the use of genetic engineering to alter specific pathways (Goswami et al., 2021). Metabolic engineering and genetic engineering are two of the most important tools for increasing the production of biological bio-H_2. Hydrogenases use ferredoxin, a photosynthetic electron transporter, to acquire electrons from PSI. While FNR plays a role in accepting electrons from photosynthesis, it is not the primary receptor. In this regard, the fact that *HYDA* is less competitive with NADPH formation catalyzed by ferredoxin-$NADP^+$-reductase represents the primary problem that needs to be addressed (FNR). Since the electron donor ferredoxin (Fd) was fused with *HYDA* and expressed in *C. reinhardtii*, the efficiency of bio-H_2 generation in living organisms was studied by Eilenberg et al. (2016). Once the Fd-*HYDA* fusion gene was expressed in *C. reinhardtii* cells, it was shown that the resulting fusion enzyme could efficiently intercept photosynthetic electrons and put them to use in bio-H_2 generation (Eilenberg et al., 2016). Several other engineering goals to avoid FNR competition were also met, including photosynthetic electron delivery of ferredoxin and FNR genetic modifications (Rumpel et al., 2014). Moreover, *C. reinhardtii* with a mutation in the *Moc1* gene has been used in a variety of applications, as detailed in an open patent (Akhlaghi and Najafpour-Darzi, 2020). In order to increase bio-H_2 production, it was suggested that electrons be transferred via a respiratory electron transfer chain (Akhlaghi and Najafpour-Darzi, 2020). In comparison to traditional chemical mutagens or UV light, atmospheric and room-temperature plasma (ARTP) is a more advantageous method of whole-cell mutagenesis. ARTP relies on helium radio-frequency atmospheric-pressure glow discharge plasma (Ottenheim et al., 2018). In light of this, Ban et al. (2019) aimed to boost microalgal bio-H_2 production by truncating the Chl antenna of *C. reinhardtii* via ARTP mutagenesis. After selection, it was found that an algal mutant produced 2.7–3.1 times (356.5–405.2 mL/L) and 1.8–5.2 times (28.5–84.1 mL/L) more bio-H_2 than the wild type did during the course of the algal subcultures grown in co-cultures and pure cultures, respectively (Ban et al., 2019).

2.6.2 MICROALGAE-BACTERIA CONSORTIUM AS A STRATEGY TO ENHANCE BIO-H_2 PRODUCTION

Even though there is a lot of interest in microalgae as a source of biofuel to replace fossil fuels, not much has been learned about how bacterial symbionts affect large-scale microalgae farming systems (Zhang et al., 2021a). Moreover, it has been suggested that microalgae and bacteria (such as *Pseudomonas*, *Escherichia coli*, *Azotobacter*, and *Bradyrhizobium*) co-cultivation is a practical method to obtain anaerobiosis via the respiration of the bacterial partners (Ban et al., 2018). Four bacterial strains and one fungal strain were isolated from *C. sorokiniana* cultures by Watanabe et al. (2005), and it was shown that the bacterium *Microbacterium trichotecenolyticum* and the fungus *Acremonium*-like hyphomycete could stimulate the growth of *Chlorella. Pseudomonas* sp. strain D, recently discovered by (Ban et al., 2018), was shown to significantly contribute to the production of bio-H_2 by the green microalgae *Chlorella* and *Scenedesmus*, as well as to the enhancement of photolysis-mediated bio-H_2 generation from *C. reinhardtii*. Microalgae *C. reinhardtii* cc124 and the bacterium *A. chroococcum* were co-cultured to produce a maximal bio-H_2 yield of 255 μmol/mg Chl (Xu et al., 2017), which was nearly 16 times higher than the control. Further, the authors hypothesized that the combined system's increased bio-H_2 yields were due to a decrease in oxygen levels, an increase in microalgal growth, an increase in H_2ase activity, and a decrease in starch content (Xu et al., 2017). The presence of many consortia led to a threefold increase in bio-H_2 productivity compared to monocultures, as co-cultures affected parameters like yield, rate, and duration (Fakhimi et al., 2020). Fakhimi and Tavakoli (2019) recently looked into how bacterial co-culture influences bio-H_2 production from *Chlamydomonas* under dim lighting conditions. The results showed that the growth of *Chlamydomonas* was reduced by 46%, 24%, and 32%, respectively, when *P. stutzeri*, *E. coli*, and *P. putida* were co-cultured in the bioreactors under low light conditions, which increased their respiration rate. The production of bio-H_2 could be increased by co-cultivating *C. reinhardtii* with isolated bacteria, as suggested by Li et al. (2013). The maximum bio-H_2 yield was discovered to be roughly four times higher than that of a single algal culture. Algal bio-H_2 generation in co-cultures may also be enhanced by the addition of sugars. When the green microalga *C. reinhardtii* was co-cultured with *E. coli* (35.1 mL bio-H_2/L), *P. putida* (40.8 mL bio-H_2/L), and *Rhizobium etli* (16.1 mL bio-H_2/L), they were able to produce bio-H_2 in sugar-containing media, as reported by Fakhimi et al. (2019). Additionally, the results show that the microalgae's ability to utilize the released bacterial acetic acid to produce bio-H_2 was crucial to the success of this collaboration (Fakhimi et al., 2019). The syntrophic interaction between microalgae and bacteria appeared to be the primary factor in increasing microalgae bio-H_2 production because it reduced Chl reduction, increased starch accumulation, and preserved protein composition (Show et al., 2019). Despite the promising potential of microalgae-bacteria consortiums for bio-H_2 production, significant research needs to be done in this area. The microalgae-bacteria consortium, on the other hand, shows great promise as a promising strategy for enhancing bio-H_2 production.

2.6.3 ELECTRO-BIO-HYDROGENATION AS A NEW APPROACH FOR MICROALGAL BIO-H_2

In addition to the aforementioned conventional methods, a novel hybrid biological bio-H_2 generation approach employing the electro-chemical process has been proposed. By oxidizing organic matter with the help of electroactive biofilms, microbial electrolysis cells (MECs) show great promise as future sources of renewable energy (Hua et al., 2019). The electrolysis based on the study and use of microbial fuel cells (MFCs) is a part of these procedures (Azwar et al., 2014). The biological oxidation of organic substances is the specific mechanism by which microorganisms generate electrons and protons. MECs have the ability to transform a wide range of organic materials into usable forms of energy (Wang et al., 2020a). The microorganisms at the anode generate protons, which travel through the reactor solution and the external circuit to the cathode. At the cathode, bio-H_2 is formed when electrons and protons bond (Hua et al., 2019). The generation of bio-H_2 does not take place within the microalgal cell, so MECs cannot be considered "traditional" bio-H_2 generation platforms (Oey et al., 2016). Bio-H_2 productivity and efficiency were shown to improve when wastewater and microalgae were combined with photosynthetic and non-photosynthetic bacteria in a fermentation process that included the use of electricity (Kim et al., 2018).

2.6.4 NANOMATERIALS FOR ENHANCING ENZYME STABILITY AND HYDROLYTIC EFFICIENCY

Since nanotechnology can influence intracellular electron transfer, microbial growth, and metalloenzyme activity—all of which contribute to bio-H_2 generation—it holds great promise for enhancing bio-H_2 production (Shanmugam et al., 2020). Since Fe^{2+} is a crucial component of the hydrogenase's active center, it has been the subject of extensive study and has been found to increase hydrogenase activity (Wu et al., 2017). Zero-valent iron (FeO) NPs were shown to increase the activity of crucial enzymes by discharging Fe^{2+}, according to research by Yang and Wang (2018). The maximum rate of bio-H_2 production in this study was 12.1 mL/h when 400 mg/L of FeO NPs were used, which is an increase of 128.3% over the rate without NPs (Yang and Wang, 2018). In a separate study, Nath et al. (2015) found that adding FeNPs increased the percentage and yield of bio-H_2 compared to the control group. The maximum bio-H_2 yield of 1.9 mol/mol glucose was seen when 100 mg/L of FeNPs were added, and the efficiency of turning glucose into bio-H_2 went up by two times. Several nickel ions (Ni^{2+}) were discovered to be important cofactors in the active site of [Ni-Fe] hydrogenase, in addition to the iron ions (Fe^{2+}) previously identified (Shanmugam et al., 2020). Dehydrogenase enzyme activity may be increased through co-cultivation with Ni^{2+}. Maximum bio-H_2 yields of 24.73 ± 1.12 mL/g COD initial were demonstrated by Elreedy et al. (2017) at a dosage of 60 mg/L for Ni NPs, which was higher than the control without the addition of NPs. Researchers have also focused more on gold NP catalysis and other novel catalytic systems for a variety of reactions. Zhang and Shen (2007), for instance, found that gold NPs significantly boosted the bioactivity of bio-H_2 producing bacteria. Moreover, the effectiveness was highly dependent on the size of the gold particles used. Since the hydrogenase involved in bio-H_2 generation is oxygen-sensitive, this limits its potential for widespread use. Recent advances in nanotechnology have made it possible to increase hydrogenase efficiency and boost the productivity of the process (Shanmugam et al., 2020). For instance, titanium (TiO_2) NPs were found to increase hydrogenase activity by Pandey et al. (2015); this property can be exploited for biophotolytic bio-H_2 generation. While there has been research into NPs for bacteria, there has been much less investigation into the role of NPs in the bio-H_2 production from microalgae. In conclusion, genetic engineering and metabolic engineering are among the essential technologies for increasing the production of biological bio-H_2. To date, there have been few studies on the effects of bacterial symbionts in commercial microalgae production systems. However, more study is needed to determine the yield and efficiency of bio-H_2 using the novel hybrid biological bio-H_2 generation approach via the electrochemical process. While nanotechnology has the potential to improve bio-H_2 production by facilitating intracellular electron transfer, microbial growth, and the activity of metalloenzymes, its widespread implementation is still in its infancy.

2.7 CHALLENGES OF BIO-H_2 PRODUCTION FROM MICROALGAE

It is undeniable that numerous microalgal species hold great promise for bio-H_2 production via a variety of mechanisms. However, there are a number of obstacles that prevent the bio-H_2 production process from running smoothly and sustainably when microalgal biomass is the driving force. This section will discuss some of the most significant difficulties that are shared by a wide variety of production mechanisms.

2.7.1 OXYGEN SENSITIVITY

Many methods of bio-H_2 production suffer from a sensitivity to oxygen, which is widely regarded as the most widespread and problematic challenge. Hydrogenase enzyme activity is inhibited by oxygen molecules due to the enzyme's extreme sensitivity to the gas. This happens because of the oxygen's binding potential in the hydrogenase's active site (El-Dalatony et al., 2020). Since these bonds are irreversible, it will be even more challenging to separate the hydrogen and oxygen molecules that make up this compound (Eroglu and Melis, 2016). It has been found that most microalgal species are inhibited from producing bio-H_2 below a saturation level of 15% in the air; however, two species have been identified that are able to express their corresponding hydrogenase enzyme at a saturation level of around 21% in the air (Razu et al., 2019). Reduced bio-H_2 production occurs when hydrogenase activities are blocked. For this reason, bioengineered oxygen-tolerant microalgal species are required to overcome this obstacle and produce hydrogenase enzymes that are effective enough to overcome their oxygen sensitivity (Mona et al., 2020b). It is hoped that by doing so, the problem of oxygen sensitivity can be solved.

2.7.2 ALTERATION OF THYLAKOID PROTON (ATP) GRADIENT

Disrupted proton transfer during bio-H_2 production from microalgae frequently results from a different thylakoid constitution. When ATP levels drop too low, electron transport stops and the proton gradient weakens, as occurs in this case (Razu et al., 2019). This problem seems to arise more often in procedures that involve using microbes capable of photosynthesis. Hydrogenase's proton flow and electron supply can be improved with the help of a leakier ATPase (Eroglu and Melis, 2016). Thus, difficulties associated with modifying the thylakoid proton gradient will be accomplished.

2.7.3 LIGHT CAPTURE DISRUPTION

Photons are taken in by the light-harvesting antenna proteins (LHCs) in photosynthetic microbes and then released as light. There is frequently interference with the functions of these antennae. It is hoped that by making changes to the LHC protein's genetic constitution, the problem can be fixed and more efficient proteins can be expressed. Minimal adjustments for some of these species have shown promising improvement (Razu et al., 2019).

2.7.4 HAMPERED PHOTOSYNTHETIC EFFICIENCY

Multiple photosynthetic processes have dismally low conversion efficiencies. Even if a substantial amount of bio-H_2 is being produced, it is not always possible to fully capture it, especially in processes involving photosynthetic stages (El-Dalatony et al., 2020). The development of an effective strategy to address this issue calls for focused effort.

2.7.5 PRODUCTION OF CONTAMINANTS

Contaminants like tar, coal, carbon dioxide, and methane are often byproducts of producing bio-H_2 from wastewater microalgae. Many methods of producing biohydrogen operate in this manner. This presents a conundrum with regards to determining the environmental viability of the relevant technology. If this keeps happening, we cannot say this process is completely eco-friendly. Therefore, all processes should incorporate some modifications to ensure source-level reduction of these contaminants, despite this being a very challenging task.

2.7.6 HIGH CONSTRUCTION AND OPERATING COST

Another huge challenge to large-scale bio-H_2 generation from wastewater microalgae is the high implementation and operating costs of many of the bioreactors. There needs to be a reduction in these costs in order to gain widespread support for these technologies from a variety of interested parties.

2.8 APPLICATIONS OF HYDROGEN IN INDUSTRY

Although H_2 is rapidly emerging as a viable alternative energy source, these uses actually make up less than 10% of the world's total H_2 consumption. The world is rethinking how H_2 is produced, transported, and used as the H_2 economy expands (Neuwirth et al., 2022). The majority of H_2 used today is used in industries like oil refining, ammonia production, and methanol production. H_2 has a long history of use in many different industries, but as a result of the pressing need to cut carbon emissions, H_2 is being creatively used to create both new and old products. This includes sectors like steel production, where H_2 is being considered as a potential substitute for coal and other carbon-emitting fuels. Unsaturated fats and oils, such as the hydrogenated vegetable oils used in margarine and butter spreads, are converted to saturated fats and oils in food. In medicine, H_2 is used to create hydrogen peroxide (H_2O_2). H_2 gas has also recently been the subject of research as a potential therapeutic gas for a number of diseases. Making concrete, a process that consumes a lot of energy and produces a lot of carbon, is another potential new application that could be explored.

2.8.1 HYDROGEN FUEL CELLS PRODUCE ELECTRICITY

Hydrogen fuel cells mix H_2 and O_2 atoms to produce energy. H_2 and O_2 react in an electrochemical cell like a battery to produce electricity, water, and a small amount of heat. There are several varieties of fuel cells available for use in various contexts. Military equipment, laptops, and mobile phones all make use of small fuel cells. Power grids, individual buildings in need of emergency or backup power, and remote places without access to power networks are all viable destinations for the electricity generated by large fuel cells (Oliveira et al., 2021).

2.8.2 BURNING HYDROGEN FOR ELECTRICITY GENERATION

H_2 is becoming increasingly common as a power plant fuel. Several US power plants have announced the installation of combustion gas turbines powered by a combination of natural gas and H_2. The 485 MW Long Ridge Energy Generation Project facility in Ohio, for example, will use a fuel blend of 95% natural gas and 5% H_2 to begin operations. Eventually, the plant will switch to 100% green H_2 obtained from renewable sources (Atilhan et al., 2021). Another example is the plan by Intermountain Power Agency to convert a coal-fired power station in Utah to a combined-cycle gas-fired plant that will use up to 30% hydrogen at first and 100% bio-H_2 in the long run.

2.8.3 AMMONIA PRODUCTION

Over the next 30 years, there will be a 40% increase in ammonia production, and H_2 will be essential to that growth. Since the ammonia industry currently accounts for 1.8% of global carbon dioxide emissions, there is a sizable window of opportunity for global carbon emissions reduction. Clean H_2 can be substituted in the industrial process by using carbon capture, utilization, and storage (CCUS) and renewable energy, resulting in a significant reduction in emissions and enabling the sustainable expansion of the petrochemical industry as a whole. Due to its ease of transportation compared to liquid or gas-based H_2, ammonia has the potential to serve as a storage medium for H_2 (Oliveira et al., 2021). As a result, initiatives that produce clean ammonia have begun to receive a lot of attention and funding.

2.8.4 METHANOL PRODUCTION

Methanol is made with the help of H_2, which reacts with carbon monoxide to create the chemical feedstock. Emissions can be decreased by replacing dirty H_2 in the hydrogenation process (Oliveira et al., 2021). Researchers are also investigating how methanol production could aid in lowering atmospheric emissions. In addition to its significant chemical uses, methanol is a viable alternative fuel for machines and vehicles because it can increase fuel efficiency and reduce emissions.

2.8.5 STEELMAKING

As mentioned in a previous FCHEA update, the steel making sector is starting to test the use of H_2 to lower emissions during the steel making process. In order to achieve the high temperatures and chemical reactions needed for the steel making process, coal is typically used (Yáñez et al., 2019). Both the necessary heat and the chemical reactions in this process can be replaced with H_2. Because steel is an important part of both industrial processes and modern buildings, using clean hydrogen could help cut emissions by a lot.

2.8.6 CONCRETE PRODUCTION

Researchers are currently conducting experimental testing to determine whether H_2 can be used in the production of concrete. Alternative methods are required because H_2 cannot be directly substituted as an ingredient or reactant in the processes used to produce cement today. To reduce emissions during the process, one strategy is to use a mixture of H_2 and other inputs, such as biomass or CCUS fossil fuels (Yáñez et al., 2019). There are already several instances of this in use, such as a demonstration at Heidelberg Cement, a business that created cement using a mixture of glycerin, biomass, and H_2 as fuel. As the desire to cut emissions grows, more businesses are likely to follow suit, and some have already done so.

2.8.7 OIL REFINING

Another application for clean H_2 that can cut carbon emissions is oil refining. Since oil is so vital to the global economy and demand is predicted to peak around 2040, there are many opportunities to cut carbon emissions during the oil refining process (Zou et al., 2022). Utilizing clean H_2, whether it is produced on-site using electrolysis or produced and shipped after being processed by CCUS, could lower emissions while still enabling advanced refining.

2.8.8 PUBLIC TRANSPORTATION

Additional applications for trains and buses using H_2 fuel cells are also being considered. H_2-powered buses have undergone trials in several major cities, including Beijing, Vancouver, London, Chicago, and Vancouver. Additional versions are expected to arrive in the United Kingdom, France, Italy, Japan, South Korea, and the United States during the following five years. In Germany, H_2-powered trains have recently gone into service (Neuwirth et al., 2022).

2.9 CONCLUSIONS AND FUTURE PERSPECTIVES

The potential benefits of producing bio-H_2 from microalgae, such as carbon neutrality and energy sustainability, have garnered significant attention in recent years. The current level of bio-H_2 generation from microalgae is still insufficient as a commercially competitive energy carrier, but there has been a lot of research on increasing bio-H_2 yield through microalgae. Although none of the methods for producing bio-H_2 are currently practical for widespread use, they each have their own unique advantages. While fermentation processes have a lower environmental impact, they are inefficient at converting solar energy and producing bio-H_2. Carbon-free fuel production and CO_2 sequestration are both greatly aided by the widespread use of microalgal biomass (as a source of carbohydrates). The amount of carbohydrate in a microalgal biomass is affected by the substrates and species being used. Aside from carbohydrate content, protein content is also important when using algae as biomass. Using genetic engineering to create microalgae with the capacity for bio-H_2 production is an attractive strategy for achieving long-term sustainability in the field of bioenergy utilization and thus easing the pressure on the world's energy supply. Bio-H_2 production from microalgae at industrial scale is preferred, but there is currently no standardized technology that uses a wide variety of nutrients to supplement potential microorganisms. It is still necessary for researchers and academics from all over the world to conduct more multidisciplinary research in order to develop methods for industrial-scale production at a cost-effective level.

KEYWORDS

- **biohydrogen**
- **biohydrogen production**
- **Chlorococcales**
- **dinitrogenase**
- **metabolic engineering**
- **microalgae**

REFERENCES

Acar, C., & Dincer, I. (2018a). 3.1 Hydrogen production. *Comprehensive Energy Systems, 3*, 1–40.

Akhlaghi, N., & Najafpour-Darzi, G. (2020). A comprehensive review on biological hydrogen production. *International Journal of Hydrogen Energy, 45*(43), 22492–22512.

Antal, T. K., Krendeleva, T. E., & Rubin, A. B. (2011). Acclimation of green algae to sulfur deficiency: Underlying mechanisms and application for hydrogen production. *Applied Microbiology and Biotechnology, 89*(1), 3–15.

Aslam, M., Ahmad, R., Yasin, M., Khan, A. L., Shahid, M. K., Hossain, S., Khan, Z., Jamil, F., Rafiq, S., Bilad, M. R., Kim, J., & Kumar, G. (2018). Anaerobic membrane bioreactors for biohydrogen production: Recent developments, challenges and perspectives. *Bioresource Technology, 269*, 452–464.

Atilhan, S., Park, S., El-Halwagi, M. M., Atilhan, M., Moore, M., & Nielsen, R. B. (2021). Green hydrogen as an alternative fuel for the shipping industry. *Current Opinion in Chemical Engineering, 31*, 100668.

Azwar, M., Hussain, M., & Abdul-Wahab, A. (2014). Development of biohydrogen production by photobiological, fermentation and electrochemical processes: A review. *Renewable and Sustainable Energy Reviews, 31*, 158–173.

Baeyens, J., Zhang, H., Nie, J., Appels, L., Dewil, R., Ansart, R., & Deng, Y. (2020). Reviewing the potential of bio-hydrogen production by fermentation. *Renewable and Sustainable Energy Reviews, 131*, 110023.

Ban, S., Lin, W., Luo, Z., & Luo, J. (2019). Improving hydrogen production of *Chlamydomonas reinhardtii* by reducing chlorophyll content via atmospheric and room temperature plasma. *Bioresource Technology, 275*, 425–429.

Ban, S., Lin, W., Wu, F., & Luo, J. (2018). Algal-bacterial cooperation improves algal photolysis-mediated hydrogen production. *Bioresource Technology, 251*, 350–357.

Batyrova, K., & Hallenbeck, P. C. (2017). Hydrogen production by a *Chlamydomonas reinhardtii* strain with inducible expression of photosystem II. *International Journal of Molecular Sciences, 18*(3), 647.

Benemann, J. (1997). Feasibility analysis of photobiological hydrogen production. *International Journal of Hydrogen Energy, 22*(10, 11), 979–987.

Bhatia, S. K., Mehariya, S., Bhatia, R. K., Kumar, M., Pugazhendhi, A., Awasthi, M. K., Atabani, A., Kumar, G., Kim, W., Seo, S., & Yang, Y. (2020). Wastewater based microalgal biorefinery for bioenergy production: Progress and challenges. *Science of the Total Environment, 751*, 141599.

Bolatkhan, K., Kossalbayev, B. D., Zayadan, B. K., Tomo, T., Veziroglu, T. N., & Allakhverdiev, S. I. (2019). Hydrogen production from phototrophic microorganisms: Reality and perspectives. *International Journal of Hydrogen Energy, 44*(12), 5799–5811.

Chader, S., Hacene, H., & Agathos, S. N. (2009). Study of hydrogen production by three strains of *Chlorella* isolated from the soil in the Algerian Sahara. *International Journal of Hydrogen Energy, 34*(11), 4941–4946.

Deviram, G., Mathimani, T., Anto, S., Ahamed, T. S., Ananth, D. A., & Pugazhendhi, A. (2020). Applications of microalgal and cyanobacterial biomass on a way to safe, cleaner and a sustainable environment. *Journal of Cleaner Production, 253*, 119770.

Eilenberg, H., Weiner, I., Ben-Zvi, O., Pundak, C., Marmari, A., Liran, O., Wecker, M. S., Milrad, Y., & Yacoby, I. (2016). The dual effect of a ferredoxin-hydrogenase fusion protein in vivo: Successful divergence of the photosynthetic electron flux towards hydrogen production and elevated oxygen tolerance. *Biotechnology for Biofuels, 9*(1), 182.

El-Dalatony, M. M., Zheng, Y., Ji, M., Li, X., & Salama, E. (2020). Metabolic pathways for microalgal biohydrogen production: Current progress and future prospectives. *Bioresource Technology, 318*, 124253.

Elreedy, A., Ibrahim, E., Hassan, N., El-Dissouky, A., Fujii, M., Yoshimura, C., & Tawfik, A. (2017). Nickel-graphene nanocomposite as a novel supplement for enhancement of biohydrogen production from industrial wastewater containing mono-ethylene glycol. *Energy Conversion and Management, 140*, 133–144.

Eroglu, E., & Melis, A. (2016). Microalgal hydrogen production research. *International Journal of Hydrogen Energy, 41*(30), 12772–12798.

Fakhimi, N., & Tavakoli, O. (2019). Improving hydrogen production using co-cultivation of bacteria with *Chlamydomonas reinhardtii* microalga. *Materials Science for Energy Technologies, 2*(1), 1–7.

Fakhimi, N., Dubini, A., Tavakoli, O., & González-Ballester, D. (2019). Acetic acid is key for synergetic hydrogen production in *Chlamydomonas*–bacteria co-cultures. *Bioresource Technology, 289*, 121648.

Fakhimi, N., Gonzalez-Ballester, D., Fernández, E., Galván, A., & Dubini, A. (2020). Algae-bacteria consortia as a strategy to enhance H_2 production. *Cells, 9*(6), 1353.

Fan, X., Wang, H., Guo, R., Yang, D., Zhang, Y., Yuan, X., Qiu, Y., Yang, Z., & Zhao, X. (2016). Comparative study of the oxygen tolerance of *Chlorella pyrenoidosa* and *Chlamydomonas reinhardtii* CC124 in photobiological hydrogen production. *Algal Research – Biomass, Biofuels and Bioproducts, 16*, 240–244.

Florin, L., Tsokoglou, A., & Happe, T. (2001). A novel type of iron hydrogenase in the green alga *Scenedesmus obliquus* is linked to the photosynthetic electron transport chain. *The Journal of Biological Chemistry, 276*(9), 6125–6132.

Foong, S. Y., Chan, Y. H., Cheah, W. Y., Kamaludin, N. H., Tengku, I. T. N. B., Sonne, C., Peng, W., Show, P. L., & Lam, S. S. (2021). Progress in waste valorization using advanced pyrolysis techniques for hydrogen and gaseous fuel production. *Bioresource Technology, 320*(Pt A), 124299.

Forestier, M., King, P. W., Zhang, L., Posewitz, M. C., Schwarzer, S., Happe, T., Ghirardi, M. L., & Seibert, M. L. (2003). Expression of two [Fe]-hydrogenases in *Chlamydomonas reinhardtii* under anaerobic conditions. *European Journal of Biochemistry, 270*(13), 2750–2758.

Gadhe, A., Sonawane, S. S., & Varma, M. N. (2015). Enhancement effect of hematite and nickel nanoparticles on biohydrogen production from dairy wastewater. *International Journal of Hydrogen Energy, 40*, 4502–4511.

Gaffron, H. (1939). Reduction of carbon dioxide with molecular hydrogen in green algæ. *Nature, 143*, 204–205.

Ge, S., Foong, S. Y., Ma, N. L., Liew, R. K., Mahari, W. A., Xia, C., Yek, P. N., Peng, W., Nam, W. L., Lim, X. Y., Liew, C. M., Chong, C. C., Sonne, C., & Lam, S. S. (2020). Vacuum pyrolysis incorporating microwave heating and base mixture modification: An integrated approach to transform biowaste into eco-friendly bioenergy products. *Renewable & Sustainable Energy Reviews, 127*, 109871.

Ghirardi, M. L., Kosourov, S. N., Tsygankov, A., & Seibert, M. L. (2000). Two-phase photobiological algal H_2-production system. In *Proceedings of the 2000 DOE Hydrogen Program Review* (pp. 9–11). Citeseer.

Godman, J. E., Molnár, A., Baulcombe, D. C., & Balk, J. (2010). RNA silencing of hydrogenase(-like) genes and investigation of their physiological roles in the green alga *Chlamydomonas reinhardtii*. *The Biochemical Journal, 431*(3), 345–351.

Goswami, R. K., Mehariya, S., Obulisamy, P. K., & Verma, P. (2021). Advanced microalgae-based renewable biohydrogen production systems: A review. *Bioresource Technology, 320*, 124301.

Hallenbeck, P. C., & Benemann, J. R. (2002). Biological hydrogen production: Fundamentals and limiting processes. *International Journal of Hydrogen Energy, 27*(11, 12), 1185–1193.

Hankamer, B., Lehr, F., Rupprecht, J., Mussgnug, J. H., Posten, C., & Kruse, O. (2007). Photosynthetic biomass and H_2 production by green algae: From bioengineering to bioreactor scale-up. *Physiologia Plantarum, 131*(1), 10–21.

He, J., Xi, L., Sun, X., Ge, B., Liu, D., Han, Z., Pu, X., & Huang, F. (2018). Enhanced hydrogen production through co-cultivation of *Chlamydomonas reinhardtii* CC-503 and a facultative autotrophic sulfide-oxidizing bacterium under sulfurated conditions. *International Journal of Hydrogen Energy, 43*(32), 15005–15013.

Horner, D. S., Heil, B., Happe, T., & Embley, T. M. (2002). Iron hydrogenases—Ancient enzymes in modern eukaryotes. *Trends in Biochemical Sciences, 27*(3), 148–153.

Hua, T., Li, S., Li, F., Zhou, Q., & Ondon, B. S. (2019). Microbial electrolysis cell as an emerging versatile technology: A review on its potential application, advance and challenge. *Journal of Chemical Technology & Biotechnology, 94*, 1697–1711.

Hwang, J. H., Kim, H. C., Choi, J. A., Abou-Shanab, R. A., Dempsey, B. A., Regan, J. M., Kim, J. R., Song, H., Nam, I. H., Kim, S. N., Lee, W., Park, D., Kim, Y., Choi, J., Ji, M. K., Jung, W., & Jeon, B. H. (2014). Photoautotrophic hydrogen production by eukaryotic microalgae under aerobic conditions. *Nature Communications, 5*, 3234.

Hwang, J., Lee, M., Kang, E. H., & Lee, W. H. (2021). Renewable algal photo H_2 production without S control using acetate-enriched fermenter effluents. *International Journal of Hydrogen Energy, 46*(2), 1740–1751.

IEA. (2020). *Energy technology perspectives 2020: Special report on carbon capture, utilization and storage*. IEA: Paris, France.

Jiménez-Llanos, J., Ramírez-Carmona, M., Rendón-Castrillón, L., & Ocampo-López, C. (2020). Sustainable biohydrogen production by *Chlorella* sp. microalgae: A review. *International Journal of Hydrogen Energy, 45*, 8310–8328.

Khetkorn, W., Rastogi, R. P., Incharoensakdi, A., Lindblad, P., Madamwar, D., Pandey, A., & Larroche, C. (2017). Microalgal hydrogen production – A review. *Bioresource Technology, 243*, 1194–1206.

Khosravitabar, F. (2019). Microalgal biohydrogen photoproduction: Scaling up challenges and the ways forward. *Journal of Applied Phycology, 32*, 277–289.

Kim, J., Jun, A., Gwon, O., Yoo, S., Liu, M., Shin, J., Lim, T., & Kim, G. (2018). Hybrid-solid oxide electrolysis cell: A new strategy for efficient hydrogen production. *Nano Energy, 44*, 121–126.

Kruse, O., Rupprecht, J., Bader, K. P., Thomas-Hall, S., Schenk, P. M., Finazzi, G., & Hankamer, B. (2005). Improved photobiological H_2 production in engineered green algal cells. *The Journal of Biological Chemistry, 280*(40), 34170–34177.

Kubas, A., Orain, C., De Sancho, D., Saujet, L., Sensi, M., Gauquelin, C., Meynial-Salles, I., Soucaille, P., Bottin, H., Baffert, C., Fourmond, V., Best, R. B., Blumberger, J., &

Léger, C. (2017). Mechanism of O_2 diffusion and reduction in FeFe hydrogenases. *Nature Chemistry, 9*(1), 88–95.

Kucharska, K., Cieśliński, H., Rybarczyk, P., Słupek, E., Łukajtis, R., Wychodnik, K., & Kamiński, M. (2019). Fermentative conversion of two-step pre-treated lignocellulosic biomass to hydrogen. *Catalysts, 9*, 858.

Kumar, G., Nguyen, D. D., Sivagurunathan, P., Kobayashi, T., Xu, K., & Chang, S. W. (2018). Cultivation of microalgal biomass using swine manure for biohydrogen production: Impact of dilution ratio and pretreatment. *Bioresource Technology, 260*, 16–22.

Kumar, S., Sharma, S., Thakur, S., Mishra, T., Negi, P., Mishra, S. K., Hesham, A. E., Rastegari, A. A., Yadav, N. R., & Yadav, A. (2019). Bioprospecting of microbes for biohydrogen production: Current status and future challenges. In *Bioprocessing for Biomolecules Production* (Vol. 2040, pp. 443–471). Springer.

Lam, M. K., Loy, A. C. M., Yusup, S., & Lee, K. T. (2019). Biohydrogen production from algae. In *Biohydrogen* (2nd ed., pp. 219–245). Elsevier.

Lee, X. J., Ong, H. C., Gan, Y. Y., Chen, W., & Mahlia, T. M. I. (2020). State of art review on conventional and advanced pyrolysis of macroalgae and microalgae for biochar, bio-oil and bio-syngas production. *Energy Conversion and Management, 210*, 112707.

Li, T., Zheng, Y., Yu, L., & Chen, S. (2014). Mixotrophic cultivation of a *Chlorella sorokiniana* strain for enhanced biomass and lipid production. *Biomass and Bioenergy, 66*, 204–213.

Li, X., Huang, S., Yu, J., Wang, Q., & Wu, S. (2013). Improvement of hydrogen production of *Chlamydomonas reinhardtii* by co-cultivation with isolated bacteria. *International Journal of Hydrogen Energy, 38*(25), 10779–10787.

Limongi, A. R., Viviano, E., De Luca, M., Radice, R., Bianco, G., & Martelli, G. (2021). Biohydrogen from microalgae: Production and applications. *Applied Sciences, 11*(4), 1616.

Lin, C., & Shei, S. (2008). Heavy metal effects on fermentative hydrogen production using natural mixed microflora. *International Journal of Hydrogen Energy, 33*, 587–593.

Lin, H. D., Liu, B. H., Kuo, T. T., Tsai, H. C., Feng, T. Y., Huang, C. C., & Chien, L. F. (2013). Knockdown of *PsbO* leads to induction of *HydA* and production of photobiological H_2 in the green alga *Chlorella* sp. DT. *Bioresource Technology, 143*, 154–162.

Liu, J., Ge, Y., Xia, S., Sun, J., & Mu, J. (2016). Photoautotrophic hydrogen production by *Chlorella pyrenoidosa* without sulfur-deprivation. *International Journal of Hydrogen Energy, 41*(20), 8427–8432.

Malek Shahkouhi, A., & Motamedian, E. (2020). Reconstruction of a regulated two-cell metabolic model to study biohydrogen production in a diazotrophic cyanobacterium *Anabaena variabilis* ATCC 29413. *PLOS ONE, 15*(3), e0227977.

Martín del Campo, J. S., Escalante, R., Robledo, D., & Patiño, R. (2014). Hydrogen production by *Chlamydomonas reinhardtii* under light-driven and sulfur-deprived conditions: Using biomass grown in outdoor photobioreactors at the Yucatan Peninsula. *International Journal of Hydrogen Energy, 39*(36), 20950–20957.

Masukawa, H., Mochimaru, M., & Sakurai, H. (2002). Disruption of the uptake hydrogenase gene, but not of the bidirectional hydrogenase gene, leads to enhanced photobiological hydrogen production by the nitrogen-fixing cyanobacterium *Anabaena* sp. PCC 7120. *Applied Microbiology and Biotechnology, 58*(5), 618–624.

Mathews, J., & Wang, G. (2009). Metabolic pathway engineering for enhanced biohydrogen production. *International Journal of Hydrogen Energy, 34*, 7404–7416.
Melis, A., & Happe, T. (2001). Hydrogen production: Green algae as a source of energy. *Plant Physiology, 127*(3), 740–748.
Meuser, J. E., D'Adamo, S., Jinkerson, R. E., Mus, F., Yang, W., Ghirardi, M. L., Seibert, M., Grossman, A. R., & Posewitz, M. C. (2012). Genetic disruption of both *Chlamydomonas reinhardtii* [FeFe]-hydrogenases: Insight into the role of *HYDA2* in H_2 production. *Biochemical and Biophysical Research Communications, 417*(2), 704–709.
Mona, S., Kaushik, A., & Kaushik, C. P. (2011). Hydrogen production and metal-dye bioremoval by a *Nostoc linckia* strain isolated from textile mill oxidation pond. *Bioresource Technology, 102*(3), 3200–3205.
Mona, S., Kumar, S., Kumar, V., Parveen, K., Saini, N., Deepak, B., & Pugazhendhi, A. (2020). Green technology for sustainable biohydrogen production (waste to energy): A review. *The Science of the Total Environment, 728*, 138481.
Mosebach, L., Heilmann, C., Mutoh, R., Gäbelein, P., Steinbeck, J., Happe, T., Ikegami, T., Hanke, G., Kurisu, G., & Hippler, M. (2017). Association of ferredoxin: $NADP^+$ oxidoreductase with the photosynthetic apparatus modulates electron transfer in *Chlamydomonas reinhardtii*. *Photosynthesis Research, 134*(3), 291–306.
Mu, D., Liu, H., Lin, W., Shukla, P., & Luo, J. (2020). Simultaneous biohydrogen production from dark fermentation of duckweed and waste utilization for microalgal lipid production. *Bioresource Technology, 302*, 122879.
Nagarajan, D., Chang, J. S., & Lee, D. J. (2020a). Pretreatment of microalgal biomass for efficient biohydrogen production: Recent insights and future perspectives. *Bioresource Technology, 302*, 122871.
Nagarajan, D., Dong, C., Chen, C., Lee, D., & Chang, J. (2020b). Biohydrogen production from microalgae: Major bottlenecks and future research perspectives. *Biotechnology Journal, 16*, 2000124.
Nagy, V., Podmaniczki, A., Vidal-Meireles, A., Tengölics, R., Kovács, L. G., Rákhely, G., Scoma, A., & Tóth, S. Z. (2018). Water-splitting-based, sustainable and efficient H_2 production in green algae as achieved by substrate limitation of the Calvin–Benson–Bassham cycle. *Biotechnology for Biofuels, 11*, 69.
Nath, D., Manhar, A. K., Gupta, K. K., Saikia, D., Das, S. K., & Mandal, M. (2015). Phytosynthesized iron nanoparticles: Effects on fermentative hydrogen production by *Enterobacter cloacae* DH-89. *Bulletin of Materials Science, 38*, 1533–1538.
Neuwirth, M., Fleiter, T., Manz, P., & Hofmann, R. (2022). The future potential hydrogen demand in energy-intensive industries: A site-specific approach applied to Germany. *Energy Conversion and Management, 252*, 115052.
Oey, M., Sawyer, A. L., Ross, I. L., & Hankamer, B. (2016). Challenges and opportunities for hydrogen production from microalgae. *Plant Biotechnology Journal, 14*(7), 1487–1499.
Oliveira, A. M., Beswick, R. R., & Yan, Y. (2021). A green hydrogen economy for a renewable energy society. *Current Opinion in Chemical Engineering, 33*, 100701.
Oncel, S. S. (2015). Biohydrogen from microalgae: Uniting energy, life, and green future. In M. A. Alam & Z. Wang (Eds.), *Microalgae* (pp. 159–196). Elsevier.

Osman, A. I., Deka, T. J., Baruah, D. C., & Rooney, D. W. (2020). Critical challenges in biohydrogen production processes from organic feedstocks. *Biomass Conversion and Biorefinery*, 1–19.

Ottenheim, C., Nawrath, M., & Wu, J. C. (2018). Microbial mutagenesis by atmospheric and room-temperature plasma (ARTP): The latest development. *Bioresources and Bioprocessing, 5*(1), 1–14.

Pandey, A., Gupta, K., & Pandey, A. (2015). Effect of nanosized TiO_2 on photofermentation by *Rhodobacter sphaeroides* NMBL-02. *Biomass and Bioenergy, 72*, 273–279.

Petrova, E., Kukarskikh, G. P., Krendeleva, T. E., & Antal, T. K. (2020). The mechanisms and role of photosynthetic hydrogen production by green microalgae. *Microbiology, 89*, 251–265.

Pongpadung, P., Liu, J., Yokthongwattana, K., Techapinyawat, S., & Juntawong, N. (2015). Screening for hydrogen-producing strains of green microalgae in phosphorus or sulfur deprived medium under nitrogen limitation. *ScienceAsia, 41*, 97–107.

Posewitz, M. C., Smolinski, S., Kanakagiri, S., Melis, A., Seibert, M. L., & Ghirardi, M. L. (2004). Hydrogen photoproduction is attenuated by disruption of an isoamylase gene in *Chlamydomonas reinhardtii. The Plant Cell, 16*, 2151–2163.

Rajesh, B. J., Kavitha, S., Yukesh, K. R., Bhosale, R. R., & Kumar, G. (2020). Industrial wastewater to biohydrogen: Possibilities towards successful biorefinery route. *Bioresource Technology, 298*, 122378.

Rashid, N., Lee, K., Han, J. I., & Gross, M. (2013). Hydrogen production by immobilized *Chlorella vulgaris*: Optimizing pH, carbon source, and light. *Bioprocess and Biosystems Engineering, 36*(7), 867–872.

Razu, M. H., Hossain, F., & Khan, M. (2019). Advancement of biohydrogen production from microalgae. In M. A. Alam & Z. Wang (Eds.), *Microalgae Biotechnology for Development of Biofuel and Wastewater Treatment* (pp. 423–462). Springer Singapore.

Ruiz-Marin, A., Canedo-López, Y., & Chávez-Fuentes, P. (2020). Biohydrogen production by *Chlorella vulgaris* and *Scenedesmus obliquus* immobilized and cultivated in artificial wastewater under different light quality. *AMB Express, 10*(1), 191.

Rumpel, S., Siebel, J. F., Farés, C., Duan, J., Reijerse, E. J., Happe, T., Lubitz, W., & Winkler, M. (2014). Enhancing hydrogen production of microalgae by redirecting electrons from photosystem I to hydrogenase. *Energy & Environmental Science, 7*, 3296–3301.

Scoma, A., Krawietz, D., Faraloni, C., Giannelli, L., Happe, T., & Torzillo, G. (2012). Sustained H_2 production in a *Chlamydomonas reinhardtii* D1 protein mutant. *Journal of Biotechnology, 157*(4), 613–619.

Sengmee, D., Cheirsilp, B., Suksaroge, T. T., & Prasertsan, P. (2017). Biophotolysis-based hydrogen and lipid production by oleaginous microalgae using crude glycerol as exogenous carbon source. *International Journal of Hydrogen Energy, 42*(4), 1970–1976.

Shanmugam, S., Hari, A., Pandey, A., Mathimani, T., Felix, L., & Pugazhendhi, A. (2020). Comprehensive review on the application of inorganic and organic nanoparticles for enhancing biohydrogen production. *Fuel, 270*, 117453.

Shetty, P., Boboescu, I. Z., Pap, B., Wirth, R., Kovács, K. L., Bíró, T., Futó, Z., White, R. A., & Maróti, G. (2019). Exploitation of algal-bacterial consortia in combined biohydrogen generation and wastewater treatment. *Frontiers in Energy Research, 7*, 1–12.

Show, K. Y., Yan, Y., Zong, C., Guo, N., Chang, J., & Lee, D. (2019). State of the art and challenges of biohydrogen from microalgae. *Bioresource Technology, 289*, 121747.
Song, W., Ding, L., Liu, M., Cheng, J., Zhou, J., & Li, Y. Y. (2020). Improving biohydrogen production through dark fermentation of steam-heated acid-pretreated *Alternanthera philoxeroides* by mutant *Enterobacter aerogenes* ZJU1. *The Science of the Total Environment, 716*, 134695.
Song, W., Rashid, N., Choi, W., & Lee, K. (2011). Biohydrogen production by immobilized *Chlorella* sp. using cycles of oxygenic photosynthesis and anaerobiosis. *Bioresource Technology, 102*(18), 8676–8681.
Swanson, K. D., Ratzloff, M. W., Mulder, D. W., Artz, J. H., Ghose, S., Hoffman, A. B., White, S., Zadvornyy, O. A., Broderick, J. B., Bothner, B., King, P. W., & Peters, J. W. (2015). [FeFe]-hydrogenase oxygen inactivation is initiated at the H cluster 2Fe subcluster. *Journal of the American Chemical Society, 137*(5), 1809–1816.
Tamagnini, P., Axelsson, R., Lindberg, P., Oxelfelt, F., Wünschiers, R., & Lindblad, P. (2002). Hydrogenases and hydrogen metabolism of cyanobacteria. *Microbiology and Molecular Biology Reviews, 66*(1), 1–20.
Tiwari, A., & Pandey, A. (2012). Cyanobacterial hydrogen production: A step towards clean environment. *International Journal of Hydrogen Energy, 37*, 139–150.
Touloupakis, E., Faraloni, C., Silva Benavides, A. M., Masojídek, J., & Torzillo, G. (2021). Sustained photobiological hydrogen production by *Chlorella vulgaris* without nutrient starvation. *International Journal of Hydrogen Energy, 46*(5), 3684–3694.
Tsygankov, A., Kosourov, S. N., Tolstygina, I. V., Ghirardi, M. L., & Seibert, M. L. (2006). Hydrogen production by sulfur-deprived *Chlamydomonas reinhardtii* under photoautotrophic conditions. *International Journal of Hydrogen Energy, 31*, 1574–1584.
Ubando, A. T., Africa, A. D. M., Maniquiz-Redillas, M. C., Culaba, A. B., Chen, W., & Chang, J. (2021). Microalgal biosorption of heavy metals: A comprehensive bibliometric review. *Journal of Hazardous Materials, 402*, 123431.
Vignais, P. M., Billoud, B., & Meyer, J. (2001). Classification and phylogeny of hydrogenases. *FEMS Microbiology Reviews, 25*(4), 455–501.
Wang, A., Wang, H., Cheng, H., Liang, B., Liu, W., Han, J., Zhang, B., & Wang, S. (2020). Electrochemistry-stimulated environmental bioremediation: Development of applicable modular electrode and system scale-up. *Environmental Science and Ecotechnology, 3*, 100050.
Wang, K., Khoo, K. S., Chew, K. W., Selvarajoo, A., Chen, W., Chang, J., & Show, P. L. (2021). Microalgae: The future supply house of biohydrogen and biogas. *Frontiers in Energy Research, 9*, 158.
Wang, Y., Ho, S., Cheng, C., Nagarajan, D., Guo, W., Lin, C., Li, S., Ren, N., & Chang, J. (2017a). Nutrients and COD removal of swine wastewater with an isolated microalgal strain *Neochloris aquatica* CL-M1 accumulating high carbohydrate content used for biobutanol production. *Bioresource Technology, 242*, 7–14.
Wang, Y., Ho, S., Yen, H., Nagarajan, D., Ren, N., Li, S., Hu, Z., Lee, D., Kondo, A., & Chang, J. (2017b). Current advances on fermentative biobutanol production using third generation feedstock. *Biotechnology Advances, 35*(8), 1049–1059.
Wang, Y., Yang, H., Zhang, X., Han, F., Tu, W., & Yang, W. (2020b). Microalgal hydrogen production. *Small Methods, 4*, 1900514.

Watanabe, K., Takihana, N., Aoyagi, H., Hanada, S., Watanabe, Y., Ohmura, N., Saiki, H., & Tanaka, H. (2005). Symbiotic association in *Chlorella* culture. *FEMS Microbiology Ecology, 51*(2), 187–196.

Wirth, R., Lakatos, G., Böjti, T., Maróti, G., Bagi, Z., Rákhely, G., & Kovács, K. L. (2018). Anaerobic gaseous biofuel production using microalgal biomass: A review. *Anaerobe, 52*, 1–8.

Wu, H., Wang, C., Chen, P., He, A., Xing, F., Kong, X., & Jiang, M. (2017). Effects of pH and ferrous iron on the coproduction of butanol and hydrogen by *Clostridium beijerinckii* IB4. *International Journal of Hydrogen Energy, 42*(10), 6547–6555.

Xu, L., Cheng, X., & Wang, Q. (2017). Effect of co-cultivation of *Chlamydomonas reinhardtii* with *Azotobacter chroococcum* on hydrogen production. *International Journal of Hydrogen Energy, 42*(36), 22713–22719.

Yáñez, M., Ortiz, A., Brunaud, B., Grossmann, I., & Ortiz, I. (2019). The use of optimization tools for the hydrogen circular economy. *Computer Aided Chemical Engineering, 46*, 1777–1782.

Yang, D., Syn, J., Hsieh, C., Huang, C., & Chien, L. (2019). Genetically engineered hydrogenases promote biophotocatalysis-mediated H_2 production in the green alga *Chlorella* sp. DT. *International Journal of Hydrogen Energy, 44*(5), 2533–2545.

Yang, G., & Wang, J. (2018). Improving mechanisms of biohydrogen production from grass using zero-valent iron nanoparticles. *Bioresource Technology, 266*, 413–420.

Yilmaz, F., Balta, M. T., & Selbaş, R. (2016). RETRACTED: A review of solar-based hydrogen production methods. *Renewable and Sustainable Energy Reviews, 56*, 171–178.

Yuan, L., & Koo, J. (2019). O_2 sensitivity and H_2 production activity of hydrogenases: A review. *Biotechnology and Bioengineering, 116*, 3124–3135.

Zhang, C., Li, S., & Ho, S. (2021). Converting nitrogen and phosphorus wastewater into bioenergy using microalgae-bacteria consortia: A critical review. *Bioresource Technology, 342*, 126056.

Zhang, T., Jiang, D., Zhang, H., Jing, Y., Tahir, N., Zhang, Y., & Zhang, Q. (2020). Comparative study on biohydrogen production from corn stover: Photo-fermentation, dark-fermentation and dark-photo co-fermentation. *International Journal of Hydrogen Energy, 45*, 3807–3814.

Zhang, Y., & Shen, J. (2007). Enhancement effect of gold nanoparticles on biohydrogen production from artificial wastewater. *International Journal of Hydrogen Energy, 32*(1), 17–23.

Zou, C., Li, J., Zhang, X., Jin, X., Xiong, B., Yu, H., Liu, X., Wang, S., Li, Y., Zhang, L., Miao, S., Zheng, D., Zhou, H., Song, J., & Pan, S. (2022). Industrial status, technological progress, challenges, and prospects of hydrogen energy. *Natural Gas Industry B, 9*(5), 427–447.

CHAPTER 3

Genetically Engineered Microbes Mediated Insulin Production: Progress and Prospects

MAHARUDRA PRATAP SINGH,[1] MANISH MISHRA,[2,*] BRIJESH PANDEY,[1] AKHILESH KUMAR SINGH,[1] PRAKASH KUMAR SARANGI,[3] and ANAND PRAKASH[1*]

[1]*Department of Biotechnology, School of Life Sciences, Mahatma Gandhi Central University, Motihari, Bihar, India*

[2]*Department of Biomedical Sciences, Mercer University School of Medicine, Macon, USA*

[3]*College of Agriculture, Central Agricultural University, Imphal, Manipur, India*

*Corresponding author

ABSTRACT

Demand for recombinant insulin is expected to upsurge in future as the epidemiology of diabetic cases increasing worldwide exponentially. There are two methods for human insulin synthesis has been developed towards its commercial production. One involves converting porcine insulin to human insulin through an amino acid substitution procedure ("semi-synthetic" human insulin), whereas the other involves developing recombinant human insulin using bacteria ("biosynthetic" human insulin). It has been determined that both forms of human insulin are secure along with comparable pharmacokinetics and efficacy. The commercial approach for producing semisynthetic human insulin was proposed

Industrial Microbiology: Current Progress and Novel Technologies.
Prakash Kumar Sarangi, Akhilesh Kumar Singh, Rajesh Kumar Srivastava, and Brijesh Pandey (Eds.)

DOI: 10.1201/9781779642219-3

through enzyme-catalyzed semi-synthesis at the C-terminus of the insulin B-chain, but it was dependent on porcine insulin supply which can be hampered by supply issues. This manufacturing approach would be unable to meet the increasing demand for affordable insulin due to restricted production capacity and high production costs. The recombinant human insulin production necessitates a suitable host organism with efficient post-translational modifications and protein folding machinery. A plant-based expression system has enormous potential for high-capacity and low-cost production, while another very appealing approach for mass production of insulin is to utilize transgenic plants.

3.1 INTRODUCTION

Insulin is a peptide hormone, synthesized from the pancreatic beta (β) cells and is crucial for maintaining a proper metabolism in the body. Insulin, along with another pancreatic hormone, glucagon, balance the glucose homeostasis in the bloodstream, and when insulin production is insufficient to properly control the glucose level in blood, then hyperglycemia occurs initially, and further, these people might develop DM (Mishra et al., 2017, 2018). DM is a chronic disease which causes a major public health problem, and its prevalence is increasing rapidly worldwide (Mishra et al., 2018; Panta et al., 2018). The International Diabetes Federation (IDF) estimated that worldwide, 463 million people were suffering from DM by 2019 (Sun et al., 2022). Insulin is a polypeptide made-up of 51 amino acids that is produced by the β cells of islets of Langerhans. Proinsulin, a single-chain precursor (86 amino acid chain), is converted to mature insulin, which consists of two polypeptide chains, A and B, respectively, which are linked by disulfide bridges. In order to manage glucose homeostasis and glucose levels in the blood, insulin has two main complementary actions: (i) promote uptake of circulating glucose into the cells/tissues, ii. reduce hepatic glycogenolysis and gluconeogenesis (Hua, 2010; Mishra et al., 2018). Von Mering and Minkowski observed at the end of the 19th century that removing the pancreas cause the onset of DM in dogs. Schafer originally postulated in 1916 that pancreatic islets secrete an anti-diabetic hormone that he termed "insulin." Barron noted in 1920 that the islets, as Langerhans referred to them in 1869, needed to be injured as well for pancreatic duct ligation with exocrine pancreas loss to cause DM.

A study conducted in the early 1920s by Banting and his colleagues identified a chemical in pancreatic extracts that displayed an amazing ability to lower blood glucose in diabetic rats. By 1923, diabetic patients were being successfully treated with pancreatic extracts (Wilcox, 2005). Certain membrane channels (Glucose transporter-4, GLUT-4) increase glucose transport into cells in response to insulin stimulation, lowering plasma glucose concentrations. After being transported, glucose would either be used as an energy source for the body's processes or stored as the dense polymer molecules known as glycogen. When there is insufficient insulin synthesis or increasing insulin demands in the body, insulin therapy may be initiated to overcome this insufficiency (Mathieu et al., 2017; Mishra et al., 2017).

3.2 BIOGENESIS OF INSULIN AND ITS RELEASE MECHANISMS

The first hormonal peptide discovered was insulin. All hormones were thought to be smaller entities until Abel's 1926 crystallization of insulin and Jensen and Evans' 1935 discovery of N-terminal phenylalanine of B-chain. In the midst of 1950s, Sanger analyzed its amino acid sequence and found that insulin is composed of 21-residue A-chain joined by two disulfide bridges formed of cysteine residues (A7–B7, A20–B19) and intrachain disulfides (A6–A11) (Weiss et al., 2014). The mechanism of insulin production and secretion was not fully understood until the late 1960s, when proinsulin was discovered. Proinsulin is a precursor protein (about 9,000 D) that contains both the A- and B-chains of insulin, which are linked by an interstitial region known as the C domain. The mature insulin and a free C-peptide (devoid of the dibasic residues) are released when a trypsin-like enzyme dissociates these connections (Steiner et al., 1969). In a study published by Chan et al., pre-proinsulin, a brand-new and bigger precursor of insulin was found. An amino-terminal hydrophobic 24-residue signal peptide lengthens this single-chain polypeptide's (12,000 D) length. Typically, such signal sequences are present in proteins that move along the secretory route. Pre-proinsulin's signal peptide is cleaved as it enters the ER, and proinsulin, in collaboration with ER's oxidative folding machinery, promotes correct disulfide pairings and three-dimensional protein folding (Chan et al., 1976) (Figure 3.1).

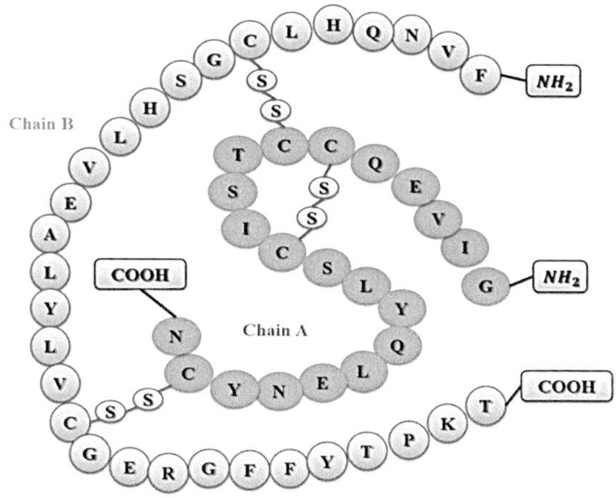

	A8	A9	A10	B30
Human	T	S	I	T
Porcine	T	S	I	A
Bovine	A	S	V	A
Sheep	A	G	V	A

FIGURE 3.1 Human insulin's primary structure, as well as the amino acid sequences of human, porcine, bovine, and sheep insulin.

Bell and colleagues discovered the human insulin gene in 1980 and used simple biosynthetic methods to produce pre-proinsulin, which is further transformed into insulin by a proteolytic process (Bell et al., 1980). The mature mRNA transcript of pre-proinsulin protein is produced by cleaving the two introns from the initial mRNA transcript, capping the 5' end using 7-methylguanosine, and polyadenylating the 3' end. Pre-proinsulin is translated by ribosomes of the rough endoplasmic reticulum (rER), and then it is translocated into the rER lumen due to interactions between the signal peptide, the signal recognition particle (SRP), and the SRP-receptor of the rER membrane. In the rER's lumen, a signal peptidase breaks down the signal peptide and produces proinsulin. This signal peptide cannot be regarded as a normal β-cell secretory product because it is immediately degraded in the rER. To produce the native structure or the original form of proinsulin, proinsulin swiftly folds and creates disulfide bonds inside the cisternae of the rER. The proinsulin is further transported to the Golgi apparatus, where the last step is processed to form native insulin and stored in secretory granules. Condensing vesicles, or early secretory granules, carry out the transformation process, which begins in the trans Golgi network (Kreil, 2014). The products are then sealed inside the mature secretory vesicles, which are released at equimolar levels of insulin and C-peptide. Glucose stimulates the insulin gene, increases the stability of the insulin mRNA, and facilitates translation in addition to boosting β-cells to release the insulin (Tokarz et al., 2018). By combining the actions of two separate proteases, proinsulin is transformed into insulin. One of the proteases has exopeptidase activity to digest the basic residue left by tryptic-like cleavage, while the other has trypsin-like endopeptidase action to cleave after dibasic residue pairs at each extremity of the C domain (Bell et al., 1980) (Figure 3.2).

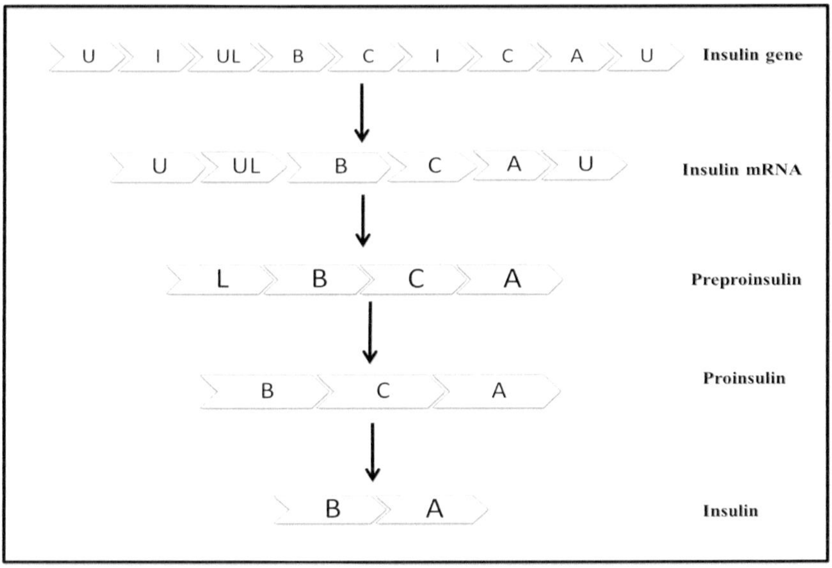

FIGURE 3.2 A diagram showing insulin processing.

3.3 INSULIN ACTION MECHANISM

It has been demonstrated that insulin increases the glucose transporters (GLUT-4) on the plasma membranes of target cells, thereby increasing glucose uptake in cells. GLUT-4 is present in several cell types but predominantly expressed in the adipocytes and skeletal muscles. When insulin came into contact with the insulin receptors of the aforementioned target cells, and due to further signaling cascade, the GLUT-4 transporters are released from intracellular storage areas to the plasma membrane to aid the transfer of glucose into the cells. Within 30 seconds of insulin stimulation, it has been observed that GLUT-4 receptors are translocating to the plasma membrane, while when the stimulus wears off, the amount of plasma membrane receptors decreases and causes a decline in glucose transfer to the cells. In contrast, the chemical potential of the glucose gradient and the transporters' Vmax control the passive nature of other types of glucose transporters (Petersen and Shulman, 2018).

Insulin binds to insulin receptors on the cell membrane. The insulin receptors are present on almost all cell types, but hepatocytes, adipocytes, and skeletal muscle cells have the highest density. The insulin receptor belongs to the receptor tyrosine kinase superfamily (RTK), and in its insulin-unbound (inactive) form, it is present on the plasma membrane as two dimeric units, each made up of one alpha (α) and one β subunit. When insulin binds to distinct sites of two α subunits, it helps the cross-linking of two receptor halves to make an active heterotetrametric insulin receptor (α2β2). This means transmembrane β subunits have tyrosine kinase activity, while extracellular α subunits have insulin binding sites (Belfiore et al., 2009; Kahn, 1985). When insulin binds to the α subunits, the receptor activates a complex intracellular signaling cascade through IRS (insulin receptor substrate) proteins along with PI3K (phosphatidylinositol 3-kinase) and ERK (extracellular signal-regulated kinase) signaling pathways. This signaling cascade upregulates the translocation of the cytoplasmic vesicles to the cell membrane (Posner, 2017). These glucose transporters assist in lowering blood glucose levels by allowing glucose from the blood to enter the cells (Jaldin-Fincati et al., 2017). Insulin promotes glycogenesis, protein biosynthesis, and lipid biosynthesis while inhibiting glycogenolysis, gluconeogenesis, and ketogenesis in hepatocytes. It enhances lipogenesis and lowers lipolysis in adipocytes. It increases glycogenesis, protein synthesis, and inhibits protein catabolism in muscle cells (Mishra et al., 2011; Rahman et al., 2021). Insulin also acts as an appetite regulator in brain tissue and suppresses glucagon release from pancreatic α cells (Rahman et al., 2021; Wilcox, 2005) (Figure 3.3).

In Type 2 DM, the insulin receptor is impaired, which affects insulin's ability to induce the translocation of GLUT4 from intracellular reserves and contributes to postprandial hyperglycemia. Insulin gives target cells a short- and long-term signal to boost glucose-transporter trafficking and keep the expression of these transporters at a basal level. Loss of regulation and transmembrane glucose transporter expression are two common

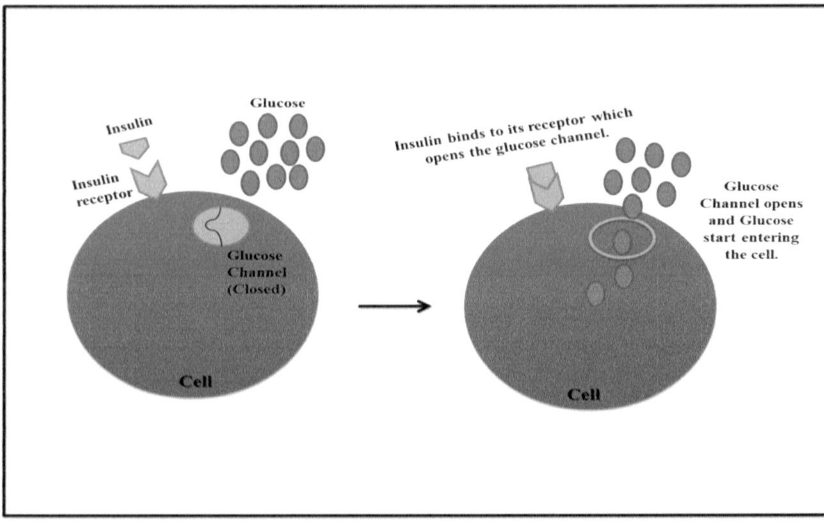

FIGURE 3.3 Insulin's mechanism of action promotes cellular absorption of glucose.

mechanisms that can lead to pathologically high plasma glucose levels in DM. Insulin release is regulated by GLUT2 expression on the surface of β-cells, which is non-insulin but glucose concentration dependent and regulates insulin secretion according to glucose level in the β-cells (Kitamura et al., 2014).

3.4 TYPES OF INSULIN

As per ADA, insulin is divided into several groups based on how quickly they work, when they peak, and how long they last: rapid-acting, regular, or short-acting, intermediate-acting, and long-acting (American Diabetes Association, 2021; Thota et al., 2022).

- Lispro, Glulisine, and Aspart are rapid-acting insulins that begin to work within 5 to 15 minutes and reach their peak within 30. The activity lasts 2 to 4 hours. To regulate blood sugar levels throughout the day, they are usually taken prior to food and are frequently taken in conjunction with either short-acting or long-acting insulin.
- Humulin R, Novolin R, and Venolin R are regular or short-acting insulins that reach their peak in 2–3 hours after injection.

The activity lasts 3 to 6 hours. Patients must take their medication as prescribed and be sure to eat a meal within the 30-minute window in order to maintain safe blood glucose levels and prevent hypoglycemia.

- Humulin N, Novolin N, and ReliOn are intermediate-acting insulins that reach their peak in 4 to 12 hours after injection. The activity lasts 12 to 18 hours. Usually used twice a day, dosing aids in regulating blood glucose levels all day long.
- Degludec, Detemir, and Glargine are long-acting insulins, like glargine and detemir, that begin working after one to two hours. Over the course of 12 to 24 hours, they produce a plateau effect. After eating, the dose is often taken at night. As a result of their prolonged duration of action, the frequency of daily dose is reduced (Figure 3.4 and Table 3.1).

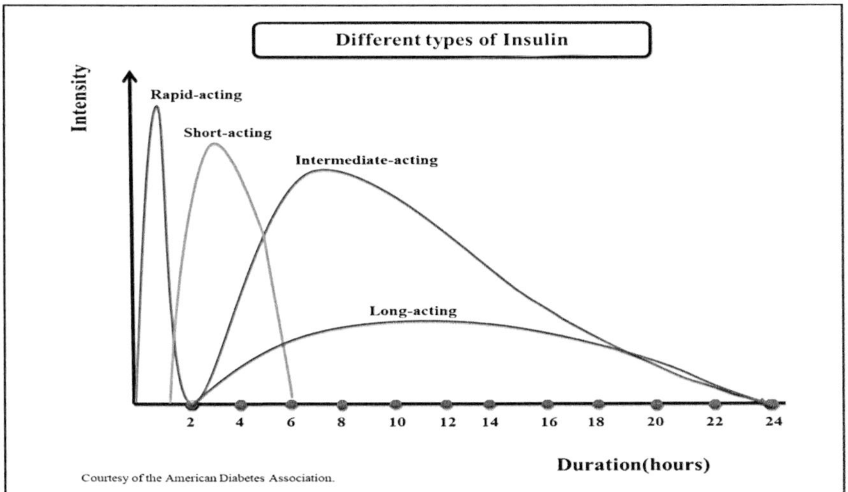

FIGURE 3.4 Graph showing the time action curves for different types of insulin.

3.5 WHY WE NEED INSULIN COMMERCIALIZATION?

It is a lengthy process for insulin to be synthesized because its chemical structure is crucial for its biological function, and any modifications may impair insulin's capacity to bind with cell receptors and control blood glucose levels. In order to create medications that can assist in

TABLE 3.1 Different Types of Insulin and How They Work

Different Types of Insulin and How They Work

Insulin Type	Onset after injection	Peaks	Duration
Rapid-acting	About 15 minutes	1 hour	2 to 4 hours
Short-acting	Within 30 minutes	2 to 3 hours	3 to 6 hours
Intermediate-acting	2 to 4 hours	4 to 12 hours	12 to 18 hours
Long-acting	Several hours	Does not peak	24 hours; some last longer

Courtesy of the American Diabetes Association.

controlling blood sugar levels, it is crucial to comprehend the synthesis as well as the structure of insulin. After years of research and development, synthetic insulin is now readily accessible. It is used as an insulin therapy for diabetics and has been demonstrated to be just as efficient as natural insulin. Researchers have also studied the impact of insulin on the body using synthetic insulin. Type 1 and type 2 DM both carry a high risk of coronary artery disease, stroke, peripheral vascular disease, and other vascular issues, and both can be managed with the help of insulin (Mishra et al., 2022; Thota et al., 2022). According to the IDF, 10.5% of the population had diabetes by 2019. Approximately 90–95% of diabetics have type 2 (or insulin-resistant), 5–10% have type 1, and 1%–2% have maturity-onset diabetes (American Diabetes Association Professional Practice Committee, 2022; Sun et al., 2022).

According to the American Diabetes Association Professional Practice Committee, there are 5 classes of diabetes, and the FDA has approved insulin treatment for them (American Diabetes Association Professional Practice Committee, 2022).

- Type 1 DM (previously known as Juvenile DM) is a disease in which the pancreatic β cells are destroyed, either by the body's own immune system or by trauma or injury to the pancreas. Patients with type 1 DM are insulin dependent and have diminished or nonexistent insulin production (Smith et al., 2017).

- The most prevalent kind of diabetes, type 2 DM, often develops in adulthood. However, due to widespread childhood obesity, it is occurring more commonly in teenagers and young adults. Type 2 DM is characterized by insulin resistance. Although the specific etiology of this is unknown, there are some potential contributors, including genetics, obesity, metabolic syndrome, and a sedentary lifestyle. The pancreas produces more insulin in an effort to get glucose into cells, but eventually this is unsuccessful, and blood glucose levels begin to rise. For type 2 DM in its later stages, insulin therapy is required (Donner et al., 2012; Kumar et al., 2013).
- As a result of a genetic defect, β cells in the pancreas are unable to secrete insulin in people with maturity-onset diabetes of the young (MODY), an inherited disorder. Patients have first-degree relatives with a similar form of diabetes, are younger, have healthy weights, lack antibodies for pancreatic islet cells, and have no other health issues. Treatment options include insulin or sulfonylureas (Anik et al., 2015).
- Gestational diabetes mellitus (GDM): In the United States, GDM complicates about 4% of all pregnancies (American Diabetes Association Professional Practice Committee, 2022). The first line of treatment for GDM is diet and exercise, and screening for the condition is done between 24 and 28 weeks of pregnancy. In order to avoid the side effects of shoulder dystocia, macrosomia, and preeclampsia, treatment is required. Insulin is recommended if diet and exercise do not manage blood sugar levels (Patti et al., 2018).

3.6 COMMERCIAL PRODUCTION OF INSULIN

By the 1970s, it had become increasingly obvious that the production of insulin from the pancreas of pigs and cows was not sufficient to meet the world's needs. Chemical synthesis was ruled out as a viable option due to the molecule's complexity and the enormous volumes necessary to fulfill global supply. Despite being able to produce human sequence hormone, the semi-synthesis process was still reliant on the availability of porcine products, which made the supply issues even worse because the conversion was not quite quantitative. With the goal of using rDNA technology to biosynthesize human insulin, Lilly Research Laboratories and Genentech launched a ground-breaking joint project (Chance et al., 1981).

3.6.1 TRADITIONAL METHODS FOR PRODUCING INSULIN

Historically, cattle or pig pancreases were diced and frozen, and insulin was extracted from these pancreas tissues using extraction and chromatographic methods (Lee and Yoon, 2021). After being extracted with ethanol and acidified with HCl or H_2SO_4 to bring the pH to 2, these steps inactivated and removed the trypsin enzyme, which has the potential to break down insulin. The extract was then concentrated after being vacuum extracted at a low temperature, adding salt to precipitate the insulin, and adding calcium carbonate to neutralize the solution. It precipitates again when the precipitate was dissolved in water and the pH was raised to insulin's isoelectric point. Gel filtration followed by ion exchange systems as the subsequent step in the technique to further purify the insulin (Keefer et al., 1981). The primary peak of insulin produced when it was subjected to a gel filtration column contains glucagon, pancreatic polypeptide, somatostatin, desamido insulin, arginine insulin, and insulin ethyl ester, among other substances. Proinsulin, proinsulin intermediates, and various covalent insulin dimers were present in the minor fraction of the higher molecular weight material that elutes prior to the insulin peak. It was necessary to remove proinsulin-like molecules during the purification procedure since they may be immunogenic. Ion exchange chromatography and gel filtering were used to extract the insulin from these other substances (Cheta, 1990).

3.6.2 SEMISYNTHETIC HUMAN INSULIN PRODUCTION VIA PORCINE INSULIN ENZYMATIC CONVERSION

In the 1970s, Novo Nordisk A/S manufactured the first semi-synthetic human insulin. A threonine residue was added to the porcine insulin's

B-30 alanine position to create it. Human mono component insulin is manufactured in five steps (Lee and Yoon, 2021). To begin with, insulin was isolated from frozen pancreatic glands of porcine. The use of typical purification methods occurs in the second step of processing. Porcine insulin was first hydrolyzed into a mixture of threonine ester, organic solvent, and trace amounts of organic solvent to remove alanine from point B-30 in the insulin molecule. Then, trypsin catalyzes the hydrolysis of insulin at LYSBBAlaBso and catalyzes the reverse process to convert it into human insulin. Using soluble trypsin to trans peptide porcine insulin into human insulin increased the yield up to 97% (Markussen et al., 1983). Although immobilized trypsin catalyzes transpeptidation, the yield was only 80%. After that, the proinsulin was reduced to concentrations that can be determined by chromatographic purification. Following formulation and sterile filling, the product was packed, distributed, and finally sold (Ueno and Morihara, 1989) (Figure 3.5).

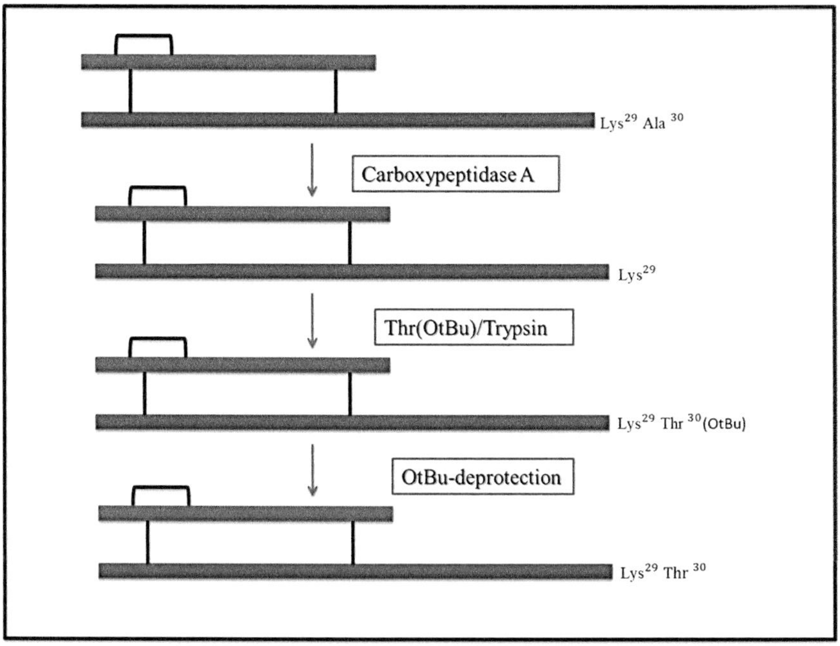

FIGURE 3.5 Semisynthetic human insulin production via porcine insulin enzymatic conversion.

3.6.3 RECOMBINANT METHODS

Human insulin turned out to be the first animal protein synthesized in bacteria that has a sequence similar to a human pancreatic peptide (Ladisch and Kohlmann, 1992). Eli Lilly and Company in Indianapolis and Genentech created this insulin (San Francisco, CA). In 1978, these companies collaborated to produce recombinant human insulin in *Escherichia coli* (*E. coli*) K-12 using genes created at the City of Hope National Medical Center. The researchers at Genentech duplicated these same genes from plasmid pBR322 and added the β-galactosidase gene before amplifying the recombinant plasmid in *E. coli* (Chance et al., 1981). In 1982, after the first successful expression had been made public in 1978, it had been scaled up and given the go-ahead from the applicable drug regulatory agencies. The decision to clone this peptide was largely influenced by the small size of insulin in both the A and B chains and the absence of methionine (Met) and tryptophan (Trp) residues. Reagents used in the insulin recovery process might digest Met and Trp residues. If these amino acids had been present, insulin would have hydrolyzed and broken down (Johnson, 1983).

3.6.3.1 TWO-CHAIN METHOD

Genentech developed the first successful technology for making recombinant human insulin on a lab scale, and Eli Lilly and Company thereafter scaled it up successfully (Baeshen et al., 2014). In the two-chain method, each chain of insulin was created independently in engineered *E. coli* as a fusion protein with β-galactosidase (β-gal). After intracellular synthesis, these fusion proteins were found in the cytoplasmic inclusion body (Williams et al., 1982). Typically, 10 to 40% of all proteins are produced via recombinant methods in *E. coli* (Burgess, 1987). CNBr dissociation at the Met residue between the A or B chain and β-gal can separate the chains A or B from the inclusion bodies. It was succeeded by purification. A small amount of mercaptan was required for the proper folding of the A and B chains in a 2:1 ratio in order to create bioactive insulin. After 24 hours, it was estimated that the yield would be 60% based on the B chain used in the procedure (Johnson, 1986).

Goeddel et al. published their findings in 1979, with 20% of all cellular proteins produced as A/B chain heterodimers with similar outcomes. The S-sulfonated chain was folded several times, and 50–80% of the time the folds were accurate. The yield was reduced because of the extreme large size of the fusion protein, which has chains A or B (21 or 30 amino acids, respectively) and β-gal (1000 amino acids), which causes early chain termination during translation. As a result, defective insulin peptides were produced. The *tryptophan (Trp) operon* is an alternative to the β-gal system (*lac operon*) for producing shorter fusion proteins with higher yield. *Trp operon* is a group of five genes that produce tryptophan-anabolic enzymes sequentially. Unlike β-gal, which has 1000 amino acids, Trp E, one of these enzymes, only has 190. Due to the strength of the Trp promoter in *E. coli*, when the Trp E gene is flanked by genes encoding the A or B chains of insulin, the production of fusion proteins increases from 5–10% to 20–30%. This results in a 10-fold or more increase in polypeptide expression over the β-gal system (Burnett, 1983). When fermentation is finished, the cells were recovered and disrupted. After separating the inclusion bodies from the cell debris, the proinsulin A and B chains are released via a CNBr cleavage of the Trp-LE'-Met-A and Trp-LE'-Met-B chains. Occasionally, inclusion bodies are dissolved with guanidine HCl (6 M) and dithiothreitol (0.1 mM) (Burgess, 1987). To create crude insulin, additional modifications to A and B chains are made via oxidative sulfitolysis, filtration, and joining. Ion exchange, size exclusion, and reversed-phase high-performance liquid chromatography are used to produce pure recombinant human insulin from crude insulin (Frank and Chance, 1983). A patent includes the chaotropic and sulfitolysis solubility of heterologous protein inclusion bodies. Thanks to the sulfitolysis and heat processes included in this approach, the protein S-sulfonate can precipitate in a high purity condition, 95% pure in contrast to the starting material. The goal of this procedure is to avoid the formation of improper disulfide bonds or intermolecular cross-linking that occur following cell lysis in traditional processing (Bobbitt and Manetta, 1990) (Figure 3.6).

3.6.3.2 PROINSULIN METHOD (INTRACELLULAR)

In this method, recombinant insulin is produced by using microorganisms that produce the full proinsulin segment instead of producing chains A

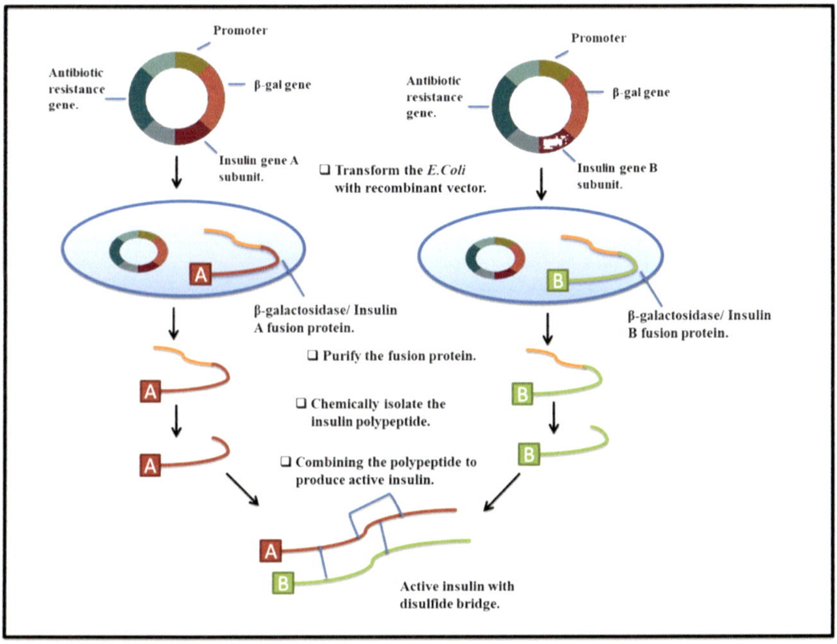

FIGURE 3.6 Using genetic engineering, a two-chain method of producing insulin is shown.

or B independently. The proinsulin method is preferred over the two-chain method for manufacturing insulin because it only requires one set of production and purification steps. In 1989, Kroeff and his colleagues summarized this technique, which is based on Watson's recombinant technology. Proinsulin mRNA is first converted into cDNA, and then a methionine codon is chemically substituted at the 5' end of the cDNA. After that, the cDNA is inserted into a plasmid and grown in *E. coli* (Kroeff et al., 1989). Trp-LE'-Met-proinsulin (Trp proinsulin) and the bacterial enzyme β-gal fragment include a methionine linker that can be broken to release the proinsulin. After release, the proinsulin chain is allowed to fold in order to release the C-peptide by an enzymatic method to produce bioactive insulin. Inter-polypeptide disulfides are then produced. In comparison, the previously mentioned two-chain technique is more difficult. After recovery, Trp proinsulin is cleaved by CNBr to produce proinsulin. To create crude insulin, the proinsulin undergoes oxidative sulfitolysis, folding under mercaptan-assisted conditions, multiple purification stages, and enzymatic processing. In order to produce pure insulin, purification of recombinant insulin was performed by size exclusion, reversed phase, and ion exchange chromatography techniques (Frank and Chance, 1983) (Figure 3.7).

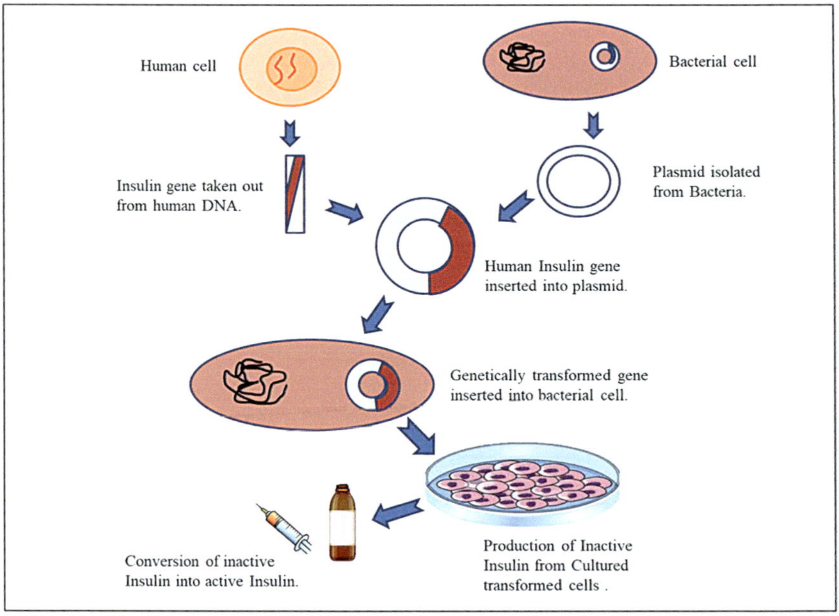

FIGURE 3.7 Using genetic engineering, a proinsulin method of producing insulin is shown.

3.6.3.3 PROINSULIN (SECRETED)

The first description of a human proinsulin secretion mechanism in *E. coli* was made by Villa-Komaroff et al. (1978). The optimum environment for the creation of recombinant proteins (RCP) would be for the bacteria to secrete vast quantities of foreign protein into the medium in an efficient manner. The RCP insulin may contain both proinsulin and j3-lactamase (an enzyme naturally released by bacteria that deactivates penicillin), but bacterium only release a small amount of their own proteins. Yeasts are also a desirable candidate for this kind of system, as it also synthesizes less unwanted proteins, and fewer unwanted proteins need to be eliminated during purification. Yeasts can help in the synthesis of disulfide bonds as well (Villa-Komaroff et al., 1978), but yeasts frequently glycosylate proteins in excess. Novo Nordisk A/S used *Saccharomyces cerevisiae (S. cerevisiae)* to manufacture proinsulin as a single-chain. In 1986, the Danish Parliament approved both the procedure and the product. This proinsulin contains an unfolded peptide which begins with a leader segment, then a B chain, a short peptide bridge, the Lys-Arg sequence, and finally the A chain. In this precursor segment, a short connecting peptide contains one basic amino acid near the A chain, which connects the 29th residue of the B chain to the 1st residue of the A chain. The final product, insulin, is created through transpeptidation and ester hydrolysis of this precursor. Further purification was achieved by chromatography techniques (Diers et al., 1991). A yeast system equipped with effective batch-fed fermentation can yield the recombinant insulin superoxide dismutase–human proinsulin (SOD–PI) up to 1500 mg/L. SOD-PI can be implemented as a raw material for the production of recombinant insulin; yields of the finished product have not been disclosed. Evidently, yeast has occasionally expressed heterologous polypeptides at levels that were less than ideal. An Eli Lilly Co. patent outlines a procedure for cultivating yeast under aerobic, anaerobic, and then aerobic conditions again. According to reports, this approach produces a high level of product expression (Beckage and Ingolia, 1988).

3.7 EXPRESSION SYSTEMS USED IN RECOMBINANT INSULIN PRODUCTION

3.7.1 E. COLI EXPRESSION SYSTEM

E. coli is a well-known bacterium enabling large-scale protein production. However, several disadvantages limit its usage in the manufacturing of recombinant insulin. *E. coli* does not perform a variety of post-translational modifications (PTMs) that are required for biological function, such as disulfide bridge formation, phosphorylation, glycosylation, and proteolytic processing (Walsh et al., 2006). The most common kind of PTM is eukaryotic N-glycosylation. *Campylobacter jejuni (C. jejuni)* has been shown to be able to glycosylate proteins, and it has also been suggested that this bacterium may provide *E. coli* with a functionally useful N-glycosylation pathway. Despite the fact that the structure of bacterial N-glycan differs from that of eukaryotic cells, *E. coli* can synthesize heterologous proteins in glycosylated form by introducing the *C. jejuni* N-linked glycosylation system.

Expression of *C. jejuni* Pglb oligosaccharyltransferase (OTase) in *E. coli* markedly boosted glycopeptide synthesis (Wacker et al., 2006). Studies have focused on producing glycosylated proteins from substrates other than those found in *C. jejuni* and *E. coli* (Pandhal et al., 2011). The utilization of codons by the heterologous protein has a substantial impact on the protein expression level. If the heterologous protein's codon usage differs significantly from that of the *E. coli* host, expression may be very low. The frequency of use of each codon normally reflects the quantity of tRNA that it correlates with, which causes substantial variations in codon usage that may lead to premature translation termination, improper amino acid incorporation, and inhibition of protein synthesis (Makrides et al., 1996). By substituting infrequently occurring minor codons with more advantageous major codons, it is possible to increase the production of heterologous proteins in *E. coli*. Similar to this, co-expression in *E. coli* of genes for numerous tRNAs for uncommon codons may improve the synthesis of heterologous proteins. Commercial *E. coli* strains can encode for tRNA for unusual codons, including Rosetta (DE3), BL21 (DE3) CodonPlus-RIL, and BL21 (DE3) CodonPlusRP (all from Stratagene, USA) (Kane, 1995). By lowering proteolytic degradation, using *E. coli* strains with protease-preventing mutations may boost the production of RCPs. Two proteases, lon and ompT, encoded by cytoplasmic and periplasmic genes, are absent from the *E. coli* strain BL-21. Specific methods, such as modifications in protease and secretion routes, may offer insight into the biology of RCPs instead of outward characteristics (Bill et al., 2011). Complex and large therapeutic proteins may be secreted in the periplasm of *E. coli* because it offers an oxidizing environment and promotes the formation of disulfide linkage that allow for the best folding of RCPs and seem likely to produce reliable N-termini of manufactured proteins (Huang et al., 2012). Because of its lower protein composition and proteolytic activity, higher manufacturing titer, and enhanced RCP solubility, periplasm has advantages over cytoplasm. Heterologous proteins typically assemble in *E. coli* as solid, misfolded protein aggregates called inclusion bodies (Overton, 2014). Molecular chaperones can assist RCPs in folding appropriately and increasing their solubility. Some chaperones help misfolded proteins to be solubilized and refolded, whereas others prevent protein aggregation. In order to increase protein solubility in *E. coli*, the most important chaperones are GrpE, DnaK, DnaJ, GroEL, and GroES, which can be used singly or in combination (de Marco, 2007).

The very first synthetic therapeutic recombinant insulin was produced by the two-chain method. In 1978, Genentech was the first to produce recombinant insulin in *E. coli*. In this method, they chemically synthesized the cDNA of the insulin A or B chain and expressed it in *E. coli* (Chance and Frank, 1993). After successful expression, individual chains are purified and incubated under optimal conditions to produce bioactive insulin with disulfide linkage. Another approach was also used to create recombinant insulin, which was derived from the single-chain production of proinsulin. Following that, the proinsulin was enzymatically treated to remove the C peptide and join the A and B chains via disulfide bonds. This strategy has been shown to be more effective and useful for production on an industrial scale (Chance et al., 1999). Using this technique, Eli Lilly created the first recombinant insulin, Humulin, which was approved for use in 1982. Because of their less allergic effect and similar amino acid arrangement to human insulin, these were the first-generation recombinant insulins preferred over animal-derived insulin (Walsh, 2005). As genetic engineering advances, properties such as the chemical synthesis of genes with altered nucleotide sequences will aid in the development of insulin derivatives with minor amino acid sequence changes (Blundell et al., 1971). Due to exceptionally high concentrations, native insulin has been found to commonly exist in the oligomerized form in commercial preparations as zinc hexamers. The amino acid residues B8, B9, B12, B13, B16, and B23–28, whose side chains are essential for oligomerization, were eliminated in the production of fast-acting insulin (Brange et al., 1990). For insulin to act in the body, the oligomeric form must be resolved to the monomeric form, which allows insulin to solubilize in the bloodstream from the administration site. Therefore, subcutaneous administration of recombinant insulin frequently results in a slow onset, with peak occurring two hours after injection and persisting for six to eight hours (Kang et al., 1991). In 1996, Lispro, a fast-acting insulin analogue manufactured by Eli Lilly, became the first one to acquire regulatory approval for therapeutic use. Lispro was created by inverting proline-lysine residues at the 28th and 29th positions of the B chain, which prevents dimer formation by reducing hydrophobic interactions. The amino acid sequence of lispro is comparable to that of native insulin. Lispro was created using a synthetic cDNA containing LysB28-ProB29. This cDNA was then introduced in *E. coli*, which produces proinsulin, which, after trypsin and carboxypeptidase treatment, finally produces lispro insulin. In 2004, the FDA approved Glulisine

(*Apidra*), a fast-acting analogue developed by Aventis Pharmaceuticals by substituting B3 asparagine with lysine and B29 lysine with glutamic acid. In order to reduce the number of injections, long-acting insulin analogues have also been created. Insulin Glargine, an Aventis Pharmaceuticals long-acting analogue, was approved by US and EU regulatory bodies in 2000. Insulin Glargine was created by substituting a glycine residue for the asparagine residue at the C-terminus of A-chain and adding two arginine residues to C-terminus of the B-chain. The isoelectric point (pI), which was 5.4 before these changes, was raised to neutral values. Glargine was produced in *E. coli* as proinsulin and expressed in the soluble form at pH 4. But after subcutaneous delivery, it precipitated because subcutaneous tissue has a neutral pH. Slow insulin re-solubilization enhances the time required for insulin to be released into the blood (Walsh, 2005).

3.7.2 YEAST EXPRESSION SYSTEM

Yeast is the most common host for the creation of many heterologous proteins, the biological activity of which is dependent on PTMs like acetylation, glycosylation, phosphorylation, and acylation. Yeast can produce RCP in a soluble state, where they undergo proper folding to yield proteins that are functionally active. It is also quite inexpensive to create biopharmaceuticals utilizing a yeast expression system, which can be scaled up using large bioreactors. One significant disadvantage of using yeast to produce therapeutic proteins for human use is that yeast N-glycosylation produces high levels of mannose residue, which reduces the protein's effectiveness by making it immunogenic and shortens its *in vivo* life. For therapeutic protein production in yeast, many trials have been performed to humanize the N-glycosylation mechanism (Wildt et al., 2005). Thus, yeast other than *S. cerevisiae* gains the focus of research for the production of biopharmaceuticals on a large scale. One example is *Pichia pastoris* (*P. pastoris*), which has a methanol-inducible alcohol oxidase 1 (AOX1) promoter, and a high production rate for RCPs with high quality and quantity. Because it cannot hyperglycosylated heterologous proteins, *P. pastoris* performs better than *S. cerevisiae* in the glycosylation of secreted proteins. Despite the fact that both yeast strains normally express proteins with significant mannose N-linked glycosylation, the oligosaccharide chains associated to the proteins in *P. pastoris* are substantially shorter than those attached to the proteins produced in *S. cerevisiae*, indicating that the glycoproteins synthesized in *P. pastoris* are better for therapeutic application in humans (Grinna et al., 1989). Additionally, compared to *S. cerevisiae*, *P. pastoris* also expresses heterologous proteins at very high quantities, which may make up about 30% of the total amount of protein in a cell (Ellis et al., 1985). As a result, *P. pastoris* may be a good choice for the manufacturing of recombinant insulin and its analogues at industrial-scale (Porro et al., 2011). Much recombinant commercial insulin has been produced since the early 1980s using *S. cerevisiae*. For successful production of recombinant proinsulin in yeast, the construct must contain chains A and B without B30 threonine at the C terminal. These two chains might be connected directly or through a shorter C peptide. When the cDNA from these constructs is combined with the *S. cerevisiae* α-factor signal sequence, it can produce up to 80 mg/mL of insulin. The proinsulin produced was extracted and converted to bioactive insulin via transpeptidation using trypsin and threonine ester (Kjeldsen, 2000). Along with natural recombinant insulin, *S. cerevisiae* also generates insulin analogues. Aspart is fast-acting insulin synthesized in *S. cerevisiae*. In 2001, the US FDA approved Aspart for therapeutic use. It is produced by an amino acid substitution from proline to aspartic acid at the 28th residue of the B chain. The amino acid substitution creates interchain repulsion, which decreases self-assembly, resulting in rapid solubilization in the bloodstream from the site of injection (Brange et al., 1988). In 2004, the European regulatory body authorized another insulin analog, Detemir, developed by Novo Nordisk, which is long-acting and produced in *S. cerevisiae*. Detemir was produced by some modifications in its B chain, like covalently joining the C14 fatty acid chain to lysine at the 29th residue and also eliminating threonine at the 30th residue. These modifications provide Detemir with the ability to bind albumin in the plasma, which provides continuous release and increases its action duration up to 24 hours (Kurtzhals et al., 1997).

3.7.3 TRANSGENIC PLANTS AS EXPRESSION SYSTEM

Transgenic plants are utilized to produce RCPs due to their advantages in terms of protein processing at a low cost. Some of the benefits of adopting transgenic plants include the lack of pathogens, its ease of manufacturing, and the presence of eukaryotic PTM. The human growth hormone was produced as RCP from a transgenic tobacco plant (Staub et al., 2000). Recombinant insulin has been produced as well as expressed with success using *Arabidopsis thaliana* oil seeds. In this method, insulin was only expressed in oil bodies, which are subcellular organelles that enable high production levels and simple recovery method of recombinant insulin. Oil bodies are storage organelles found in oil seeds and are made up of an exterior oleosin-coated protein wall and a hydrophobic triacylglycerol core inside a phospholipid membrane. RCP that is specifically directed to oil bodies in the form of an oleosin fusion has been genetically engineered into oil seeds (Nykiforuk et al., 2006). The number of chromatography steps required to obtain pure inulin was reduced because only the oil body could be separated from the other seed components through liquid–liquid phase separation. Insulin has been found to accumulate in significant amounts (0.13% of total protein in transgenic seeds). Recombinant insulin was isolated from its oleosin fusion partner and matured via trypsin digestion to yield physiologically active insulin after oil body purification. This study unambiguously shows that insulin expression as an oleosin fusion protein in plants allowed for the accumulation of considerable levels of recombinant insulin within the seed as well as the provision of simple downstream purification through centrifugation or oil body separation (von Rooijen et al., 1995). Standard enzymatic processes that are currently employed for the commercial synthesis of insulin from yeast and *E. coli* can be used for subsequent maturation to get biologically active insulin. Recombinant insulin can also be kept on hand in oil seeds, which serve as a kind of natural cellular storage (Deckers et al., 1999). Another method involved creating transgenic plants by fusing the cholera toxin B component with human proinsulin, which was utilized to modify the chloroplasts of lettuce and tobacco. Proinsulin has been found to accumulate in old tobacco leaves up to 47% of the total protein in the leaf, and in old lettuce leaves up to 53% of the total protein. As much as 40% of proinsulin was identified in even dried, senescent lettuce leaves, proving that proinsulin contained in lettuce leaves is quite durable. Proinsulin was purified to 98% from tobacco leaves and then released as insulin peptides by Furin protease. Oral administration of unprocessed proinsulin encapsulated in plant cell or injection into animals resulted in levels of blood glucose similar to commercially available insulins. Based on the production (3 mg of proinsulin/gm of leaves), it was calculated that one acre of tobacco plantation might produce up to 20 million daily doses of insulin annually.

Very high levels of biologically active proinsulin expression and long-term stability in dried leaves provide a dependable, affordable technique for proinsulin injection and oral administration (Boyhan et al., 2011).

3.8 CONCLUSION

Since Sanger's description of insulin's primary structure and distinct disulfide cross-linking pattern, along with the first account of a hormone's chemical synthesis more than half a century ago, there has been significant interest in the development of more effective chemical techniques for the manufacturing of this fascinating and crucial peptide. There have been two main strategies: synthesis using two different chains, then merging and forming disulfide bonds; or synthesis using a single-chain pro-hormone precursor, then folding and forming disulfide bonds. Many researchers have shown incredible creativity in addressing a number of significant issues connected to the intrinsic complexity of the two chains, notably the A-chain, the appropriate pairing of the three disulfide bonds, and the full chemical characterization of the synthetic products. Synthetic insulin as well as its analogues currently has overall yields that are comparable to those obtained using recombinant DNA synthesis techniques. Finally, genetically modified insulin offers the potential to give individuals with diabetes with a new and enhanced source of insulin. Due to its similarity to human insulin, it offers the benefit of allowing for the treatment of diabetes. Furthermore, genetically modified insulin has the capacity to be generated in vast amounts with cheaper price, which would increase the potential to transform the diabetes treatment and enhance the quality of life for millions of people worldwide.

Over the next 20 years, the WHO estimates that global sales of insulin will increase from $12 billion to $54 billion due to expected increase in diabetes cases worldwide due to changes in dietary patterns and lifestyle. Both type I and type II diabetics may need insulin, but those with late-stage type II diabetes need to take a higher dose due to insulin resistance. For the production of recombinant insulin, the primary organisms used are *S. cerevisiae* and *E. coli*. There is an urgent need to scale up the production of bioactive insulin and its analogs from *S. cerevisiae* and *E. coli* using the most advanced and efficient technology. Another strategy, a plant-based expression system for recombinant insulin production with high-efficiency and low-cost has great potential. A low-cost method for injectable and oral proinsulin delivery is made possible by the development of new transgenic plants which produce very high levels of bioactive proinsulin and exhibit long-term stability. Transgenic seeds can also store recombinant insulin until needed.

The significant rise in diabetic cases and research into other insulin administration techniques like oral or nasal or spray inhalation are expected to result in a rise in the demand for recombinant insulin in the near future. Current manufacturing techniques might be unable to meet the world's growing demand for insulin due to capacity constraints and expensive production costs, so we need to keep looking for new technologies which can produce large amounts of insulin with lower cost.

KEYWORDS

- **diabetic mellitus**
- **disulfide linkage**
- **fast-acting insulin**
- **long-acting insulin**
- **recombinant insulin**
- ***Saccharomyces cerevisiae***
- **short-acting insulin**

REFERENCES

American Diabetes Association Professional Practice Committee. (2022). Classification and diagnosis of diabetes: Standards of medical care in diabetes—2022. *Diabetes Care, 45*(Suppl 1), S17–S38. https://doi.org/10.2337/dc22-S002.

American Diabetes Association. (2021). 9. Pharmacologic approaches to glycemic treatment: Standards of medical care in diabetes—2021. *Diabetes Care, 44*(Suppl 1), S111–S124. https://doi.org/10.2337/dc21-S009.

Anık, A., Çatlı, G., Abacı, A., & Böber, E. (2015). Maturity-onset diabetes of the young (MODY): An update. *Journal of Pediatric Endocrinology & Metabolism, 28*(3, 4), 251–263. https://doi.org/10.1515/jpem-2014-0384.

Baeshen, N. A., Baeshen, M. N., Sheikh, A., Bora, R. S., Ahmed, M. M., Ramadan, H. A., Saini, K. S., & Redwan, E. M. (2014). Cell factories for insulin production. *Microbial Cell Factories, 13*, 141. https://doi.org/10.1186/s12934-014-0141-0.

Beckage, C. A., & Ingolia, T. D. (1988). *Method, vectors and transformants for high expression of heterologous polypeptides in yeast.* Publication of Patent US4745057A. https://patents.google.com/patent/US4745057A (accessed on 15th November 2025).

Belfiore, A., Frasca, F., Pandini, G., Sciacca, L., & Vigneri, R. (2009). Insulin receptor isoforms and insulin receptor/insulin-like growth factor receptor hybrids in physiology and disease. *Endocrine Reviews, 30*(6), 586–623. https://doi.org/10.1210/er.2008-0046.

Bell, G. I., Pictet, R. L., Rutter, W. J., Cordell, B., Tischer, E., & Goodman, H. M. (1980). Sequence of the human insulin gene. *Nature, 284*(5751), 26–32. https://doi.org/10.1038/284026a0.

Bill, R. M., Henderson, P. J., Iwata, S., Kunji, E. R., Michel, H., Neutze, R., Newstead, S., Poolman, B., Tate, C. G., & Vogel, H. (2011). Overcoming barriers to membrane protein structure determination. *Nature Biotechnology, 29*(4), 335–340. https://doi.org/10.1038/nbt.1833.

Blundell, T. L., Cutfield, J. F., Dodson, G. G., Dodson, E., Hodgkin, D. C., & Mercola, D. (1971). The structure and biology of insulin. *The Biochemical Journal, 125*(3), 50P–51P. https://doi.org/10.1042/bj1250050p.

Bobbitt, J. L., & Manetta, J. V. (n.d.). *Purification and Refolding of Recombinant Proteins*. U.S. Patent No. 4,109,092,3967.

Boyhan, D., & Daniell, H. (2011). Low-cost production of proinsulin in tobacco and lettuce chloroplasts for injectable or oral delivery of functional insulin and C-peptide. *Plant BiotechnologyJournal, 9*(5),585–598.https://doi.org/10.1111/j.1467-7652.2010.00577.x.

Brange, J., Owens, D. R., Kang, S., & Vølund, A. (1990). Monomeric insulins and their experimental and clinical implications. *Diabetes Care, 13*(9), 923–954. https://doi.org/10.2337/diacare.13.9.923.

Brange, J., Ribel, U., Hansen, J. F., Dodson, G., Hansen, M. T., Havelund, S., Melberg, S. G., Norris, F., Norris, K., & Snel, L. (1988). Monomeric insulins obtained by protein engineering and their medical implications. *Nature, 333*(6174), 679–682. https://doi.org/10.1038/333679a0.

Burgess, R. R. (2009). Refolding solubilized inclusion body proteins. In *Methods in Enzymology* (Vol. 463, pp. 259–282). Academic Press. https://doi.org/10.1016/S0076-6879(09)63017-2.

Burnett, J. P. (1983). Chapter 14 - Commercial production of recombinant DNA-derived products. In M. Inouye (Ed.), *Experimental Manipulation of Gene Expression* (pp. 259–277). Academic Press.

Chan, S. J., Keim, P., & Steiner, D. F. (1976). Cell-free synthesis of rat preproinsulins: Characterization and partial amino acid sequence determination. *Proceedings of the National Academy of Sciences, 73*(6), 1964–1968. https://doi.org/10.1073/pnas.73.6.1964.

Chance, R. E., & Frank, B. H. (1993). Research, development, production, and safety of biosynthetic human insulin. *Diabetes Care, 16*(Suppl 3), 133–142.

Chance, R. E., Glazer, N. B., & Wishner, K. L. (1999). Insulin lispro (Humalog). In *Insulin* (pp. 339–354). Springer.

Chance, R. E., Kroeff, E. P., Hoffmann, J. A., & Frank, B. H. (1981). Chemical, physical, and biologic properties of biosynthetic human insulin. *Diabetes Care, 4*(2), 147–154. https://doi.org/10.2337/diacare.4.2.147.

Cheţa, D. (1990). Immunological aspects of diabetes. *Recenti Progressi in Medicina, 81*(2), 72–78.

de Marco, A. (2007). Protocol for preparing proteins with improved solubility by co-expressing with molecular chaperones in *Escherichia coli*. *Nature Protocols*, *2*(10), 2632–2639.

Deckers, H. M., Moloney, M. M., & Baum, A. S. (1999). Chapter 24. The case for recombinant production of pharmaceutical proteins in plants. *Annual Reports in Medicinal Chemistry*, *34*, 237–245.

Diers, I. V., Rasmussen, E., Larsen, P. H., & Kjaersig, L. (1991). In Y. H. Chiu & D. L. Gueriguian (Eds.), *Drug Biotechnology Regulations (Scientific Basis and Practices)* (pp. 167–177). Marcel Dekker, Inc.

Donner, T., & Muñoz, M. (2012). Update on insulin therapy for type 2 diabetes. *The Journal of Clinical Endocrinology and Metabolism*, *97*(5), 1405–1413.

Ellis, S. B., Brust, P. F., Koutz, P. J., Waters, A. F., Harpold, M. M., & Gingeras, T. R. (1985). Isolation of alcohol oxidase and two other methanol regulatable genes from the yeast *P. pastoris*. *Molecular and Cellular Biology*, *5*(5), 1111–1121.

Frank, B. H., & Chance, R. E. (1983). Two routes for producing human insulin utilizing recombinant DNA technology. *Münchener Medizinische Wochenschrift*, *Suppl 1*, S14–S20.

Goeddel, D. V., Kleid, D. G., Bolivar, F., Heyneker, H. L., Yansura, D. G., Crea, R., Hirose, T., Kraszewski, A., Itakura, K., & Riggs, A. D. (1979). Expression in *Escherichia coli* of chemically synthesized genes for human insulin. *Proceedings of the National Academy of Sciences of the United States of America*, *76*(1), 106–110.

Grinna, L. S., & Tschopp, J. F. (1989). Size distribution and general structural features of N-linked oligosaccharides from the methylotrophic yeast, *P. pastoris*. *Yeast*, *5*(2), 107–115.

Hua, Q. (2010). Insulin: A small protein with a long journey. *Protein & Cell*, *1*(6), 537–551.

Huang, C. J., Lin, H., & Yang, X. (2012). Industrial production of recombinant therapeutics in *Escherichia coli* and its recent advancements. *Journal of Industrial Microbiology & Biotechnology*, *39*(3), 383–399.

Jaldin-Fincati, J. R., Pavarotti, M., Frendo-Cumbo, S., Bilan, P. J., & Klip, A. (2017). Update on GLUT4 vesicle traffic: A cornerstone of insulin action. *Trends in Endocrinology and Metabolism*, *28*(8), 597–611.

Johnson, I. S. (1983). Human insulin from recombinant DNA technology. *Science*, *219*(4585), 632–637.

Johnson, R. D. (1986). The processing of biomacromolecules: A challenge for the eighties. *Fluid Phase Equilibria*, *29*, 109–123.

Kahn, C. R. (1985). The molecular mechanism of insulin action. *Annual Review of Medicine*, *36*, 429–451.

Kane, J. F. (1995). Effects of rare codon clusters on high-level expression of heterologous proteins in *Escherichia coli*. *Current Opinion in Biotechnology*, *6*(5), 494–500.

Kang, S., Creagh, F. M., Peters, J. R., Brange, J., Vølund, A., & Owens, D. R. (1991). Comparison of subcutaneous soluble human insulin and insulin analogues (AspB9,

GluB27; AspB10; AspB28) on meal-related plasma glucose excursions in type I diabetic subjects. *Diabetes Care*, *14*(7), 571–577.

Keefer, L. M., Piron, M. A., & De Meyts, P. (1981). Human insulin prepared by recombinant DNA techniques and native human insulin interact identically with insulin receptors. *Proceedings of the National Academy of Sciences of the United States of America*, *78*(3), 1391–1395.

Kitamura, T., Kahn, C. R., & Accili, D. (2003). Insulin receptor knockout mice. *Annual Review of Physiology*, *65*, 313–332.

Kjeldsen, T. (2000). Yeast secretory expression of insulin precursors. *Applied Microbiology and Biotechnology*, *54*(3), 277–286.

Kreil, G. (1981). Transfer of proteins across membranes. *Annual Review of Biochemistry*, *50*, 317–328.

Kroeff, E. P., Owens, R. A., Campbell, E. L., Johnson, R. D., & Marks, H. I. (1989). Production scale purification of biosynthetic human insulin by reversed-phase high-performance liquid chromatography. *Journal of Chromatography*, *461*, 45–61.

Kumar, H., Mishra, M., Bajpai, S., Pokhria, D., Arya, A. K., Singh, R. K., & Tripathi, K. (2013). Correlation of insulin resistance, beta cell function and insulin sensitivity with serum sFas and sFasL in newly diagnosed type 2 diabetes. *Acta Diabetologica*, *50*(4), 511–518.

Kurtzhals, P., Havelund, S., Jonassen, I., & Markussen, J. (1997). Effect of fatty acids and selected drugs on the albumin binding of a long-acting, acylated insulin analogue. *Journal of Pharmaceutical Sciences*, *86*(12), 1365–1378.

Ladisch, M. R., & Kohlmann, K. L. (1992). Recombinant human insulin. *Biotechnology Progress*, *8*(6), 469–478.

Lee, S. H., & Yoon, K. H. (2021). A century of progress in diabetes care with insulin: A history of innovations and foundation for the future. *Diabetes & Metabolism Journal*, *45*(5), 629–640.

Makrides, S. C. (1996). Strategies for achieving high-level expression of genes in *Escherichia coli*. *Microbiological Reviews*, *60*(3), 512–538.

Markussen, J., Damgaard, U., Pingel, M., Snel, L., Sørensen, A. R., & Sørensen, E. (1983). Human insulin (Novo): Chemistry and characteristics. *Diabetes Care*, *6*(Suppl 1), 4–8.

Mathieu, C., Gillard, P., & Benhalima, K. (2017). Insulin analogues in type 1 diabetes mellitus: Getting better all the time. *Nature Reviews. Endocrinology*, *13*(7), 385–391.

Mishra, M., Kumar, H., Bajpai, S., Singh, R. K., & Tripathi, K. (2011). Level of serum IL-12 and its correlation with endothelial dysfunction, insulin resistance, proinflammatory cytokines and lipid profile in newly diagnosed T2DM. *Diabetes Research and Clinical Practice*, *94*(2), 255–261.

Mishra, M., Nichols, L., Dave, A. A., Pittman, E. H., Cheek, J. P., Caroland, A. J. V., Lotwala, P., Drummond, J., & Bridges, C. C. (2022). Molecular mechanisms of cellular injury and role of toxic heavy metals in chronic kidney disease. *International Journal of Molecular Sciences*, *23*(19), 11105.

Mishra, M., Panta, R., Miyares, M., & Solanki, R. (2018). Association of diabetes mellitus and thyroid disorders: An adipocytokines prospective. *Journal of Endocrinology and Thyroid Research*, *3*(3), 555612.

Mishra, M., Sotto, M. R., Panta, R., Miyares, M., & Solanki, R. (2017). Association of diabetes mellitus and thyroid disorders: A metabolic prospective. *Asian Pacific Journal of Health Sciences*, *4*(3), 253–262.

Nykiforuk, C. L., Boothe, J. G., Murray, E. W., Keon, R. G., Goren, H. J., Markley, N. A., & Moloney, M. M. (2006). Transgenic expression and recovery of biologically active recombinant human insulin from *Arabidopsis thaliana* seeds. *Plant Biotechnology Journal*, *4*(1), 77–85.

Overton, T. W. (2014). Recombinant protein production in bacterial hosts. *Drug Discovery Today*, *19*(5), 590–601.

Pandhal, J., Ow, S. Y., Noirel, J., & Wright, P. C. (2011). Improving N-glycosylation efficiency in *Escherichia coli* using shotgun proteomics, metabolic network analysis, and selective reaction monitoring. *Biotechnology and Bioengineering*, *108*(4), 902–912.

Panta, R., Paudel, K., Mishra, M., Solanki, R., Shrestha, B., & Ibrahim, J. (2018). Effectiveness of carbohydrate diet restriction in type 2 diabetes mellitus on insulin and incretin-based therapies. *Asian Pacific Journal of Health Sciences*, *5*(1), 1–8.

Patti, A. M., Giglio, R. V., Pafili, K., Rizzo, M., & Papanas, N. (2018). Pharmacotherapy for gestational diabetes. *Expert Opinion on Pharmacotherapy*, *19*(13), 1407–1414.

Petersen, M. C., & Shulman, G. I. (2018). Mechanisms of insulin action and insulin resistance. *Physiological Reviews*, *98*(4), 2133–2223.

Porro, D., Gasser, B., Fossati, T., Maurer, M., Branduardi, P., Sauer, M., & Mattanovich, D. (2011). Production of recombinant proteins and metabolites in yeasts: When are these systems better than bacterial production systems. *Applied Microbiology and Biotechnology*, *89*(4), 939–948.

Posner, B. I. (2017). Insulin signaling: The inside story. *Canadian Journal of Diabetes*, *41*(1), 108–113.

Rahman, M. S., Hossain, K. S., Das, S., Kundu, S., Adegoke, E. O., Rahman, M. A., Hannan, M. A., Uddin, M. J., & Pang, M. G. (2021). Role of insulin in health and disease: An update. *International Journal of Molecular Sciences*, *22*(12), 6403.

Smith, M. J., Simmons, K. M., & Cambier, J. C. (2017). B cells in type 1 diabetes mellitus and diabetic kidney disease. *Nature Reviews. Nephrology*, *13*(11), 712–720.

Staub, J. M., Garcia, B., Graves, J., Hajdukiewicz, P. T., Hunter, P., Nehra, N., Paradkar, V., Schlittler, M., Carroll, J. A., Spatola, L., Ward, D., Ye, G., & Russell, D. A. (2000). High-yield production of a human therapeutic protein in tobacco chloroplasts. *Nature Biotechnology*, *18*(3), 333–338.

Steiner, D. F., Clark, J. L., Nolan, C., Rubenstein, A. H., Margoliash, E., Aten, B., & Oyer, P. E. (1969). Proinsulin and the biosynthesis of insulin. *Recent Progress in Hormone Research*, *25*, 207–282.

Sun, H., Saeedi, P., Karuranga, S., Pinkepank, M., Ogurtsova, K., Duncan, B. B., Stein, C., Basit, A., Chan, J. C. N., Mbanya, J. C., Pavkov, M. E., Ramachandaran, A., Wild, S. H., James, S., Herman, W. H., Zhang, P., Bommer, C., Kuo, S., Boyko, E. J., & Magliano, D. J. (2022). IDF Diabetes Atlas: Global, regional and country-level diabetes prevalence estimates for 2021 and projections for 2045. *Diabetes Research and Clinical Practice*, *183*, 109119.

Thota, S., & Akbar, A. (2022). *Insulin*. In StatPearls. StatPearls Publishing. https://www.ncbi.nlm.nih.gov/books/NBK519007/ (accessed on 15th November 2025).

Tokarz, V. L., MacDonald, P. E., & Klip, A. (2018). The cell biology of systemic insulin function. *The Journal of Cell Biology*, *217*(7), 2273–2289.

Ueno, Y., & Morihara, K. (1989). Use of immobilized trypsin for semisynthesis of human insulin. *Biotechnology and Bioengineering*, *33*(1), 126–128.

van Rooijen, G. J., & Moloney, M. M. (1995). Plant seed oil-bodies as carriers for foreign proteins. *Biotechnology (Nature Publishing Company)*, *13*(1), 72–77.

Villa-Komaroff, L., Efstratiadis, A., Broome, S., Lomedico, P., Tizard, R., Naber, S. P., Chick, W. L., & Gilbert, W. (1978). A bacterial clone synthesizing proinsulin. *Proceedings of the National Academy of Sciences of the United States of America*, *75*(8), 3727–3731.

Wacker, M., Feldman, M. F., Callewaert, N., Kowarik, M., Clarke, B. R., Pohl, N. L., Hernandez, M., Vines, E. D., Valvano, M. A., Whitfield, C., & Aebi, M. (2006). Substrate specificity of bacterial oligosaccharyltransferase suggests a common transfer mechanism for the bacterial and eukaryotic systems. *Proceedings of the National Academy of Sciences of the United States of America*, *103*(18), 7088–7093.

Wacker, M., Linton, D., Hitchen, P. G., Nita-Lazar, M., Haslam, S. M., North, S. J., Panico, M., Morris, H. R., Dell, A., Wren, B. W., & Aebi, M. (2002). N-linked glycosylation in *Campylobacter jejuni* and its functional transfer into *E. coli*. *Science*, *298*(5599), 1790–1793.

Walsh, G. (2005). Therapeutic insulins and their large-scale manufacture. *Applied Microbiology and Biotechnology*, *67*(2), 151–159.

Walsh, G., & Jefferis, R. (2006). Post-translational modifications in the context of therapeutic proteins. *Nature Biotechnology*, *24*(10), 1241–1252.

Weiss, M., Steiner, D. F., & Philipson, L. H. (2014). Insulin biosynthesis, secretion, structure, and structure-activity relationships. In K. R. Feingold et al. (Eds.), *Endotext*. MDText.com, Inc. https://www.ncbi.nlm.nih.gov/books/NBK279027/ (accessed on 15th November 2025).

Wilcox, G. (2005). Insulin and insulin resistance. *The Clinical Biochemist. Reviews*, *26*(2), 19–39.

Wildt, S., & Gerngross, T. U. (2005). The humanization of N-glycosylation pathways in yeast. *Nature Reviews. Microbiology*, *3*(2), 119–128.

Williams, D. C., Van Frank, R. M., Muth, W. L., & Burnett, J. P. (1982). Cytoplasmic inclusion bodies in *Escherichia coli* producing biosynthetic human insulin proteins. *Science*, *215*(4533), 687–689.

CHAPTER 4

Cyanobacteria for the Production of Polyhydroxyalkanoate Bioplastic Factories: Progress and Prospects

SANDEEP SIROHI,[1] RITU VISHNOI,[1] D. V. SURYA PRAKASH,[2] NITIKA VATS,[2] PRITI PAL,[3] AKHILESH KUMAR SINGH,[4,*] and LAXUMAN SHARMA[5,*]

[1]*Department of Botany, Hariom Saraswati P.G. College, Dhanauri, Haridwar (affiliated to Sri Dev Suman Uttarakhand University, Badshahithaul, Tehri Garhwal) Uttarakhand, India*

[2]*Department of Biotechnology, Meerut Institute of Engineering and Technology, Meerut, Uttar Pradesh, India*

[3]*Amity Institute of Biotechnology, Amity University Uttar Pradesh, Lucknow Campus, Lucknow, Uttar Pradesh, India*

[4]*Department of Biotechnology, School of Life Sciences, Mahatma Gandhi Central University, Motihari, Bihar, India*

[5]*Department of Horticulture, Sikkim University, Sikkim, India*

*Corresponding author

ABSTRACT

Recently, a lot of scientific attention is being paid towards polyhydroxyalkanoates (PHA), commonly known as "bioplastics." Basically, PHA is produced through bacterial fermentation, incurring high production costs due to high and costly carbon substrates, making the products much costlier than conventional plastics. The most ideal proposition is to use

Industrial Microbiology: Current Progress and Novel Technologies.
Prakash Kumar Sarangi, Akhilesh Kumar Singh, Rajesh Kumar Srivastava, and Brijesh Pandey (Eds.)

DOI: 10.1201/9781779642219-4

photoautotrophic organisms like cyanobacteria for reducing the production cost of bioplastics. Nutrient limitation, exogenous application of low levels of carbon in growing substrates, manipulation of culture conditions, development of recombinant cyanobacteria, and metabolic engineering are the major strategies so far taken for increasing the PHA content in cyanobacteria. There are significant efforts in understanding the mechanisms of production of PHA in cyanobacterial cells.

4.1 INTRODUCTION

Plastic, being the most affordable polymer, is a crucial part of daily human life, particularly for residential and industrial uses (Arino et al., 1995; Brandl et al., 1988). Due to the indispensable nature of plastic in the modern world and its affordability, its usage has grown to an unfathomable scale. The non-degradability of conventional petrochemical plastics poses serious environmental problems with dire consequences. It is estimated that about 100 million tons of plastic are produced annually. Plastic waste is usually managed by recycling, particularly to avoid the load of its disposal in the environment, but the quality of plastics degrades over time (Asada et al., 1999) and is ultimately disposed of into the environment, where it persists for generations. Some forms of plastic, such as Bakelite, cannot be recycled. Improvement in manufacturing processes, raw material selection, recycling, and conversion of wastes to acceptable forms are the better proposition to prevent the accumulation of material waste in the environment. Therefore, from a comprehensive perspective, recycling is a time-consuming and expensive alternative and cannot be a management strategy for non-degradable polymers. The only solution for managing the plastic menace is replacement.

In the quest of searching for an alternative to plastics, an environment-friendly polymer or biodegradable plastic has found its place in the industries and is now produced in many nations. The governments of different nations and the industries are also considering the issue seriously and have endorsed the green policy for replacing non-degradable plastics with biodegradable ones.

The production of environmental-friendly bioplastics is one such development that can help us get over the pollution issue brought on by non-biodegradable plastics (Dephillippis et al., 1992; Gomes et al., 2020). New sources of biodegradable polymers become necessary due to the projected decline of oil supply (Brandl et al., 1990, 1995, 2011). The petroleum crisis, the threat of petroleum oil price increases (Poirier, 1995), and awareness of the scientific community about plastic pollution (Law, 2017) led to the production of commercial biodegradable polymers. Among all the biopolymers, polyhydroxyalkanoates (PHAs) are proven to be the best substitute for conventional plastics due to their thermoplastic properties (Senior and Dawes, 1973). PHAs are of microbial origin, composed of various hydroxy alkanoic acids (HA) which are eco-friendly, renewable, with analogous material properties to conventional thermoplastics like polypropylene (Singh et al., 2017).

After the disposal of PHAs thermoplastics, they tend to degrade when they come into contact with compost, soil, or marine trash and liberate CO_2 and water as breakdown byproducts. PHAs, with many desirable material properties, find a wide range of applications. The possible applications of bacterial PHAs are closely related to desirable attributes like thermoplasticity, piezoelectricity, biological degradability, and depolymerization of PHB to monomeric D(-)-3-hydroxybutyric acid. Three main industries—commodity packaging, agriculture, and pharmaceuticals—have seen the most application of bacterial PHAs. Despite the fact that bacterial PHAs show great promise as consumer goods, it is believed that their greatest level of development is in the pharmaceutical industry (Doi et al., 1990, 1994; Flechter et al., 1993).

The discovery of PHA inclusion in bacterial cells in 1888 by Beijerinck, the extraction of PHB in hot chloroform and hypothesizing it as an intracellular storage molecule by Lemoigne (1923), understanding the metabolic role of PHB in 1958, the first patent of PHB as a substitute for plastic in 1962, the production of PHA at an industrial scale in 1981, and PHA production in phototrophic organisms and extremophiles are the major milestones that encourage extensive research (Palmeiro-Sánchez et al., 2022). Despite tremendous effort in research and upscaling of production of PHA, the replacement of plastics is still not achievable due to the high cost of production of bacterial PHA. The high cost is mainly attributed to the high requirement of carbon sources for bacterial growth. In order to reduce the production cost by lowering the cost of nutritional inputs and carbon sources, cyanobacteria are now considered as a cheaper source of biodegradable plastic. Being photoautotrophic, cyanobacteria have minimal nutrient requirements and produce PHAs through oxygenated photosynthesis, and they can be grown in mixotrophic growth conditions with small quantities of organic substrates such as acetate, glucose, propionate, valerate, and so on (Singh and Mallick, 2017).

Organisms like cyanobacteria can be grown in agro-industrial waste like animal manure (Byrom et al., 1990, 1994), and this way there will be a drastic reduction in production costs. Other advantages would be to recycle waste water, reduce CO_2, promote bioremediation, and properly utilize cyanobacterial metabolites in other sectors like the food industry and fermentation processes. However, the design and establishment of an acceptable protocol by optimizing all the relevant parameters for the production of PHA is of utmost importance for reducing the cost of production and to meet the desirable quality of materials.

4.2 POLYHYDROXYALKANOATES (PHAS) AS A MATERIAL

The majority of PHAs are aliphatic polyesters of carbon, oxygen, and hydrogen and constitute a family of natural, water-insoluble, stereospecific polyesters of a wide range of different HAs, being characterized by the general chemical structure as presented in Figure 4.1, where 'X' can reach up to 30,000 and the R-pendant group includes an H-atom or a large variety of C-chains. A side chain R group is present on each PHA monomeric unit and can take the form of a hydrogen atom, a methyl group, or a tridecyl group. This normally saturated alkyl side chain can be accepted by the unusual molecular structure of unsaturated, branched, aromatic, heterocyclic, and substituted alkyl groups. The side chains of PHA thermoplastics could become chemically modified as a result of the cross-linking of unsaturated bonds (Carr et al., 1966). Changes in the side chains and the capacity to alter their substituents have made the PHA thermoplastics family more versatile with a wide range of potential applications.

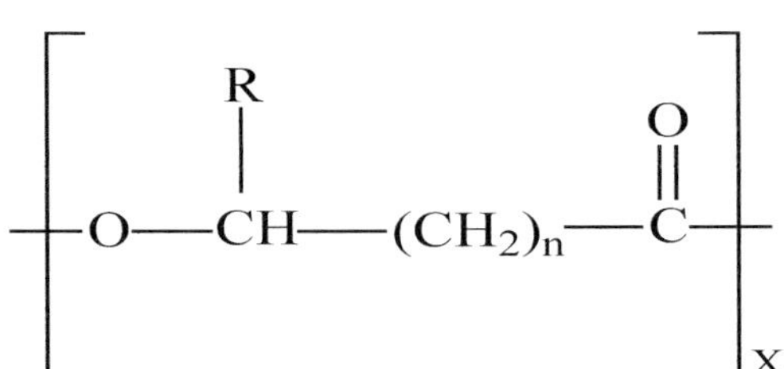

FIGURE 4.1 General structure of polyhydroxyalkanoates.

For example, when

$n = 1$	R = hydrogen	Poly(3-hydroxypropionate)
	methyl	Poly(3-hydroxybutyrtate)
	ethyl	Poly(3-hydroxyvalerate)
	propyl	Poly(3-hydroxyhexanoate)
	pentyl	Poly(3-hydroxyoctanoate)
	nonyl	Poly(3-hydroxydodecanoate)
$n = 2$	R = hydrogen	Poly(4-hydroxybutyarte)
$n = 3$	R = hydrogen	Poly(5-hydroxyvalerate)

PHAs are basically classified into three classes (Anderson and Dawes, 1990; Steinbuchel, 1992), depending on the number of carbon atoms in the monomer units: short chain length (SCL), medium chain length (MCL), and long chain length polyhydroxyalkanoates (LCL-PHAs) composed by hydroxy acids with 3–5, 6–14, or more than 14 carbon atoms, respectively. Poly-β-hydroxybutyrate (PHB), Poly (3-hydroxyvalerate), and their copolymer poly (3-hydroxybutyrate-*co*-3-hydroxyvalerate) (PHBV) are the most common SCL PHA found in various bacteria and cyanobacteria. PHB is a stereoregular polymer with 100% of the asymmetric carbon atoms in R-stereochemical configuration, which makes it isotactic, highly crystalline (crystallinity ranges between 55–80%), and relatively stiff. Its glass-to-rubber transition temperature (Tg) and its melting temperature (Tm) are comparable with isotactic polypropylene, as well as some mechanical properties like Young's modulus and tensile strength. Other properties of PHB useful for specific applications are resistance to humidity, piezoelectricity, and optical purity (Lee et al., 2001). The material properties like elongation and elasticity characteristics improve with copolymerization of 3-hydroxybutyrate (HB) and 3-hydroxyvalerate (HV).

MCL-PHAs are fully isotactic, semicrystalline elastomers characterized by low Tm values, low tensile strength, and high values of elongation to break (De Koning, 1995; Lee, 1996). poly(3-hydroxyoctanoate) (PHO) and poly(3-hydroxynonanoate) (PHN), and the copolymers with 3-hydroxyhexanoate (HHx), 3-hydroxyheptanoate (HH), and/or 3-hydroxydecanoate (HD), are MCL-PHAs. This family of polyesters usually softens at temperatures around 40°C and has low crystallization rates. However, both these limitations can be overcome by cross-linking the polymers (De Koning, 1995). So far, more than 150 different PHAs are reported, and it is the largest group of natural polyesters.

4.3 COMMERCIAL STATUS OF POLYHYDOXYALKANOATES

The most attractive features of PHAs are their biodegradability and hydrophobicity. There are other commercially available, cheaper biodegradable plastics in the market, but only PHAs have good moisture resistance, which finds its application particularly in sectors like packaging of liquids and food, paper coating, and so forth. The possible applications of PHAs are in different fields like disposables (razors, trays for food, utensils, etc.), packaging (bottles, films), agriculture (controlled release of pesticides, fertilizers, growth regulators, seed encapsulation), medical (absorbable sutures, surgical pins, staples), pharmaceutical (retarded drug release, drug carrier), and hygiene products (diapers, feminine hygiene products). This wide range of uses of PHA has led to the expansion and growth of the global market, particularly in the packaging and food services sectors. The market demand and the economic interest in PHA have propelled due to growing public awareness of biodegradable materials, governments' pursuit to prioritize environmentally friendly regulations, and the rising price of fossil fuels. The PHA bioplastic industry spans many countries in North America, Europe, Asia-Pacific, and the rest of the world. Substantial investment in the industry has been seen over the decades due to the large amounts of capital available in many enterprises in Europe and the Asia-Pacific area, as well as the high levels of consumer disposable income (Campbell et al., 1982; Capon et al., 1983). As a consequence of a mutually beneficial relationship between Metabolic Inc. and Archer Daniels Midland, the first US commercial plant for the manufacturing of P(3HB-co-3HV) co-polymer was established. However, this collaboration did not endure long and was broken up in 2012. At the moment, various other companies across the globe are producing different types of PHA under different brand names. Some of the important brands (Koller and Mukherjee, 2022) in the market are Biocycle (PHB Industrial SA, Brazil), Minerv-PHA (Bio-On Italy), BIOMER (Biomer Schwalbach, Germany), YOOP (Mango Material red wood, USA), ENMAT (Tianan Biologic Material Co., China), and Tepha Flex (Tepha Medical device Inc. USA). Large-scale PHA polymer production is still in its early stages despite being commercialized. Due to high recovery costs and the low polymer content of bacterial PHAs, production cost is higher than standard plastics. PHA bioplastics are 18 times more expensive than polypropylene, a common plastic (Gibson et al., 1981). The best possible combination of technological breakthroughs, economic viability, and integrated scale-up for commercial manufacturing and marketing is the need of the hour for the promotion and popularization of PHA bioplastics. As a consequence, an expanded and enhanced research initiative is currently concentrating on the bacterial system with the ultimate objective of making PHAs affordable. This endeavor aims to leverage photoautotrophic hosts to combat the constraints. Plants and agricultural crops that produce PHA are thought to be a viable option among photoautotrophic hosts. Using the transgenic approach, PHA production in diverse plant systems has been studied. The expression of the bacterial enzymes necessary for PHA synthesis in the various subcellular regions of plant cells, including the cytoplasm, plastids, and peroxisomes, has been attempted on numerous occasions. Low differential expression, lengthy growth times, and difficulties separating the PHA from other cellular components pose hurdles in the manufacture of plant-based PHA bioplastic. As a result, cyanobacteria are becoming increasingly significant as photoautotrophic hosts for the effective generation of PHAs.

4.4 PHB METABOLISM IN CYANOBACTERIA

The regulation of PHAs in the microbial cell has been a recent topic of importance, as it may help researchers in developing strains with improved PHA-producing potential through engineering the regulatory factors (Mitra et al., 2021). In a continuous approach for reducing the cost of production and making PHAs an affordable polymer, genetic manipulation of the microorganism for producing a high-yielding producer has been one of the strategies (Ye et al., 2018). The synthesis of PHAs in microbial cells is controlled by complex regulatory circuits (Mitra et al., 2021), where the carbon flow plays a major role in balancing the synthesis of PHA and microbial growth.

Basically, the synthesis process of PHB begins from the acetyl CoA monomer and enters the tricarboxylic acid (TCA) cycle. The fixation of carbon dioxide occurs via the Calvin-Benson-Bassham pathway in cyanobacteria, and the carbon is stored in the form of intracellular polymers like glycogen, polyglucans, and cyanophycean, including PHB (Hai et al., 2001; Wu et al., 2001), and these compounds are carbon and energy reserves. Several studies have revealed that the glycogen biosynthetic pathway has a notional link to PHA production in cyanobacteria. Glycogen, a major carbon and energy reserve, is accumulated during light phases in cyanobacteria. The biosynthesis pathway of glycogen is reported to have competition with the PHA pathway at the level of the 3-phosphoglycerate (3PG) pool (Wu et al., 2002). A large pool of 3PG is directed to glycogen synthesis, affecting PHA biosynthesis. The suppression of glycogen biosynthesis without affecting the growth of cyanobacteria can be a better strategy for improving PHA production (Singh et al., 2017).

3-phosphoglycerate (PGA) finally converts to Acetyl CoA through different pathways, viz. the Entner–Doudoroff (ED) pathway, glycolysis, and the pentose phosphate pathway (Singh and Mallick, 2017). Acetyl CoA ultimately enters the TCA cycle, where the biosynthesis process of PHA starts. However, the TCA cycle is incomplete in cyanobacteria, which is due to the lack of 2-oxoglutarate dehydrogenase complex, an essential enzyme that converts 2-oxoglutarate (2-OG) to succinyl-CoA. The incompleteness of the TCA cycle has led to many studies, and there has been continuous development in understanding the TCA cycle of cyanobacteria. Some of the studies have gone a long way in proving the completeness of the TCA cycle in cyanobacteria. The genes encoding 2-oxoglutarate decarboxylase and succinic semialdehyde dehydrogenase are now identified (Zhang and Bryant, 2011), and these genes are reported to be present in all types of cyanobacterial genomes except a few, like *Prochlorococcus* and marine *Synechococcus* species. Another variant of the cycle includes the glyoxylate shunt for the assimilation of acetate and γ-amino-butyric acid (GABA), which occurs only in a few cyanobacteria (Zhang and Bryant, 2015). The two enzymes, isocitrate lyase and malate synthase, that complete the cycle were also characterized in *Chlorogloeopsis fritschii* PCC 9212. In addition to the above, ambiguity due to the missing GABA aminotransferase GABA shunt was also studied. Zhang et al. (2016) isolated and characterized the enzymes of the GABA shunt and showed that N-acetylornithine aminotransferase (ArgD) can function as a GABA aminotransferase and help in completing the GABA shunt together with glutamate decarboxylase (GadA). Thus, the cyanobacterial TCA cycle is not incomplete and is possibly closed with an alternative shunt. Such scientific advances have led to an understanding of not only the metabolism of TCA but also the PHA biosynthetic pathway (Singh and Mallick, 2017).

Enzymes encoded as Pha A (acetyl-CoA acetyltransferase), Pha B (acetoacetyl-CoA reductase), and the synthase dimers Pha C and E (PHB

Synthase) regulate the synthesis of PHB (Afreen et al., 2021; Koch and Forchhammer, 2020). The first step of condensation of two units of acetyl CoA to acetoacetyl-CoA is catalyzed by PhaA. In the second step, acetoacetyl-CoA is reduced to 3-hydroxyacetyl-CoA, which is catalyzed by Pha B. During this step, one molecule of NADPH is oxidized to $NADP^+$. Finally, Pha C and E connect a 3-hydroxyacetyl-CoA to an elongated PHB polymer.

PHA synthase in cyanobacteria is classified as type-III PHA synthases, and the biosynthesis of only short-chain-length PHAs is reported (Hai et al., 2001). So far, the presence of type –III PHA synthase has been studied on only a few strains of cyanobacteria, viz. *Synechococcus* sp. strains MA19 and PCC 6715, *Chlorogloeopsis fritschii* PCC 6912, *Anabaena cylindrica* SAG 1403-2, *Cyanothece* sp. strains PCC 7424, PCC 8303, and PCC 8801, and *Gloeocapsa* sp. strain PCC 7428; however, more diversity may be discovered from diverse groups of cyanobacteria.

The presence of PHA synthase in heterotrophic bacteria differs from that of cyanobacteria. In heterotrophic bacteria, four classes of PHA synthase are reported instead of just two in cyanobacteria (photoautotrophic growth), thereby enabling the synthesis of substrate-specific short-chain-length and medium-chain-length polymers of PHA. The synthases are specific to substrate (3-HA) C-chain length. The classes of PHA comprise Class I (PhaC), responsible for the biosynthesis of short-chain-length 3-hydroxy acid (scl-3HA); Class II (PhaC), responsible for medium-chain-length 3-hydroxy acid (mcl-3HA); Class III (PhaEC), which enables the synthesis of two subunits (EC) (scl-3HA); and Class IV (PhaRC), which enables two subunits (RC) (scl-3HA) (Afreen et al., 2021). The addition of different substrates to the growing media, like sugars (glucose, sucrose, maltose, fructose, and lactose), glycerol, starch, methanol, lignin, agricultural wastewater, industrial wastewater, and dairy by-products, has been reported as a potential substrate for PHA production (Bhatia et al., 2021; Kalia et al., 2021; Kumar et al., 2019). The combination of the substrate and the feeding regime has been found to decide the PHA composition in terms of scl and mcl-3 HA (Ray and Kalia, 2017; Singh et al., 2015).

4.5 TRENDS IN PRODUCTION OF PHA IN CYANOBACTERIA

Since the first report of the presence of PHB in the cyanobacterium, *Chlorogloea fritschii* (Carr, 1966), there has been no looking back in the understanding of cyanobacterial PHA. Tremendous scientific efforts have been seen during the last 4–5 decades to establish cyanobacteria as a cell factory to produce bioplastics. Cyanobacteria are indigenously the sole prokaryotes that accumulate PHA by oxygenic photosynthesis. Hundreds of strains are reported to accumulate PHAs, particularly Poly-B-hydroxybutyrate, either in photoautotrophic or mixotrophic growth conditions coupled with nutrient limitations or supplementation of carbon dioxide. A plethora of publications are also available for having the added advantage of cyanobacteria used for wastewater treatment along with the production of PHAs.

Cyanobacteria, being photoautotrophic organisms, carry out oxygenic photosynthesis using H_2O as an electron donor in the presence of chlorophyll *a* and phycobiliproteins as photosynthetic pigments (Stal and Moezelar, 1997). The minimal requirement of nutrients for optimal growth of cyanobacteria attributes to the short generation time of cyanobacteria. Further, these organisms can successfully be cultivated in wastewaters due to their ability to use inorganic nitrogen and phosphorus for their growth, and wastewaters such as effluents of farm-yards, fish-farms, rubber industries, sewage treatment plants, and so forth, are rich sources of N and P.

Several workers have screened hundreds of cyanobacterial species for the presence of PHAs (Allen, 1984; Bhati et al., 2010; Campbell et al., 1982; Carpine et al., 2017; Carr, 1966; Jensen and Baxter, 1981; Miyake et al., 1996; Nishioka et al., 2001; Porta et al., 2000; Rippka et al., 1971; Rueda et al., 2022; Shrivastav and Mishra, 2010; Sicko-Goad, 1982; Stal, 1992; Stal et al., 1990; Sudesh et al., 2002; Suzuki et al., 1996; Tarawat et al., 2020; Vincenzini et al., 1990; Wu et al., 2001). In most of the studies, it was found that initially several cyanobacterial species/strains were screened for the presence of PHAs, particularly PHB and PHV. Strategically, efforts were given to increasing the PHA contents through manipulation of growing conditions, creating a deficiency of nitrogen or phosphorus, gas exchange limitation, and the addition of exogenous carbon sources like acetate, valerate, propionate, sucrose, glucose, and fructose. Efforts were also made to study the role of physical factors such as pH, temperature, incubation period, spectral quality, and so forth. Other strategies, so far, tried are the addition of elicitors, manipulation of the synthetic pathway, metabolic engineering, or using recombinant cyanobacteria. The initiation of research with just the detection of PHB in cyanobacteria has now taken a better status with reports of a tremendous increase in the accumulation of the homopolymer of 3-hydroxybutyric acid (PHB) and co-polymers 3-hydroxybutyrate-co-3-hydroxyvalerate (PHBV) (Summarized in Tables 4.1–4.3).

Irrespective of the studies carried out so far, there has been a clear observation that cyanobacteria synthesize PHB or PHV in a small quantity in a natural photoautotrophic condition. Upon the addition of exogenous CO_2 or carbon sources or the imposition of N- and P- deficiency, the PHB or PHV pool tends to increase manifold. Trends of PHB research and the use of cyanobacterial cells for PHAs synthesis are summarized in Tables 4.1–4.3.

TABLE 4.1 Accumulation of Homopolymer PHB in Photoautotrophic Growth Conditions By *Cyanobacterial* Species/Strains

Cyanobacterial Species/ Strains	Growing Conditions	PHB (% Dry Weight) or Productivity	References
Spirulina platensis	Carbon dioxide	4.5 mg l^{-1} h^{-1}	Campbell et al. (1982)
Gloeothece PC6909	Photoautotrophic	2.2%	Stal (1992)
Anabaena cylindrica 10C	Photoautotrophic	0.2%	Lama et al. (1996)
Synechococcus MA19	Photoautotrophic	5.0%	Miyake et al. (1996)
Synechococcus MA19	photoautotrophic + 2% CO_2 bubbling	20%	Miyake et al. (1996)
Nostoc muscorum	photoautotrophic	8.56	Sharma and Mallick (2005)
Nostoc muscorum	Photoautotrophic + P-Deficiency	23%	Sharma (2005)
Arthrospira subsalsa	Photoautotrophic	5.9	Shrivastav et al. (2010)
Aulosira fertilissima	Photoautotrophic	6.5	Bhati et al. (2010)
Recombinant Synechocystis sp. PCC 6803	Photoautotrophic + N-deficiency	26.0	Khetkorn et al. (2016)
Calothrix scytonemicola TISTR 8095	CO_2 + N-deficiency	25.0	Kaewbai et al. (2016)
Nostoc muscorum NCCU-442	Photoautotrophic	6.44	Ansari and Fatma (2016)
Recombinant Synechocystis sp. PCC6803	Photoautotrophic condition + CO_2	12	Carpine et al. (2017)
Synechocystis sp. PCC 6714 (Mutant generation)	Photoautotrophic + CO_2	37	Kamravamanesh et al. (2018)
Tolypothrix distorta TISTR 8985, Nostoc muscorum TISTR 8871, Chlorogloea fritschii TISTR 8547	CO_2	2.2, 0.8, 0.3 (mg gDW^{-1} day^{-1}) productivity	Tarawat et al. (2020)
Synechocystis sp.	Inorganic carbon 2g C/L in form of NAHCO3	14.3%	Rueda et al. (2022)

It is visualized from Table 4.1 that the cyanobacterial species are able to accumulate PHB even up to 8.56% (dcw) in a photoautotrophic condition with a normal light and dark cycle. Supplementation of growing media with CO_2 has been shown to increase the PHB content in different cyanobacterial strains like *Tolypothrix distorta TISTR 8985, Nostoc muscorum TISTR 8871, Chlorogloea fritschii TISTR 8547* (Tarawat et al., 2020), including recombinant *Synechocystis sp. PCC6803* (Carpine et al., 2017) and a high PHB yielding mutant of sp. PCC 6714 (Kamravamanesh et al., 2018). In all the studies, supplementation of gaseous (CO_2) and inorganic sources of carbon dioxide ($NaHCO_3$) has increased the PHB content.

TABLE 4.2 Accumulation of Homopolymer PHB with Supplementation of Carbon Sources and Nutrient Limitation Conditions By Cyanobacterial Species/Strains

Cyanobacterial Species	Growth Conditions	Accumulation (Percent Dry Cell Weight or Productivity)	References
Nostoc muscorum	Acetate + glucose + P-limitation + dark incubation	45.6	Sharma et al. (2007)
Synechocystis sp. PCC 6803	Phosphate deficiency + gas exchange limitation + acetate + fructose	38.0	Panda and Mallick (2007)
Aulosira fertilissima	Citrate + acetate + P limitation	85.1	Samantaray and Mallick (2012)
Nostoc muscorum	Glucose + acetate + P-deficiency	16.5	Haase et al. (2012)
Aulosira fertilissima CCC 444	Fructose + Valerate	76.9	Samantaray and Mallick (2014)
Synechococcus	Acetate + N-deficiency	35.0	Mendhulkar and Shetye (2017)

Several workers have reported that the PHB content can be increased by supplementation of culture media with carbon sources. The increase is more pronounced under nitrogen and phosphorus deficit conditions (Table 4.2). The PHB accumulation was reported to be as high as 85.1% in *Aulosira fertilissima* under specific culture conditions (Citrate + acetate + P limitation).

Besides, some earlier works have also reported the enhancement of PHB content with supplementation of carbon sources to the media. Almost all the studies have shown that the mixotrophic condition, achieved by adding carbon sources to the media, has increased the PHB yield. In *Gloeothece* PCC 6909, Stal (1992) made an effort to standardize the conditions for maximum PHB production in a N_2-fixing cyanobacterium, where the effects of growth phase, light-dark cycles, and mixotrophy on PHB accumulation were evaluated. The initial PHB accumulation of 2.2% (dcw) under photoautotrophic conditions increased to 2.5% after 48 h of dark incubation, and further addition of acetate (10 mM) increased the content to 3.3% in the presence of nitrogen sources. Under nitrogen-free conditions, however, the addition of the same amount of acetate yielded PHB up to 4.7% (dcw).

In a case study with *Oscillatoria limosa* for PHA accumulation, Stal (1992) did not find any effect of acetate (0.5–2.0 mM). The culture age was shown to be an important factor, and maximum accumulation was recorded at the stationary phase of the culture. In photoautotrophic growth conditions under continuous illumination, PHB content in *Spirulina maxima* SOSA18 was undetectable (De Philippis et al., 1992), but when these cultures were grown under mixotrophic conditions in the presence of acetate (8.7 mM), an accumulation of up to 3% (dcw) was recorded after 200 h of incubation, which was shown to be the stationary phase of the culture. The cell density was 1.2 g l^{-1} with an average production rate of 0.19 mg l^{-1} h^{-1}.

Accumulation of PHB in *Anabaena cylindrica* 10 C was studied by Lama et al. (1996). In the absence of combined nitrogen sources, the photo autotrophically grown cultures accumulated PHB up to 0.2% (dcw) after 21 days of growth (initial accumulation potential in nitrogen-supplemented medium was 0.005% dcw). Addition of acetate (8.8 mM) to the nitrogen-free BG-11 medium enhanced intracellular PHB concentration by 10-fold, *ca.* 2% (dcw). Addition of glucose (3.3 mM) to the above growth medium resulted in only a 3-fold increase in the PHB pool, whereas citrate (3.1 mM) supplementation caused a strong decrease in PHB content. Interestingly, *Anabaena cylindrica* 10C showed the possibility of producing the copolyester P(3HB-co-3HV) following supplementation of propionate or valerate to the culture medium.

The most interesting results, however, were obtained by Miyake and his group with the thermophilic cyanobacterium, *Synechococcus* MA19, isolated from the surface of a wet volcanic rock in Japan. When the cyanobacterium was transferred to nitrogen-free medium, PHB accumulation reached 21% (dcw) from an initial accumulation of 5% (dcw) when grown in usual BG-11 medium. Accumulation of PHB was found to reach 20% (dcw) in cultures bubbled with 2% CO_2 at a rate of 0.4 l min^{-1}. The cellular PHB content reached 27% (dcw) when the above culture was subjected to dark incubation (Miyake et al., 1996). Interestingly, under phosphate-limited conditions at 50°C, when the intracellular phosphate concentration was 0.043–0.076 mmol per gram of cellular components, this strain accumulated PHB up to 55% (dcw) after 260 h of incubation (Nishioka et al., 2001). To maintain the above internal phosphate pool, $Ca_3(PO_4)_2$ was used as the phosphate source. The accumulation of PHB with the addition of $Ca_3(PO_4)_2$ was twice that without $Ca_3(PO_4)_2$ under complete depletion of phosphate. This condition yielded a cell density of 4.4 g l^{-1} and an average production rate of 9.2 mg l^{-1} h^{-1} PHB, the highest value recorded for cyanobacterial species so far.

Synechocystis sp. PCC 6803, the unicellular non-nitrogen-fixing form, is being used as a model cyanobacterium around the world. The distinguishing feature of spontaneous transformability, coupled with a short generation time, made it one of the most popular organisms for genetic and physiological studies. Under mixotrophic growth conditions with 5 mM glucose, this strain accumulated PHB only up to 0.2% (dcw), whereas the cells in nitrogen-starved conditions could accumulate up to 4% (dcw) (Wu et al., 2001). This was explained as being due to the reduction of growth in nitrogen-deprived conditions, but not due to the enhanced biosynthesis of PHB. The simultaneous effects of acetate and nitrogen deprivation on the stimulation of PHB biosynthesis were later demonstrated by Wu et al. (2002). The addition of sodium acetate (15 mM) in nitrogen-sufficient conditions yielded 9.9% PHB, while under nitrogen-starved conditions, the PHB pool reached up to 15.2% (dcw).

Studies conducted by Sudesh et al. (2002) with various carbon sources viz. acetate, 4-hydroxybutyric acid, 1,4-butanediol, γ-butyrolactone, pentatonic acid, and propionic acid with the wild type *Synechocystis* sp. PCC 6803 reported PHB accumulation up to 7% (dcw) in the cultures supplemented with 10 mM sodium acetate. The vessels supplemented with other carbon sources mentioned above were not found to affect PHB accumulation noticeably.

The results of a thorough study of these data showed that the activation of PHB synthesis is promoted in cells with an unbalanced carbon:nitrogen (C:N) and NADPH:ATP ratio, as well as nitrogen and phosphorus deficiency. Phosphorus/phosphate limitation may promote the increased synthesis of the PHA pool. It is probably due to additional availability of reducing power, when $NADP^+$ reduction is encouraged by ongoing noncyclic ATP production, while the photosynthetic electron flow is still present. A similar improvement in the PHA pool under N_2 restriction may be possible if one considers the past findings. These observations suggest that the abundance of NADPH or the NADPH: NADP ratio increases intracellular PHA polymer synthesis. Cells were raised in the control culture under standard conditions, and it was found that the rate of flow of reducing energy into the Citric acid cycle remained essentially constant over the course of the culture period. Inhibition of amino acid biosynthesis pathways, specifically the reaction that results from nitrogen pool depletion during cell culture involves -ketoglutaric acid ($C_5H_6O_5$) conversion to glutamic acid ($C_5H_9NO_4$), which in turn results in an increased NADPH pool. As a result, fewer NADPH molecules are used as reducing power. PHA thermoplastics may build more intracellularly when there is an insufficient or lack of nitrogen. This lingering NADPH could be the root of the problem.

4.6 PRODUCTION OF CO-POLYMER PHBV IN CYANOBACTERIA

PHB as a material has similar tensile strength and thermal properties to that of polypropylene, but has lower elongation. Elongation and elasticity characteristics improve with copolymerization of 3-hydroxybutyrate (HB) and 3-hydroxyvalerate (HV) yielding copolymer poly (3-hydroxybutyrate-co-3-hydroxyvalerate) (PHBV). The co-polymerization of HB and HV in varying proportions yields PHBV with a higher elongation ability (Chanprateep, 2010; Verlinden et al., 2007). Increased mole ratio (mol%) of HV in PHBV enhances the polymer elongation but reduces the melting temperature, glass transition temperature, and tensile strength (Chanprateep and Kulpreecha, 2006; Lee, 1996). Alternation of the HV mol% in PHBV polymer help in diversifying its material properties, useful for divergent material usages (Chanprateep, 2010). So far, only limited species and strains are reported to produce PHBV with a maximum range of HV ratio of 5–38 mol% when cultured with one or two type(s) of organic compound(s) viz. fructose, acetate, propionate, and/or valerate (Bhati and Mallick, 2012; Samantaray and Mallick, 2015; Taepucharoen et al., 2017). Most recently Tarawat et al. (2020), examined 20 cyanobacteria species from diverse group for PHBV production. Under photoautotrophy with carbon dioxide supply, only one strain produced PHBV, but with supplementation of different organic substrate (acetate or propionate or valerate), three other strains were found to accumulate PHBV.

TABLE 4.3 Accumulation of Copolymer PHBV By Cyanobacteria

Cyanobacterial Species	Culture Conditions	% PHB-HV	References
Nostoc muscorum Agardh	Acetate + glucose + valerate + N-deficiency	78	Bhati and Mallick (2015)
Nostoc muscorum Agardh	Acetate + valerate + N-deficiency	60	Bhati and Mallick (2012)
	Acetate + glucose + valerate + poultry litter	67.0 (PHB-HV)	Bhati and Mallick (2016)
Oscillatoria okeni TISTR 8549	N-Deficiency	14 ± 4%	Taepucharoen (2017)
	Acetate + N-deficiency + Dark condition	42 ± 8%	Taepucharoen (2017)
Anabaena spiroides TISTR 8075	Photoautotrophic condition	0.5 ± 0.0 (mg gDW^{-1} day^{-1})	Tarawat et al. (2020)
Anabaena ambigua TISTR 8001	CO_2 + Acetate	7.2 ± 3.0 (mg gDW^{-1} day^{-1})	Tarawat et al. (2020)

4.7 GENETICALLY ENGINEERED/RECOMBINANT CYANOBACTERIA AND PHA PRODUCTION

It is well known that several companies around the globe are interested in creating PHA thermoplastics for commercial usage. The strategies employed for PHA thermoplastic synthesis, as well as the media composition, which includes carbon substrates, are some of the factors that lead to the high production costs of PHAs (Holmes et al., 1985; Huisman et al., 1989). Another strategy is to construct high-yielding recombinant cyanobacteria.

Introduction of heterologous enzymes and a heterologous pathway is a noble strategy for the production of biofuels and biochemicals, including PHB (Choi et al., 2020). Genetically modified cyanobacterial strains with a high level of PHA synthesis are considered the most sought-after approach. Well-designed metabolic engineering, along with genetic engineering, recombinant DNA technology, and other relevant techniques, are to be holistically employed to achieve the desired objectives. Several researchers have already made breakthroughs in engineering cyanobacteria for better PHB production. The production of acetoacetic-CoA reductase, PHB synthase, and 3-ketothiolase has been genetically engineered in cyanobacteria. The ability of a non-modified species to accumulate bioplastics promotes the growth of a suitable genetically modified PHA accumulator. PHA production increased from 4 to 24% (dcw) after transformation of *Synechococcus* PCC 7942 with the *C. necator* PHA operon. Additionally, the *Synechococcus* PCC 7002 *RuBisCO* gene (rbc) possess a potent promoter that is activated by CO_2 concentration (Lafferty et al., 1988; Lama et al., 1996).

The critical factors known so far for enhanced PHB accumulation in cyanobacteria, like nitrogen limitation and supplementation of acetate, were also investigated in recombinant cyanobacteria. The genetic modification of cyanobacteria for PHB production was first attempted in a PHA non-producing strain, *Synechococcus* sp. PCC 7942 (Suzuki et al., 1996). *Synechococcus* sp. PCC 7942 was transformed using the gene from *W. eutropha* encoding PHB-synthesizing enzymes, PHB synthase, *β*-ketothiolase, and acetoacetyl-CoA reductase (Takahashi et al., 1998). Under photoautotrophic conditions, the transformant in nitrogen-free BG-11 medium accumulated 1.4% (dcw) PHB, and upon the addition of acetate, there was accumulation up to 10%, and under nitrogen-starved conditions, up to 25% (dcw). PHB synthase activity in *Synechocystis* sp. PCC 6803 was increased two-fold by introducing the PHA biosynthetic genes of *W. eutropha*, where the number of PHB granules in the recombinant cells ranged from 4–6, and their sizes were in the range of 70–240 nm. In comparison to the wild-type, the recombinant cells with increased synthase activity showed only a marginal increase in PHB content, suggesting that PHB synthase is not the limiting enzyme of PHB biosynthesis in *Synechocystis* sp. PCC 6803.

Increasing PHB content in *Synechocystis* 6803 was tried by the introduction of heterologous PHA synthase from *Ralstonia eutropha*, wherein 11% PHB (dcw) was achieved under nitrogen deficit conditions and supplementation of acetate (Sudesh et al., 2002). Modification of carbon partitioning can also enhance PHB production (Wu et al., 2002). ADP-glucose pyrophosphorylase (Agp) was eliminated in the pathway to weaken glycogen synthesis and to increase carbon flux to produce PHB. In that condition, PHB content increased to 19% with acetate supplementation. Carpine et al. (2017) tried to engineer carbon metabolism to increase acetyl-CoA and to link carbon flux toward the PHB pathway. PHB accumulation was increased up to 12% dcw in *Synechocystis* 6803 by the removal of phosphotransacetylase (Pta) and acetyl-CoA hydrolase (Ach) and overexpression of heterologous phosphoketolase (Xpka) in a photoautotrophic growing condition. N-deprived conditions and overexpression of PhaA/B genes have been shown to increase PHB accumulation up to 26% (Khetkorn et al., 2016), and with the addition of acetate, it was further increased up to 35%.

Synechocystis sp. PCC 6803 promoter was successfully used as a regulatory sequence for PHA synthesis and the regulation of PHA biosynthetic genes. By employing carbon dioxide as a substrate, this creates a brand-new, intriguing prospect for the commercial-scale, economical aggregation of PHA in cyanobacteria. Contrarily, the technique of metabolic engineering can be widely applied to introduce new metabolic pathways in order to increase the range of substrates that can be employed and to make unique PHA polymers, including increased PHA aggregation in blue green algae. A thorough review of the literature showed that, despite the fact that there are only a few studies accessible, this field of study on the generation of PHA by cyanobacteria is relatively new. The continued improvement and development of cyanobacterial genetic and metabolic engineering techniques, combined with the expanding accessibility of genome data, has generally enabled the emergence of cyanobacteria as suitable and proficient construction of commercial cell factories for PHA production. Future research will concentrate on modifying the metabolism and genetic makeup of cyanobacteria to increase PHA thermoplastic synthesis (Lee et al., 2001). Recent work on recombinant cyanobacteria for increased PHB content can be seen in *Synechocystis* sp. PCC 6803 strain. Koch et al. (2020a) reported the increase in PHB content in nutrient limitation in new strain *Synechocystis* sp. PCC 6803 that lacks the regulatory protein PirC (product of sll0944). Pir C exhibits a higher activity of the phosphoglycerate mutase that results in increasing PHB content under nutrient limitation. A high yielding PHB strain PPT 1 (ΔpirC-REphaAB) was developed by introducing Pha A and Pha B under the control of promoter PpsbA2 from *Cupriavidus necator* to improve the intracellular PHB content. PHB up to 63% dcw was obtained in nitrogen and phosphorus depleted medium, which upon addition of acetate yielded 81% PHB in dcw basis (Table 4.4).

TABLE 4.4 Production of PHA By Recombinant Cyanobacteria

Cyanobacterial Species	Gene Regulation	Culture Conditions	References
Synechocystis 6803	Introduction of PHA biosynthetic genes from *Ralstonia eutropha*	+Acetate 11%	Sudesh et al. (2002)
Synechocystis 6803	Δagp (glycogen synthesis inhibition)	+Acetate 19%	Wu et al. (2002)
Synechocystis 6803	Ppsab2: phaAB -Nitrogen and ± acetate	26% (without acetate addition) and 35% (with acetate))	Khetkorn et al. (2016)
Synechocystis 6803	Ptrc: xfpk and Δpta Δach	12%	Carpine et al. (2017)
Synechocystis 6803	Ptac: phaA-(RBSopt)-phaB1, PpsbA12-Ptac: tesB, and ΔphaEC	5% CO_2 1.84 g L^{-1}	Wang et al. (2018)
Synechococcus 7942	Ptrc: nphT7, tesB, phaAB, and PLlacO1: pptesB	50 mM $NaHCO_3$ 840 mg L^{-1}	Ku and Lan (2018)
Synechocystis 6803	PPT 1 (ΔpirC-REphaAB)	63% (N- P-) 83% (Acetate + N-P-)	Koch et al. (2020b)

4.8 INCREASING THE PHA CONTENT IN CYANOBACTERIA BY METABOLIC ENGINEERING

An approach of metabolic engineering to increase the amount of PHBs in different cyanobacteria species has been tried. Particularly in *Synechocystis* 6803, it has been attempted for almost two decades. The initial attempt was to introduce the PHA synthase gene from the bacterial species *Ralstonia eutropha (Cupriavidus necator)* in view of increasing the PHA synthase activity and in turn the PHA content. In one of the mutants of *Synechocystis* 6803, Tyo et al. (2009) reported an increased PHB level through disruption of a gene encoding gamma-glutamyl phosphate reductase (proA, sll0461) or sll0565 (encoding a hypothetical protein). The nitrogen deficiency induced gene sll0783 (with unknown function), when knocked out, resulted in no PHB accumulation in nitrogen starvation conditions, which indicates the posttranslational regulation of PHB biosynthesis by Sll0783 (Schlebusch and Forchammen, 2010). In N-deprived conditions, overexpression of the sigE gene, which in fact increases levels of glycogen phosphorylase, isoamylases, glucose-6-phosphate dehydrogenase, and 6-phosphogluconate dehydrogenase respectively, encoded by glgPs, glgXs, zwf, and gnd (Osanai et al., 2013), and the protein levels of different genes associated with PHB synthesis viz. PhaA, PhaB, PhaC, and PhaE also increases with overexpression of sigE. PHB levels increase from 5.7 mg/L in the wild-type to 14.0 mg/L in sigE-overexpressing strains (Osanai et al., 2013).

Overexpression of Rre37, an OmpR-type response regulator, upregulates the genes related to sugar catabolism and nitrogen metabolism. It was found that the protein levels of GlgP(slr1367), one of the two glycogen phosphorylases, was higher in the rre37-overexpressing strain than the wild-type under both nitrogen-replete and nitrogen-depleted conditions, and in turn decreased the level of glycogen under nitrogen-replete conditions. Overall, it was demonstrated that Rre37 is a pathway-level regulator that activates the metabolic flow from glycogen to polyhydroxybutyrate (Osanai, 2014). PhaA, PhaB, PhaC, and PhaE were also shown to increase in overexpressed rre37-overexpresing *Synechocystis* 6803. An additive effect in increasing the PHB levels was also seen in the case of the co-overexpression of rre37 and sigE. Overexpression of the *phaABEC* gene cluster in *Synechocystis* 6803 cells has been shown to increase PHB levels by 12 times (Hondo et al., 2015). The increase in acetyl-CoA, acetoacetyl-CoA, and 3-hydroxybutyryl-CoA with overproduction of the *Synechocystis* 6803 phaABEC gene cluster indicates that the pathway-level regulation is important for increasing PHB (Hondo et al., 2015).

One of the noteworthy advancements in the regulatory mechanism of PHB synthesis was the identification of novel regulatory proteins in *Synechocystis* sp. PCC 6803. During nitrogen depletion, this strain undergoes chlorosis in which the cells first accumulate glycogen and later synthesize PHB by metabolizing the intracellular glycogen pool (Koch et al., 2019). Two novel proteins, Slr0058 and PirC, are known to regulate PHB synthesis in this strain (Koch et al., 2020c; Orthwein et al., 2021).

4.9 PRODUCTION OF CYANOBACTERIAL PHAS USING A TWO-STAGE TECHNIQUE THAT OPERATES IN ONE

Any techniques that lead to rapid production of cyanobacterial biomass with the synthesis of a high concentration of PHA molecules would be helpful in the economic manufacturing of biodegradable plastics in cyanobacteria. The well-documented two-stage cultivation method is widely used to maximize biomass production and the synthesis of PHA thermoplastics. A high biomass concentration is the goal of the first stage, or growth stage, which is continued by a second stage of aggregation where a particular nutrient is restricted to promote PHA synthesis. In the aggregation stage, PHA is produced intracellularly, which alone causes the residual cell concentration to rise (Martin et al., 2020). A technique like this is appealing and might be economically feasible for the manufacturing of PHAs on a wide scale.

4.10 CONCLUSION

In pursuit of developing eco-friendly substitutes to conventional plastics, PHA thermoplastics have gained popularity. PHA production in photoautotrophic organisms like cyanobacteria is currently regarded as a viable alternative to heterotrophic production due to their ability to grow biomass on non-arable land with minimal nutrients and reduced carbon levels. There is a high possibility of growing cyanobacteria using wastewater and photosynthesizing carbon dioxide into PHA thermoplastics.

Cyanobacteria are just beginning to be studied as potential cell factories for PHA synthesis. Numerous studies conducted to date for enhancing PHA development in cyanobacteria indicate that there is still a great deal of untapped potential. The genetic modification of cyanobacterial cells to include genes that efficiently synthesize PHAs, using metabolic inhibitors, blocking the glycogen biosynthesis pathway, switching from a two-step culture procedure to a single stage, improving biomass harvesting and cultivation practices together with efficient and low-cost extraction techniques, and the exploitation of inexpensive substrates and mixed cultures are all promising techniques for producing affordable cyanobacterial PHAs. Therefore, scientific progress, as well as financial sustainability and scale-up, are required for the commercial manufacturing and marketing of cyanobacterial PHA thermoplastics.

KEYWORDS

- **bioplastic**
- **carbon source optimization**
- **cyanobacteria**
- **metabolic engineering**
- **PHB accumulation**
- **polyhydroxyalkanoate**

REFERENCES

Afreen, R., Tyagi, S., Singh, G. P., & Singh, M. (2021). Challenges and perspectives of polyhydroxyalkanoate production from microalgae/cyanobacteria and bacteria as microbial factories: An assessment of hybrid biological system. *Frontiers in Bioengineering and Biotechnology*, *9*, 624885. https://doi.org/10.3389/fbioe.2021.624885 (accessed on 15th November 2025).

Allen, M. M. (1984). Cyanobacterial cell inclusions. *Annual Review of Microbiology*, *38*, 1–25. https://doi.org/10.1146/annurev.mi.38.100184.000245 (accessed on 15th November 2025).

Anderson, A. J., & Dawes, E. A. (1990). Occurrence, metabolism, metabolic role, and industrial uses of bacterial polyhydroxyalkanoates. *Microbiological Reviews*, *54*(4), 450–472. https://doi.org/10.1128/mr.54.4.450-472.1990 (accessed on 15th November 2025).

Ansari, S., & Fatma, T. (2016). Cyanobacterial polyhydroxybutyrate (PHB): Screening, optimization and characterization. *PLOS ONE*, *11*(6), e0158168. https://doi.org/10.1371/journal.pone.0158168 (accessed on 15th November 2025).

Arino, X., Ortega-Calvo, J. J., Hernandez-Mavine, M., & Saiz-Jimenez, C. (1995). Effect of Sulfur starvation on the morphology and structure of the cyanobacterium *Gleotheoceae* sp. PCC 6909. *Archives of Microbiology*, *163*, 447–453.

Asada, Y., Miyake, M., Miyake, J., Kurane, R., & Tokiwa, Y. (1999). Photosynthetic accumulation of poly-(hydroxybutyrate) by cyanobacteria—The metabolism and potential for CO_2 recycling. *International Journal of Biological Macromolecules*, *25*(1–3), 37–42. https://doi.org/10.1016/S0141-8130(99)00013-6 (accessed on 15th November 2025).

Atlas, R. M. (1993). *Microbial ecology: Fundamentals and applications* (pp. 39–43). Kluwer Academic Publishers.

Bhati, R., & Mallick, N. (2012). Production and characterization of poly(3-hydroxybutyrate-co-3-hydroxyvalerate) co-polymer by a N_2-fixing cyanobacterium, *Nostoc muscorum* Agardh. *Journal of Chemical Technology & Biotechnology*, *87*, 505–512.

Bhati, R., & Mallick, N. (2015). Poly(3-hydroxybutyrate-co-3-hydroxyvalerate) copolymer production by the diazotrophic cyanobacterium *Nostoc muscorum* Agardh: Process optimization and polymer characterization. *Algal Research-Biomass Biofuels and Bioproducts*, *7*, 78–85.

Bhati, R., Samantaray, S., Sharma, L., & Mallick, N. (2010). Poly-β-hydroxybutyrate accumulation in cyanobacteria under photoautotrophy. *Biotechnology Journal*, *5*(11), 1181–1185.

Bhatia, S. K., Otari, S., Jeon, J., Gurav, R. G., Choi, Y., Bhatia, R. K., Pugazhendhi, A., Kumar, V., Rajesh Banu, J., Yoon, J., Choi, K., & Yang, Y. (2021). Biowaste-to-bioplastic (polyhydroxyalkanoates): Conversion technologies, strategies, challenges, and perspective. *Bioresource Technology*, *326*, 124733.

Brandi, H., Bachofen, R., Mayer, J., & Wintermantel, E. (1995). Degradation and applications of polyhydroxyalkanoates. *Canadian Journal of Microbiology*, *41*(13), 143–153.

Brandl, H., Gross, R. A., Lenz, R. W., & Fuller, R. C. (1988). *Pseudomonas oleovorans* as a source of poly(beta-hydroxyalkanoates) for potential applications as biodegradable polyesters. *Applied and Environmental Microbiology*, *54*(8), 1977–1982. https://doi.org/10.1128/aem.54.8.1977-1982.1988 (accessed on 15th November 2025).

Brandl, H., Gross, R. A., Lenz, R. W., & Fuller, R. C. (1990). Plastics from bacteria and for bacteria: Poly(beta-hydroxyalkanoates) as natural, biocompatible, and biodegradable polyesters. *Advances in Biochemical Engineering/Biotechnology*, *41*, 77–93. https://doi.org/10.1007/BFb0010232 (accessed on 15th November 2025).

Byrom, D. (1987). Polymer synthesis by microorganisms: Technology and economics. *Trends in Biotechnology*, *5*(9), 246–250. https://doi.org/10.1016/0167-7799(87)90100-4 (accessed on 15th November 2025).

Byrom, D. (1990). Industrial production of copolymer from *Alcaligenes eutrophus*. In E. A. Dawes (Ed.), *Novel Biodegradable Microbial Polymers* (pp. 113–117). Kluwer Academic Publishers.

Byrom, D. (1994). Polyhydroxyalkanoates. In D. P. Mobley (Ed.), *Plastics from Microbes: Microbial Synthesis of Polymers and Polymer Precursors* (pp. 5–33). Munich Hanser.

Campbell, J. 3rd., Stevens, S. E. Jr., & Balkwill, D. L. (1982). Accumulation of poly-beta-hydroxybutyrate in *Spirulina platensis*. *Journal of Bacteriology*, *149*(1), 361–363. https://doi.org/10.1128/jb.149.1.361-363.1982 (accessed on 15th November 2025).

Capon, R. J., Dunlop, R. W., Ghisalberti, E. L., & Jefferies, P. R. (1983). Poly-3-hydroxyalkanoates from marine and freshwater cyanobacteria. *Phytochemistry*, *22*(5), 1181–1184. https://doi.org/10.1016/0031-9422(83)80217-9 (accessed on 15th November 2025).

Carpine, R., Du, W., Olivieri, G., Pollio, A., Hellingwerf, K. J., Marzocchella, A., & Santos, F. B. (2017). Genetic engineering of *Synechocystis* sp. PCC6803 for poly-β-hydroxybutyrate overproduction. *Algal Research-Biomass Biofuels and Bioproducts*, *25*, 117–127.

Carr, N. G. (1966). The occurrence of poly-beta-hydroxybutyrate in the blue-green alga, *Chlorogloea fritschii*. *Biochimica et Biophysica Acta*, *120*(2), 308–310. https://doi.org/10.1016/0926-6585(66)90353-0 (accessed on 15th November 2025).

Chanprateep, S. (2010). Current trends in biodegradable polyhydroxyalkanoates. *Journal of Bioscience and Bioengineering*, *110*(6), 621–632.

Chanprateep, S., & Kulpreecha, S. (2006). Production and characterization of biodegradable terpolymer poly(3-hydroxybutyrate-co-3-hydroxyvalerate-co-4-hydroxybutyrate) by *Alcaligenes* sp. A-04. *Journal of Bioscience and Bioengineering*, *101*(1), 51–56.

Choi, Y., Lee, J. W., Kim, J. W., & Park, J. (2020). Acetyl-CoA-derived biofuel and biochemical production in cyanobacteria: A mini review. *Journal of Applied Phycology*, *32*, 1643–1653.

Dawes, E. A. (1992). *Prokaryotic structure and function: A new perspective* (pp. 88–122). Cambridge University Press.

De Koning, G. (1995). Physical properties of bacterial poly ((R)-3-hydroxyalkanoates). *Canadian Journal of Microbiology*, *41*(13), 303–309.

De Philippis, R., Sili, C., & Vincenzini, M. (1992). Glycogen and poly-β-hydroxybutyrate synthesis in *Spirulina maxima*. *Microbiology*, *138*(8), 1623–1628.

Doi, Y. (1990). *Microbial Polyesters* (pp. 562–564). VCH Publishers.

Doi, Y., Mukai, K., Kasuya, K., & Yamada, K. (1994). Biodegradation of biosynthetic and chemosynthetic polyhydroxyalkanoates. *Studies in Polymer Science*, *12*, 39–51. https://doi.org/10.1016/B978-0-444-81708-2.50010-5 (accessed on 15th November 2025).

Doi, Y., Segawa, A., & Kunioka, M. (1990). Biosynthesis and characterization of poly(3-hydroxybutyrate-co-4-hydroxybutyrate) in *Alcaligenes eutrophus*. *International Journal of Biological Macromolecules*, *12*(2), 106–111.

Flechter, A. (1993). In *Plastics from bacteria and for bacteria: PHA as natural, biodegradable polyesters* (pp. 77–93). J Springer Verlag.

Gibson, J. L. (1981). Movement of acetate across the cytoplasmic membrane of the unicellular cyanobacteria *Synechococcus* and *Aphanocapsa*. *Archives of Microbiology*, *130*, 175–179. https://doi.org/10.1007/BF00411073 (accessed on 15th November 2025).

Gomes, G. D., Pereira, X. L., & Valadares, S. A. (2020). Cyanobacterial polyhydroxyalkanoates: A sustainable alternative in circular economy. *Molecules*, *25*(18), 4331. https://doi.org/10.3390/molecules25184331 (accessed on 15th November 2025).

Haase, S. M., Huchzermeyer, B., & Rath, T. (2012). PHB accumulation in *Nostoc muscorum* under different carbon stress situations. *Journal of Applied Phycology*, *24*, 157–162.

Hai, T., Hein, S., & Steinbüchel, A. (2001). Multiple evidence for widespread and general occurrence of type-III PHA synthases in cyanobacteria and molecular characterization of the PHA synthases from two thermophilic cyanobacteria: *Chlorogloeopsis fritschii* PCC 6912 and *Synechococcus* sp. strain MA19. *Microbiology*, *147*(11), 3047–3060.

Holmes, P. A. (1985). Applications of PHB – A microbially produced biodegradable thermoplastic. *Physics in Technology*, *16*, 32–36. https://doi.org/10.1088/0305-4624/16/1/305 (accessed on 15th November 2025).

Hondo, S., Takahashi, M., Osanai, T., Matsuda, M., Hasunuma, T., Tazuke, A., & Asayama, M. (2015). Genetic engineering and metabolite profiling for overproduction of polyhydroxybutyrate in cyanobacteria. *Journal of Bioscience and Bioengineering*, *120*(5), 510–517.

Huisman, G., de Leeuw, O. S., Eggink, G., & Witholt, B. (1989). Synthesis of poly-3-hydroxyalkanoates is a common feature of fluorescent pseudomonads. *Applied and Environmental Microbiology*, *55*, 1949–1954. https://doi.org/10.1128/aem.55.8.1949-1954.1989 (accessed on 15th November 2025).

Jensen, T. E. (1980). Polymorphic bodies of various isolates of *Nostoc* (Cyanophyceae). *Letters in Microbiology*, *11*, 117–125.

Jensen, T. E., & Baxter, M. (1981). Vesicles and a vesicular crystalloid in *Microcystis aeruginosa* (Cyanophyceae). *Cytobios*, *32*, 129–137.

Jensen, T. E., & Sicko, L. M. (1971). Fine structure of poly-β-hydroxybutyric acid granules in a blue-green alga, *Chlorogloea fritschii*. *Journal of Bacteriology*, *106*, 683–686. https://doi.org/10.1128/jb.106.2.683-686.1971 (accessed on 15th November 2025).

Kaewbai-Ngam, A., Incharoensakdi, A., & Monshupanee, T. (2016). Increased accumulation of polyhydroxybutyrate in divergent cyanobacteria under nutrient-deprived photoautotrophy: An efficient conversion of solar energy and carbon dioxide to polyhydroxybutyrate by *Calothrix scytonemicola* TISTR 8095. *Bioresource Technology*, *212*, 342–347.

Kalia, V. C., Patel, S. K. S., Shanmugam, R., & Lee, J. K. (2021). Polyhydroxyalkanoates: Trends and advances toward biotechnological applications. *Bioresource Technology*, *326*, 124737.

Kalia, V. C., Raizada, N., & Sonakya, V. (2000). Bioplastics. *Journal of Scientific & Industrial Research*, *59*, 433–445.

Kamravamanesh, D., Kovacs, T., Pflügl, S., Druzhinina, I., Kroll, P., Lackner, M., & Herwig, C. (2018). Increased poly-β-hydroxybutyrate production from carbon dioxide in randomly mutated cells of cyanobacterial strain *Synechocystis* sp. PCC 6714: Mutant generation and characterization. *Bioresource Technology*, *266*, 34–44.

Kamravamanesh, D., Pflügl, S., Nischkauer, W., Limbeck, A., Lackner, M., & Herwig, C. (2017). Photosynthetic poly-β-hydroxybutyrate accumulation in unicellular cyanobacterium *Synechocystis* sp. PCC 6714. *AMB Express*, *7*(1), 1–12.

Kessler, B., & Witholt, B. (2001). Factors involved in the regulatory network of polyhydroxyalkanoate metabolism. *Journal of Biotechnology*, *86*(2), 97–104. https://doi.org/10.1016/s0168-1656(00)00404-1 (accessed on 15th November 2025).

Khetkorn, W., Incharoensakdi, A., Lindblad, P., & Jantaro, S. (2016). Enhancement of poly-3-hydroxybutyrate production in *Synechocystis* sp. PCC 6803 by overexpression of its native biosynthetic genes. *Bioresource Technology*, *214*, 761–768.

Koch, M., Berendzen, K. W., & Forchhammer, K. (2020a). On the role and production of polyhydroxybutyrate (PHB) in the cyanobacterium *Synechocystis* sp. PCC 6803. *Life*, *10*(4), 47.

Koch, M., Bruckmoser, J., Scholl, J., Hauf, W., Rieger, B., & Forchhammer, K. (2020b). Maximizing PHB content in *Synechocystis* sp. PCC 6803: A new metabolic engineering strategy based on the regulator PirC. *Microbial Cell Factories*, *19*(1), 231. https://doi.org/10.1186/s12934-020-01491-1 (accessed on 15th November 2025).

Koch, M., Doello, S., Gutekunst, K., & Forchhammer, K. (2019). PHB is produced from glycogen turnover during nitrogen starvation in *Synechocystis* sp. PCC 6803. *International Journal of Molecular Sciences*, *20*, 1942–1956. https://doi.org/10.3390/ijms20081942 (accessed on 15th November 2025).

Koch, M., Orthwein, T., Alford, J. T., & Forchhammer, K. (2020c). The Slr0058 protein from *Synechocystis* sp. PCC 6803 is a novel regulatory protein involved in PHB granule formation. *Frontiers in Microbiology*, *11*, 1–13. https://doi.org/10.3389/fmicb.2020.00809 (accessed on 15th November 2025).

Koller, M., & Mukherjee, A. (2022). A new wave of industrialization of PHA biopolyesters. *Bioengineering (Basel, Switzerland)*, *9*(2), 74. https://doi.org/10.3390/bioengineering9020074 (accessed on 15th November 2025).

Ku, J. T., & Lan, E. I. (2018). A balanced ATP driving force module for enhancing photosynthetic biosynthesis of 3-hydroxybutyrate from CO_2. *Metabolic Engineering*, *46*, 35–42.

Kumar, P., Maharjan, A., Jun, H. B., & Kim, B. S. (2019). Bioconversion of lignin and its derivatives into polyhydroxyalkanoates: Challenges and opportunities. *Biotechnology and Applied Biochemistry*, *66*(2), 153–162.

Lafferty, R. M., Korsatko, B., & Korsatko, W. (1988). Microbial production of poly-β-hydroxybutyric acid. In *Biotechnology: Special Microbial Processes* (pp. 136–176). UCH.

Lama, L., Nicolaus, B., Calandrelli, V., Manca, M. C., Romano, I., & Gambacorta, A. (1996). Effect of growth conditions on endo- and exopolymer biosynthesis in *Anabaena cylindrica* 10 C. *Phytochemistry*, *42*(3), 655–659. https://doi.org/10.1016/0031-9422(95)00985-X (accessed on 15th November 2025).

Law, K. L. (2017). Plastics in the marine environment. *Annual Review of Marine Science*, *9*, 205–229.

Lee, S. Y. (1996). Bacterial polyhydroxyalkanoates. *Biotechnology and Bioengineering*, *49*(1), 1–14.

Lee, S. Y., Hong, S. H., Park, S. J., Van Wegen, R., & Middelberg, A. P. J. (2001). Metabolic flux analysis on the production of poly(3-hydroxybutyrate). In *J Wiley-VCH* (pp. 249–261).

Lemoigne, M. (1926). Products of dehydration and of polymerization of β-hydroxybutyric acid. *Bulletin of the Chemical Society of Japan*, *8*, 770–782.

Martin, K. (2020). Bioplastics from microalgae: Polyhydroxyalkanoate production by cyanobacteria. In *Handbook of Microalgae-Based Processes and Products* (pp. 597–645).

Mendhulkar, V. D., & Shetye, L. A. (2017). Synthesis of biodegradable polymer polyhydroxyalkanoate (PHA) in cyanobacteria *Synechococcus elongatus* under mixotrophic nitrogen- and phosphate-mediated stress conditions. *Industrial Biotechnology*, *13*(2), 85–93.

Mitra, R., Xu, T., Chen, G., Xiang, H., & Han, J. (2021). An updated overview on the regulatory circuits of polyhydroxyalkanoates synthesis. *Microbial Biotechnology*, *15*, 1446–1470. https://doi.org/10.1111/1751-7915.13915 (accessed on 15th November 2025).

Miyake, M., Erata, M., & Asada, Y. (1996). A thermophilic cyanobacterium, *Synechococcus* sp. MA19, capable of accumulating poly-β-hydroxybutyrate. *Journal of Fermentation and Bioengineering*, *82*(5), 512–514.

Nishioka, M., Nakai, K., Miyake, M., Asada, Y., & Taya, M. (2001). Production of poly-β-hydroxybutyrate by thermophilic cyanobacterium, *Synechococcus* sp. MA19, under phosphate-limited conditions. *Biotechnology Letters*, *23*, 1095–1099.

Orthwein, T., Scholl, J., Spät, P., Lucius, S., Koch, M., Macek, B., & Forchhammer, K. (2021). The novel PII-interactor PirC identifies phosphoglycerate mutase as key control point of carbon storage metabolism in cyanobacteria. *Proceedings of the National Academy of Sciences*, *118*(6), e2019988118.

Osanai, T., Numata, K., Oikawa, A., Kuwahara, A., Iijima, H., Doi, Y., Tanaka, K., Saito, K., & Hirai, M. Y. (2013). Increased bioplastic production with an RNA polymerase sigma factor SigE during nitrogen starvation in *Synechocystis* sp. PCC 6803. *DNA Research*, *20*(6), 525–535. https://doi.org/10.1093/dnares/dst028 (accessed on 15th November 2025).

Osanai, T., Oikawa, A., Numata, K., Kuwahara, A., Iijima, H., Doi, Y., & Hirai, M. Y. (2014). Pathway-level acceleration of glycogen catabolism by a response regulator in the cyanobacterium *Synechocystis* species PCC 6803. *Plant Physiology*, *164*(4), 1831–1841.

Palmeiro-Sánchez, T., O'Flaherty, V., & Lens, P. N. (2022). Polyhydroxyalkanoate bio-production and its rise as biomaterial of the future. *Journal of Biotechnology*, *348*, 10–25.

Panda, B., & Mallick, N. (2007). Enhanced poly-β-hydroxybutyrate accumulation in a unicellular cyanobacterium, *Synechocystis* sp. PCC 6803. *Letters in Applied Microbiology*, *44*(2), 194–198.

Porta, D., Rippka, R., & Hernández-Mariné, M. (2000). Unusual ultrastructural features in three strains of *Cyanothece* (cyanobacteria). *Archives of Microbiology*, *173*, 154–163.

Ray, S., & Kalia, V. C. (2017). Co-metabolism of substrates by *Bacillus thuringiensis* regulates polyhydroxyalkanoate co-polymer composition. *Bioresource Technology*, *224*, 743–747.

Rippka, R., Neilson, A., Kunisawa, R., & Cohen-Bazire, G. (1971). Nitrogen fixation by unicellular blue-green algae. *Archiv für Mikrobiologie*, *76*, 341–348.

Rueda, E., Álvarez-González, A., Vila, J., Díez-Montero, R., Grifoll, M., & García, J. (2022). Inorganic carbon stimulates the metabolic routes related to the polyhydroxybutyrate production in a *Synechocystis* sp. strain (cyanobacteria) isolated from wastewater. *Science of the Total Environment*, *829*, 154691.

Samantaray, S., & Mallick, N. (2012). Production and characterization of poly-β-hydroxybutyrate (PHB) polymer from *Aulosira fertilissima*. *Journal of Applied Phycology*, *24*, 803–814.

Samantaray, S., & Mallick, N. (2014). Production of poly(3-hydroxybutyrate-co-3-hydroxyvalerate) co-polymer by the diazotrophic cyanobacterium *Aulosira fertilissima* CCC 444. *Journal of Applied Phycology*, *26*, 237–245.

Samantaray, S., & Mallick, N. (2015). Role of cultural variables in tailoring poly(3-hydroxybutyrate-co-3-hydroxyvalerate) copolymer synthesis in the diazotrophic cyanobacterium *Aulosira fertilissima* CCC 444. *Journal of Applied Phycology*, *27*, 197–203.

Schlebusch, M., & Forchhammer, K. (2010). Requirement of the nitrogen starvation-induced protein Sll0783 for polyhydroxybutyrate accumulation in *Synechocystis* sp. strain PCC 6803. *Applied and Environmental Microbiology*, *76*(18), 6101–6107.

Senior, P. J., & Dawes, E. A. (1973). The regulation of poly-β-hydroxybutyrate metabolism in *Azotobacter beijerinckii*. *Biochemical Journal*, *134*(1), 225–238.

Sharma, L. (2005). *Studies on accumulation of poly-β-hydroxybutyrate in N_2-fixing cyanobacterium Nostoc muscorum* (PhD thesis). Indian Institute of Technology, Kharagpur.

Sharma, L., & Mallick, N. (2005). Accumulation of poly-β-hydroxybutyrate in *Nostoc muscorum*: Regulation by pH, light–dark cycles, N and P status and carbon sources. *Bioresource Technology*, *96*(11), 1304–1310.

Sharma, L., Panda, B., Singh, A. K., & Mallick, N. (2006). Studies on poly-β-hydroxybutyrate synthase activity of *Nostoc muscorum*. *The Journal of General and Applied Microbiology*, *52*(4), 209–214. https://doi.org/10.2323/jgam.52.209 (accessed on 15th November 2025).

Sharma, L., Singh, A. K., Panda, B., & Mallick, N. (2007). Process optimization for poly-β-hydroxybutyrate production in a nitrogen-fixing cyanobacterium, *Nostoc muscorum* using response surface methodology. *Bioresource Technology*, *98*(5), 987–993.

Shrivastav, A., Mishra, S. K., & Mishra, S. (2010). Polyhydroxyalkanoate (PHA) synthesis by *Spirulina subsalsa* from Gujarat coast of India. *International Journal of Biological Macromolecules*, *46*(2), 255–260.

Sicko-Goad, L. (1982). A morphometric analysis of algal response to low dose, short-term heavy metal exposure. *Protoplasma*, *110*, 75–86.

Singh, A. K., & Mallick, N. (2017). Advances in cyanobacterial polyhydroxyalkanoates production. *FEMS Microbiology Letters*, *364*(20).

Singh, A. K., Sharma, L., Mallick, N., & Mala, J. (2017). Progress and challenges in producing polyhydroxyalkanoate biopolymers from cyanobacteria. *Journal of Applied Phycology*, *29*, 1213–1232.

Singh, M., Kumar, P., Ray, S., & Kalia, V. C. (2015). Challenges and opportunities for customizing polyhydroxyalkanoates. *Indian Journal of Microbiology*, *55*, 235–249.

Sirohi, S., Mallick, N., Sirohi, S. P., Tyagi, P. K., Pg, K., & Simbhaoli, C. (2015). A novel approach to enhancement of poly-β-hydroxybutyrate accumulation in *Aulosira fertilissima* by mixotrophy and chemoheterotrophy. *Research Journal of Pharmaceutical, Biological and Chemical Sciences*, *6*(2), 1266–1270.

Stal, L. J. (1992). Poly(hydroxyalkanoate) in cyanobacteria: An overview. *FEMS Microbiology Reviews*, *9*(2–4), 169–180.

Stal, L. J., & Moezelaar, R. (1997). Fermentation in cyanobacteria. *FEMS Microbiology Reviews*, *21*(2), 179–211. https://doi.org/10.1016/S0168-6445(97)00056-9 (accessed on 15th November 2025).

Stal, L. J., Heyer, H., & Jacobs, G. (1990). Occurrence and role of poly-hydroxyalkanoates in the cyanobacterium *Oscillatoria limosa*. In E. A. Dawes (Ed.), *Novel Biodegradable Microbial Polymers* (pp. 435–438). NATO ASI Series, E 186. Kluwer Academic Publishers.

Steinbüchel, A. (1992). Biodegradable plastics. *Current Opinion in Biotechnology*, *3*, 291–297.

Sudesh, K., Taguchi, K., & Doi, Y. (2002). Effect of increased PHA synthase activity on polyhydroxyalkanoates biosynthesis in *Synechocystis* sp. PCC6803. *International Journal of Biological Macromolecules*, *30*(2), 97–104. https://doi.org/10.1016/s0141-8130(02)00010-7 (accessed on 15th November 2025).

Suzuki, T., Miyake, M., Tokiwa, Y., Saegusa, H., Saito, T., & Asada, Y. (1996). A recombinant cyanobacterium that accumulates poly-(hydroxybutyrate). *Biotechnology Letters*, *18*, 1047–1050.

Taepucharoen, K., Tarawat, S., Puangcharoen, M., Incharoensakdi, A., & Monshupanee, T. (2017). Production of poly(3-hydroxybutyrate-co-3-hydroxyvalerate) under photoautotrophy and heterotrophy by non-heterocystous N_2-fixing cyanobacterium. *Bioresource Technology*, *239*, 523–527.

Takahashi, H., Miyake, M., Tokiwa, Y., & Asada, Y. (1998). Improved accumulation of poly-3-hydroxybutyrate by a recombinant cyanobacterium. *Biotechnology Letters*, *20*(2), 183–186.

Tarawat, S., Incharoensakdi, A., & Monshupanee, T. (2020). Cyanobacterial production of poly(3-hydroxybutyrate-co-3-hydroxyvalerate) from carbon dioxide or a single organic substrate: Improved polymer elongation with an extremely high 3-hydroxyvalerate mole proportion. *Journal of Applied Phycology*, *32*, 1095–1102.

Tyo, K. E., Jin, Y. S., Espinoza, F. A., & Stephanopoulos, G. (2009). Identification of gene disruptions for increased poly-3-hydroxybutyrate accumulation in *Synechocystis* PCC 6803. *Biotechnology Progress*, *25*(5), 1236–1243.

Verlinden, R. A., Hill, D. J., Kenward, M. A., Williams, C. D., & Radecka, I. (2007). Bacterial synthesis of biodegradable polyhydroxyalkanoates. *Journal of Applied Microbiology*, *102*(6), 1437–1449.

Vincenzini, M., Sili, C., De Philippis, R., Ena, A., & Materassi, R. (1990). Occurrence of poly-beta-hydroxybutyrate in *Spirulina* species. *Journal of Bacteriology*, *172*(5), 2791–2792.

Wang, B., Xiong, W., Yu, J., Maness, P. C., & Meldrum, D. R. (2018). Unlocking the photobiological conversion of CO_2 to (R)-3-hydroxybutyrate in cyanobacteria. *Green Chemistry*, *20*(16), 3772–3782.

Wu, G. F., Wu, Q. Y., & Shen, Z. Y. (2001). Accumulation of poly-β-hydroxybutyrate in cyanobacterium *Synechocystis* sp. PCC6803. *Bioresource Technology*, *76*(2), 85–90.

Ye, J., Hu, D., Che, X., Jiang, X., Li, T., Chen, J., & Chen, G. Q. (2018). Engineering of *Halomonas bluephagenesis* for low-cost production of poly(3-hydroxybutyrate-co-4-hydroxybutyrate) from glucose. *Metabolic Engineering*, *47*, 143–152.

Zhang, S., & Bryant, D. A. (2011). The tricarboxylic acid cycle in cyanobacteria. *Science*, *334*(6062), 1551–1553.

Zhang, S., & Bryant, D. A. (2015). Biochemical validation of the glyoxylate cycle in the cyanobacterium *Chlorogloeopsis fritschii* strain PCC 9212. *Journal of Biological Chemistry*, *290*(22), 14019–14030.

Zhang, S., Qian, X., Chang, S., Dismukes, G. C., & Bryant, D. A. (2016). Natural and synthetic variants of the tricarboxylic acid cycle in cyanobacteria: Introduction of the GABA shunt into *Synechococcus* sp. PCC 7002. *Frontiers in Microbiology*, *7*, 1972.

CHAPTER 5

Microalgae as a Promising Platform for the Production of Eco-Friendly Biodiesel: Recent Advances, Bottlenecks, and Solutions

PRITI PAL,[1] SAURABH SINGH RATHORE,[2,3] BRIJESH PANDEY,[3] ANAND PRAKASH,[3] RAM PRASAD,[4] PRAKASH KUMAR SARANGI,[5,*] and AKHILESH KUMAR SINGH[3,*]

[1]*Amity Institute of Biotechnology, Amity University Uttar Pradesh, Lucknow Campus, Lucknow, Uttar Pradesh, India*

[2]*Department of Life Sciences, Inter University Centre for Teacher Education, Bananas Hindu University, Varanasi, Uttar Pradesh, India*

[3]*Department of Biotechnology, School of Life Sciences, Mahatma Gandhi Central University, Motihari, Bihar, India*

[4]*Department of Botany, School of Life Sciences, Mahatma Gandhi Central University, Motihari, Bihar, India*

[5]*College of Agriculture, Central Agricultural University, Imphal, Manipur, India*

*Corresponding author

ABSTRACT

Biodiesel made from microalgae is now being mass-produced and sold as a viable alternative to petroleum-based feedstock due to advances in industrial production. Microalgae have a high growth rate and high carbon sequestration, and they can be grown without the need for arable

Industrial Microbiology: Current Progress and Novel Technologies.
Prakash Kumar Sarangi, Akhilesh Kumar Singh, Rajesh Kumar Srivastava, and Brijesh Pandey (Eds.)

DOI: 10.1201/9781779642219-5

soil in both fresh and salt water. Recent improvements in the cultivation, harvesting, pretreatment, lipid extraction, and transesterification subsystems are bringing microalgal biodiesel closer to being able to be used as a fuel. Although promising, microalgal platforms have yet to be adopted for bioenergy production on an industrial scale due to a number of obstacles in the form of technical and economic hurdles. Challenges in microalgae production processes for bioenergy purposes are outlined in this review, along with techno-economic assessments of the cultivation, harvesting, and downstream processes necessary to produce biodiesel. Microalgae cultivation, lipid accumulation, microalgae harvesting, and lipid extraction have all been summed up in the section on influencing parameters for biodiesel production. The goal of this analysis is to improve the economics and technology of making biodiesel from microalgae and to help come up with efficient and marketable technology to do this.

5.1 INTRODUCTION

Biodiesel production is one of the most impressive applications of micro-algae, a type of biomass feedstock. They are well-known for their high lipid content, ability to reduce carbon dioxide emissions, quick rate of growth, and cultivation using non-arable land (Zhang et al., 2020). Due to these characteristics, microalgae stand out as a superior feedstock option. Also, unlike food crops, microalgae cannot be eaten, so it is a promising alternate feedstock to food crops (Hou et al., 2021). Algae have many benefits as a biofuel source, including rapid growth (a biomass doubling time of 1–6 days) and oil production (10–20 times more oil per hectare per year) (Li and Guo, 2017). The solar energy conversion capacity is about 4.5%, and the photosynthetic rate is about 6.9 x 10^4 cells/ml/h, with up to 50% algal biomass constituents of carbon (Zeng et al., 2021). Pyrolysis (thermal cracking), micro-emulsification, and transesterification are the three primary processing techniques for transforming oils from feedstock into biodiesel (Sanghamitra et al., 2020). Transesterification is the most common and widely used production method for biodiesel, and it has been shown to work effectively with a wide range of feedstocks (Zhang et al., 2020). Renewable energy production is very appealing and a promising source of energy due to the extensive biomass resource and its advantages over fossil fuels. Microalgae have many possible engineering applications. Wastewater treatment is one example. Using bioremediation techniques to apply algae to waste water treatment is an effective method of pollution control that can get rid of nitrates, phosphates, and metals like cadmium, zinc, nickel, and lead. Low concentrations of nitrogen and phosphorus in effluent can be achieved by combining algae and wastewater treatment. Biomass and oils derived from wastewater treatment sludge can be harvested for further processing (Su et al., 2022). This method does not generate any unnecessary byproducts that could harm the ecosystem. Sewage treatment, preventing eutrophication, creating bio-fertilizers, and reducing Carbon dioxide (CO_2) emissions from power plants are all examples of its many uses (Hou et al., 2021). The capture of CO_2 is one of the most pressing issues facing governments and businesses around the world today. CO_2 emissions can be reduced more sustainably through the use of microalgae because of their unique ability to absorb CO_2 during the growth process (Sanghamitra et al., 2020). Microalgae can be used to capture CO_2 from industrial processes that produce significant quantities of gas flue, and they can do so while also producing other, beneficial byproducts. We will look into the many ways of producing and processing microalgal biomass in order to create functional biodiesel production. Several methods now being used by scientists and industry to cultivate microalgal biomass for biodiesel production are depicted in Figure 5.1. Due to their high carbohydrate and lipid content, quick growth rate, and tolerance to variable climatic conditions, microalgae are attracting a lot of attention as a potential feedstock (Li and Guo, 2017). Technology advancements in the areas of genome sequencing, genetic transformation, gene targeting, promoters, selection markers, and molecular biology tools like clustered regularly interspaced short palindromic repeats (CRISPR), transcription activator-like effectors (TALEs), and zinc-finger nucleases (ZFN) have paved the way for the discovery of previously unknown metabolic pathways within algal cells, allowing for the design and synthesis of novel biological systems. Moreover, it has been experimentally demonstrated that a wide variety of microalgal species produce over 50% of their biomass as lipids, primarily triglycerides, the raw material for biodiesel production (Jung et al., 2021). Lipid types and fatty acid profiles are also significantly influenced by the microalgal strain and the cultivation method (Asadi et al., 2020). For biodiesel production, for example, the simplest and most widely used method for inducing the lipid accumulation on the microalgae cells follows an initial period of growth under optimal conditions (Renuka et al., 2018). However, there are still obstacles to be overcome before strategies can be developed to increase biodiesel production and commercialization, despite the promising nature of these organisms (Zhang et al., 2018). Growth-related characteristics, biomass yield, and nutrient competition ability are all examples of areas where tradeoffs exist. A sustainable microalgal biodiesel production system also requires a constant supply of nutrients and the recycling of water after biomass dewatering (Sakai et al., 2009). In this chapter, we give an overview of the current state of making biodiesel from microalgae and talk about the main problems that need to be solved. The review also summarizes various approaches to valorizing microalgal biomass with the goal of increasing the commercial viability of microalgae for biodiesel production.

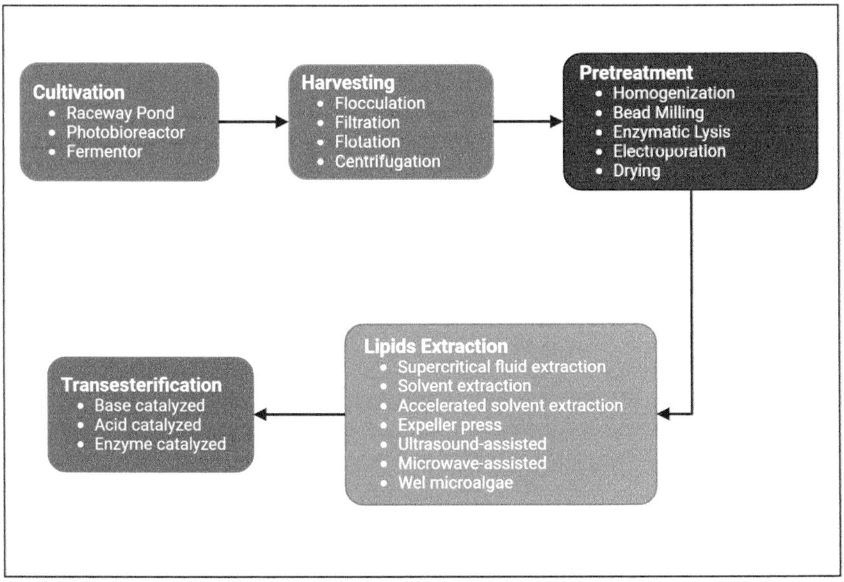

FIGURE 5.1 Overview of the microalgae to biodiesel production process.

5.2 MICROALGAE AS SUITABLE SOURCES OF LIPID PRODUCTION: PRECURSOR FOR BIODIESEL

Microalgae are unicellular, microscopic microorganisms that have been present on Earth for centuries. Microalgae's photosynthetic process is analogous to plants. Since it is so easily able to absorb water, carbon dioxide, and other nutrients when submerged in water, it can efficiently convert solar energy into biomass resources. Freshwater microalgae can be found in a variety of natural conditions, such as rivers, lakes, reservoirs, creeks, ponds, marshes, and soils, while marine microalgae are found in saltwater environments (Singh et al., 2020). Many applications of microalgae include food, cosmetics, aquaculture, environmental testing, wastewater treatment, and energy production. Microalgae have a high lipid content, so they can be harvested and converted (esterified) into useful motor fuel. Due to their increased photosynthetic efficiency and biomass production compared to alternative sources (Zhu et al., 2016), microalgae have been proposed as a viable fuel feedstock. In order to survive in their natural environments, microorganisms have developed a wide variety of strategies. Microscopic plants called microalgae convert water, carbon dioxide, and sunlight into chemical energy and produce essential organic compounds like carbohydrates, lipids, and proteins. Large quantities of triacylglycerols (TAGs) used as biofuel storage can be produced by microalgae under photooxidative stress or other extreme environmental circumstances (Hu et al., 2008). Also, they produce many different kinds of hydrocarbons. It has been discovered that certain algal species grow rapidly, produce large amounts of TAG or oil, and are thus classified as oleaginous. Some researchers have proposed using these algae as a potential source of cell factories. Since microalgae are responsible for 40% of global carbon fixation and can contain up to a maximum of 70% oil (dry weight) in some species shown in Table 5.1 (Chu, 2017), biofuels derived from microalgae have the potential to reduce greenhouse gas emissions. Several strategies and technologies have been explored to increase microalgae's fatty acid production rates and make the process more sustainable and scalable (Chen et al., 2021). Since lipids are used to construct cellular membranes and are essential for the development of other biomolecules (Leu et al., 2014), it is crucial to understand that lipid production in microalgal cells serves a purpose beyond energy storage. Despite these benefits, commercial microalgal exploitation for biodiesel production is still not feasible. Much research and development has been carried out on these species in both academic and commercial contexts. Non-polar lipids, such as triglycerides (TAGs), are essential for energy storage in microalgae. Polar lipids, such as glycerophospholipids, play an important role in cell structure. Structural lipids contain long chains of fatty acids, and these fatty acids are converted into polyunsaturated fatty acids

(PUFAs) like docosapentaenoic acid (DPA), eicosapentaenoic acid (EPA), and docosahexaenoic acid (DHA) (Shokravi et al., 2020). These lipids are directly involved in intracellular membrane fusion events and play a dynamic role in optimizing biosynthetic processes and membrane fluidity. Their role in signaling pathways and cellular responses to environmental changes is also crucial. Among storage lipids, TAGs have the highest ratio of saturated to unsaturated fatty acids (Thuy et al., 2022). The metabolic energy needs are met by the efficient catabolization of TAGs. Polar lipids are synthesized primarily in the presence of light, stored in cytosolic lipid bodies, and used when light is unavailable. The chloroplast contains a large number of phosphoglycerides and a large number of glycosylglycerides, including mono-galactosyldiacylglycerol, digalactosyldiacylglycerol, and sulfoquinovosyldiacylglycerol (Ahorsu et al., 2018). These proteins reside in membranes, including the plasma membrane and several endoplasmic reticulum membrane systems (Shokravi et al., 2020).

TABLE 5.1 Microalgae with Lipid Contents >50% (Dry Cell Weight) Have Been Reported Under Controlled Laboratory Conditions

Microalgae Species	Culture Condition	Lipid Content (wt.%)	Lipid Productivity (mg Lipids L^{-1} d^{-1})	References
Scenedesmus obliquus	Nitrogen and phosphorus limitations in presence of thiosulphate	58	5.6–58.6	Mandal and Mallick (2009)
Choricystis minor	Nitrate and phosphate limitations	60	14.8–45.3	Sobczuk and Chisti (2010)
Nannochloropsis sp.	15% CO_2 sparging under nitrogen starvation and high light intensity	62	9.0–19.0	Jiang et al. (2011)
Chlorella vulgaris	Nitrogen, iron, and phosphorus limitations	57	16.6–50.0	Mallick et al. (2012)
Scenedesmus sp. strain R-16	Nitrogen deficiency	53	–	Ren et al. (2013)
Nannochloropsis oceanica	High light intensity and nitrogen depleted culture conditions	49.0–53.2	100.2–158.8	Ma et al. (2014)
Chlorella vulgaris	Municipal wastewater utilization with nitrogen-limitation approach	63	33	Robles-Heredia et al. (2015)
Trentepohila arborum	High light while nitrogen-replete conditions	26.3	43.6	Chen et al. (2016)

TABLE 5.1 *(Continued)*

Microalgae Species	Culture Condition	Lipid Content (wt.%)	Lipid Productivity (mg Lipids L^{-1} d^{-1})	References
Chlamydomonas reinhardtii	Nitrogen depletion	63–94	14.4–33.8	Tan and Lee (2017)
Acutodesmus obliquus	Nitrogen-limitation	49.07–77.20	11.3–23.8	Renuka et al. (2018)
Chlamydomonas reinhardtii.	Nitrogen-deficient medium	64.25	80.92	Shin et al. (2019)
Chlorella sorokiniana strain pa.91	Nutrients-rich media of dairy wastewater	36.67	41.3–50.9	Asadi et al. (2020)
Euglena gracilis	High dose of nutrients (glucose, nitrogen, and phosphorus) under the oxygen-free environment	24.81	131	Jung et al. (2021)
Chlorella sorokiniana strain P21	Cassava wastewater	54.11–61.52	21–25	Padri et al. (2022)

5.3 MICROALGAL PATHWAY INVOLVED IN PRODUCTION OF LIPID

Triacylglycerol (TAG) synthesis in microalgal cells occurs within the chloroplasts and the endoplasmic reticulum, whereas fatty acid synthesis takes place in the plastids (Figure 5.2). TAGs are lipids utilized for energy and carbon storage but, unlike phospholipids in biological membranes, do not fulfill structural roles in cells. Microalgal cells employ either the Kennedy pathway or the monoacylglycerol pathway to synthesize TAG. Both pathways involve the esterification of acetyl-CoA with the hydroxyl groups on glycerol, which is the fundamental process of TAG synthesis. Fatty acid biosynthesis commences with the carboxylation of acetyl-CoA, a reaction catalyzed by acetyl-CoA carboxylase (ACCase), resulting in malonyl-CoA. The activity of ACCase is a key regulatory point in fatty acid synthesis. Acyl chain elongation in microalgal cells begins with the transfer of malonyl-CoA to the acyl carrier protein (ACP). Multiple fatty acid synthase subunits catalyze the synthesis of C16 or C18 products. The fatty acid elongation process is terminated by two distinct enzymatic activities. First, acyl groups are removed from ACP by chloroplast acyltransferase; subsequently, ACP transfers the newly synthesized fatty acid directly to glycerol-3-phosphate (G-3-P). Additionally, free fatty acids (FFA) are produced via the hydrolysis of acyl-ACP by acyl-ACP thioesters. Prior to glyceride production, FFA must be exported from the chloroplast. In the Kennedy pathway, the initial acyl group is esterified to glycerol-3-phosphate (Reddy et al., 2010). The second reaction is phospholipid synthesis. The dephosphorylation of phospholipid to diacylglycerol (DAG) by phosphatase is the initial step in the pathway leading to triglyceride (TG) formation. The monoacylglycerol pathway is initiated by two monoacylglycerols (sn2-MAG), and monoacylglycerol 2 acyltransferase converts sn2-MAG to diacylglycerol (DAG) (MGAT). TAG biosynthesis depends on acyl-CoA: DAG acyltransferase, an enzyme that catalyzes the conversion of newly synthesized DAG to TAG. In the diacylglycerol acyltransferase (PDAT)-catalyzed pathway for TAG synthesis, phospholipids or galactolipids serve as potential substrates, with the acyl donor being a phospholipid (Dahlqvist et al., 2000). While PDAT activity has been observed in certain plant species, its role in TAG synthesis varies considerably among species. The identification of the gene encoding PDAT has revealed that microalgae possess a mechanism for TAG production independent of acetyl-CoA. Fatty acid methyl esters (biodiesel) are generated through a process known as TAG transesterification (Figure 5.2).

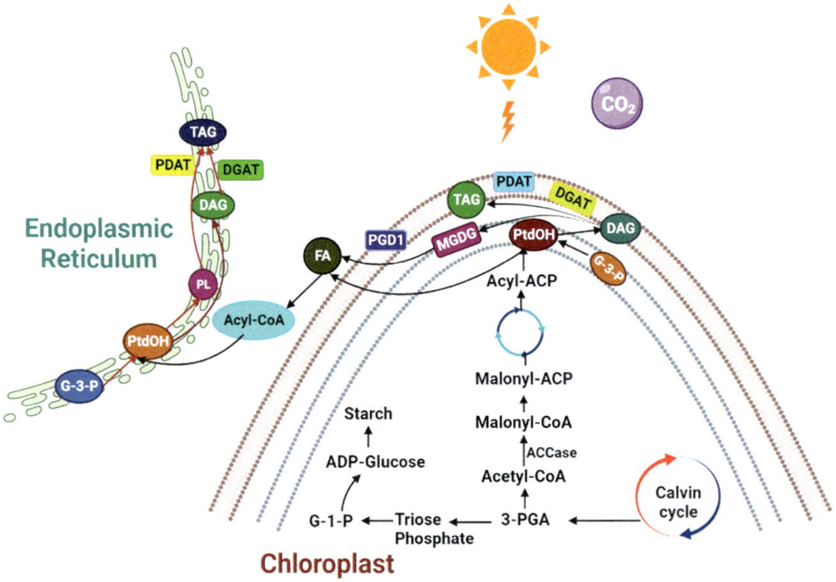

FIGURE 5.2 Triacylglyceride (TAG) synthetic pathway in microalgae cell.

5.4 MICROALGAL BIODIESEL PRODUCTION PROCESSES

As microalgae biodiesel production research has progressed, developments in microalgae cultivation systems, cell disruption, transesterification, and lipid extraction technology have emerged as central concerns (Chhandama et al., 2021).

5.4.1 CELL DISRUPTION

To maximize the wet extraction recovery of intracellular products, the algal biomass must be treated in the cell disruption unit with the right technology. In this study, we considered the possibility of using high-pressure homogenization. This approach was chosen because it is a mature technology that has been tested and refined at both the laboratory and industrial scales (Theegala, 2015). Cell disruption and lysis using pressure homogenization should be performed on the biomass with a 25 wt.% from the harvesting processes before sending it on to extraction (Davis et al., 2011). It was estimated that pressure homogenization would use 0.20 kWh/kg of dry biomass with an efficiency of 90% (equivalent to a 25% wt.% input) (Davis et al., 2012). It is hypothesized that the homogenizer's undisturbed algae flow through the extraction (without lipid recovery) and into the digester, where they are broken down along with the rest of the residue.

5.4.2 LIPID EXTRACTION

Extracting lipids from algae is typically done by processing either wet algal paste or dry algal cake, with or without cell disruption (Theegala, 2015). In this research, wet algal paste is subjected to solvent extraction following pretreatment or cell disruption in order to remove lipid (Figure 5.3). Ethanol is the standard solvent for lipid extraction. Ethanol's polarity makes it ideal for this purpose, as it allows it to dissolve polar lipids in the cell membrane and thus release more cell material for extraction (Lohrey and Kochergin, 2012). Moreover, ethanol is readily available in the production facility and has a low toxicity (ethanol is produced from cane molasses in Metahara factory). The present study assumed a solvent to dry biomass ratio of 5:1 (w/w), as was done in previous extraction studies (Nappa et al., 2015). The literature Halim et al. (2011) reports a lipid recovery rate of up to 97%. However, an 80% lipid recovery is assumed here for the sake of the study. Centrifugation in a disc stack is assumed to separate the lipid-rich solvent from the algae residue slurry (Davis et al., 2011). After the anaerobic digester (AD) has finished producing biogas from the algal waste, the oil-solvent solution can be sent to a stripping column, where the ethanol can be recovered from the oil and reused to produce a lipid stream that is 99.50% pure (Davis et al., 2011). To extract the biomass, 0.28 kWh/kg of dry biomass in electricity is needed according to some estimates (Davis et al., 2012), while 1.30 kWh/kg of dry biomass in thermal energy is taken into account (Delrue et al., 2012). Lipid loss in the stripper is estimated to be 5 wt.% and solvent loss in circulation to be 5.20 g ethanol/kg of oil (Nappa et al., 2015).

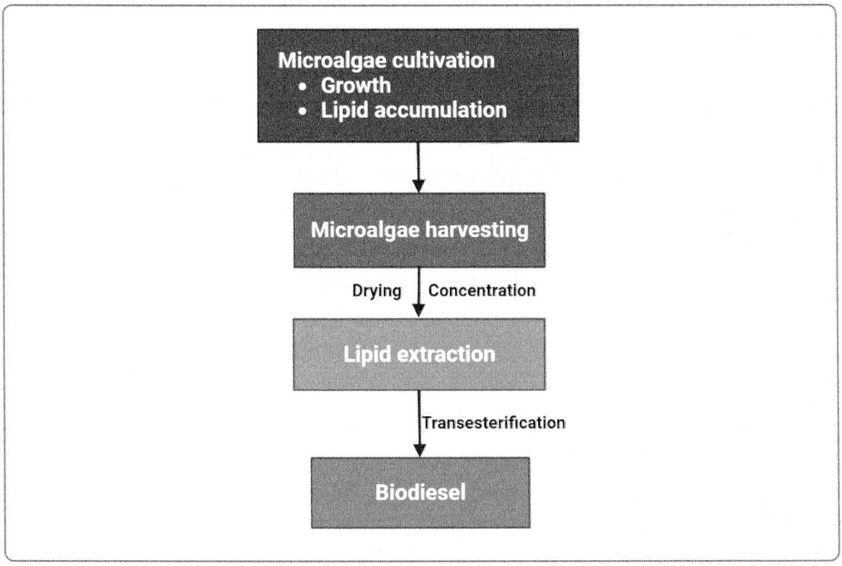

FIGURE 5.3 Production process of microalgal biodiesel.

5.4.3 TRANSESTERIFICATION

Once the lipids have been extracted, they can be transported and converted into biodiesel via transesterification using methanol and sodium hydroxide as catalysts (1 wt.%) (Delrue et al., 2012). It was assumed that the mass ratio of methanol to fatty acids in the reactor would be 1:10 (Delrue et al., 2012). We assumed a base conversion rate of 80% (wt.%) in the reactor from oil to biodiesel (Pokoo-Aikins et al., 2010). Other research shows that the microalgae have an extremely low free fatty acid content, at around 0.05%. As a result of this consideration, prior treatment is unnecessary for the current investigation. The glycerol produced during transesterification was supposed to be sent to the AD after being decanted to a purity of 85% glycerol and 15% methanol (wt.%) (Figure 5.4) (Andersson et al., 2011). Approximately 0.10 kg of glycerol was predicted to be produced for every 1 kg of biodiesel (Pokoo-Aikins et al., 2010). It was assumed that the unused methanol was distilled off and reused in the reactor. The final fatty acid methyl ester (FAME) purity was determined to be 96.50% (wt%). It was predicted that the FAME would contain 0.50 wt.% water, 0.24 wt.% glycerol, and 0.20 wt.% methanol (Atadashi et al., 2011). It was estimated that 3.80 x 10^4 joules of electrical energy and 0.68 kWh/kg of converted oil would be needed for the transesterification (Delrue et al., 2012). Biodiesel was calculated to have a density of 900 kg/m^3 and an energy content of 42 MJ/kg (Delrue et al., 2012).

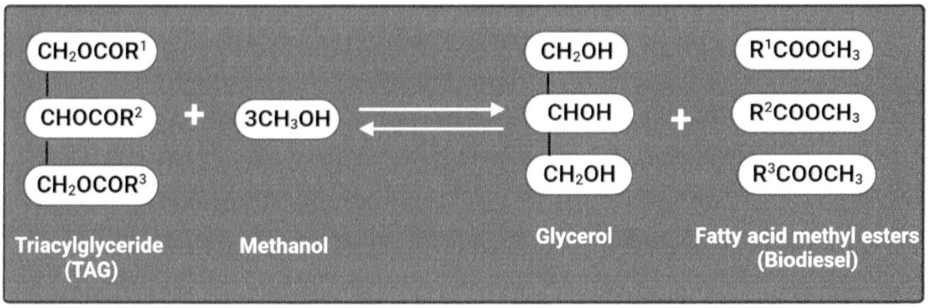

FIGURE 5.4 Triacylglyceride (TAG) is used to synthesize biodiesel through transesterification.

5.5 VARIOUS STRATEGIES TO INCREASE LIPID ACCUMULATION IN MICROALGAE

Upstream microalgal cultivation (growth and lipid accumulation) and downstream microalgal harvesting and lipid extraction are central to studies of microalgal biodiesel production. Because of how well established the technology is to transform lipids into biodiesel, it will not be discussed in detail here. Lipid, biodiesel's starting point, is a hot topic in the scientific community right now. Microalgal lipids' fatty acid components are pivotal to biodiesel quality, which in turn affects the engine's efficiency at burning fuel and producing heat. To confirm the usefulness of microalgal biodiesel and its potential to partially replace fossil fuels, microalgae under stress produce lipids primarily consisting of neutral fatty acids with a low degree of saturation (no more than two unsaturated bonds in most cases) (Song et al., 2022). In addition to being readily available and renewable, biodiesel also has excellent transportability, a high combustion efficiency, a low sulfur and aromatic content, a high cetane number, and sufficient biodegradability. When compared to standard diesel fuels, it provides superior performance in terms of both safety and energy balance. Below, we will go over the variables that come into play during each phase of making biodiesel from microalgae (Chhandama et al., 2021).

5.5.1 MICROALGAE CULTIVATION

The amount of lipid harvested from microalgae can be altered by changing how they are grown. Microalgal growth is affected by a number of environmental variables. These include the availability of light, carbon sources, nutrients (such as nitrates, phosphates, and carbohydrates), and trace elements (such as Mn, Co, Zn, and Mo) (Aziz et al., 2020). Temperature, pH, salinity, oxygen removal, and carbon dioxide uptake are additional factors.

5.5.1.1 LIGHT

Both light and dark reactions are necessary for photosynthesis to occur. During the photosynthesis process, oxygen is produced as a byproduct of capturing and converting light energy into an energy carrier (Sung et al., 2022). Electrons from oxygen can be transferred to photosystems I (PSI) and II (PSII) (Ranadheer et al., 2019). The productivity of microalgae can be lowered by photo-oxidation of PSII components brought on by intense light (Arun et al., 2021). Microalgae typically absorb and use natural light with wavelengths ranging from 400 to 700 nm (43% of solar energy), so an optimal light intensity is one of the crucial parameters of microalgal growth (Teo et al., 2014).

5.5.1.2 TEMPERATURE

The growth rate of microorganisms is significantly influenced by environmental temperature. Microalgae benefit from an increase in temperature for CO_2 uptake and fixation, but at very high temperatures, their respiratory metabolism is stifled (Barten et al., 2021). Temperatures between 20 and 30°C are generally optimal for growing most microalgae species. Some freshwater microalgae are able to thrive in a wide temperature range (from 5–35°C), but in industrial cultivation, the ideal temperature range of 25 to 30°C must be maintained (Zhong et al., 2022). Therefore, the ideal temperature for microalgal growth should be maintained to avoid negative effects on microalgal cell metabolism and a decrease in lipid accumulation.

5.5.1.3 SALINITY, NUTRIENTS, AND PH

Various species of microalgae have different salinity, nutrient, and pH requirements. Microalgae need water with just the right amount of salt in it to thrive. While high salinity typically inhibits microalgal growth, it is possible to increase the salinity of the culture medium by adding NaCl and Na_2SO_4. It is important to note that microalgae species vary greatly in their salinity preferences. Marine microalgae, for instance, can thrive in saltier water than their freshwater counterparts (Maity and Mallick, 2022). Microalgae require a wide variety of nutrients for growth, including carbon, oxygen, hydrogen, nitrogen, potassium, magnesium, calcium, iron, sulfur, and phosphorus (Chew et al., 2018). Microalgae rely on nitrogen and phosphorus (N and P) for growth, with some species requiring more specialized nutrients, like silicon (Si) for diatoms. Enzyme activity, phosphorus availability, ammonia toxicity, and inorganic carbon availability are all controlled by the pH of the culture medium, making it an essential factor in microalgal growth. Growing microalgae causes the pH of the culture medium to rise, making it more alkaline. Microalgae readily capture CO_2 from the air and use it to produce biomass as photosynthesis accelerates and OH-ions accumulate (Isiramen et al., 2022). The permeability of microalgal cells and the shape of hydrated hydrogen ions in inorganic compounds are both affected by shifts in pH.

5.5.1.4 CULTIVATION MODES

Microalgae can use light and inorganic materials (CO_2, water, and inorganic salts) for photosynthesis and organic material synthesis in a process called "autotrophic mode" (Ruiz et al., 2022). Microalgae can grow in the dark using external carbon sources when they are in heterotrophic mode. When set to mixotrophic mode, microalgae are given access to both inorganic and organic sources of carbon. Both biomass and lipid yields were dramatically increased in mixotrophic mode compared to autotrophic mode. Microalgae also exhibit quick cell division rates when growing in mixotrophic mode. Microalgae's photosynthetic efficiency is affected by a number of factors, one of which is carbon dioxide (CO_2). An increase in CO_2 is good for enhancing photosynthetic efficiency, which in turn increases biomass yield (Cheng et al., 2022).

5.5.2 LIPID ACCUMULATION

Lipid accumulation in microalgae is affected by a wide variety of environmental factors including nutrient availability, carbon source, salinity, light intensity, temperature, and metal ions (Ma et al., 2022).

5.5.2.1 NUTRIENTS

In microalgae, nitrogen (N) plays a crucial role in cellular structure because it is required for the production of proteins, amino acids, nucleic acids, enzymes, and photosynthetic pigments (Gao et al., 2019). The nutrient phosphorus (P) plays a crucial role in controlling cell growth and metabolism. It is crucial to the processes of energy transfer, signal transduction, macromolecule biosynthesis, photosynthesis, and cellular respiration, among many others (Wu et al., 2021). Microalgae take in phosphorus in the form of inorganic phosphate, which is then converted to orthophosphate by phosphatase on the cell surface. Microalgae store phosphorus in the form of polyphosphate particles when there is an abundance of the nutrient, allowing them to continue growing even when phosphorus is scarce (Zhang et al., 2021).

5.5.2.2 SALINITY

Salinity affects both the growth of microalgae and their resistance to infection by microorganisms like bacteria. This is because microalgae accumulate lipids. Increased intracellular osmotic pressure caused by high salinity triggers physiological and biochemical responses (Yang et al., 2022). The fluidity and permeability of the microalgal cell membrane are affected by the high salinity, allowing for the controlled entry and exit of ions through the cell membrane and the activation of the accumulation of osmoprotective solutes and stress proteins. Salinity affects the cellular structure and metabolic processes of microalgae, which also promote lipid accumulation (Zafar et al., 2021).

5.5.2.3 METAL IONS

The accumulation of lipids and carbohydrates in microalgae, in addition to the growth rate, are significantly impacted by trace metal elements. Abd El Baky et al. (2012) discovered that adding bioavailable Fe^{3+} to the culture medium can aid in the growth and storage of fats in microalgae, although they are stymied by both high and low concentrations. Intense amounts of Mg^{2+} can stimulate acetyl coenzyme A carboxylase (ACCase), leading to increased lipid accumulation (Qian et al., 2021). Lipid synthesis in microalgae is more tightly regulated by Ca^{2+} than by other cations involved in signal transduction (Di Caprio et al., 2018). Heavy metals like Cd, Cu, and Zn can affect the lipid metabolism in microalgae, which can result in increased lipid accumulation (Urrutia et al., 2019).

5.5.2.4 TEMPERATURE

Inevitably, microalgae's growth conditions will shift as a result of seasonal temperature fluctuations, daily temperature changes, and sudden temperature changes for a variety of reasons (Xin et al., 2011). When microalgae are under stress from high or low temperatures, changes in their metabolic pathways cause them to store lipids (Rehman et al., 2022).

5.5.2.5 LIGHT

Since microalgae are photosynthetic organisms, sunlight is crucial to their development. However, microalgae can only take in a small amount of natural light, specifically between 400 and 700 nm (about 43% of solar energy) (Yan et al., 2016). Consequently, the most important factor influencing or even controlling microalgae growth and lipid accumulation is light, specifically light of the right wavelength and intensity.

5.5.3 MICROALGAE HARVESTING

The most challenging part of making biodiesel from microalgae is the harvesting process, which entails isolating the microalgae from the culture medium. Over-shading occurs when there are so many microalgae in a given area that not enough light can reach the ground, resulting in a decrease in productivity. This means that a lot of water needs to be removed in order to harvest microalgae, as the biomass concentration of microalgae in culture medium is typically kept low (0.5–5 g/L) (Liu et al., 2018). The small cell size (2–20 m) and stable colloid in the suspension of microalgae further complicate the harvesting process. Thus, the sedimentation efficiency of microalgae cells depends on microalgae cell size, cation concentration in the culture medium, and the microalgal cell's own extracellular secretions (Ortiz et al., 2021).

5.5.3.1 CELL SIZE

As it stands, most research into microalgal flocculation is limited to single species grown in controlled environments. The surface properties of the microalgae cell, which change with species and cultivation conditions, are the primary determinants of flocculation. As microalgal cells get smaller, their specific surface area goes up (Min et al., 2022). This means that the dose of flocculant needed to effectively remove microalgae increases in direct proportion to the size of the cells involved. The relatively large microalgae cell, meanwhile, aids in the sedimentation process.

5.5.3.2 CATION CONCENTRATION

There is a correlation between the flocculation efficiency of microalgae and the concentrations of divalent cations (such as Ca^{2+} and Mg^{2+}) and trivalent cations (such as Al^{3+} and Fe^{3+}). Both types of molecules contribute to microalgal self-flocculation, but the latter primarily binds to the negatively charged extracellular polymeric substance (EPS) on the microalgal surface cells (Choi et al., 2020).

5.5.3.3 pH

The value of the pH must be carefully considered. Microalgae cell surface charge and the charge of chemical flocculants are both influenced by the pH. As the pH of the surrounding environment rises, the negative charge on the surface of the microalgae cell decreases. Microalgae flocculate and settle out when the pH rises above 10, because their "zero charge" surface transitions from a homogeneous stable phase to an inhomogeneous unstable phase (Liu et al., 2014a). At higher pH levels, Mg(OH), $CaCO_3$, and $Ca_3(PO_4)_2$ begin to precipitate.

5.5.3.4 CO_2

Self-flocculation may occur if microalgae, which consume CO_2 through photosynthesis, raise the pH of their environment. Commercial microalgae culture units commonly use carbon dioxide to increase microalgae yield. An interesting new microalgal technology is the ability to control microalgal self-flocculation with CO_2. The efficiency with which microalgae absorb HCO_3- depends heavily on their CO_2 concentration mechanism (CCM) (Liu et al., 2014b). Self-flocculation of microalgae occurs when the pH rises above 10, allowing most microalgae to continue photosynthesis in this range (Tran et al., 2017).

5.5.3.5 EXTRACELLULAR POLYMERIC SUBSTANCE

Proteins, polysaccharides, humic substances, and nucleic acids make up the majority of EPS. They are made when microalgae sporulate. Self-flocculation describes the process by which microalgal cells connect to one another and form large aggregates with the help of EPS. Additionally, microalgal cell extracellular secretions and metal cations in the culture medium can interact to promote microalgal sedimentation (Wang et al., 2022). By binding to the negatively charged carboxyl groups in the cell wall and the carboxyl and hydroxyl groups in the extracellular secretion, the positively charged metal cations form flocs. Microalgal extracellular secretions vary according to the type of nutrients present, the microalgae's developmental stage, and their external environment (Taghavijeloudar et al., 2021).

5.5.4 LIPID EXTRACTION

Proteins, cellulose, hemicellulose, and pectin make up the bulk of a microalgae cell wall. They present the primary difficulty in extracting lipids from microalgae because they are thick and hard. The resistance of microalgal cells to mechanical fragmentation is directly proportional to the mechanical strength and thickness of the cell wall (e.g., homogenization under high pressure). As a result, environmental factors like salinity, nutrients, and light all have an effect on the microalgae cell wall's properties.

5.5.4.1 SALINITY

According to Munns et al. (1983), when grown in a high salinity culture medium, up to 70% of a *Chlorella* cell was found to be its cell wall, implying that cell wall thickness varied with the microalgae's cultivation conditions. Cellulose is the primary component of their cell wall, and the enzymes involved in cellulose synthesis are profoundly impacted by salinity (Menegazzo and Fonseca, 2019). But no one has looked into how salinity affects the parts of microalgal cell walls and their mechanical strength.

5.5.4.2 NUTRIENTS

Cell size remains a significant determinant of cell resistance to rupture in high-pressure homogenizers, despite species-specific variations in the strength and elasticity of the cell walls of microalgal species. A greater resistance to rupture may be a stress response in N-deficient microalgae. Their average size was also greater than that of N sufficient microalgae, and their cell walls were thicker (Enamala et al., 2018). Growing larger made N-deficient microalgal cells more vulnerable to rupturing. The increased susceptibility of microalgae to mechanical fragmentation may be due to factors other than cell size and cell wall thickness (Ahmed and Kumar, 2022).

5.5.4.3 LIGHT

The most damaging to cells are the short-wavelength ultraviolet B (UV-B) and ultraviolet C (UV-C) rays. Damage to DNA, proteins, and lipids may be caused by UV-generation A's of reactive oxygen species, making it a less direct and effective carcinogen (Kuluncsics et al., 1999). Photon absorption by DNA bases in the visible spectrum (UV-B and UV-C) leads to chemical quenching and the formation of pyridine dimers in the DNA sequence, both of which are considered direct DNA damage (Rastogi et al., 2010). The structural damage and signaling/metabolic disruption in UV-induced cell lysis can be traced back to the uptake of UV radiation by DNA, RNA, proteins, and lipids. Photo repair of DNA damage in algae occurs when the enzyme cyclobutene pyrimidine dimer photolyase is exposed to UV light (Rastogi et al., 2010). Since photo repair prevents cell lysis, microalgae are typically kept in the dark after UV irradiation for as long as possible. Microalgae can be irreparably damaged by exposure to strong ultraviolet light on an industrial scale. Damage to DNA in the form of pyrimidine dimers is a potential outcome of exposure to UV-B radiation. When an algal cell suffers a severe mutation, it may lyse due to necrosis or apoptosis, resulting in the uncontrolled release of intracellular components. Due to its tightly controlled nature, this is also known as "programmed cell death" and is triggered when certain mutations occur in DNA. Intracellular components, such as lipids, are "packaged" into apoptotic bodies and then released into the culture medium (Sydney et al., 2018).

5.6 SYSTEMIC APPROACHES FOR INTEGRATING MICROALGAL OMICS AND GENETIC ENGINEERING FOR CUSTOM DESIGNING MICROALGAE FOR BIODIESEL PRODUCTION

An accurate depiction of the changes in the biosynthetic processes of potential microalgal strains cultivated under a given set of conditions will be possible due to the integration of the omics studies, including transcriptomics, proteomics, and metabolomics (Akula et al., 2009). As a result, a hierarchical map of the regulation of genes, proteins, and metabolites will be made available, enabling more precise and effective manipulation of microalgal strains to boost lipid productivity. The ability to increase lipid accumulation through genetic engineering of microalgal strains may significantly improve the economic viability of biodiesel made from microalgae. Several genome editing tools, such as nuclear transformation (conventional technologies), adaptive laboratory evolution (ALE), RNA silencing (RNAi), transcriptional factor engineering, zinc finger nucleases (ZFNs), transcription activator like effector nucleases (TALENs), and CRISPR-Cas9 have been developed for microalgal systems in the past

(Beacham et al., 2017). Each of the aforementioned methods for editing the genome is covered separately in the sections that follow.

5.6.1 CONVENTIONAL GENETIC ENGINEERING

One of the direct routes to increasing lipid yield in microalgal strains is through the overexpression of particular lipid biosynthetic genes. According to Xue et al. (2015), overexpressing ACCase did not increase the total lipid content of the diatoms *Cyclotella cryptica* and *Navicula saprophila*, but rather only led to a twofold increase in the transcript level. On the other hand, *P. tricornutum* and *C. reinhardtii*'s acyl-ACP esterase, which is in charge of stopping chain elongation during fatty acid synthesis, increased C12–C14 fatty acids without affecting the overall lipid yield. Additionally, TAG accumulation was increased in *Chlorella minutissima* when yeast-derived LPAT, PAP, GPAT, DGAT, and G3PDH were present. Finally, when the nitrogen-deprived medium was supplemented with acetate, acetyl-CoA synthetase (ACS) under nitrogen deprivation in *C. vulgaris* led to a 1.5-fold and 50% increase in lipid content (Rengel et al., 2018). In engineered strains of *Haematococcus pluvialis*, overexpression of glycerol kinase, aldolase, PCC 6803, and NAB 1/Lci A decreased chlorophyll content and increased biomass and carotenoid production. The commercialization of algal biodiesel depends on this rise in biomass and accumulation of lipids. Also, a smaller antenna can stop too much sunlight from being soaked up, which lets more light reach dense algal cultures.

5.6.2 ADAPTIVE LABORATORY EVOLUTION

An approach known as ALE modifies the genetic code of microorganisms to produce single-nucleotide polymorphisms (SNPs), smaller insertions and deletions (indels), or larger deletions and insertions that improve growth and biochemical properties (Dragosits and Mattanovich, 2013). To create a strain with the desired properties, ALE is conducted over an extended period of time (weeks or years) in a controlled laboratory environment under clearly defined conditions (Shin et al., 2017). In recent years, ALE has been used to create microalgae strains with novel biological and phenotypic traits. For instance, the adapted strain of *C. reinhardtii* CC-124H demonstrated reduced flagellar mobility and an increased palaelodid state, which increased cell size, biomass productivity, and total lipids (66% in normal media and 116% in nitrogen-deprived medium as compared to the wild-type strain) (Shin et al., 2017). Also, a whole-genome sequencing analysis of the adapted strain found 44 CDS (coding DNA sequence) changes, of which 34 were caused by non-synonymous substitutions in 336 genes, most of which were important for the cell cycle to continue.

Additionally, adaptive strains of *C. reinhardtii* cc4326 and cc4334 low-starch mutants demonstrated an increase in lipid accumulation (36.67% and 44.67%) in comparison to the wild type (13%), when nitrogen was depleted (Yu et al., 2013). In a similar vein, under nitrogen-depleted conditions, an adaptive strain of *C. reinhardtii starchless* mutants (sta6-1 cells) demonstrated a 175% increase in total lipid content as compared to the wild type (50%) (Velmurugan et al., 2014). ALE has also been used to improve CO_2 tolerance in *Chlorella* sp., increase resistance to phenol in *Chlorella* sp., reduce light sensitivity in *C. reinhardtii*, and reduce salinity stress in *Chlamydomonas* sp. (Li et al., 2015).

5.6.3 ENGINEERING TRANSCRIPTIONAL FACTORS

Transcriptional factors (TFs) bind to specific DNA motifs within cis elements to control the expression profiles of numerous metabolic pathway components (Bajhaiya et al., 2017). The expression of numerous enzymes is altered when TFs are engineered, increasing the likelihood that genetically engineered strains will succeed. To this end, about 147 putative transcriptional regulators (TRs) and 147 putative transcription factors (TFs) have been discovered in *C. reinhardtii* (Manuelle et al., 2009). A few of the TFs, such as *PSR1* (Pi starvation response), *SNO3*, *GmDOF4* (Glycine maxDNA binding with one), *CHT7* (compromised hydrolysis of triacylglycerols 7), WRI 1 (Wrinkled 1), *bHLH* (basic helix loop-helix), *ROC40* (rhythm of chloroplast 40), *NRR-1* (nitrogen response regulator 1), *TAR1* (TAG accumulation (Ajjawi et al., 2017). PSR1 is crucial for controlling the upregulation of phosphatases and Pi transporters, which regulate the acquisition of phosphorus (Bajhaiya et al., 2015). Under phosphorus-deficient conditions, the mutants of *C. reinhardtii* lacking *PSR1* displayed inhibition of lipid and starch accumulation, while over-expression of this TF increased starch accumulation but decreased neutral lipid content (Bajhaiya et al., 2015). The *TAZ* zinc finger protein-related mRNA encoding protein (*SNO3*), also known as CREB-binding protein/P300, was the subject of a recent patent application by Sapphire Energy

Inc. (Yohn et al., 2016). Without significantly affecting the microalga's growth, its overexpression in *C. reinhardtii* led to an increase in lipid accumulation under nitrogen deprivation (Yohn et al., 2016). Similarly, in nitrogen-deficient conditions, *C. reinhardtii* mutants of *NRR-1* accumulated half as much TAG as the parent strain (Boyle et al., 2012).

On the other hand, under mixotrophic conditions, overexpression of *GmDOF4* from soya bean increased the lipid content by 46–52% without affecting the growth rate in *Chlorella ellipsoidea* (Zhang et al., 2014). *GmDOF4* directly binds to the cis-DNA elements found in the promoter region, which activates ACCase. According to a theory, when microalga are nitrogen-deficient, they accumulate TAG as a stress response to achieve cellular quiescence, but when their growth media are nitrogen-rich, the condition is reversed and TAG is degraded (Tsai et al., 2014). When nitrogen was replenished, the TAG degradation took longer in C. reinhardtii mutants of *CHT7* that had sluggish growth (Tsai et al., 2014). Compared to non-transformed cells, heterologous overexpression of *WRI* recently demonstrated a 44% increase in the total lipid content (Kang et al., 2017). Furthermore, *Nannochloropsis salina* exhibited an increased growth rate (60%) and lipid content increases of 24.5%, 46%, and 32.5% under normal, nitrogen-depleted, and osmotic conditions (50 g/L sea salts), respectively, when compared to the parental strains (21.7%, 42.3%, and 30.1%) (Kang et al., 2017). Furthermore, under nitrogen and sulfur starvation, TAG accumulation was increased (0.5–1-fold) in *C. reinhardtii* mutants with *TAR1* defects (Kajikawa et al., 2015). In *N. gaditana*, Zn(II)2Cys6 was recently knocked out, which improved total carbon portioning into lipids and increased total lipid content to 40–55% in nutrient-rich conditions (Ajjawi et al., 2017). In a different study, the proteomics of nitrogen-starved *Chlorella* sp. cells revealed that *ROC40* was the most strongly induced protein, playing a crucial role in the circadian rhythm that is essential for the survival of microalga under fluctuating conditions (Goncalves et al., 2016). This target, though, has not yet been overexpressed, so more research is necessary.

5.6.4 RNAi SILENCING APPROACHES

Gene silencing, or RNAi interference/silencing, occurs in eukaryotes when double-stranded RNA (dsRNA) triggers the degradation of homologous mRNA (Cerutti, 2003). RNAi silencing can be used to learn more about gene expression, which can then be used to identify promising engineering targets for increasing TAG accumulation in microalgae. Two high lipid (25%) accumulating lines (PEPCK19 and PEPCK21) were generated in *P. tricornutum* by knockdown of phosphoenolpyruvate carboxykinase (PEPC), in comparison to the control (21%) (Yang et al., 2016). Oxaloacetate is produced by phosphoenolpyruvate carboxylase (PEPC), a crucial enzyme in the TCA cycle and protein synthesis (Kao and Ng, 2017). For this reason, inhibiting PEPC will shift the metabolic emphasis from glucose to lipid synthesis. Also, two classes of the PEPC complex, PEPC1 and PECP2, have been found in microalgae (Deng et al., 2014). The former is a rare homotetramer (the p109 subunit), while the latter is a common heteromeric (p109 and p131 subunits). Knocking down *PEPC1* led to significantly higher lipid accumulation (20% higher TAG compared to wild type), while silencing *PEPC2* had the opposite effect (Deng et al., 2014).

Because DGAT catalyzes the final step in TAG formation, its silencing will result in less TAG accumulation, as noted in the TAG synthesis section. There are two distinct DGAT families found in microalgae, *DGAT1* and *DGAT2*, and they are highly dissimilar to one another. To determine which DGAT is crucial for increasing TAG accumulation, Dang et al. (2012) silenced five homologous genes in *C. reinhardtii* (Cr DGAT1, Cr DGAT2, Cr DGAT3, Cr DGAT4, and Cr DGAT5). Knocking out DGAT1 or DGAT5 led to a 16–24% and 28–37% decrease in lipid content, respectively, indicating that these genes were crucial for TAG accumulation. Similarly, in *N. oceanica*, silencing DGAT1A reduced TAG content by 25%, whereas overexpressing it increased TAG accumulation by 39% in nitrogen-starved conditions compared to untransformed cell lines (Wei et al., 2017). Silencing genes in alternative pathways may also increase total TAG accumulation, though this approach is less direct than going after the lipid-enhancer genes themselves. Transgenic lines of *C. reinhardtii* in which citrate synthase (CS) was silenced had a 169% increase in total lipid content compared to the wild type (Deng et al., 2013). The mitochondrial enzyme CS is responsible for the oxidation of acetyl-CoA to citryl-CoA.

5.6.5 ADVANCED GENOME EDITING TOOLS

There is an additional step of optimizing the codon or specific promoters needed for a given set of physiological states that is necessary for all of the aforementioned genetic engineering approaches because they all regulate genes in cis and require replacement of native expression cassettes, which alters the native regulation (Gordon et al., 2016). The optimization of native regulation and efficient transgenic cell lines are both eliminated in favor of engineering trans acting tools. CRISPR-Cas9, an RNA-guided DNA-cutting system, is a powerful, yet easy-to-use tool for creating desirable engineered strains (Kao and Ng, 2017). The Cas9 endonuclease, derived from the CRISPR system of Streptococcus pyogenes, is guided to the target gene on the host chromosome by a single guide RNA (sgRNA) containing a sequence complementary to the target (Kao and Ng, 2017). CRISPR interference (CRISPRi) is a modification of CRISPR-Cas9 that is accomplished by altering the Cas9 protein to produce a version with a mutated catalytic active site, which is referred to as "dead Cas9" (dCas9) (Posewitz, 2017). The sgRNA-dCas9 complex can still bind to the target sequence and modulate gene expression (repress or activate) (Gordon et al., 2016).

CRISPR-Cas9 has only been used to edit a small number of genes in a variety of microalgal species so far. Silencing PEPC in the cell wall-deficient *C. reinhardtii* CC-400 cw 15 mt^+ strain led to a 74% increase in lipid content (28.5% of dry cell weight) compared to the wild-type strain (Kao and Ng, 2017). Furthermore, CRISPRi attenuation of Zn(II)2Cys6 at the 5' UTR increased lipid productivity in *N. gaditana* by twofold (Ajjawi et al., 2017). Non-lipid enhancing targets have also been investigated, with examples including the replacement of the silacidin gene with a resistance cassette (FCP: NAT) leading to 85% NAT-resistant colonies in *T. pseudonana* (Belshaw et al., 2017), and the knockout of the CpFTSY gene in *C. reinhardtii,* leading to a reduction in the size of the chlorophyll antenna and an increase in the plant's overall growth rate (Baek et al., 2016).

TALEN is a trans editing tool that uses TAL effector proteins, which are secreted by *Xanthomonas* bacteria through the type III secretion system (Christian et al., 2010). Multiple highly conserved 34-amino-acid (AA) repeat modules, each containing two divergent amino acids known as repeat variable di-residues (RVD), make up the DNA-binding domains of TAL proteins (Serif et al., 2017). Together, RVD is part of what is called a targeting sequence and helps with sequence recognition (Serif et al., 2017). To create a double-strand break at the designated site, the nuclease (Fok1) acts as a dimer and needs two TAL protein constructs (Miller et al., 2010).

The host machinery can then use either homologous recombination (HR) or non-homologous recombination (NHR) to fix the strand beak (NHER). The former has a low error rate and can be used to introduce foreign sequences that have string homology to the DNA sequence next to the target site, while the latter has a high error rate and can be used to generate random insertions and deletions (Serif et al., 2017). TALE nucleases were shown to be effective in *P. tricornutum* by demonstrating a 45% increase in TAG content relative to the wild type through the inactivation of UDP-glucose pyro phosphorylase, an enzyme involved in carbohydrate storage (Daboussi et al., 2014).

5.6.6 OMICS DATABASES

Integrating the findings of various omics studies with mathematical models and computational tools, system biology has the potential to generate comprehensive models depicting the cellular metabolism of any given microalga (Rodríguez-Moyá and Gonzalez, 2010). There are currently a handful of integrated platforms available, such as ChlamyCys (for *C. reinhardtii*), Greenhouse (an online tool developed by Las Alamos National laboratory), the Diatom EST database (for *T. pseudonana* and *P. tricornutum*), the Alga-PrAs (algal protein annotation suite), and the Cyan-Omics (omics database for cyanobacteria) (Zhang et al., 2017).

5.6.7 GENOME-SCALE MODELS

Creation of computer simulations that mimic biological processes, including flux balance analysis (FBA) that includes all of an organism's genes that code for enzymes (Orth et al., 2010). *C. reinhardtii*, *Chlorella sorokiniana*, *Chlorella protothecoides*, and *Chlorella sp.* FC2 IITG, as well as *P. tricornutum* and *Synechocystis*, are just a few examples of the microalgal strains for which FBA models have been developed (Wu et al., 2015). To better illustrate primary metabolism, we reconstructed *C. reinhardtii's* metabolic network, which results in 458 metabolites and 484 metabolic reactions distributed across the cytosol, mitochondria, and chloroplasts (Boyle and Morgan, 2009). In addition, the FBA model helped researchers locate 16 previously uncharacterized genes and a previously unknown gene (fructose 1,6-bisphosphatase). Light-driven respiration and incident photon flux density were highlighted in another FBA model of *C. reinhardtii* that used a constraint-based approach to investigate microalgal response under photoautotrophic conditions (Cogne et al., 2011). Because of the significance of light-driven metabolism in green microalgae, Chang et al. (2011) built an FBA model for *C. reinhardtii* called iRC1080, which includes 1080 genes, 2190 reactions, and 1068 unique metabolites.

C. sorokiniana's metabolic network under heterotrophic conditions was built out of 34 reactions and 37 metabolites, and it included the Embden Meyerhof-Parnas pathway (EMP), hexose monophosphates (HMP), tricarboxylic acid (TCA), and fatty acid synthesis (Zhu and Huang, 2017). Comparable to the reconstruction of the FBA model, which was partitioned into four organelles based on the *C. protothecoides* genome, namely the cytosol, chloroplasts, mitochondria, and peroxisomes (Wu et al., 2015). When the microalga was grown under both autotrophic and heterotrophic conditions, the model included 272 reactions, 270 enzymes, and 461 encoding genes. An FBA model of *Chlorella sp.* F22 IITG was built by (Muthuraj et al., 2013b), which predicted an intracellular flux change from nutrient-rich to nutrient-poor growth (Muthuraj et al., 2013b). Under heterotrophic conditions, the phosphate pentose pathway, which leads to carbohydrate synthesis, received 50% of the total carbon flux and glycolysis received the other 50%. But in autotrophic conditions, carbon flux is channeled into the Calvin cycle for CO_2 fixation, where it produces dihydroxyacetone phosphate, which then branches off to serve as a substrate in both gluconeogenesis and the glycolytic pathway (Muthuraj et al., 2013b). Using data from the diatom genome, biochemical, and online bioinformatics databases, an FBA model of *P. tricornutum* was developed (Kim et al., 2016). There are a total of 587 metabolites and 849 reactions that take place inside the cell, and these are split between the cytosol, the mitochondrial matrix, the chloroplasts, and the peroxisomes, depending on whether the cell is autotrophic, mixotrophic, or heterotrophic.

Several genome-scale models have been created for different types of algae, such as AlgaGEM (for *C. reinhardtii*), DRUM (for *T. lutea* and *C. sorokiniana*), iLB1027 lipid (for *P. tricornutum*), iCZ843 (for *C. vulgaris* UTEX 395), iN934 (for *N. salina*), iRJL321 (for *N. gaddii*). These improvements in genetic engineering and system biology have helped us learn more about algal biology and made microalgal strains produce more lipids. This makes it possible to make biodiesel that is good for the environment.

5.7 CURRENT CHALLENGES IN MICROALGAL BIODIESEL PRODUCTION

In the current scenario, rising temperatures can be traced back to an increase in both energy demand and serious environmental issues brought on by the extensive use of fossil fuels. As a result, scientists have been diligently looking for a long-term solution to these problems. For these reasons, microalgae are being looked at as a viable alternative fuel source and as a great way to clean up the environment. Previous research has estimated that the price per gallon of biodiesel made from microalgae using a photobioreactor (PBR) is \$20.53, while the price using an open raceway pond cultivation method is \$9.84. This demonstrates the potential of microalgal biodiesel as a green energy source. Based on the reviewed literature, it is clear that many different types of microalgae exist, but only a select few are considered optimal for biodiesel production due to their high lipid content and high yields. More than 25,000 species of microalgae have been described so far (Raja et al., 2008), but currently only a small fraction of these are utilized.

There are still many obstacles in the way of using microalgae as a feedstock for the production of bioenergy and bioproducts, but we can work to overcome them by bringing laboratory-scale technologies up to commercial standards. Increasing pretreatment and extraction yields, lowering water content in biomass, and boosting biodiesel production are the most pressing issues. Microalgal biodiesel is too expensive due to the cultivation system design requiring temperature and growth limiting condition control, despite the availability of several cutting-edge technologies for massive biomass production and lipid conversion into biodiesel (viz., CO_2, water sources, nutrients, and optimization). The other major challenge is the high cost and energy requirements of dewatering biomass.

In large-scale algal cultivation, the closed PBR system is typically more expensive than open raceway ponds. Overheating and fouling are also significant operational difficulties for the PBR system because of the restrictions on gas exchange. Since open ponds, and especially mixed raceway ponds, can be easily scaled up to several hectares and cost much less to build and operate, they are the optimal choice for industrial-scale biomass production of microalgae. Even for high value-added bioproducts that sell for over 100 or 1000 dollars, about 95% of commercial microalgal biomass is produced using open raceway ponds. However, open cultivation methods have a few drawbacks, most notably contamination from other microalgal species, algal grazers, fungi, amoeba, and so forth, and temperature. Despite the hundreds of research papers that have been published, there is still a lack of information on commercial-scale cultivation designs, operations, yields, and other crucial factors, according to a literature review (Zhu et al., 2017). The high upfront and ongoing costs of producing biofuel from microalgae have been a major obstacle. Despite the fact that microalgae-based biofuels have been the subject of a number of studies, a significant technological gap was discovered during commercialization. Water, CO_2, N, and P consumption are thought to be significant barriers to large-scale biomass production. While wastewater can be used as a nutrient source, there is a significant contamination risk due to the presence of bacteria, pathogens, and chemical compounds in the water. According to previous research, 1 liter of algal oil requires 0.16 kg of nitrogen and 0.022 kg of phosphorus (Venkata et al., 2015). Additionally, microalgae require 3.5–9.3 kg of CO_2 for every 1 liter of algal oil produced; this indicates that algae use a considerable amount of CO_2 for their growth and biomass production.

The removal of algal lipids as a prerequisite to making biodiesel presents another significant difficulty. Here, after the biomass has been dried, the process of lipid extraction employs high-priced solvents, which add significantly to the overall cost of production. To lower the cost of biomass pretreatment, many researchers are looking for very advanced technologies that do not require drying or solvent extraction of the algal slurry.

The current methods for producing biodiesel are costly due to the need for pure water and a lipid feedstock that is free of FFA. The biomass must be dried for this extraction method, but this is an additional crucial step that comes at a higher cost. The esterification process is carried out either via an acidic or enzymatic route, with the goal of decreasing the FFA content of the lipid feedstock; however, this process is still in the experimental phase. An additional benefit of the enzymatic esterification using lipases is that it can be carried out at room temperature. However, the main problem with this approach is the by-product glycerol, which can inhibit the lipase activity. In order to prevent glycerol formation and lipase inhibition, some researchers have shown that methyl acetate can be used as a substrate instead of methanol (Baadhe et al., 2014).

5.8 ECONOMIC ASPECTS AND IMPACTS OF MICROALGAL BIODIESEL PRODUCTION

This is because the process of making biodiesel requires a lot of power. The economic viability of biodiesel is enhanced by every step of the production process, from sourcing raw materials to processing, manufacturing, warehousing, and selling the finished product. Initially, the cost-effectiveness of biodiesel will determine whether or not it is used. Microalgal biodiesel is currently more expensive than diesel made from petroleum, so researchers are working to reduce the price. Thus, a few crucial strategies are generally followed to guarantee the microalgal biodiesel's cost effectiveness. These include: (i) increasing the amount of energy captured from the atmosphere; (ii) increasing the amount of energy harvested from the microalgal biomass; (iii) increasing the biomass yield of the resource; (iv) increasing the number of co-products produced; (v) decreasing the energy input during downstream processes; and (vi) increasing the ability of the product to compete with other fuels.

Microalgal biodiesel production systems have come a long way in terms of technological and economic development in recent years. Microalgal biodiesel was produced in 2010–2011 at a cost of more than $100/GGE (gallons of gasoline equivalent) in a paddle-wheel driven microalgal pond cultivation system (National Research Council [NRC], 2012). However, as technology has improved, the cost of producing algal biodiesel has decreased to $7.50/GGE (National Biodiesel Board [NBB], 2009) and is projected to decrease to $3.00/GGE by 2030 (Office of Energy Efficiency and Renewable Energy [EERE], 2017). Reduced prices for biodiesel can be traced back to advancements in cultivation, strain selection, harvesting and extraction technologies, and co-product creation. Many groups and businesses have put forth diverse proposals for cutting costs. University of Utah researchers have used jet mixer technology for direct extraction of lipids from wet algal biomass (Mohanty, 2019), Missing Link Technology and Algae Venture Systems have used their patented harvesting and algae oil extraction systems (Lane, 2014), and Origin Oil Co. has used its patented quantum fracture technology for efficient and innovative single step oil extraction from microalgae (Eckelberry and Eckelberry, 2008). Using their new invention, Origin Oil Co. claims they can bring the price of microalgal biodiesel down to $2.00/GGE, which is quite a reduction. For algal oil to reach commercial viability, many companies must compete, including Sapphire Energy, Muradel, Solazyme, and Algae. Tec, Cellana, and Neste Oil, BioProcess Algae, and Algenol, are constructing large-scale production units in addition to innovating the production and conversion processes (European Technology and Innovation Platform Bioenergy [ETIP], 2014). As a result of concerted efforts in the realm of scientific inquiry and technological advancement, the production of microalgal biodiesel is making the transition from a theoretical to a practical endeavor.

Research into the production of microalgal biodiesel is ongoing around the world, but there are few reports on a detailed cost analysis of the final product (as an effect of the entire production system). This gap has been filled, however, by production models that include cost reduction calculations in anticipation of a competitive market for microalgal biodiesel (Harun et al., 2011). With time and effort, microalgal biodiesel can become cheaper than petroleum diesel, but this has not yet been achieved. The global average price of diesel is \$3.08 per gallon as of 2021 (Diesel Prices), so an equivalent market price for algal diesel is necessary. Given that all integrated systems based on algal biomass processing are limited by high cultivation variability, increasing algal biomass yields from 12 to >30 gdw/m^2/day on a sustained basis is necessary to meet this goal. Furthermore, the energy-return-on-investment (EROI) for harvesting algae from ponds should ideally be >20, meaning that no more than 5% of the energy content of the algae should be spent during harvesting and the lipid extraction and co-product fermentation processes. Additional cost savings for microalgal biodiesel can be obtained by valorizing algal biomass and optimizing lipid content in high biomass yielding microalgal strains (Zhu, 2015). High biomass and lipid productivity can also be achieved in photobioreactors; however, this approach appears less viable for commercial use due to construction and operation challenges such as overheating, fouling, improper gas exchange, and so forth. Veeramuthu and Ngamcharusrivichai (2020) found that the price of producing algal biodiesel with a PBR and an open raceway pond was \$20.53 and \$9.84 per gallon, respectively. Several other groups of researchers and industry, including the National Renewable Energy Laboratory (Davis, 2017) and the Solix Corporation, have published reports corroborating this claim. Another emerging potential strategy for the future is to reduce the price of algal biodiesel by combining economic and environmental sustainability.

Microalgae use the CO_2 released into the air from burning fossil fuels to fuel their own growth and development. The International Panel on Climate Change has identified BECCS technology as the most technologically and economically promising solution for reducing the effects of greenhouse gas emissions (National Biodiesel Board [NBB], 2009). Tax incentives and/or carbon credit programs that promote carbon capture for bioenergy production can lead to even greater cost savings (Sayre, 2010). In what comes next, we will look at the details of the proposed plan for achieving environmental sustainability while also making sure that the whole process is cost-effective.

5.9 CONCLUSIONS AND FUTURE PERSPECTIVES

Since microalgae are capable of utilizing the carbon resources present in water and soil in nature, biodiesel can be produced from microalgae, making it a more sustainable and environmentally friendly alternative to fossil fuel. However, a great deal of work is needed to find solutions to a variety of problems, especially with regards to producing biodiesel at a low cost while maintaining a high level of efficiency, treating wastewater, and reducing carbon dioxide emissions. Research shows that the main barrier to mass-scale adoption of microalgal biodiesel production is financial. Further research is needed to perfect the zero-nutrient-cost technology for biomass production, low-cost harvesting on a massive scale, and the biodiesel conversion process. However, there was still a bare need for more study and novel ideas in the area of sustainability analysis of microalgae biofuel's iterative production rounds. Microalgal biodiesel production on a commercial scale has the potential to make a significant contribution to the current global energy scenario and the attendant environmental concerns. Microalgae are seen as a third-generation candidate for meeting the world's growing energy needs and the difficulties associated with doing so. Consequently, a novel, low-cost technology based on biorefineries should be developed to address these issues. Moreover, it aids in the development of cost-free technologies for treating wastewater, reducing CO_2 emissions, and creating bioproducts with desirable added value. Adapting various cost-cutting activities, such as recycling wastewater and using industrial flue gas as nutrients and carbon sources, could enhance these economic processes.

KEYWORDS

- **biodiesel production**
- **cultivation**
- **lipid extraction**
- **microalgae**
- **sustainable bioenergy**
- **techno-economics**

REFERENCES

Abd El Baky, H. H., El-Baroty, G. S., Bouaid, A., Martinez, M., & Aracil, J. (2012). Enhancement of lipid accumulation in *Scenedesmus obliquus* by optimizing CO_2 and Fe^{3+} levels for biodiesel production. *Bioresource Technology*, *119*, 429–432. https://doi.org/10.1016/j.biortech.2012.05.104 (accessed on 15th November 2025).

Ahmed, J., & Kumar, V. (2022). Effect of high-pressure treatment on oscillatory rheology, particle size distribution and microstructure of microalgae *Chlorella vulgaris* and *Arthrospira platensis*. *Algal Research*, *62*, 102617. https://doi.org/10.1016/j.algal.2021.102617 (accessed on 15th November 2025).

Ahorsu, R., Medina, F., & Constantí, M. (2018). Significance and challenges of biomass as a suitable feedstock for bioenergy and biochemical production: A review. *Energies*, *11*, 3366. https://doi.org/10.3390/en11123366 (accessed on 15th November 2025).

Ajjawi, I., Verruto, J., Aqui, M., Soriaga, L. B., Coppersmith, J., Kwok, K., Peach, L., Orchard, E., Kalb, R., Xu, W., Carlson, T. J., Francis, K., Konigsfeld, K., Bartalis, J., Schultz, A., Lambert, W., Schwartz, A. S., Brown, R., & Moellering, E. R. (2017). Lipid production in *Nannochloropsis gaditana* is doubled by decreasing expression of a single transcriptional regulator. *Nature Biotechnology*, *35*(7), 647–652. https://doi.org/10.1038/nbt.3865 (accessed on 15th November 2025).

Akula, S. P., Miriyala, R. N., Thota, H., Rao, A. A., & Gedela, S. (2009). Techniques for integrating -omics data. *Bioinformation*, *3*(6), 284–286. https://doi.org/10.6026/97320630003284 (accessed on 15th November 2025).

Andersson, V., Broberg, S., & Hackl, R. (2011). *Integrated algae cultivation for biofuels production in industrial clusters.* https://doi.org/10.13140/RG.2.1.4473.4889 (accessed on 15th November 2025).

Arun, S., Ramasamy, S., & Pakshirajan, K. (2021). Mechanistic insights into nitrification by microalgae–bacterial consortia in a photo-sequencing batch reactor under different light intensities. *Journal of Cleaner Production*, *321*, 128752. https://doi.org/10.1016/j.jclepro.2021.128752 (accessed on 15th November 2025).

Asadi, P., Rad, H. A., & Qaderi, F. (2020). Lipid and biodiesel production by cultivation isolated strain *Chlorella sorokiniana* pa.91 and *Chlorella vulgaris* in dairy wastewater treatment plant effluents. *Journal of Environmental Health Science & Engineering*, *18*(2), 573–585. https://doi.org/10.1007/s40201-020-00483-y (accessed on 15th November 2025).

Atadashi, I., Aroua, M., Aziz, A. A., & Sulaiman, N. (2011). Refining technologies for the purification of crude biodiesel. *Applied Energy*, *88*(12), 4239–4251. https://doi.org/10.1016/j.apenergy.2011.05.029 (accessed on 15th November 2025).

Aziz, M. M. A., Kassim, K. A., Shokravi, Z., Jakarni, F. M., Liu, H. Y., Zaini, N., Tan, L. S., Islam, A. S., & Shokravi, H. (2020). Two-stage cultivation strategy for simultaneous increases in growth rate and lipid content of microalgae: A review. *Renewable and Sustainable Energy Reviews*, *119*, 109621. https://doi.org/10.1016/j.rser.2019.109621 (accessed on 15th November 2025).

Baadhe, R. R., Potumarthi, R., & Gupta, V. K. (2014). Lipase-catalyzed biodiesel production: Technical challenges. In *Bioenergy Research: Advances and Applications* (pp. 119–129). https://doi.org/10.1016/B978-0-444-59561-4.00008-5 (accessed on 15th November 2025).

Baek, K., Kim, D. H., Jeong, J., Sim, S. J., Melis, A., Kim, J. S., Jin, E., & Bae, S. (2016). DNA-free two-gene knockout in *Chlamydomonas reinhardtii* via CRISPR-Cas9 ribonucleoproteins. *Scientific Reports*, *6*, 30620. https://doi.org/10.1038/srep30620 (accessed on 15th November 2025).

Bajhaiya, A. K., Dean, A. P., Zeef, L. A., Webster, R. E., & Pittman, J. K. (2016). PSR1 is a global transcriptional regulator of phosphorus deficiency responses and carbon storage metabolism in *Chlamydomonas reinhardtii*. *Plant Physiology*, *170*(3), 1216–1234. https://doi.org/10.1104/pp.15.01907 (accessed on 15th November 2025).

Bajhaiya, A. K., Ziehe, M. J., & Pittman, J. K. (2017). Transcriptional engineering of microalgae: Prospects for high-value chemicals. *Trends in Biotechnology*, *35*(2), 95–99. https://doi.org/10.1016/j.tibtech.2016.06.001 (accessed on 15th November 2025).

Barten, R., Djohan, Y., Evers, W., Wijffels, R., & Barbosa, M. (2021). Towards industrial production of microalgae without temperature control: The effect of diel temperature fluctuations on microalgal physiology. *Journal of Biotechnology*, *336*, 56–63. https://doi.org/10.1016/j.jbiotec.2021.06.017 (accessed on 15th November 2025).

Beacham, T. A., Sweet, J. B., & Allen, M. J. (2017). Large-scale cultivation of genetically modified microalgae: A new era for environmental risk assessment. *Algal Research*, *25*, 90–100. https://doi.org/10.1016/j.algal.2017.04.028 (accessed on 15th November 2025).

Belshaw, N., Grouneva, I., Aram, L., Gal, A., Hopes, A., & Mock, T. (2022). Efficient gene replacement by CRISPR/Cas-mediated homologous recombination in the model diatom *Thalassiosira pseudonana*. *The New Phytologist*. Advance online publication. https://doi.org/10.1111/nph.18587 (accessed on 15th November 2025).

Boyle, N. R., & Morgan, J. A. (2009). Flux balance analysis of primary metabolism in *Chlamydomonas reinhardtii*. *BMC Systems Biology*, *3*, 4. https://doi.org/10.1186/1752-0509-3-4 (accessed on 15th November 2025).

Cerutti, H. (2003). RNA interference: Traveling in the cell and gaining functions? *Trends in Genetics*, *19*(1), 39–46. https://doi.org/10.1016/s0168-9525(02)00010-0 (accessed on 15th November 2025).

Chang, R. L., Ghamsari, L., Manichaikul, A., Hom, E. F., Balaji, S., Fu, W., Shen, Y., Hao, T., Palsson, B. Ø., Salehi-Ashtiani, K., & Papin, J. A. (2011). Metabolic network reconstruction of *Chlamydomonas* offers insight into light-driven algal metabolism.

Molecular Systems Biology, *7*, 518. https://doi.org/10.1038/msb.2011.52 (accessed on 15th November 2025).

Chen, H., & Wang, Q. (2021). Regulatory mechanisms of lipid biosynthesis in microalgae. *Biological Reviews of the Cambridge Philosophical Society*, *96*(5), 2373–2391. https://doi.org/10.1111/brv.12759 (accessed on 15th November 2025).

Chen, L., Zhang, L., & Liu, T. (2016). Concurrent production of carotenoids and lipid by a filamentous microalga *Trentepohlia arborum*. *Bioresource Technology*, *214*, 567–573. https://doi.org/10.1016/j.biortech.2016.05.017 (accessed on 15th November 2025).

Cheng, P., Huang, J., Song, X., Yao, T., Jiang, J., Zhou, C., Yan, X., & Ruan, R. (2022). Heterotrophic and mixotrophic cultivation of microalgae to simultaneously achieve furfural wastewater treatment and lipid production. *Bioresource Technology*, *349*, 126888. https://doi.org/10.1016/j.biortech.2022.126888 (accessed on 15th November 2025).

Chew, K. W., Chia, S. R., Show, P. L., Yap, Y. J., Ling, T. C., & Chang, J. (2018). Effects of water culture medium, cultivation systems and growth modes for microalgae cultivation: A review. *Journal of the Taiwan Institute of Chemical Engineers*, *91*, 332–344. https://doi.org/10.1016/j.jtice.2018.05.039 (accessed on 15th November 2025).

Chhandama, M. V. L., Satyan, K. B., Changmai, B., Vanlalveni, C., & Rokhum, S. L. (2021). Microalgae as a feedstock for the production of biodiesel: A review. *Bioresource Technology Reports*, *15*, 100771. https://doi.org/10.1016/j.biteb.2021.100771 (accessed on 15th November 2025).

Choi, O. K., Hendren, Z., Kim, G. D., Dong, D., & Lee, J. W. (2020). Influence of activated sludge derived-extracellular polymeric substance (ASD-EPS) as bio-flocculation of microalgae for biofuel recovery. *Algal Research*, *45*, 101736. https://doi.org/10.1016/j.algal.2019.101736 (accessed on 15th November 2025).

Christian, M., Cermak, T., Doyle, E. L., Schmidt, C., Zhang, F., Hummel, A., Bogdanove, A. J., & Voytas, D. F. (2010). Targeting DNA double-strand breaks with TAL effector nucleases. *Genetics*, *186*(2), 757–761. https://doi.org/10.1534/genetics.110.120717 (accessed on 15th November 2025).

Chu, W. L. (2017). Strategies to enhance production of microalgal biomass and lipids for biofuel feedstock. *European Journal of Phycology*, *52*, 419–437. https://doi.org/10.1080/09670262.2017.1379100 (accessed on 15th November 2025).Cogne, G., Rügen, M., Bockmayr, A., Titica, M., Dussap, C. G., Cornet, J. F., & Legrand, J. (2011). A model-based method for investigating bioenergetic processes in autotrophically growing eukaryotic microalgae: Application to the green algae *Chlamydomonas reinhardtii*. *Biotechnology Progress*, *27*(3), 631–640. https://doi.org/10.1002/btpr.596 (accessed on 15th November 2025).

Courchesne, N. M., Parisien, A., Wang, B., & Lan, C. Q. (2009). Enhancement of lipid production using biochemical, genetic and transcription factor engineering approaches. *Journal of Biotechnology*, *141*(1, 2), 31–41. https://doi.org/10.1016/j.jbiotec.2009.02.018 (accessed on 15th November 2025).

Daboussi, F., Leduc, S., Maréchal, A., Dubois, G., Guyot, V., Perez-Michaut, C., Amato, A., Falciatore, A., Juillerat, A., Beurdeley, M., Voytas, D. F., Cavarec, L., & Duchateau, P. (2014). Genome engineering empowers the diatom *Phaeodactylum tricornutum* for biotechnology. *Nature Communications*, *5*, 3831. https://doi.org/10.1038/ncomms4831.

Dahlqvist, A., Stahl, U., Lenman, M., Banas, A., Lee, M., Sandager, L., Ronne, H., & Stymne, S. (2000). Phospholipid: diacylglycerol acyltransferase: An enzyme that catalyzes the acyl-CoA-independent formation of triacylglycerol in yeast and plants. *Proceedings of the National Academy of Sciences of the United States of America*, *97*(12), 6487–6492. https://doi.org/10.1073/pnas.120067297 (accessed on 15th November 2025).

Davis, R. (2017). *Algal biofuels techno-economic analysis*. Project Peer Review. Denver, CO: DOE Bioenergy Technologies Office (BETO), National Renewable Energy Laboratory. https://doi.org/10.2172/1044475 (accessed on 15th November 2025).

Davis, R., Aden, A., & Pienkos, P. T. (2011). Techno-economic analysis of autotrophic microalgae for fuel production. *Applied Energy*, *88*(10), 3524–3531. https://doi.org/10.1016/j.apenergy.2011.04.018 (accessed on 15th November 2025).

Davis, R., Fishman, D. B., Frank, E. D., Wigmosta, M. S., Aden, A., Coleman, A. M., Pienkos, P. T., Skaggs, R. J., Venteris, E. R., & Wang, M. (2012). *Renewable diesel from algal lipids: An integrated baseline for cost, emissions, and resource potential from a harmonized model* (No. ANL/ESD/12-4; NREL/TP-5100-55431; PNNL-21437). National Renewable Energy Laboratory (NREL); Golden, CO (United States). https://doi.org/10.2172/1044475 (accessed on 15th November 2025).

Delrue, F., Setier, P., Sahut, C., Cournac, L., Roubaud, A., Peltier, G., & Froment, A. (2012). An economic, sustainability, and energetic model of biodiesel production from microalgae. *Bioresource Technology*, *111*, 191–200. https://doi.org/10.1016/j.biortech.2012.02.020 (accessed on 15th November 2025).

Deng, X. D., Cai, J. J., & Fei, X. W. (2013). Involvement of phosphatidate phosphatase in the biosynthesis of triacylglycerols in *Chlamydomonas reinhardtii*. *Journal of Zhejiang University. Science. B*, *14*(12), 1121–1131. https://doi.org/10.1631/jzus.B1300180.

Deng, X. D., Gu, B., Li, Y. J., Hu, X. W., Guo, J. C., & Fei, X. W. (2012). The roles of acyl-CoA:diacylglycerol acyltransferase 2 genes in the biosynthesis of triacylglycerols by the green algae *Chlamydomonas reinhardtii*. *Molecular Plant*, *5*(4), 945–947. https://doi.org/10.1093/mp/sss040 (accessed on 15th November 2025).

Deng, X., Cai, J., Li, Y., & Fei, X. (2014). Expression and knockdown of the *PEPC1* gene affect carbon flux in the biosynthesis of triacylglycerols by the green alga *Chlamydomonas reinhardtii*. *Biotechnology Letters*, *36*(11), 2199–2208. https://doi.org/10.1007/s10529-014-1593-3 (accessed on 15th November 2025).

Di Caprio, F., Altimari, P., & Pagnanelli, F. (2018). Effect of Ca^{2+} concentration on *Scenedesmus* sp. growth in heterotrophic and photoautotrophic cultivation. *New Biotechnology*, *40*(Pt B), 228–235. https://doi.org/10.1016/j.nbt.2017.09.003 (accessed on 15th November 2025).

Dragosits, M., & Mattanovich, D. (2013). Adaptive laboratory evolution—Principles and applications for biotechnology. *Microbial Cell Factories*, *12*, 64. https://doi.org/10.1186/1475-2859-12-64 (accessed on 15th November 2025).

Eckelberry, N., & Eckelberry, T. R. (2008). *Algae growth system for oil production* (U.S. Patent No. US20090029445A1). Washington, DC: U.S. Patent and Trademark Office. https://patents.google.com/patent/US20090029445A1 (accessed on 15th November 2025).

Enamala, M. K., Enamala, S., Chavali, M., Donepudi, J., Yadavalli, R., Kolapalli, B., Aradhyula, T. V., Velpuri, J., & Kuppam, C. (2018). Production of biofuels from microalgae—A review on cultivation, harvesting, lipid extraction, and numerous applications of microalgae. *Renewable and Sustainable Energy Reviews*, *94*, 49–68. https://doi.org/10.1016/j.rser.2018.05.012 (accessed on 15th November 2025).

Gao, F., Yang, H. L., Li, C., Peng, Y. Y., Lu, M. M., Jin, W. H., Bao, J. J., & Guo, Y. M. (2019). Effect of organic carbon to nitrogen ratio in wastewater on growth, nutrient uptake and lipid accumulation of a mixotrophic microalgae *Chlorella* sp. *Bioresource Technology*, *282*, 118–124. https://doi.org/10.1016/j.biortech.2019.03.011 (accessed on 15th November 2025).

Goncalves, E. C., Wilkie, A. C., Kirst, M., & Rathinasabapathi, B. (2016). Metabolic regulation of triacylglycerol accumulation in the green algae: Identification of potential targets for engineering to improve oil yield. *Plant Biotechnology Journal*, *14*(8), 1649–1660. https://doi.org/10.1111/pbi.12523 (accessed on 15th November 2025).

Gordon, G. C., Korosh, T. C., Cameron, J. C., Markley, A. L., Begemann, M. B., & Pfleger, B. F. (2016). CRISPR interference as a titratable, trans-acting regulatory tool for metabolic engineering in the cyanobacterium *Synechococcus* sp. strain PCC 7002. *Metabolic Engineering*, *38*, 170–179. https://doi.org/10.1016/j.ymben.2016.07.007 (accessed on 15th November 2025).

Halim, R., Gladman, B., Danquah, M. K., & Webley, P. A. (2011). Oil extraction from microalgae for biodiesel production. *Bioresource Technology*, *102*(1), 178–185. https://doi.org/10.1016/j.biortech.2010.06.136 (accessed on 15th November 2025).

Harun, R., Davidson, M., Doyle, M., Gopiraj, R., Danquah, M., & Forde, G. (2011). Technoeconomic analysis of an integrated microalgae photobioreactor, biodiesel and biogas production facility. *Biomass and Bioenergy*, *35*(1), 741–747. https://doi.org/10.1016/j.biombioe.2010.10.007 (accessed on 15th November 2025).

Hou, Q., Qi, X., Zhen, M., Qian, H., Nie, Y., Bai, C., Zhang, S., Bai, X., & Ju, M. (2021). Biorefinery roadmap based on catalytic production and upgrading 5-hydroxymethylfurfural. *Green Chemistry*, *23*, 119–123. https://doi.org/10.1039/D0GC02770G (accessed on 15th November 2025).

Hu, Q., Sommerfeld, M., Jarvis, E., Ghirardi, M., Posewitz, M., Seibert, M., & Darzins, A. (2008). Microalgal triacylglycerols as feedstocks for biofuel production: Perspectives and advances. *The Plant Journal: For Cell and Molecular Biology*, *54*(4), 621–639. https://doi.org/10.1111/j.1365-313X.2008.03492.x (accessed on 15th November 2025).

Isiramen, O. E., Bahri, P. A., Moheimani, N. R., Vadiveloo, A., Shayesteh, H., & Parlevliet, D. A. (2022). Improving pH control and carbon dioxide utilization efficiency in microalgae cultivation systems with the use of a proportional-integral + dead-zone control strategy. *Bioresource Technology Reports*, *17*, 100917. https://doi.org/10.1016/j.biteb.2021.100917 (accessed on 15th November 2025).

Jiang, L., Luo, S., Fan, X., Yang, Z., & Guo, R. (2011). Biomass and lipid production of marine microalgae using municipal wastewater and high concentration of CO_2. *Applied Energy*, *88*(10), 3336–3341. https://doi.org/10.1016/j.apenergy.2011.03.043 (accessed on 15th November 2025).

Jung, J., Kim, J. Y., Jung, S., Choi, Y., & Kwon, E. E. (2021). Quantitative study on lipid productivity of *Euglena gracilis* and its biodiesel production according to the cultivation conditions. *Journal of Cleaner Production*, *291*, 125218. https://doi.org/10.1016/j.jclepro.2020.125218 (accessed on 15th November 2025).

Kajikawa, M., Sawaragi, Y., Shinkawa, H., Yamano, T., Ando, A., Kato, M., Hirono, M., Sato, N., & Fukuzawa, H. (2015). Algal dual-specificity tyrosine phosphorylation-regulated kinase, triacylglycerol accumulation regulator1, regulates accumulation of triacylglycerol in nitrogen or sulfur deficiency. *Plant Physiology*, *168*(2), 752–764. https://doi.org/10.1104/pp.15.00319 (accessed on 15th November 2025).

Kang, N. K., Kim, E. K., Kim, Y. U., Lee, B., Jeong, W. J., Jeong, B. R., & Chang, Y. K. (2017). Increased lipid production by heterologous expression of *AtWRI1* transcription factor in *Nannochloropsis salina*. *Biotechnology for Biofuels*, *10*, 231. https://doi.org/10.1186/s13068-017-0919-5 (accessed on 15th November 2025).

Kao, P., & Ng, I. (2017). CRISPRi mediated phosphoenolpyruvate carboxylase regulation to enhance the production of lipid in *Chlamydomonas reinhardtii*. *Bioresource Technology*, *245*, 1527–1537. https://doi.org/10.1016/j.biortech.2017.04.111 (accessed on 15th November 2025).

Kim, G., Mujtaba, G., & Lee, K. (2016). Effects of nitrogen sources on cell growth and biochemical composition of marine chlorophyte *Tetraselmis* sp. for lipid production. *Algae*, *31*, 257–266. https://doi.org/10.4490/ALGAE.2016.31.8.18.

Kuluncsics, Z., Perdiz, D., Brulay, E., Muel, B., & Sage, E. (1999). Wavelength dependence of ultraviolet-induced DNA damage distribution: Involvement of direct or indirect mechanisms and possible artefacts. *Journal of Photochemistry and Photobiology B: Biology*, *49*(1), 71–80. https://doi.org/10.1016/S1011-1344(99)00034-2 (accessed on 15th November 2025).

Lane, J. (2014). Where are we with algae biofuels? *Biofuels Digest*. Available at: https://www.biofuelsdigest.com/bdigest/2014/10/13/where-are-we-with-algae-biofuels/ (accessed on 15th November 2025).

Leu, S., & Boussiba, S. (2014). Advances in the production of high-value products by microalgae. *Industrial Biotechnology*, *10*, 169–183. https://doi.org/10.1089/IND.2013.0039 (accessed on 15th November 2025).

Li, J., & Guo, Z. (2017). Structure evolution of synthetic amino acids-derived basic ionic liquids for catalytic production of biodiesel. *ACS Sustainable Chemistry & Engineering*, *5*, 1237–1247. https://doi.org/10.1021/ACSSUSCHEMENG.6B02732 (accessed on 15th November 2025).

Li, X., Hu, H. Y., & Zhang, Y. P. (2011). Growth and lipid accumulation properties of a freshwater microalga *Scenedesmus* sp. under different cultivation temperature. *Bioresource Technology*, *102*(3), 3098–3102. https://doi.org/10.1016/j.biortech.2010.10.055 (accessed on 15th November 2025).

Li, Y., Mu, J., Chen, D., Xu, H., Han, F., Feng, B., & Zeng, H. (2015). Proteomics analysis for enhanced lipid accumulation in oleaginous *Chlorella vulgaris* under a heterotrophic-Na^+ induction two-step regime. *Biotechnology Letters*, *37*(5), 1021–1030. https://doi.org/10.1007/s10529-014-1758-0 (accessed on 15th November 2025).

Liu, G., Qiao, L., Zhang, H., Zhao, D., & Su, X. (2014b). The effects of illumination factors on the growth and HCO_3^- fixation of microalgae in an experiment culture system. *Energy*, *78*, 40–47. https://doi.org/10.1016/j.energy.2014.05.043 (accessed on 15th November 2025).

Liu, H., Chen, H., Wang, S., Liu, Q., Li, S., Song, X., Huang, J., Wang, X., & Jia, L. (2018). Optimizing light distribution and controlling biomass concentration by continuously pre-harvesting *Spirulina platensis* for improving the microalgae production. *Bioresource Technology*, *252*, 14–19. https://doi.org/10.1016/j.biortech.2017.12.046 (accessed on 15th November 2025).

Liu, J., Tao, Y., Wu, J., Zhu, Y., Gao, B., Tang, Y., Li, A., Zhang, C., & Zhang, Y. (2014a). Effective flocculation of target microalgae with self-flocculating microalgae induced by pH decrease. *Bioresource Technology, 167*, 367–375. https://doi.org/10.1016/j.biortech.2014.06.036 (accessed on 15th November 2025).

Lohrey, C., & Kochergin, V. (2012). Biodiesel production from microalgae: Co-location with sugar mills. *Bioresource Technology, 108*, 76–82. https://doi.org/10.1016/j.biortech.2011.12.035 (accessed on 15th November 2025).

Ma, X., Mi, Y., Zhao, C., & Wei, Q. (2022). A comprehensive review on carbon source effect of microalgae lipid accumulation for biofuel production. *The Science of the Total Environment, 806*(Pt 3), 151387. https://doi.org/10.1016/j.scitotenv.2021.151387 (accessed on 15th November 2025).

Ma, Y., Wang, Z., Yu, C., Yin, Y., & Zhou, G. (2014). Evaluation of the potential of 9 *Nannochloropsis* strains for biodiesel production. *Bioresource Technology, 167*, 503–509. https://doi.org/10.1016/j.biortech.2014.06.047 (accessed on 15th November 2025).

Maity, S., & Mallick, N. (2022). Trends and advances in sustainable bioethanol production by marine microalgae: A critical review. *Journal of Cleaner Production, 345*, 131153. https://doi.org/10.1016/j.jclepro.2022.131153 (accessed on 15th November 2025).

Mallick, N., Mandal, S., Singh, A. K., Bishai, M., & Dash, A. (2012). Green microalga *Chlorella vulgaris* as a potential feedstock for biodiesel. *Journal of Chemical Technology & Biotechnology, 87*, 137–145. https://doi.org/10.1002/jctb.2694 (accessed on 15th November 2025).

Mandal, S., & Mallick, N. (2009). Microalga *Scenedesmus obliquus* as a potential source for biodiesel production. *Applied Microbiology and Biotechnology, 84*(2), 281–291. https://doi.org/10.1007/s00253-009-1935-6 (accessed on 15th November 2025).

Menegazzo, M. L., & Fonseca, G. G. (2019). Biomass recovery and lipid extraction processes for microalgae biofuels production: A review. *Renewable and Sustainable Energy Reviews, 107*, 87–107. https://doi.org/10.1016/j.rser.2019.01.064 (accessed on 15th November 2025).

Miller, R., Wu, G., Deshpande, R. R., Vieler, A., Gärtner, K., Li, X., Moellering, E. R., Zäuner, S., Cornish, A. J., Liu, B., Bullard, B., Sears, B. B., Kuo, M. H., Hegg, E. L., Shachar-Hill, Y., Shiu, S. H., & Benning, C. (2010). Changes in transcript abundance in *Chlamydomonas reinhardtii* following nitrogen deprivation predict diversion of metabolism. *Plant Physiology, 154*(4), 1737–1752. https://doi.org/10.1104/pp.110.165159 (accessed on 15th November 2025).

Min, K. H., Kim, D. H., Ki, M. R., & Pack, S. P. (2022). Recent progress in flocculation, dewatering, and drying technologies for microalgae utilization: Scalable and low-cost harvesting process development. *Bioresource Technology, 344*(Pt B), 126404. https://doi.org/10.1016/j.biortech.2021.126404 (accessed on 15th November 2025).

Mohanty, S. (2019). Engineers lower energy costs of algae biofuel production. *Biofuels International*. Available at: https://biofuels-news.com/news/engineers-lower-energy-costs-of-algae-biofuel-production/ (accessed on 15th November 2025).

Munns, R., Greenway, H., Setter, T. L., & Kuo, J. (1983). Turgor pressure, volumetric elastic modulus, osmotic volume and ultrastructure of *Chlorella emersonii* grown at high and low external NaCl. *Journal of Experimental Botany, 34*(2), 144–155. https://doi.org/10.1093/jxb/34.2.144 (accessed on 15th November 2025).

Muthuraj, M., Palabhanvi, B., Misra, S., Kumar, V., Sivalingavasu, K., & Das, D. (2013). Flux balance analysis of *Chlorella* sp. FC2 IITG under photoautotrophic and heterotrophic growth conditions. *Photosynthesis Research*, *118*(1, 2), 167–179. https://doi.org/10.1007/s11120-013-9943-x (accessed on 15th November 2025).

Nappa, M., Karinen, P., & Hytönen, E. (2015). Producing lipids, biogas and fertilizer from microalgae – Conceptual design and techno-economic analysis. *Carbon Capture and Storage Program (CCSP): Deliverable D*, 605.

Orth, J. D., Thiele, I., & Palsson, B. Ø. (2010). What is flux balance analysis? *Nature Biotechnology*, *28*(3), 245–248. https://doi.org/10.1038/nbt.1614 (accessed on 15th November 2025).

Ortiz, A., García-Galán, M. J., García, J., & Díez-Montero, R. (2021). Optimization and operation of a demonstrative full-scale microalgae harvesting unit based on coagulation, flocculation and sedimentation. *Separation and Purification Technology*, *259*, 118171. https://doi.org/10.1016/j.seppur.2020.118171 (accessed on 15th November 2025).

Padri, M., Boontian, N., Teaumroong, N., Piromyou, P., & Piasai, C. (2022). Co-culture of microalga *Chlorella sorokiniana* with syntrophic *Streptomyces thermocarboxydus* in cassava wastewater for wastewater treatment and biodiesel production. *Bioresource Technology*, *347*, 126732. https://doi.org/10.1016/j.biortech.2022.126732 (accessed on 15th November 2025).

Pokoo-Aikins, G., Nadim, A. H., El-Halwagi, M. M., & Mahalec, V. (2010). Design and analysis of biodiesel production from algae grown through carbon sequestration. *Clean Technologies and Environmental Policy*, *12*, 239–254. https://doi.org/10.1007/s10098-009-0215-6 (accessed on 15th November 2025).

Posewitz, M. C. (2017). Algal oil productivity gets a fat bonus. *Nature Biotechnology*, *35*(7), 636–638. https://doi.org/10.1038/nbt.3920 (accessed on 15th November 2025).

Qian, J., Shimotori, K., Liu, X., Ban, S., Akizuki, S., Fujiwara, M., Kodera, T., Toda, T., & Imai, A. (2021). Enhancement of algal growth by Mg^{2+} released from anaerobic digestion effluent of aquatic macrophytes through photolysis. *Biochemical Engineering Journal*, *172*, 108065. https://doi.org/10.1016/j.bej.2021.108065 (accessed on 15th November 2025).

Raja, R., Hemaiswarya, S., Kumar, N. A., Sridhar, S., & Rengasamy, R. (2008). A perspective on the biotechnological potential of microalgae. *Critical Reviews in Microbiology*, *34*(2), 77–88. https://doi.org/10.1080/10408410802086783 (accessed on 15th November 2025).

Ranadheer, P., Kona, R., Sreeharsha, R. V., & Venkata, M. S. (2019). Non-lethal nitrate supplementation enhances photosystem II efficiency in mixotrophic microalgae towards the synthesis of proteins and lipids. *Bioresource Technology*, *283*, 373–377. https://doi.org/10.1016/j.biortech.2019.03.089 (accessed on 15th November 2025).

Rastogi, R. P., Richa, Kumar, A., Tyagi, M. B., & Sinha, R. P. (2010). Molecular mechanisms of ultraviolet radiation-induced DNA damage and repair. *Journal of Nucleic Acids*, *2010*, 592980. https://doi.org/10.4061/2010/592980 (accessed on 15th November 2025).

Reddy, V. S., Rao, D. K., & Rajasekharan, R. (2010). Functional characterization of lysophosphatidic acid phosphatase from *Arabidopsis thaliana*. *Biochimica et Biophysica Acta*, *1801*(4), 455–461. https://doi.org/10.1016/j.bbalip.2009.12.005 (accessed on 15th November 2025).

Rehman, M., Kesharvani, S., Dwivedi, G., & Gidwani, S. K. (2022). Impact of cultivation conditions on microalgae biomass productivity and lipid content. *Materials Today: Proceedings*, *56*, 282–290. https://doi.org/10.1016/j.matpr.2022.01.152 (accessed on 15th November 2025).

Ren, H. Y., Liu, B. F., Ma, C., Zhao, L., & Ren, N. Q. (2013). A new lipid-rich microalga *Scenedesmus* sp. strain R-16 isolated using Nile red staining: Effects of carbon and nitrogen sources and initial pH on the biomass and lipid production. *Biotechnology for Biofuels*, *6*(1), 143. https://doi.org/10.1186/1754-6834-6-143 (accessed on 15th November 2025).

Rengel, R., Smith, R. T., Haslam, R. P., Sayanova, O., Vila, M., & León, R. (2018). Overexpression of acetyl-CoA synthetase (ACS) enhances the biosynthesis of neutral lipids and starch in the green microalga *Chlamydomonas reinhardtii*. *Algal Research*, *31*, 183–193. https://doi.org/10.1016/j.algal.2018.02.009 (accessed on 15th November 2025).

Renuka, N., Guldhe, A., Singh, P., & Bux, F. (2018). Combined effect of exogenous phytohormones on biomass and lipid production in *Acutodesmus obliquus* under nitrogen limitation. *Energy Conversion and Management*, *168*, 522–528. https://doi.org/10.1016/j.enconman.2018.05.029 (accessed on 15th November 2025).

Robles-Heredia, J. C., Sacramento-Rivero, J. C., Canedo-López, Y., Ruiz-Marín, A., & Vilchiz-Bravo, L. E. (2015). A multistage gradual nitrogen reduction strategy for increased lipid productivity and nitrogen removal in wastewater using *Chlorella vulgaris* and *Scenedesmus obliquus*. *Brazilian Journal of Chemical Engineering*, *32*(2), 335–345. https://doi.org/10.1590/0104-6632.20150322s00003304 (accessed on 15th November 2025).

Rodríguez-Moyá, M., & Gonzalez, R. (2010). Systems biology approaches for the microbial production of biofuels. *Biofuels*, *1*, 291–310. https://doi.org/10.4155/bfs.10.5 (accessed on 15th November 2025).

Ruiz, J., Wijffels, R. H., Dominguez, M., & Barbosa, M. J. (2022). Heterotrophic vs autotrophic production of microalgae: Bringing some light into the everlasting cost controversy. *Algal Research*, *64*, 102698. https://doi.org/10.1016/j.algal.2022.102698 (accessed on 15th November 2025).

Sakai, T., Kawashima, A., & Koshikawa, T. (2009). Economic assessment of batch biodiesel production processes using homogeneous and heterogeneous alkali catalysts. *Bioresource Technology*, *100*(13), 3268–3276. https://doi.org/10.1016/j.biortech.2009.02.010 (accessed on 15th November 2025).

Sanghamitra, S., Deshmukh, S., & Narayan, K. P. (2020). Effects of alternate nutrient medium on microalgae biomass and lipid production as a bioenergy source for fuel production. *Materials Today: Proceedings*, *28*, 659–664. https://doi.org/10.1016/j.matpr.2019.12.238 (accessed on 15th November 2025).

Sayre, R. (2010). Microalgae: The potential for carbon capture. *Engineering, Biology, Environmental Science*, *60*, 722–727. https://doi.org/10.1525/bio.2010.60.9.9 (accessed on 15th November 2025).

Serif, M., Lepetit, B., Weißert, K., Kroth, P., & Rio Bartulos, C. (2017). A fast and reliable strategy to generate TALEN-mediated gene knockouts in the diatom *Phaeodactylum tricornutum*. *Algal Research*, *23*, 186–195. https://doi.org/10.1016/j.algal.2017.02.005 (accessed on 15th November 2025).

Shin, S. E., Koh, H. G., Kang, N. K., Suh, W. I., Jeong, B. R., Lee, B., & Chang, Y. K. (2017). Isolation, phenotypic characterization and genome wide analysis of a *Chlamydomonas reinhardtii* strain naturally modified under laboratory conditions: Towards enhanced microalgal biomass and lipid production for biofuels. *Biotechnology for Biofuels*, *10*, 308. https://doi.org/10.1186/s13068-017-1000-0 (accessed on 15th November 2025).

Shin, Y. S., Jeong, J., Nguyen, T. H. T., Kim, J. Y. H., Jin, E., & Sim, S. J. (2019). Targeted knockout of phospholipase A2 to increase lipid productivity in *Chlamydomonas reinhardtii* for biodiesel production. *Bioresource Technology*, *271*, 368–374. https://doi.org/10.1016/j.biortech.2018.09.121 (accessed on 15th November 2025).

Shokravi, Z., Shokravi, H., Chyuan, O. H., Lau, W. J., Koloor, S. S., Petrů, M., & Ismail, A. (2020). Improving 'lipid productivity' in microalgae by bilateral enhancement of biomass and lipid contents: A review. *Sustainability*, *12*, 9083. https://doi.org/10.3390/su12219083 (accessed on 15th November 2025).

Singh, A., Ummalyma, S. B., & Sahoo, D. (2020). Bioremediation and biomass production of microalgae cultivation in river water contaminated with pharmaceutical effluent. *Bioresource Technology*, *307*, 123233. https://doi.org/10.1016/j.biortech.2020.123233 (accessed on 15th November 2025).

Sobczuk, T. M., & Chisti, Y. (2010). Potential fuel oils from the microalga *Choricystis minor*. *Journal of Chemical Technology & Biotechnology*, *85*, 100–108. https://doi.org/10.1002/jctb.2272 (accessed on 15th November 2025).

Song, X., Liu, B., Kong, F., Ren, N., & Ren, H. (2022). Overview on stress-induced strategies for enhanced microalgae lipid production: Application, mechanisms and challenges. *Resources, Conservation and Recycling*, *183*, 106355. https://doi.org/10.1016/j.resconrec.2022.106355 (accessed on 15th November 2025).

Su, G., Ong, H. C., Gan, Y. Y., Chen, W., Chong, C. T., & Ok, Y. S. (2022). Co-pyrolysis of microalgae and other biomass wastes for the production of high-quality bio-oil: Progress and prospective. *Bioresource Technology, 344*, 126096. https://doi.org/10.1016/j.biortech.2021.126096 (accessed on 15th November 2025).

Sung, Y. J., Lee, J. S., & Sim, S. J. (2022). Accelerated sunlight-driven conversion of industrial flue gas into biofuels by microfluidic high-throughput screening towards improving photosynthesis in microalgae under fluctuating light. *Chemical Engineering Journal, 443*, 136487. https://doi.org/10.1016/j.cej.2022.136487 (accessed on 15th November 2025).

Sydney, T., Marshall-Thompson, J. A., Kapoore, R. V., Vaidyanathan, S., Pandhal, J., & Fairclough, J. P. A. (2018). The effect of high-intensity ultraviolet light to elicit microalgal cell lysis and enhance lipid extraction. *Metabolites, 8*(4), 65. https://doi.org/10.3390/metabo8040065 (accessed on 15th November 2025).

Taghavijeloudar, M., Kebria, D. Y., & Yaqoubnejad, P. (2021). Simultaneous harvesting and extracellular polymeric substances extrusion of microalgae using surfactant: Promoting surfactant-assisted flocculation through pH adjustment. *Bioresource Technology, 319*, 124224. https://doi.org/10.1016/j.biortech.2020.124224 (accessed on 15th November 2025).

Tan, K. W., & Lee, Y. K. (2017). Expression of the heterologous *Dunaliella tertiolecta* fatty acyl-ACP thioesterase leads to increased lipid production in *Chlamydomonas reinhardtii*. *Journal of Biotechnology, 247*, 60–67. https://doi.org/10.1016/j.jbiotec.2017.03.004 (accessed on 15th November 2025).

Teo, C. L., Atta, M., Bukhari, A., Taisir, M., Yusuf, A. M., & Idris, A. (2014). Enhancing growth and lipid production of marine microalgae for biodiesel production via the use of different LED wavelengths. *Bioresource Technology, 162*, 38–44. https://doi.org/10.1016/j.biortech.2014.03.113 (accessed on 15th November 2025).

Theegala, C. S. (2015). Algal cell disruption and lipid extraction: A review on current technologies and limitations. *Environmental Science, 2*, 422–438. https://doi.org/10.1007/978-3-319-20200-6_13 (accessed on 15th November 2025).

Thuy Lan Chi, N., Mathimani, T., Manigandan, S., Shanmugam, S., Thi Ha, N., Cam Nhung, T., Ali Alharbi, S., Chinnathambi, A., Brindhadevi, K., Chanasut, U., & Whangchai, K. (2022). Small scale photobioreactor, outdoor open pond cultivation of *Chlorella* sp. and harvesting at log and stationary growth phase towards lipids and methyl ester production. *Fuel, 319*, 123813. https://doi.org/10.1016/j.fuel.2022.123813 (accessed on 15th November 2025).

Tran, N. T., Seymour, J. R., Siboni, N., Evenhuis, C. R., & Tamburic, B. (2017). Photosynthetic carbon uptake induces autoflocculation of the marine microalga *Nannochloropsis oculata. Algal Research, 26*, 302–311. https://doi.org/10.1016/j.algal.2017.08.005 (accessed on 15th November 2025).

Tsai, C. H., Warakanont, J., Takeuchi, T., Sears, B. B., Moellering, E. R., & Benning, C. (2014). The protein Compromised Hydrolysis of Triacylglycerols 7 (CHT7) acts as a repressor of cellular quiescence in *Chlamydomonas. Proceedings of the National Academy of Sciences of the United States of America, 111*(44), 15833–15838. https://doi.org/10.1073/pnas.1414567111 (accessed on 15th November 2025).

Urrutia, C., Yañez-Mansilla, E., & Jeison, D. (2019). Bioremoval of heavy metals from metal mine tailings water using microalgae biomass. *Algal Research, 43*, 101659. https://doi.org/10.1016/j.algal.2019.101659 (accessed on 15th November 2025).

Veeramuthu, A., & Ngamcharussrivichai, C. (2020). Potential of microalgal biodiesel: Challenges and applications. In T. Taner (Ed.), *Renewable Energy – Technologies and Applications*. IntechOpen. https://doi.org/10.5772/intechopen.91651 (accessed on 15th November 2025).

Velmurugan, N., Sung, M., Yim, S. S., Park, M. S., Yang, J. W., & Jeong, K. J. (2014). Systematically programmed adaptive evolution reveals potential role of carbon and nitrogen pathways during lipid accumulation in *Chlamydomonas reinhardtii. Biotechnology for Biofuels, 7*(1), 117. https://doi.org/10.1186/s13068-014-0117-7 (accessed on 15th November 2025).

Venkata, M. S., Rohit, M. V., Chiranjeevi, P., Chandra, R., & Navaneeth, B. (2015). Heterotrophic microalgae cultivation to synergize biodiesel production with waste remediation: Progress and perspectives. *Bioresource Technology, 184*, 169–178. https://doi.org/10.1016/j.biortech.2014.10.056 (accessed on 15th November 2025).

Wang, Q., Shen, Q., Wang, J., Zhao, J., Zhang, Z., Lei, Z., Yuan, T., Shimizu, K., Liu, Y., & Lee, D. (2022). Insight into the rapid biogranulation for suspended single-cell microalgae harvesting in wastewater treatment systems: Focus on the role of extracellular polymeric substances. *Chemical Engineering Journal, 430*, 132631. https://doi.org/10.1016/j.cej.2021.132631 (accessed on 15th November 2025).

Wei, L., Xin, Y., Wang, Q., Yang, J., Hu, H., & Xu, J. (2017). RNAi-based targeted gene knockdown in the model oleaginous microalgae *Nannochloropsis oceanica. The Plant*

Journal: For Cell and Molecular Biology, 89(6), 1236–1250. https://doi.org/10.1111/tpj.13411 (accessed on 15th November 2025).

Wu, Q., Guo, L., Li, X., & Wang, Y. (2021). Effect of phosphorus concentration and light/dark condition on phosphorus uptake and distribution with microalgae. *Bioresource Technology, 340*, 125745. https://doi.org/10.1016/j.biortech.2021.125745 (accessed on 15th November 2025).

Wu, S., Zhou, J., Xin, Y., & Xue, S. (2015). Nutritional stress effects under different nitrogen sources on the genes in microalga *Isochrysis zhangjiangensis* and the assistance of *Alteromonas macleodii* in releasing the stress of amino acid deficiency. *Journal of Phycology, 51*(5), 885–895. https://doi.org/10.1111/jpy.12328 (accessed on 15th November 2025).

Xue, J., Niu, Y., Huang, T., Yang, W., Liu, J., & Li, H. (2015). Genetic improvement of the microalga *Phaeodactylum tricornutum* for boosting neutral lipid accumulation. *Metabolic Engineering, 27*, 1–9. https://doi.org/10.1016/j.ymben.2014.10.002 (accessed on 15th November 2025).

Yan, C., Zhu, L., & Wang, Y. (2016). Photosynthetic CO_2 uptake by microalgae for biogas upgrading and simultaneously biogas slurry decontamination by using of microalgae photobioreactor under various light wavelengths, light intensities, and photoperiods. *Applied Energy, 178*, 9–18. https://doi.org/10.1016/j.apenergy.2016.06.012 (accessed on 15th November 2025).

Yang, J., Pan, Y., Bowler, C., Zhang, L., & Hu, H. (2016). Knockdown of phosphoenolpyruvate carboxykinase increases carbon flux to lipid synthesis in *Phaeodactylum tricornutum. Algal Research, 15*, 50–58. https://doi.org/10.1016/j.algal.2016.02.004 (accessed on 15th November 2025).

Yang, Z. Y., Gao, F., Liu, J. Z., Yang, J. S., Liu, M., Ge, Y. M., Chen, D. Z., & Chen, J. M. (2022). Improving sedimentation and lipid production of microalgae in the photobioreactor using saline wastewater. *Bioresource Technology, 347*, 126392. https://doi.org/10.1016/j.biortech.2021.126392 (accessed on 15th November 2025).

Yohn, C., Brand, A., Michael, M., & Behnke, C. A. (2016). *Transformation of algae for increasing lipid production.* U.S. Patent No. 9,428,779 B2, filed February 1, 2011, and issued August 30, 2016.

Yu, S., Zhao, Q., Miao, X., & Shi, J. (2013). Enhancement of lipid production in low-starch mutants *Chlamydomonas reinhardtii* by adaptive laboratory evolution. *Bioresource Technology, 147*, 499–507. https://doi.org/10.1016/j.biortech.2013.08.069 (accessed on 15th November 2025).

Zafar, A. M., Javed, M. A., Aly Hassan, A., Mehmood, K., & Sahle-Demessie, E. (2021). Recent updates on ions and nutrients uptake by halotolerant freshwater and marine microalgae in conditions of high salinity. *Journal of Water Process Engineering, 44*, 102382. https://doi.org/10.1016/j.jwpe.2021.102382 (accessed on 15th November 2025).

Zeng, J., Wang, Z., & Chen, G. (2021). Biological characteristics of energy conversion in carbon fixation by microalgae. *Renewable and Sustainable Energy Reviews, 152*, 111661. https://doi.org/10.1016/j.rser.2021.111661 (accessed on 15th November 2025).

Zhang, C., Li, S., & Ho, S. H. (2021). Converting nitrogen and phosphorus wastewater into bioenergy using microalgae-bacteria consortia: A critical review. *Bioresource Technology, 342*, 126056. https://doi.org/10.1016/j.biortech.2021.126056 (accessed on 15th November 2025).

Zhang, R., Patena, W., Armbruster, U., Gang, S. S., Blum, S. R., & Jonikas, M. C. (2014). High-throughput genotyping of green algal mutants reveals random distribution of mutagenic insertion sites and endonucleolytic cleavage of transforming DNA. *The Plant Cell, 26*(4), 1398–1409. https://doi.org/10.1105/tpc.114.124099 (accessed on 15th November 2025).

Zhang, S., Zhou, S., Yang, X., Xi, W., Zheng, K., Chu, C., Ju, M., & Liu, L. (2020). Effect of operating parameters on hydrothermal liquefaction of corn straw and its life cycle assessment. *Environmental Science and Pollution Research International, 27*(6), 6362–6374. https://doi.org/10.1007/s11356-019-07267-4 (accessed on 15th November 2025).

Zhang, Y., Liu, X., White, M. A., & Colosi, L. M. (2017). Economic evaluation of algae biodiesel based on meta-analyses. *International Journal of Sustainable Energy, 36*, 682–694. https://doi.org/10.1080/14786451.2015.1086766 (accessed on 15th November 2025).

Zhang, Y., Xia, C., Lu, M., & Tu, M. (2018). Effect of over liming and activated carbon detoxification on inhibitors removal and butanol fermentation of poplar prehydrolysates. *Biotechnology for Biofuels, 11*, 178. https://doi.org/10.1186/s13068-018-1182-0 (accessed on 15th November 2025).

Zhong, Y., Su, Y., Zhang, D., She, C., Chen, N., Chen, J., Yang, H., & Balaji-Prasath, B. (2022). The spatiotemporal variations in microalgae communities in vertical waters of a subtropical reservoir. *Journal of Environmental Management, 317*, 115379. https://doi.org/10.1016/j.jenvman.2022.115379 (accessed on 15th November 2025).

Zhu, L. (2015). Biorefinery as a promising approach to promote microalgae industry: An innovative framework. *Renewable and Sustainable Energy Reviews, 41*, 1376–1384. https://doi.org/10.1016/j.rser.2014.09.040 (accessed on 15th November 2025).

Zhu, L. D., Li, Z. H., & Hiltunen, E. (2016). Strategies for lipid production improvement in microalgae as a biodiesel feedstock. *BioMed Research International, 2016*, 8792548. https://doi.org/10.1155/2016/8792548 (accessed on 15th November 2025).

Zhu, L., Nugroho, Y., Shakeel, S., Li, Z., Martinkauppi, B., & Hiltunen, E. (2017). Using microalgae to produce liquid transportation biodiesel: What is next? *Renewable and Sustainable Energy Reviews, 78*, 391–400. https://doi.org/10.1016/j.rser.2017.04.089 (accessed on 15th November 2025).

Zhu, Y., & Huang, Y. (2017). Use of flux balance analysis to promote lipid productivity in *Chlorella sorokiniana*. *Journal of Applied Phycology, 29*, 889–902. https://doi.org/10.1007/s10811-016-0973-6 (accessed on 15th November 2025).

CHAPTER 6

Bacterial-Based Industrial Amylase Enzyme: Progress and Prospects

BABLI KUMARI,[1] PRITI PAL,[2] AKHILESH KUMAR SINGH,[1] PRAKASH KUMAR SARANGI,[3] ANAND PRAKASH,[1] SAURABH SINGH RATHORE,[1,4] and BRIJESH PANDEY[1,*]

[1]*Department of Biotechnology, School of Life Sciences, Mahatma Gandhi Central University, Motihari, Bihar, India*

[2]*Amity Institute of Biotechnology, Amity University Uttar Pradesh, Lucknow Campus, Lucknow, Uttar Pradesh, India*

[3]*College of Agriculture, Central Agricultural University, Imphal, Manipur, India*

[4]*Department of Life Sciences, Inter University Center for Teacher Education, Bananas Hindu University, Varanasi, Uttar Pradesh, India*

*Corresponding author

ABSTRACT

Amylases are one of the most crucial and ancient industrial enzymes. Enzymes are widely employed in several industries in the period of global industrialization. The family of α-amylases and their key traits, microbiological origins, manufacturing, and industrial uses make them a highly desirable industrial enzyme in a variety of industries, including textiles, food, detergents, pharmaceuticals, and so forth. The goal of this chapter is to promote new industrial technologies based on α-amylase while also examining the possibilities of these enzymes. Researchers from all around the globe are very interested in the biotechnological potential of α-amylases

Industrial Microbiology: Current Progress and Novel Technologies.
Prakash Kumar Sarangi, Akhilesh Kumar Singh, Rajesh Kumar Srivastava, and Brijesh Pandey (Eds.)

DOI: 10.1201/9781779642219-6

from microorganisms as imaginable biological catalysts in a number of industrial developments. As opposed to submerged fermentation (SmF), solid-state fermentation (SSF) for the production of α-amylase has been approved because of its straightforward process, minimal capital outflow, reduced levels of catabolite suppression, and greater product recovery.

6.1 INTRODUCTION

Amylases are the most predominant type of endo-acting enzyme. There are still enzymes that hydrolyze starch polysaccharides to produce products with the α-anomeric configuration. These enzymes are exo-acting hydrolases rather than endo-acting hydrolases, and they are claimed to act differently at one end of a polysaccharide chain to produce largely one size of small oligosaccharide. Some of these enzymes, such as maltogenic "amylase," now have Enzyme Commission numbers that are different from those of α-amylase (MacGregor et al., 2001; Wang et al., 2021). The inner portion of the amylopectin or amylose chain contains α,1–4 glycosidic linkages that endo amylases are able to break. One well-known endo amylase is α-amylase. It may be found in an inclusive range of Archaea and bacteria related species. Oligosaccharides of varied length with α-configuration and α-limit destrins, which make up branching oligosaccharides, are the end products of α-amylase activity (Tiwari et al., 2015).

Amylases are enzymes that break down starch molecules into glucose-based polymers. α-Amylases are generally spread, and the main sources include plants, bacteria, and fungus. The bulk of microbial α-amylases belong to the family 13 glycosyl hydrolases and share a lot of characteristics with one another. However, different reaction specificities amongst family members have been identified. According to their structural makeup, α-amylases have (β/α) 8 or TIM barrel structures, and novel "beta strands" they are in charge of hydrolyzing or forming glycosidic linkages in the α-conformation (Reddy et al., 2003; Wang et al., 2020). Today enzymes are employed in many different industrial fields like detergents, the paper, textile, and food industries, among many other industrial applications. Since the beginning of time, enzymes have been employed for the saccharification of starch, the brewing of drinks like beer, the treatment of gastrointestinal diseases, and the making of cheese from milk (Sundaram et al., 2014). Amylases are enzymes that break down starch molecules into glucose-based polymers. However, varied response details have been noted across the family members (Reddy et al., 2003). α-Amylase (EC 3.2.1.1) is the most well-known amylolytic enzyme. It facilitates the process by catalyzing the hydrolysis of the 1,4-glucosidic bonds in starch and related α-glucans. The α-amylase enzyme has a broad range of preferred substrates and product specificity in general. It is one of the most commonly occurring glycoside hydrolases in the sequence-based categorization scheme of all carbohydrate-active enzymes (Janacek et al., 2014). In the industries that process starch, microbial amylases have effectively displaced chemical starch hydrolysis. They have potential use in a variety of industrial processes, including those in the food, baking, brewing, detergent, textile, and paper sectors. They are also used in the starch saccharification process. The range of amylase use has grown into many other sectors, including medicinal, clinical, and analytical chemistry, with the emergence of new frontiers in biotechnology (Bhavya et al., 2007). A reduction in the viscosity of a starch solution, a decrease in the turbidity of a starch suspension, a decrease in the intensity of the starch-iodine reaction, and an increase in the number of reducing groups may all be used to evaluate the amylase activity (Rick et al., 1947). Additionally, samples are screened for α-amylase inhibitors using the starch-iodine assay technique to measure α-amylase activity. There are signs, though, that this approach might not be suitable for filtering certain groups of compounds or bacterial source. The current research looked into the method's potential drawbacks when analyzing bacterial source and substances for α-amylase inhibition. For this research, quercetin (QC), aqueous methanol (AM), and hexane (HX) were used as solvent portions of a crude methanol extract (CR) of Dacryodes edulis, along with benzoic acid (BA) and quercetin (QC). Using spectrophotometry, the phytochemical composition and antioxidant activity were evaluated. Starch-iodine technique approaches I and II (ST-ID I and ST-ID II, respectively) and dinitro salicylic acid (DNSA) as the reference method were used to measure α-amylase inhibition (expressed in percentage and as IC50). When the results of the ST-ID I and II methods were compared to those of the DNSA method, the results showed that the extracts/compounds (AM, EA, and QC) with significantly high polyphenolic content, antioxidant activity, and starch-iodine complex decolorization effect yielded contrary results of α-amylase inhibition (Ononamadu et al., 2020).

Using X-ray diffraction methods, the structure of human pancreatic α-amylase has been identified to 1.8 resolution. It has been discovered that this enzyme consists of three structural domains. The biggest is Domain A, which consists of residues 1–99 and 169–404 and forms a core eight-stranded parallel β-barrel with the active site residues Asp 197, Glu 233, and Asp 300 at one end. A bonded chloride ion that interacts with Arg 195, Asn 298, and Arg 337 is also present in this area. The smallest domain, B (residues 100–168), is used to create a calcium-binding site against the β-barrel of Domain A wall. The protein groups Asn 100, Arg 158, Asp 167, and His 201 interact with this calcium as ligands. Anti-parallel β-structure makes up Domain C (residues 405–496), which is only tangentially related to Domains A and B (Brayer et al., 1995). The management of postprandial blood glucose levels in type 2 diabetic patients and borderline patients may benefit from the inhibition of α-glucosidase (EC 3.2.1.20) and α-amylase (EC 3.2.1.1), enzymes involved in the digestion of carbohydrates. This is because these inhibitors can significantly reduce the postprandial increase of blood glucose following a mixed carbohydrate diet (Nair et al., 2013). Due to their ease of separation and improved heat or pH stability, immobilized enzymes are increasingly being exploited as catalysts in a variety of industrial applications. Numerous inorganic and organic supports, including zeolite, amorphous aluminum phosphate, clay and modified clays, have been investigated for the immobilization of enzymes. Conventional silica has been observed to have a wide range of pore sizes, a low surface area, and poor heat stability when used to immobilize particular enzymes. In addition, the wide range of hole sizes in typical silica increases the likelihood that smaller pores would be blocked during enzyme immobilization, which may impact how substrate molecules approach the enzyme's active sites (Pandya et al., 2005). The capacity of α-amylase to generate mid chain random degradation of starch has been of essential relevance to various different businesses, including the manufacture of limit dextrin and its application in starch-processing industries. The paper, textile, and brewing sectors stand out among these. It makes it easier to employ highly concentrated barley syrup for brewing while hydrolyzing starch to produce a product with the right viscosity for paper coating and surface sizing. It is utilized for desizing in the textile industry. Enzymic hydrolysis of starch is a crucial step that profoundly affects subsequent unit actions in each of these scenarios (Hussain et al., 2021). The biological compound amylase, and particularly α-amylase, is significant and has numerous uses in both business and therapeutic settings (Partelli et al., 2014; Voskuhl et al., 2014). Because of this, bacteria—particularly a Gram-positive bacterium known as *Bacillus subtilis*—are regarded as an essential cell factory for the production of α-amylase (Miralles et al., 2013). This is because *Bacillus subtilis* produces designed proteins at a high rate (Liu et al., 2013; Zweers et al., 2008), lacks endotoxins, and is straightforward because a viable cell only contains a small number of proteins (van et al., 2013).

6.2 AMYLASE AND ITS SUBSTRATE

The three subtypes of amylases are typically α, β, and γ, with the first two garnering the majority of research. α-Amylase, an enzyme, responds more quickly than β-amylase. Because they disintegrate α-1–4 glycosidic bonds, the amylases are also known as glycoside hydrolases. Amylases are commonly distributed in biological systems and have specific substrates (Gupta et al., 2003; Guzman et al., 1995). The most well-known substrates that microorganisms can employ as a carbon source to produce α-amylase are maltose, glucose, and sucrose. The ideal carbon source for the formation of α-amylase is starch, and the suggested concentration for *Aspergillus oryzae* S2 is 10% (Far et al., 2020). Amylases interact with their polymeric substrates at a number of different binding sites on the surface of the enzyme (starch and related substances). One of these places is the extended substrate-binding site in the catalytic center (Machius et al., 1998; Prongjit et al., 2022). The abundance of inexpensive plant sources for amylase substrates makes the potential uses of the enzyme more cost-effective (Gupta et al., 2003). When the substrate is transformed into the product, the anomeric conformation is maintained, thus the enzymes act on the α-linkages in glucans or glucosides to produce α-linked products (Brayer et al., 2000). This interaction takes place between the somewhat positively charged hydrogen atom of the substrate and aromatic residues (tyrosine, tryptophan, and occasionally phenylalanine) (Baroroh et al., 2017).

6.3 BACTERIAL AMYLASE

All of the domains of animals, plants, and microorganisms have amylases. Many studies have been performed using the extracellular α-amylase generated by a variability of microbes during the past few periods. The main benefit of employing microbes to produce amylases is their inexpensive mass production capability and ease of manipulation to produce enzymes with the necessary properties. Although numerous fungi, yeasts, bacteria, and *actinomycetes* have produced amylase, enzymes from bacterial and fungal sources have predominated in commercial uses (Naidu et al., 2013). Amylase produced by bacteria is more affordable and efficient than that produced by other microorganisms. Many different bacterial species produce amylases, which are subsequently isolated. These include *Bacillus subtilis, B. cereus, B. amyloliquefaciens* (Rakaz et al., 2021; Welker et al., 1967), *B. coagulans, B. polymyxa, B. acidocaldarius, B. licheniformis, Lactobacillus* spp., *Micrococcus* spp., *Pseudomonas* spp., *Arthrobacter* spp., *Escherichia* spp., *Proteus* spp., *Thermomonospora* spp., *Serratia* spp., *Bacillus stearothermophilus, Geobacillus* spp., *Esterenkonia* spp., and *Pseudoalteromonas* spp. (Far et al., 2020). The kind of strain, content of the medium, culture technique, cell growth, food needs, incubation duration, pH, temperature, metal ions, and thermostability all affect the production of microbial amylases from bacteria (Pandey et al., 2000). To manufacture thermostable α-amylase for industrial usage, *Bacillus* sp. is a kind of bacteria that is widely used. *Bacillus licheniformis*, *Bacillus subtilis*, *Bacillus stearothermophilus*, and *Bacillus amyloliquefaciens* are well-known for producing thermostable (Prakash et al., 2010). Prior to being tested for their ability to produce certain enzymes, it is crucial to isolate promising and effective bacterial or fungal strains. Microbes are everywhere and may be found in all kinds of places, as has already been mentioned. The most effective strains, however, are typically derived from habitats with abundant substrates, from which the bacteria may be adapted to utilize a specific substrate in Table 6.1 (Anbu et al., 2014; Gopinath et al., 2005). Serial dilution is a typical technique for isolating strains since it reduces the number of colonies and makes selection easier (Gopinath et al., 2005). Substrate selection is a different technique where effective strains are identified based on their affinity for a certain substrate (Anbu et al., 2004). Several bacteria and fungi have been identified and researched for its capacity to generate amylase using these techniques. *Bacillus stearothermophilus* US100 has been linked to amylases that can produce MOS from starch (Ali et al., 2001), as have *Bacillus subtilis* (Messaoud et al., 2004), *Brachybacterium* sp. strain LB25 (Doukyu et al., 2007), and *Bacillus acidicola* (Albert et al., 2005). Only a few strains, viz., *B. subtilis* (Takasaki, 1985) and *Bacillus* sp. GM8901, synthesize maltotriose and maltotetraose.. As a result, efforts to find effective amylases that produce maltotriose and maltotetraose continue (Kim et al., 1995). Table 6.1 displays some characteristics of several bacteria.

TABLE 6.1 Bacterial α-Amylases and Their Characteristics

Bacteria	Strain	Temperature	pH	Molecular Weight	Method	Inhibitor	References
Chromohalobacter sp.	TVSP101	65°C	7–9	72KDa	SSF	–	Prakash et al. (2009)
Bacillus sp.	BKL20	70°C	11	–	SMF	NH4Cl	Kubrak et al. (2010)
Bacillus licheniformis	MTCC 2617 and 2618,	37°C	7.0	–	–	EDTA	Divakaran et al. (2011)
Bacillus subtilis	DM-03	52–55°C	9.0	42.8 kDa	SSF	–	Das et al. (2014)
Bacillus sp.	MTCC 121	40°C	7.1	-	SMF	$(NH_4)_2SO_4$	Raul et al. (2014)
Marinobacter sp.	EMB8	80°C	7.0	72KDa	-	PCMB and Hg^{2+}	Kumar et al. (2015)
Brevibacillus borstelensis	R1	37°C	7.0	–	SMF	–	Suribabu et al. (2016)
Haloarcula hispânica	D61	40–45°C	7	–	–	–	Siroosi et al. (2021)
Bacillus subtilis	QM3	70°C	6.0	–	SMF	Fe^{2+}, Zn^{2+}, Cu^{2+}, Mn^{2+} and EDTA	Hu and Liu (2021).

6.4 APPROACHES FOR THE PRODUCTION OF BACTERIAL AMYLASE

Bacterial strain, culture technique, cultivation medium composition, cell growth, nutritional needs, incubation time, pH, temperature, metal ions, and thermostability all affect the production of microbial amylases. Since they grow quickly, resulting in short fermentation cycles, have the ability to secrete proteins into the extracellular media, and are safe to handle, such industrially significant bacteria of the genus *Bacillus* can really be used commercially (Tiwari et al., 2015).

6.4.1 SCREENING AND SELECTING AMYLASE-PRODUCING BACTERIA

Amylase-producing microbes may be isolated from a broad range of environments, although soil is generally considered to be the greatest source. Because soil has a wide variety of bacteria, it was chosen as the source of bacterial isolation and is occasionally thought of as the rcpository of amylolytic microorganisms (Yassin et al., 2021). For this investigation, soil samples were taken in Afdera, Ethiopia's Afar Region. A total of 49 different colonies were identified based on colony morphology. About 31% of these isolates were seen to provide a zone of clearing surrounding their colonies using the iodine solution. There were 15 amylase manufacturers (M1–M15). Only three isolates were chosen and designated as M2, M8, and M13 based on their effectiveness in degrading starch and the short amount of time needed to decolorize iodine in starch broth. They were discovered using a test for starch hydrolysis and the addition of iodine solution to look for any clear zone production surrounding the colonies. Due to the absence of starch, which was digested by the amylase enzyme generated by the bacteria, distinct zones were created. This demonstrates the potential for the isolates to manufacture amylase, which may be employed in a variety of industrial applications (Yassin et al., 2021).

6.5 FERMENTATION STUDIES ON BACTERIAL AMYLASE PRODUCTION METHOD

The isolated microbial strains used in the starch processing procedure must have the capacity to generate the enzyme on an industrial scale.

There are primarily two techniques: (i) Solid-state fermentation (SSF) and (ii) Submerged fermentation (SMF).

6.5.1 SUBMERGED FERMENTATION SYSTEM

Bio-products made from broth medium, such as molasses, are produced using SMF. For the development of microorganisms (mainly bacteria) in the medium to manufacture α-amylase, this process requires extraordinary wetness. The consumption of substrate is rapid, so a continuous supply of substrate is required (Barragán et al., 2016). This high moisture also makes it simple to apply methods for purification, production, temperature control, nutrition, pH, and other things. For instance, *Bacillus* sp. may be used to produce amylase as a microbiological source (Vidyalakshmi et al., 2009). In SmF, genetically modified bacteria can be cultured with ease. In bioreactors, the bacteria are given a controlled environment and can be sterilized and purified quickly, resulting in a high production of finished goods. Industries regulate fermenters in batch, fed-batch, or continuous modes depending on the need and kind of microbe utilized (Paulová et al., 2013). To manufacture α-amylase by SSF, *Bacillus thuringiensis* and *Bacillus cereus* are typically co-cultured together (Abdullah et al., 2018).

6.5.2 SOLID-STATE FERMENTATION SYSTEM

SSF is extensively used in the commercial production of α-amylase. For bacteria that succeed in low-moisture environments, SSF is used. Paper pulp, polysaccharides, bran, and other solid substrates are used (Dasari et al., 2019). The need to continually replace the substrate is eliminated since the nutrient-rich substrates are recycled and consumed at a very slow pace. The medium used in this process has a very low water content and just a few uses (Paulová et al., 2013). Fungi may easily grow in SSF since it resembles their natural habitat, which makes it ideal for fungi fermentation (Belmessikh et al., 2013). To produce α-amylase by SSF, *Bacillus thuringiensis* and *Bacillus cereus* are naturally co-cultured together (Abdullah et al., 2018). The development of fermentation processes depends heavily on the optimization of fermentation situations, mainly physical and chemical parameters, because of their influence on the effectiveness and capability of the process. The impact of several variables, including temperature, pH, metal ions, nitrogen sources, carbon surface-acting agents, phosphate, and agitation, have been evaluated for the production of α-amylase (Sundarram et al., 2014) (Figure 6.1).

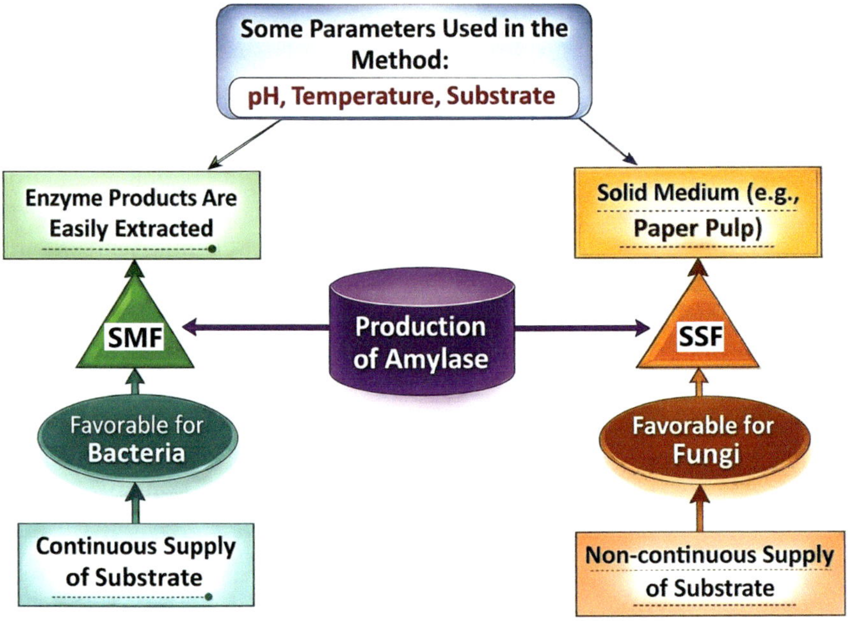

FIGURE 6.1 Production techniques for amylase.

6.6 FACTORS AFFECTING THE PRODUCTION OF MICROBIAL AMYLASE

6.6.1 CARBON SOURCE

Normally, glucose, starch, fructose, sucrose, lactose, galactose, date syrup, and molasses are used as substrates (Simair et al., 2017). *B. licheniformis* utilized the carbon sources, such as groundnut oil cake and wheat bran, at a ratio of 1:1 to create α-amylase (Razdan and Kocher, 2018). *B. subtilis* also utilized banana trash as a source of carbon for the production of α-amylase (Unakal et al., 2012). These carbon sources can be substituted with less expensive agricultural wastes including cottonseed meal, rice bran, soybean meal, wheat bran, and rice husk because they are quite expensive, similar to other media content (Pinotti et al., 2017).

6.6.2 EFFECT OF NITROGEN SOURCE

Numerous organic or inorganic nitrogenous sources are employed in the production of α-amylase. There are few inorganic nitrogen sources: $NaNO_3$, $(NH)_4$ SO_4, NH_4Cl, NH_4 NO_3, and KNO_3 (Suganthi et al., 2011). Organic nitrogen sources used generally are soya bean meal, peptone, and yeast extract (Jujjavarapu and Dhagat, 2019).

6.6.3 ROLE OF PHOSPHATE

In microorganisms, phosphate plays a crucial regulatory function in the synthesis of primary and secondary metabolites. It also has an impact on the organism's growth and α-amylase production. The synthesis of enzymes and the condition of *Aspergillus oryzae* both significantly increased above 0.2 M phosphate levels, according to reports. Low phosphate levels in *amyloliquefaciens* led to extremely low cell mass and no synthesis of amylase. High phosphate concentrations, on the other hand, prevented *Bacillus amyloliquefaciens* from producing enzymes (Naidu et al., 2013).

6.6.4 EFFECT OF pH

The most amylase at a maximum concentration of 7.0 (11 U/ml). The activity of amylase decreased as the pH level continued to rise. However, it was discovered that 7.0 was the ideal pH for the fermentation medium. The activity is reduced or denatured when the pH is changed below or above the ideal range (Vidyalakshmi et al., 2009).

6.6.5 EFFECT OF TEMPERATURE

Two temperatures must be inside the ideal range during manufacturing. They are the most suitable temperature for the growth of microbial sources and the temperature at which enzyme production is greatest. The optimal temperatures for both *B. licheniformis* and *B. subtilis* strains were examined by researchers. It was observed that the optimum temperatures for growth and the production of α-amylase were 45–46°C and 50°C, respectively (Sundarram et al., 2014).

6.6.7 AGITATION

Agitation intensity affects mixing and oxygen transfer rates, which in turn affects mycelial shape and product production. A greater agitation speed has been shown to occasionally be harmful to mycelial development, which might reduce enzyme synthesis (Naidu et al., 2013).

6.6.8 ROLE OF OTHER IONS

While Mn^{2+}, Mo^{2+}, Na^{+}, Fe^{2+}, Cl^{+}, K^{+}, and SO_4^{2+} had no impact on *Aspergillus oryzae*'s ability to produce amylase, Ca^{2+} was an inhibitor. When Mg^{2+} was removed from the medium, output fell by 50% and had a significant effect. The coordinated stimulation of enzyme synthesis by *Bacillus* sp. by Na^{+} and Mg^{2+} (Naidu et al., 2013).

6.7 INDUSTRIAL APPLICATION OF AMYLASE

Amylases play a vital role in a variety of biotechnological applications, including those in the detergent, fermentation, food, pharmaceutical, brewing, textile, and paper sectors. To meet the escalating needs of these firms, amylase manufacture at a reasonable price is required.

6.7.1 FOOD INDUSTRY

The conversion of starch into dextrin's, sugar, and syrups are the main activity in the starch processing industry. In order to enhance the digestibility of fiber, amylases are also used to clarify beer, fruit juices, and animal feed before processing (Gavrilescu et al., 2005). The food industry has seen tremendous transformation due to the current biotechnology revolution. There are several papers on the subject of using genetically modified enzymes safely in the food industry. Amylases are commonly utilized in the production of processed foods, such as bread, beer, cakes, fruit juices, digestive aids, and starch syrups (Mobini-Dehkordi et al., 2012). The enzymes are added to processed food items as processing agents upstream from the end product, which is a practice that is common to nearly all food applications. Numerous improvements have been achieved in the utilization of RCP synthesis to create effective mono-component enzymes without potential side effects and in the optimization of enzymes for current applications (Kirk et al., 2002).

6.7.2 BIO FUEL PRODUCTION

The fact that fossil fuels contribute to greenhouse gases is one of their key disadvantages. In comparison to other fossil fuels, coal is the most destructive to the environment since it produces the worst combustion by products. Natural gas, on the other hand, burns far cleaner than other fossil fuels, making it the most environmentally friendly. Due to environmental concerns and high gasoline costs over the past few decades, biofuels have attracted a lot of attention. The majority of bio fuels are ethanol fuel (Saini et al., 2017). Fuels containing ethanol may be produced using renewable resources like agricultural crops and by-products. Starch is the most common substrate for ethanol synthesis since it is a cheap and widely accessible raw material in most parts of the world (Chi et al., 2009). The bioconversion of starch into ethanol includes the processes of saccharification and liquification, in which an amylolytic microbial agent or enzymes like α-amylase convert starch into sugar (Mojsov et al., 2012).

6.7.3 PAPER INDUSTRY

It has been claimed that α-amylase is used to create low viscosity, high molecular weight starch for coating paper (Bruinenberg et al., 1996). In the paper and pulp business, amylases are used to change starches for coated paper. Similar to how textiles are sized, paper is sized to prevent processing-related mechanical damage. Additionally, it raises the standard of the final product. The size improves the paper's strength and stiffness. Additionally, it makes the paper more erasable and provides a wonderful coating. An excellent sizing agent for paper finishing is starch. In the size press, starch is introduced to the paper, and as it passes through two rollers carrying the starch slurry, the paper absorbs the starch. This procedure operates at a temperature between 45 and 60°C. Replicable outcomes at this point depend on the starch's consistent viscosity. The mill also has the freedom to alter the viscosity of the starch for various paper grades. Natural starch has a viscosity that is too high for paper size, thus in batch or continuous operations, α-amylases are used to partly degrade the polymer. The circumstances vary depending on the type of starch and α-amylase utilized (Tolan et al., 1996). Because of its embrittlement, starch paste used as a mounting adhesive and supplemented with additives like protein alum or glue commonly damages paper (Saini et al., 2017). The goal of modifying coated paper's starch using α-amylases in the pulp and paper industry is to produce low-viscosity, high molecular weight starch. The coating procedure makes the paper's surface sufficiently durable and smooth, which enhances the paper's writing performance. Although the natural starch utilized in this application has a viscosity that is too high for paper size, this may be adjusted using batch or continuous processes that partly breakdown the polymer with amylases (Mojsov et al., 2012). Starch not only serves as a fantastic coating for the paper, but it also works as a decent sizing agent for the finishing of the paper, improving its quality. The size makes the paper stronger and more rigid (Sindhu et al., 2017).

6.7.4 BREAD AND CHAPATTI INDUSTRY

To produce bread products, amylases are essential. Amylases may degrade starch and produce minute amounts of dextrin's that yeast may use. During dough fermentation, proving, and the early stages of baking, yeast may function because α-amylases convert wheat flour's damaged starch into small dextrin's (Saini et al., 2017). These enzymes have been utilized by the baking industry for hundreds of years to produce a wide range of premium goods. Since many years ago, the baking business has made extensive use of enzymes like malt and microbial α-amylases (Hamer et al., 1995; Ragab et al., 2022).

6.7.5 CHOCOLATE INDUSTRY

Chocolate syrup is created by treating amylases with cocoa slurries. Because chocolate starch is dextrinizing, the syrup does not thicken. It is possible to create chocolate-flavored syrups with a high cocoa content, excellent stability, and flow characteristics at room temperature by combining an amylolytic enzyme with a quantity of Dutch process cocoa that results in a syrup pH range of 5.5 to 7.5. The syrup is created by adding sugar and cocoa alternately to water until the solids content is between 58 and 65 weight percent, adding an amylolytic enzyme, heating the mixture to a temperature between 175 and 185°F for at least 10 to 15 minutes, raising the temperature to 200° F, and then cooling (Saini et al., 2017).

6.7.6 DETERGENT INDUSTRY

Modern compressed detergents now include enzymes among their components. The major benefit of using enzymes in detergents is the significantly gentler conditions compared to detergents without enzymes. Early automated dishwashing detergents were extremely strong, dangerous to consume, and unsuitable for use with delicate China and wooden flatware. The detergent industry was compelled by this to look for kinder and more effective remedies (Van et al., 1992). Additionally, enzymes enable decrease of washing temperatures. Since 1975, α-amylases have been a component in powder laundry detergents. α-amylase is now present in 90% of all liquid detergents (Abedi et al., 2022; Kottwitz et al., 1994).

In terms of volume and cost, detergent additives continue to be the main use for industrial enzymes. Although additional, very different hydrolases are added for a range of benefits, including the efficient removal of certain stains, proteases make up the majority of the component (Kirk al., 2002). The fact that α-amylases are extremely sensitive to calcium and oxidants, which are also present in detergents, is one of their principal drawbacks. However, by substituting different amino acids for the oxidation-sensitive ones, it has been possible to construct genetically altered strains for the generation of α-amylases to increase their bleach stability. Leucine has taken the place of methionine at position 197 in *B. amyloliquefaciens* amylase, improving the enzyme's tolerance to oxidative substances (Sindhu et al., 2017).

6.7.7 TEXTILE INDUSTRY

The warp is put under a lot of stress during weaving by modern textile production methods. Therefore, it is necessary to stop the yarn from breaking. The threads are covered in a removably protective covering for this reason. For a layer of this size, entirely different materials are employed. Starch is a very enticing size since it is affordable, widely available throughout most of the world, and easy to remove. Application of α-amylases, which target just the size and spare the fibers, results in effective desiring of starch-sized textiles. Additionally, it randomly separates the starch into water-soluble dextrin's that may be washed off. It has been claimed that α-amylases are used in the warp sizing of textile fibers to create fibers with a high degree of strength (Hendriksen et al., 1999). For textile desizing, α-amylases are used in the textile industry. The threads are covered with a removable layer of protection to prevent yarn from breaking. Because it is inexpensive, conveniently accessible, and removable, starch is a perfect desizing agent (Sindhu et al., 2017). The starch is taken out of the fabric after it has been woven, and then it is scrubbed and dyed. Usually, α-amylase is used to remove the starch from textiles (Hendriksen et al., 1999). Starch is a very attractive size due to its low cost, easy removal, and widespread accessibility throughout most of the world. The textile finishing sector subsequently removes starch from the woven fabric using a wet procedure (Mojsov et al., 2012).

6.7.8 OTHER APPLICATIONS

Clinical and medicinal chemistry analysis of the various amylase uses has broadened into many other domains, including pharmaceutical, clinical, and analytical chemistry, with the appearance of new biotechnological boundaries. The use of amylases is utilized in a number of clinical and medical procedures. It has been detailed how to use a liquid stable reagent based on α-amylase with the Ciba Corning Express clinical chemistry system (Becks et al., 1995). Additionally, an amylase-based method for the exposure of higher oligosaccharides was devised (Giri et al., 1990). It was said that this approach was more effective than the silver nitrate test. Additionally, electrolyte isolator semiconductor capacitor (EIS-CAP) transducer-equipped biosensors for process monitoring were created (Menzel et al., 1998). The liquefaction of starch to create glucose and fructose syrups is one of the most significant uses of α-amylase. Due to its strong sweetening capability, starch is transformed to high-fructose corn syrup, which is extensively utilized in the beverage sector as a sweetener for soft drinks (Sindhu et al., 2017). The range of amylase's uses has expanded in several additional industries, including clinical, medicinal, and analytical chemistry (Cherry et al., 2004; Pandey et al., 2000). To some extent, amylases are also utilized as digestive aids to enhance the digestibility of some of the constituents in animal feed and to complement the diastatic activity of wheat (Kumar et al., 1995). The use of amylases is utilized in a number of clinical and medical procedures. It has been documented how to use a liquid stable reagent based on α-amylase with the Ciba Corning Express clinical chemistry system (Becks et al., 1995). Additionally, an amylase-based method for the recognition of higher oligosaccharides was devised (Giri et al., 1990).

6.8 COMMERCIAL STATUS OF AMYLASE: BOTTLENECKS AND SOLUTIONS

6.8.1 IMMOBILIZATION OF A THERMOSTABLE AMYLASE

Food technology, biotechnology, biomedicine, and analytical chemistry all employ immobilized enzymes. They have a number of benefits over free enzymes, including easy reactant, product, and medium separation, simple enzyme recovery, and repeated or continuous recycle (Varavinit et al., 2002). The characteristics of the enzyme were studied before and after immobilization, and the generated α-amylase was immobilized on several carriers using various techniques. The immobilized enzyme's ideal pH was altered to an acidic range. The immobilized enzyme's ideal reaction temperature was changed slightly to between 70 and 80°C. The immobilized enzyme was shown to have greater Km values, Vmax values, and thermal stability than the free enzyme. $CaCl_2$ was one of the studied metals that had an influence on α-amylase's activity by activating it (Banna et al., 2007). Chemical testing was done to determine the effectiveness of immobilized enzyme in removing starch stains from cotton fabric using different commercial detergents. In the presence of immobilized amylase, all detergents provided cleaner washing than when used alone. Furthermore, washing with less costly (nonenzymic) detergents when immobilized amylase was present was nearly identical to washing with more expensive (enzymic) detergents. About 100 uses of the immobilized enzyme were made without any appreciable loss of activity (Dhingra et al., 2006). By cross-linking with chromium acetate, chromium sulphate, potassium chrome sulphate, and formaldehyde, α-amylase was immobilized in photographic gelatin. Immobilized α-amylase was coated on polyester film strips (Bayramoglu et al., 2007). Amylase (diastase) from *Aspergillus niger* has been attached to zirconia-coated alkylamine glass beads that are secured within a plastic beaker using glutaraldehyde coupling (Dhingra et al., 2006) (Table 6.2).

TABLE 6.2 Immobilization Method Advantage and Disadvantage

Immobilization Method	Disadvantage	Advantage
Adsorption method	The weak interactions keep native structure of the enzyme and its activity	Diffusional limitations and conformational changes
Cross-linking method	Increased specific activity, greater volumetric activity per biocatalyst mass, more simple production, higher purity, less production costs, and less contaminations by the support	Low mechanical stability, low activity, poor reproducibility, and difficulties with handling
Covalent attachment method	Strong attachment of enzyme with carrier	Rigorous preparation condition and loss of enzyme activity due to reaction with toxic crosslinking reagents
Entrapment method	Affordable and fast method requiring mild conditions, and protects the enzyme from mechanical shear, hydrophobic solvents, and gas bubbles	Loss of enzyme activity and limitation in mass transfer and low-level enzyme loading

6.8.2 CHAPERONE

First, the extracellular target protein is produced with the help of intracellular chaperones, and then, with the help of parts of the secretory route, the protein is exported from the cell. At the N-terminus of the target protein precursor is a unique amino acid sequence called a signal peptide, which plays a crucial role in the secretion of the mature protein. Given that the folding and degradation of the target protein between the cell membrane and the cell wall affects the efficacy of signal peptide-mediated transport, optimizing the signal peptide may be an efficient technique for achieving overall secretion balance (Yao et al., 2021).

6.8.3 PLASMID CONSTRUCTION AND TRANSFORMATION

Plasmids and primers were utilized entirely. Plasmid pBE-SPaprE-amyS was created by amplifying the α-amylase gene amyS from plasmid pET-20b-amy with primers pBE1 and pBE2, then cloning it into the vector pBE-S DNA with restriction enzymes SacI and HindIII. Primers pBE3 and pBE4 were used to isolate the pBE-amyS fragment from pBE-SPaprE-amyS, which serves as the structural backbone of pBE-SPmix-amyS. Using the Takara In-Fusion HD Cloning Kit, we fused the purified pBE-amyS fragment with a panel of 173 secretory signal peptides generated from *B. subtilis*. Improved extracellular expression of α-amylase from *Bacillus stearothermophilus* in *Bacillus subtilis* via signal peptide optimization, chaperone overexpression, and α-amylase mutant selection (Yao et al., 2021).

6.8.4 FOOD WASTE DEGRADATION

The amylase-producing bacterial strain FW2 was studied and characterized after being isolated from discarded food. *Bacillus* genus member FW2 was discovered to share a 99.86% resemblance to *Bacillus* amyloliquefaciens DSM 7T. These microorganisms were able to break down organic molecules over a wide pH range (4.5–12) and in high-salinity environments containing up to 35% NaCl, with growth occurring between 6°C and 75°C (although weak at 80°C). After 4 days of incubation at pH 7–7.5, 40–45°C, and 0%–10% NaCl, maximum enzyme synthesis was achieved, reaching 1,200 ± 23.4 U/mL for protease and 2,400 ± 45.8 U/mL for amylase. Purified forms of protease (AprM and NprM) had molecular weights of 28 and 44 kDa, respectively, as determined by SDS-PAGE, while α-amylase had a molecular weight of 55 kDa. Over the course of 15 days, we measured the volume reduction caused by food waste degradation, and we found that it was 69%. Extremozyme-producing bacteria with economic potential, including strain FW2, show promise for waste breakdown in harsh environments (Pham et al., 2021).

6.8.5 PAPER-BASED A-AMYLASE

When starch was present or absent, there was a noticeable difference in the amount of α-amylase found on the chip's top and bottom halves made of paper. We used Western blotting (WB) and enzyme-linked immunosorbent assay (ELISA) to determine that the starch-coated paper-based chip had a very high concentration of α-amylase solely in its anterior region. Using saliva and minute quantities of other biofluids like blood, tears, pancreatic juice, and urine, we believe our paper-based chip will aid α-amylase detection methods (Hyung et al., 2021).

6.8.6 STRESS BIOMARKER

In humans, salivary α-amylase (sAA) is measured as a stress biomarker in the presence of disease, but this has not been done with dogs. With the help of a kinetic enzyme test, researchers in this study determined the stress levels of 16 healthy and 31 ill canines by measuring their sAA and serum α-amylase (AA) levels. In dogs with disease, both sAA and serum AA levels were significantly higher than those in healthy dogs ($p < 0.05$), although no correlation was observed between the two groups ($r = 0.251$, $p = 0.089$). This provides supporting evidence for the potential value of sAA as a stress biomarker in dog disease (Hong et al., 2019).

6.9 CONCLUSION

One of the greatest vital enzymes utilized in industrial processes is amylase. Despite the fact that there are numerous microbial sources for the effective production of this enzyme and that amylases, particularly α-amylases, are used in starch liquefaction and other starch-based industries, the commercial production of this enzyme has only been possible through a few carefully chosen strains of bacteria and fungi. Amylases are among the most significant enzymes employed in a variety of sectors, as is seen from several publications. Over the past 50 years, amylase research has advanced incredibly quickly, and prospective industrial uses of the enzyme, particularly in solid waste organization, have been found. Microbes can be used by SSF, which uses waste materials and by-products from other processes, to make amylase. The creation of fructose syrup, ecologically friendly detergents, and baked goods are just a few of the essential uses for the enzyme. This enzyme is a glimmer of hope for people who live in a period where fossil fuels are running out and there is a great need to create another source of energy. Starch is utilized as a basic ingredient in the manufacturing of biofuels. More studies are concentrating on the microbial generation of the enzyme since it produces an enzyme that is crucial for industry and aids in maintaining a clean environment.

6.10 FUTURE PROSPECTIVE

One of the most crucial enzymes utilized in industrial processes is amylase. Although there are various microbiological sources for the actual synthesis of this enzyme, only a few chosen strains of bacteria and fungi match the requirements for commercial production despite the fact that α-amylase has been used in starch-based businesses for many decades. For instance, the ongoing discovery of novel enzymes through contemporary biotechnology may result in enzyme products with enhanced cleaning capabilities at low temperatures. In nations where hot washing is still practiced, this may enable wash temperatures to be lowered and save energy. The synthesis of α-amylase has developed into a fascinating biotechnology issue as a result of the increasing demand for α-amylase in the industrial and medicinal fields for a range of applications. Several techniques and criteria are employed to synthesis the affordable and easily accessible α-amylase from microbial sources. Even though the biosynthesis of α-amylase has advanced significantly, alternative genetic engineering techniques, such as cloning, gene expression.

KEYWORDS

- ***Bacillus* species**
- **enzyme characteristics**
- **enzyme optimization**
- **fermentation techniques**
- **industrial applications**

REFERENCES

Abdullah, R., Naeem, N., Aftab, M., Kaleem, A., Iqtedar, M., Iftikhar, T., et al. (2018). Enhanced production of alpha amylase by exploiting novel bacterial co-culture technique employing solid state fermentation. *Iranian Journal of Science and Technology, Transactions A: Science, 42*(2), 305–312. https://doi.org/10.1007/s40995-016-0015-x (accessed on 15th November 2025).

Abedi, G., Talebpour, Z., Aliahmadi, A., & Mashhadi, I. S. (2022). Identification of industrial detergent enzymes by SDS-PAGE and MALDI-TOF mass spectrometry. *New Journal of Chemistry, 46*(8), 3939–3947.

Albert, R. A., Archambault, J., Rossello-Mora, R., Tindall, B. J., & Matheny, M. (2005). *Bacillus acidicola* sp. Nov., a novel mesophilic, acidophilic species isolated from acidic *Sphagnum* peat bogs in Wisconsin. *International Journal of Systematic and Evolutionary Microbiology, 55*(5), 2125–2130.

Anbu, P., Hilda, A., & Gopinath, S. C. B. (2004). Keratinophilic fungi of poultry farm and feather dumping soil in Tamil Nadu, India. *Mycopathologia, 158*(3), 303–309. https://doi.org/10.1007/s11046-004-3465-1 (accessed on 15th November 2025).

Baroroh, U., Yusuf, M., Rachman, S. D., Ishmayana, S., Syamsunarno, M. R. A., Levita, J., & Subroto, T. (2017). The importance of surface-binding site towards starch-adsorptivity level in α-amylase: A review on structural point of view. *Enzyme Research, 2017*, Article ID 8717460.

Barragán, L. P., Figueroa, J. J. B., Durán, L. R., González, C. A., & Hennigs, C. (2016). Fermentative production methods. In *Biotransformation of Agricultural Waste By-Product* (pp. 189–217). Elsevier.

Bayramoglu, Z., Akbulut, U., & Sungur, S. (1992). Immobilization of alpha-amylase into gelatin films with various cross-linkers. *Bioorganic & Medicinal Chemistry Letters, 2*(5), 427–432.

Belmessikh, A., Boukhalfa, H., Mechakra-Maza, A., Gheribi-Aoulmi, Z., & Amrane, A. (2013). Statistical optimization of culture medium for neutral protease production by *Aspergillus oryzae*: Comparative study between solid and submerged fermentations on tomato pomace. *Journal of the Taiwan Institute of Chemical Engineers, 44*(3), 377–385.

Bhavya, D. (2007). Production and characterization of fungal amylase enzyme isolated from *Aspergillus* sp. JGI 12 in solid state culture. *African Journal of Biotechnology, 6*(5), 576–581.

Brayer, G. D., Luo, Y., & Withers, S. G. (1995). The structure of human pancreatic α-amylase at 1.8 Å resolution and comparisons with related enzymes. *Protein Science, 4*(9), 1730–1742.

Brayer, G. D., Sidhu, G., Maurus, R., Rydberg, E. H., Braun, C., Wang, Y., & Withers, S. G. (2000). Subsite mapping of the human pancreatic α-amylase active site through structural, kinetic, and mutagenesis techniques. *Biochemistry, 39*(16), 4778–4791.

Bruinenberg, P. M., Hulst, A. C., Faber, A., & Voogd, R. H. (1996). *A process for surface sizing or coating of paper.* European Patent Application EP 0,690,170 A1, filed and published in 1996.

Burhan, A., Nisa, U., Gökhan, C., Ömer, C., Ashabil, A., & Osman, G. (2003). Enzymatic properties of a novel thermostable, thermophilic, alkaline and chelator resistant amylase from an alkaliphilic *Bacillus* sp. isolate ANT-6. *Process Biochemistry, 38*(10), 1397–1403.

Das, K., Doley, R., & Mukherjee, A. K. (2004). Purification and biochemical characterization of a thermostable, alkaliphilic, extracellular α-amylase from *Bacillus subtilis* DM-03, a strain isolated from the traditional fermented food of India. *Biotechnology and Applied Biochemistry, 40*(3), 291–298.

Dasari, P. R., Ramteke, P. W., Kesri, S., & Kongala, P. R. (2019). Comparative study of cellulase production using submerged and solid-state fermentation. In *Approaches to Enhance Industrial Production of Fungal Cellulases* (pp. 37–52).

Dhingra, S., Khanna, M., & Pundir, C. S. (2006). Immobilization of alpha-amylase onto alkylamine glass beads affixed inside a plastic beaker: Kinetic properties and application. *Indian Journal of Chemical Technology, 13*(2), 119–121.

Divakaran, D., Chandran, A., & Pratap, C. R. (2011). Comparative study on production of α-amylase from *Bacillus licheniformis* strains. *Brazilian Journal of Microbiology, 42*, 1397–1404.

El-Banna, T. E., Abd-Aziz, A. A., Abou-Dobara, M. I., & Ibrahim, R. I. (2007). Production and immobilization of alpha-amylase from *Bacillus subtilis*. *Pakistan Journal of Biological Sciences: PJBS, 10*(12), 2039–2047.

Far, B. E., Ahmadi, Y., Khosroshahi, A. Y., & Dilmaghani, A. (2020). Microbial alpha-amylase production: Progress, challenges and perspectives. *Advanced Pharmaceutical Bulletin, 10*(3), 350.

Ferrer-Miralles, N., & Villaverde, A. (2013). Bacterial cell factories for recombinant protein production; expanding the catalogue. *Microbial Cell Factories, 12*(1), 1–4.

Gangadharan, D., Sivaramakrishnan, S., Nampoothiri, K. M., & Pandey, A. (2006). Solid culturing of *Bacillus amyloliquefaciens* for alpha-amylase production. *Food Technology & Biotechnology, 44*(2), 215–220.

Gavrilescu, M., & Chisti, Y. (2005). Biotechnology—A sustainable alternative for chemical industry. *Biotechnology Advances, 23*(7, 8), 471–499.

Gopinath, S. C. B., Anbu, P., & Hilda, A. (2005). Extracellular enzymatic activity profiles in fungi isolated from oil-rich environments. *Mycoscience, 46*(2), 119–126. https://doi.org/10.1007/s10267-004-0221-9 (accessed on 15th November 2025).

Gupta, R., Gigras, P., Mohapatra, H., Goswami, V. K., & Chauhan, B. (2003). Microbial α-amylases: A biotechnological perspective. *Process Biochemistry, 38*(11), 1599–1616. https://doi.org/10.1016/s0032-9592(03)00053-0 (accessed on 15th November 2025).

Guzman-Maldonado, H., Paredes-Lòpez, O., & Biliaderis, C. G. (1995). Amylolytic enzymes and products derived from starch: A review. *Critical Reviews in Food Science and Nutrition, 35*(5), 373–403. https://doi.org/10.1080/10408399509527706 (accessed on 15th November 2025).

Hamer, R. J. (1995). Enzymes in the baking industry. In G. A. Tucker & L. F. J. Woods (Eds.), *Enzymes in Food Processing* (pp. 190–222). Blackie Academic and Professional.

Hong, H. R., Oh, Y. I., Kim, Y. J., & Seo, K. W. (2019). Salivary alpha-amylase as a stress biomarker in diseased dogs. *Journal of Veterinary Science, 20*(5), Article e52.

Hu, Q., & Liu, J. (2021). Production of α-amylase by *Bacillus subtilis* QM3 and its enzymatic properties. *Open Access Library Journal, 8*(3), 1–8.

Hussain, M., & Jabbar, B. (2021). Applications of alpha (α)-amylase in biotechnology: A review. *GU Journal of Phytosciences, 1*(1), 98–105.

Hyung, S., Karima, G., Shin, K., Kim, K. S., & Hong, J. W. (2021). A simple paper-based α-amylase separating system for potential application in biological sciences. *BioChip Journal, 15*(3), 252–259.

Janeček, Š., Svensson, B., & MacGregor, E. (2014). α-Amylase: An enzyme specificity found in various families of glycoside hydrolases. *Cellular and Molecular Life Sciences, 71*(7), 1149–1170.

Jujjavarapu, S. E., & Dhagat, S. (2019). Evolutionary trends in industrial production of α-amylase. *Recent Patents on Biotechnology, 13*, 4–18.

Kirk, O., Borchert, T. V., & Fuglsang, C. C. (2002). Industrial enzyme applications. *Current Opinion in Biotechnology, 13*(4), 345–351.

Konsoula, Z., & Liakopoulou-Kyriakides, M. (2007). Co-production of alpha-amylase and beta-galactosidase by *Bacillus subtilis* in complex organic substrates. *Bioresource Technology, 98*, 150–157.

Kottwitz, B., Upadek, H., & Carrer, G. (1994). Applications and benefits of enzymes in detergent. *Chimica Oggi, 12*, 21–24.

Kubrak, O. I., Storey, J. M., Storey, K. B., & Lushchak, V. I. (2010). Production and properties of α-amylase from *Bacillus* sp. BKL20. *Canadian Journal of Microbiology, 56*(4), 279–288.

Kumar, S., & Khare, S. K. (2015). Chloride activated halophilic α-amylase from *Marinobacter* sp. EMB8: Production optimization and nanoimmobilization for efficient starch hydrolysis. *Enzyme Research, 2015*, Article ID 494219.

Kundu, A. K., & Das, S. (1970). Production of amylase in liquid culture by a strain of *Aspergillus oryzae*. *Applied Microbiology, 19*(4), 598–603.

Liu, L., Liu, Y., Shin, H. D., Chen, R. R., Wang, N. S., Li, J., & Chen, J. (2013). Developing *Bacillus* spp. as a cell factory for production of microbial enzymes and industrially important biochemicals in the context of systems and synthetic biology. *Applied Microbiology and Biotechnology, 97*, 6113–6127.

MacGregor, E. A., Janeček, Š., & Svensson, B. (2001). Relationship of sequence and structure to specificity in the α-amylase family of enzymes. *Biochimica et Biophysica Acta (BBA) - Protein Structure and Molecular Enzymology, 1546*(1), 1–20.

Machius, M., Declerck, N., Huber, R., & Wiegand, G. (1998). Activation of *Bacillus licheniformis* α-amylase through a disorder→order transition of the substrate-binding site mediated by a calcium–sodium–calcium metal triad. *Structure, 6*(3), 281–292.

Mobini-Dehkordi, M., & Javan, F. A. (2012). Application of alpha-amylase in biotechnology. *Journal of Biology and Today's World, 1*(1), 39–50.

Mojsov, K. (2012). Microbial alpha-amylases and their industrial applications: A review. *International Journal of Management, IT and Engineering (IJMIE), 2*(10), 583 609.

Naidu, M. A., & Saranraj, P. (2013). Bacterial amylase: A review. *International Journal of Pharmaceutical and Biological, 4*(2), 274–287.

Nair, S. S., Kavrekar, V., & Mishra, A. (2013). In vitro studies on alpha amylase and alpha glucosidase inhibitory activities of selected plant extracts. *European Journal of Experimental Biology, 3*(1), 128–132.

Nielsen, J. E., & Borchert, T. V. (2000). Protein engineering of bacterial α-amylases. *Biochimica et Biophysica Acta, 1543*(2), 253–274. https://doi.org/10.1016/s0167-4838(00)00240-5 (accessed on 15th November 2025).

Ononamadu, C. J., Ezeigwe, O. C., Owolarafe, T. A., Ihegboro, G. O., Lawal, T. A., Salawu, K., & Aminu, I. (2020). Starch-iodine assay method underestimates alpha-amylase inhibitory potential of antioxidative compounds and extracts. *BioTechnologia: Journal of Biotechnology, Computational Biology and Bionanotechnology, 101*(1), 45–54.

Pandey, A., Nigam, P., Soccol, C. R., Soccol, V. T., Singh, D., & Mohan, R. (2000). Advances in microbial amylases. *Applied Biochemistry and Biotechnology, 31*(2), 135–152.

Pandya, P. H., Jasra, R. V., Newalkar, B. L., & Bhatt, P. N. (2005). Studies on the activity and stability of immobilized α-amylase in ordered mesoporous silicas. *Microporous and Mesoporous Materials, 77*(1), 67–77.

Partelli, S., Tamburrino, D., Crippa, S., Facci, E., Zardini, C., & Falconi, M. (2014). Evaluation of a predictive model for pancreatic fistula based on amylase value in drains after pancreatic resection. *The American Journal of Surgery, 208*, 634–639.

Paulová, L., Patáková, P., & Brányik, T. (2013). Advanced fermentation processes. In *Engineering aspects of food biotechnology* (Vol. 29, pp. 89–110). CRC Press.

Pham, V. H. T., Kim, J., Shim, J., Chang, S., & Chung, W. (2021). Purification and characterization of strong simultaneous enzyme production of protease and α-amylase from an extremophile-*Bacillus* sp. FW2 and its possibility in food waste degradation. *Fermentation, 8*(1), 12.

Pinotti, L. M., Lacerda, J. X., Oliveira, M. M., Teixeira, R. D., Rodrigues, C., & Cassini, S. T. (2017). Production of lipolytic enzymes using agroindustrial residues. *Chemical Engineering Transactions, 56*, 1897–1902.

Prakash, B., Vidyasagar, M., Madhukumar, M., Muralikrishna, G., & Sreeramulu, K. (2009). Production, purification, and characterization of two extremely halotolerant, thermostable, and alkali-stable α-amylases from *Chromohalobacter* sp. TVSP 101. *Process Biochemistry, 44*, 210–215.

Prakash, O., & Jaiswal, N. (2010). α-Amylase: An ideal representative of thermostable enzymes. *Applied Biochemistry and Biotechnology, 160*, 2401–2414.

Prongjit, D., Lekakarn, H., Bunterngsook, B., Aiewviriyasakul, K., Sritusnee, W., & Champreda, V. (2022). Functional characterization of recombinant raw starch degrading α-amylase from *Roseateles terrae* HL11 and its application on cassava pulp saccharification. *Catalysts, 12*(6), 647.

Ragab, M. M., Othman, H., & Hassabo, A. G. (2022). Various extraction methods of different enzymes and their potential applications in various industrial sectors (a review). *Egyptian Journal of Chemistry, 65*(10), 495–508.

Rakaz, M. A., Hussien, M. O., & Ibrahim, H. M. (2021). Isolation, extraction, purification, and molecular characterization for thermostable α-amylase from locally isolated *Bacillus* species in Sudan. *Biochemistry Research International, 2021*, 1–8.

Raul, D., Biswas, T., Mukhopadhyay, S., Das, S. K., & Gupta, S. (2014). Production and partial purification of alpha-amylase from *Bacillus subtilis* (MTCC 121) using solid state fermentation. *Biochemistry Research International, 2014*. https://doi.org/10.1155/2014/568141 (accessed on 15th November 2025).

Razdan, N., & Kocher, G. S. (2018). Utilization of damaged and spoiled wheat grains for bioethanol production. *Bioscience Biotechnology Research Communications, 11*, 658–673.

Reddy, N. S., Nimmagadda, A., & Rao, K. S. (2003). An overview of the microbial α-amylase family. *African Journal of Biotechnology, 2*(12), 645–648.

Rick, W., & Stegbauer, H. P. (1974). α-Amylase measurement of reducing groups. In *Methods of Enzymatic Analysis* (pp. 885–890). Academic Press.

Saboury, A. A. (2002). Stability, activity and binding properties study of α-amylase upon interaction with Ca^{2+} and Co^{2+}. *Biologia – Section Cellular and Molecular Biology, 57*(11), 221–228.

Saini, R., Saini, H. S., & Dahiya, A. (2017). Amylases: Characteristics and industrial applications. *Journal of Pharmacognosy and Phytochemistry, 6*(4), 1865–1871.

Sharma, N., Vamil, R., Ahmad, S., & Agarwal, R. (2012). Effect of different carbon and nitrogen sources on α-amylase production from *Bacillus amyloliquefaciens*. *International Journal of Pharmaceutical, 3*(4), 1161–1163.

Simair, A. A., Qureshi, A. S., Khushk, I., Ali, C. H., Lashari, S., Bhutto, M. A., Mangrio, G. S., & Lu, C. (2017). Production and partial characterization of α-amylase enzyme from *Bacillus* sp. BCC 01-50 and potential applications. *BioMed Research International, 2017*.

Sindhu, R., Binod, P., & Pandey, A. (2017). α-Amylases. In *Current developments in biotechnology and bioengineering* (pp. 3–24). Elsevier. https://doi.org/10.1016/B978-0-444-63662-1.00001-4 (accessed on 15th November 2025).

Siroosi, M., Borujeni, F. B., Amoozegar, M. A., Babavalian, H., & Hassanshahian, M. (2021). Halophilic amylase production and purification from *Haloarcula* sp. strain D61. *Biointerface Research in Applied Chemistry, 11*(1), 7382–7392.

Suganthi, R., Benazir, J. F., Santhi, R., Ramesh, K. V., Hari, A., Meenakshi, N., & Lakshmi, R. (2011). Amylase production by *Aspergillus Niger* under solid state fermentation using agroindustrial wastes. *International Journal of Engineering Science and Technology, 3*(2), 1756–1763.

Sundarram, A., & Murthy, T. P. K. (2014). α-Amylase production and applications: A review. *Journal of Applied & Environmental Microbiology, 2*(4), 166–175.

Suribabu, K., & Hemalatha, K. P. (2016). Thermal optimization of α-amylase production in *Brevibacillus* sp. *International Journal of Current Microbiology and Applied Sciences, 5*, 909–916.

Thippeswamy, S., Girigowda, K., & Mulimani, V. H. (2006). Isolation and identification of alpha-amylase producing *Bacillus* sp. from dhal industry waste. *Indian Journal of Biochemistry and Biophysics, 43*(5), 295.

Tiwari, S. P., Srivastava, R., Singh, C. S., Shukla, K., Singh, R. K., Singh, P., & Sharma, R. (2015). Amylases: An overview with special reference to alpha-amylase. *Journal of Global Biosciences, 4*(1), 1886–1901.

Tolan, J. S. (1996). Pulp and paper. In T. Godfrey & S. West (Eds.), *Industrial enzymology* (2nd ed., pp. 327–338). Stockton Press.

Unakal, C., Kallur, R. I., & Kaliwal, B. B. (2012). Production of α-amylase using banana waste by *Bacillus subtilis* under solid state fermentation. *European Journal of Experimental Biology, 2*, 1044–1052.

Van Dijl, J., & Hecker, M. (2013). *Bacillus subtilis*: From soil bacterium to super-secreting cell factory. *Microbial Cell Factories, 12*(1), 1–6.

Van Ee, J. H., Van Rijswijk, W. C., & Bollier, M. (1992). Enzymatic automatic dishwash detergents. *Chimica Oggi, 10*(8–9), 21–24.

Varavinit, S., Chaokasem, N., & Shobsngob, S. (2002). Immobilization of a thermostable alpha-amylase. *ScienceAsia, 28*(3), 247–251.

Vidyalakshmi, R., Paranthaman, R., & Indhumathi, J. (2009). Amylase production on submerged fermentation by *Bacillus* spp. *World Journal of Chemistry, 4*(1), 89–91.

Voskuhl, M. D., Mittal, S., Sharples, E. J., Vaidya, A., Gilbert, J., Friend, P. J., & Ploeg, R. J. (2014). Improving monitoring after pancreas transplantation alone: Fine-tuning of an old technique. *Clinical Transplantation, 28*, 1047–1053.

Wang, C., Lu, L., Huang, C., He, B., & Huang, R. (2020). Simultaneously improved thermostability and hydrolytic pattern of alpha-amylase by engineering central beta strands of TIM barrel. *Applied Biochemistry and Biotechnology*, 1–14.

Wang, Y., Bai, Y., Ji, H., Dong, J., Li, X., Liu, J., & Jin, Z. (2021). Insights into rice starch degradation by maltogenic α-amylase: Effect of starch structure on its rheological properties. *Food Hydrocolloids, 117*, 106724.

Welker, N. E., & Campbell, L. L. (1967). Unrelatedness of *Bacillus amyloliquefaciens* and *Bacillus subtilis*. *Journal of Bacteriology, 94*(4), 1124–1130.

Yao, D., Zhang, K., Su, L., Liu, Z., & Wu, J. (2021). Enhanced extracellular *Bacillus stearothermophilus* α-amylase production in *Bacillus subtilis* by balancing the entire secretion process in an optimal strain. *Biochemical Engineering Journal, 168*, 107948.

Yassin, S. N., Jiru, T. M., & Indracanti, M. (2021). Screening and characterization of thermostable amylase-producing bacteria isolated from soil samples of Afdera, Afar Region, and molecular detection of amylase-coding gene. *International Journal of Microbiology, 2021*, 5592885.

Zweers, J. C., Barák, I., Becher, D., Driessen, A. J., Hecker, M., Kontinen, V. P., Saller, M. J., Vavrová, L., & van Dijl, J. M. (2008). Towards the development of *Bacillus subtilis* as a cell factory for membrane proteins and protein complexes. *Microbial Cell Factories, 7*, 10.

CHAPTER 7

Recent Trends and Industrial Applications of Microbially Fabricated 2,3-Butanediol

PRITI PAL,[1] RAM PRASAD,[2] BRIJESH PANDEY,[3] ANAND PRAKASH,[3] SAURABH SINGH RATHORE,[3,4] MANISH MISHRA,[5] PRAKASH KUMAR SARANGI,[6,*] and AKHILESH KUMAR SINGH[3,*]

[1]*Amity Institute of Biotechnology, Amity University Uttar Pradesh, Lucknow Campus, Lucknow, Uttar Pradesh, India*

[2]*Department of Botany, School of Life Sciences, Mahatma Gandhi Central University, Motihari, Bihar, India*

[3]*Department of Biotechnology, School of Life Sciences, Mahatma Gandhi Central University, Motihari, Bihar, India*

[4]*Department of Life Sciences, Inter University Centre for Teacher Education, Bananas Hindu University, Varanasi, Uttar Pradesh, India*

[5]*Department of Biomedical Sciences, Mercer University School of Medicine, Macon, USA*

[6]*College of Agriculture, Central Agricultural University, Imphal, Manipur, India*

*Corresponding author

ABSTRACT

The non-oxidative metabolism of microorganisms produces 2,3-butanediol (BD), which has a wide range of industrial applications and a high market value. A particular focus has been placed on its microbial production as

Industrial Microbiology: Current Progress and Novel Technologies.
Prakash Kumar Sarangi, Akhilesh Kumar Singh, Rajesh Kumar Srivastava, and Brijesh Pandey (Eds.)

DOI: 10.1201/9781779642219-7

a substitute for petroleum-based production. Many 2,3-BD-producing microorganisms, including yeast and bacteria, have been segregated and their metabolisms extensively changed in order to technically manufacture 2,3-BD. In comparison to chemical-based processes, microbial production of 2,3-BD is thought to be more environmentally sustainable. The viability of the microbial production process on a commercial scale is constrained by this high energy consumption. The selection of the top strains and methodical optimization of the fermentation conditions are necessary for increased 2,3-BD production. Additionally, metabolic pathway engineering is necessary to optimize outcomes and reduce production costs by making the strains adept at utilizing inexpensive substrates. Currently, renewable energy sources, whose greatest benefit is their environmental friendliness, are used as raw materials in microbial fermentation, which is how most 2,3-BD is produced. A number of fermentation techniques have been put to the test using raw materials like lignocellulose, starch, sugar, and glycerol as feed-stocks. This chapter will present current details on the parameters and strategies for increased microbial 2,3-BD production.

7.1 INTRODUCTION

The production of 2,3-butanediol (2,3-BD) by microorganisms is a more environmentally responsible method that uses less energy and produces less CO_2. As a result, it has recently received more attention (Hakizimana et al., 2020). For the production of 2,3-BD, a variety of microorganisms, including yeast and bacteria, have been employed (Lee and Seo, 2019). About 2,3-BD is a valuable bulk chemical with chiral properties that is used to make butadiene or methyl ethyl ketone (MEK), as well as printing inks, perfumes, fumigants, insecticides, explosives, plasticizers, chemicals, food, and medicine (Erian et al., 2018). In addition to being a popular antifreeze additive and valuable fuel additive, 2,3-BD has a heating value of 27.2 kJ/g, which is comparable to those of methanol (22.1 kJ/g), n-butanol (33.1 kJ/g), and ethanol (29.1 kJ/g). Because of its high-octane rating, 2,3-BD is now used as a gasoline octane booster. Additionally, 2,3-BD can be dehydrated to form 1,3-butadiene, a crucial monomer in the production of synthetic rubber, or the powerful fuel additive methyl-ethyl-ketone (MEK), an industrial solvent with higher combustion values. About 2,3-BD is expected to reach US $ 220 million in the global market by 2027, growing at a compound annual growth rate of 3% from 2019 to 2027. Its current market price is higher than that of 1,4-butanediol (Maina et al., 2022; Tinco et al., 2021). The potential global market for 2,3-BD downstream products is 32 million metric tons per year, or US$43 billion (Amraoui et al., 2022; Mailaram et al., 2022). As a result of rising populations and better living conditions, energy demand is rising globally. According to reports, between 2012 and 2040, the world's energy needs could increase by as much as 48%. Harden and Walpole 1906 published the first report on the microbial synthesis of 2,3-BD. They used *Klebsiella pneumoniae*. Studies on 2,3-BD fermentation then started to flourish during World War II because of a high demand for 1,3-butadiene, which was first proposed by Fulmer et al. (1933). Since then, numerous initiatives have been made to find microbial strains that perform better and develop enhanced fermentation techniques. 2,3-BD can be produced by a wide variety of microbial strains. *K. pneumoniae*, *K. oxytoca*, and *Serratia marcescens* are all effective producer strains (Song et al., 2019). Numerous investigations into the production of 2,3-BD using nonpathogenic strains have been conducted (e.g., *Paenibacillus polymyxa, Bacillus licheniformis, B. subtilis*, and *B. amyloliquefaciens*). In a flask cultivation, the mutant strain of *E. aerogenes* KCTC 2190 converted 60 g/L of glucose into 12.52 g/L of 2,3-BD, which is the highest reported glucose conversion rate possible (Choi et al., 2016). In order to produce 2,3-BD industrially, chemical synthesis is preferred over microbial fermentation due to environmental issues and the depletion of fossil fuel resources. This is also quite possible because numerous bacterial strains produce large amounts of 2,3-BD on their own. Currently, the following species have been determined to be the most appropriate for 2,3-BD industrial production: *Serratia marcescens, Klebsiella pneumoniae, Klebsiella oxytoca, Enterobacter aerogenes, E. ludwigii, Paenibacillus polymyxa*, and *Saccharomyces cerevisiae* (Yang and Zhang, 2019). Despite the fact that these Gram-negative bacteria produce more than 150 g/L 2,3-BD using fully optimized industrial production methods, glucose, sucrose, hexoses, pentoses, disaccharides, and sugarcane molasses were found to be the most useful substrates for the microbial producers that are currently available (Banoon et al., 2020). However, because the cost of the substrate accounts for the majority of the production costs, efforts are being made to create bioprocesses that use other affordable, renewable non-food sources. The participation of *Paenibacillus polymyxa, Bacillus licheniformis, B. subtilis,* and *B. amyloliquefaciens* has led to the development of such successful laboratory-level processes (Song et al., 2019). Most of the bacterial isolates used to make 2,3-BD can use the C6 and C5 sugars found in the biomass of lignocellulose as substrates (Vivek et al., 2021). There has been a lot of work done on producing 2,3-BD using bio-based materials in recent decades. *Klebsiella* species have a higher 2,3-BD production rate due to their ability to use a variety of substrates and higher cultural adaptability. Xylose or glucose, which are used as substrates, starch, molasses, water hyacinth and Jerusalem artichoke tubers, sweet potatoes, and other sources are also available (Mailaram and Maity, 2022). Jatropha waste also provides a favorable raw material for the production of 2,3-BD and sugar (Li et al., 2015). However, ethanol, succinic, lactic, and acetic acids are among the liquid byproducts produced by this pathway (Wang et al., 2021). In addition to boosting 2,3-BD production, bio-engineered microorganisms have been developed to grow on a wider variety of substrates than before, especially less expensive ones. Other substrates, such as plant biomass, agricultural waste, and industrial waste, can also be used by microorganisms to produce biofuel and useful byproducts (Shaoqu et al., 2022). The choice of suitable high-yielding strains and methodical optimization of the culture conditions, including temperature, aeration, and byproduct control, are crucial for increasing 2,3-BD yield. Furthermore, picking a successful cultivation strategy is essential for getting the best yields, particularly batch, fed-batch, and continuous fermentation (Ioannidou et al., 2020). The most recent techniques have been examined and reviewed in this chapter to enhance microbial 2,3-butanediol production.

7.2 THE METABOLIC PATHWAY OF PRODUCTION 2,3-BD

Numerous strains of bacteria, yeasts, and even algae are examples of microbes that can produce 2,3-BD (Sabra et al., 2016). 2,3-BD is created through a mixed-acid fermentation process. Based on the microorganism and fermentation techniques, various end products are typically manufactured. The mechanism of 2,3-BD production and metabolic regulation remains a mystery. However, it is hypothesized that 2,3-BD biosynthesis serves an important physiological role by aiding in the regulation of the NADH/NAD+ ratio in cells, stopping acidification, and storing carbon and energy for microbial growth (Van Houdt et al., 2007). The *Klebsiella budB*, *budA*, and *budC* genes, respectively, encode these enzymes (Bialkowska, 2016). Alpha-acetolactate can spontaneously transform into diacetyl, but this reaction needs oxygen to occur. The produced diacetyl is then converted into acetoin as a byproduct of the reducing power. Based on the pH and oxygen levels, pyruvate can also undergo decrease or oxidation, resulting in a number of organic acids and alcohols like lactate, format, acetate, succinate, ethanol, and acetoin (Figure 7.1). Some techniques for increasing 2,3-BD production in order to reduce such undesirable byproducts include mutant screening, genetic engineering, and fermentation future technologies (Yang et al., 2015b). Additionally, two different enzymes convert pyruvate into acetolactate. The first enzyme, catabolic acetolactate synthase, is a component of the butanediol pathway and functions best in acetate at a pH of 5.8. Acetolactate decarboxylase, the second enzyme in the butanediol-producing process, has an ideal pH of 6.3 and is in charge of catalyzing the conversion of acetolactate into acetoin. Diacetyl (acetoin) reductase is a third enzyme that catalyzes the irreversible conversion of diacetyl to acetoin as well as the reversible conversion of acetoin to 2,3-BD. NADH serves as a coenzyme for the tetrameric enzyme dioleoyl (acetoin) reductase. Based on the stereospecificity's of the active acetoin reductase enzymes or whether a cyclic pathway is being used, different isomers of 2,3-BD are generated from acetoin. The dextro- and Levo-forms, which are optically active, and a misinform, which is optically inactive, make up the three 2,3-BD stereoisomers. The enzymes acetoin racemase, L(+) 2,3-BD dehydrogenase, and D(-) 2,3-BD dehydrogenase produce these three 2,3-BD stereoisomers. Different 2,3-BD stereoisomer formation mechanisms were developed in *K. pneumonia.* Additionally, a single enzyme, butanediol dehydrogenase, catalyzes the conversion of diacetyl to S-acetoin and then to dextro-butanediol (Chen et al., 2014). To maximize the production of various isomers, it is necessary to alter a number of culture conditions, including pH, temperature, and shaking rate (Chen et al., 2013). In theory, 2,3-BD can be made by fermenting a variety of monosaccharides, including hexoses and pentoses. In this instance, pyruvate is the key intermediate, and the synthesis of 2,3-BD necessitates the activity of three enzymes: acetolactate synthase, decarboxylase, and 2,3-BD dehydrogenase. However, some by-products, such as ethanol, acetate, lactate, format, and succinate, are frequently produced in addition to 2,3-BD. The fermentation yields are significantly impacted by these by-products.

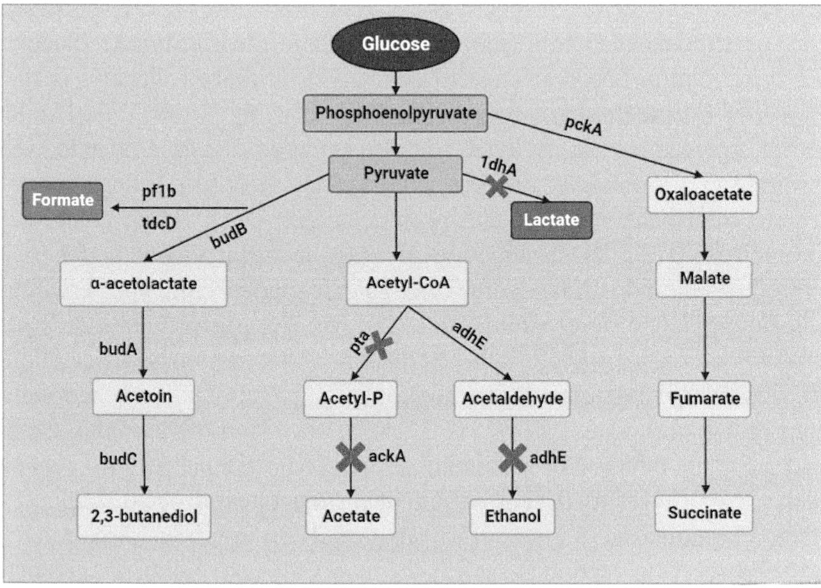

FIGURE 7.1 The metabolic pathways for 2,3-BD synthesis in microbes. Gene deletions for enhancing 2,3-BD synthesis are represented by the red crosses.

7.3 OPTIMIZATION OF SELECTIVE PARAMETERS FOR THE IMPROVED 2,3- BD PRODUCTION

To increase 2,3-BD yields, several researchers have concentrated on optimizing strains and production processes. The substrate, culture conditions, and microorganisms used all affect how much 2,3-BD is produced. Better strain selection, improved culture medium composition, and metabolic engineering of the biosynthetic pathway are all necessary for increased 2,3-BD production (Li et al., 2013a). The latter proved to be very effective, particularly by changing the production routes for ethanol, lactate, and acetate (Cho et al., 2015). For instance, higher yields of 2,3-BD by *Klebsiella* sp. were observed when lactate dehydrogenase was eliminated or acetaldehyde dehydrogenase was inactivated (Guo et al., 2014). Today, a variety of cultivation methods are employed, including fed-batch cultivation in environments that limit growth, batch cultivation in environments that promote resting cells, semi-continuous productions using immobilized cells, continuous cultivation and production using free cells, and solid-state fermentation. Hence, for increased 2,3-BD production, selecting the appropriate cultivation techniques is essential. The production of 2,3-BD is significantly influenced by the modulation of fermentation parameters. Numerous studies have found a link between 2,3-BD yields and oxygen concentration to be negative. Higher oxygen concentration influences cell mass formation while negatively affecting 2,3-BD formation. Briefly, a decrease in oxygen content increases 2,3-BD yield (Chan et al., 2016). However, in this situation, the lower cell concentration hinders the overall conversion rate (Ji et al., 2009). The regulation of metabolism also involves other elements like culture temperature, pH, and acetic acid supplementation (Celinska and Grajek, 2009). In addition to those cultural conditions' effects, genetic engineering has been successfully used to increase 2,3-BD production (Lee et al., 2013).

7.3.1 METABOLIC ENGINEERING FOR STRAINS WITH HIGH PRODUCTION

The ability to genetically modify microbes so they function as effective bio-factories has only recently become possible. However, a thorough understanding of metabolic characteristics is required in order to optimize the metabolic pathways for higher yields of the desired product (Park et al., 2013a). Some 2,3-BD pathway genes, especially those that need NADH to form the byproducts, can be changed or mutated to produce new, improved mutants (Jantama et al., 2015). Additionally, recent studies showed that over-expressing the genes responsible for the production of 2,3-BD and preventing the synthesis of byproducts increased 2,3-BD yields (Kim et al., 2012). For instance, when a short-chain acyl dehydrogenase from *K. pneumoniae* was used, high levels of L-2,3-BD were produced. Diacetyl and acetoin are reduced by the L-2,3-BD dehydrogenase, and L-2,3-BD itself is oxidized (Table 7.1) (Park et al., 2014). *K. oxytoca* GSC 12206, a 2,3-BD native hyperproducing and non-pathogenic bacterium, was isolated from cattle. The ldhA gene in this strain was disrupted, resulting in a mutant with yield and productivity of 0.41 g/g and 2.27 g/l.h and fabrication of 2,3-BD up to 115 g/l but less lactic acid (Kim et al., 2013).

Similar modifications were made to the genes for lactate dehydrogenase and phosphotransacetylase, which are involved in the pathways that result in the formation of lactic acid and acetic acid, in a wild-type *K. oxytoca* ME-303 to increase 2,3-BD production. While drastically lowering the byproducts of lactic acid and acetic acid by as much as 88% and 92%, respectively, the mutant ME-UD-3 produced 7.8% more than the wild type (Ji et al., 2008).

The same enzyme was altered in *K. pneumoniae*, and the resulting lactate dehydrogenase-deficient mutant (ΔldhA) was efficient enough to generate 2,3-BD at a rate of 2.48 g/Lh while using glucose as a substrate. When sugarcane was used as the substrate, productivity was 2.31 g/Lh and production was 76.24 g/L (Lee et al., 2013). Interfering with ethanol's biosynthesis is another way to increase 2,3-BD production. The *K. oxytoca* aldA gene, which codes for aldehyde dehydrogenase, was substituted with a tetracycline resistance cassette to create an ethanol-deficient mutant. It has been demonstrated that other microorganisms besides *Klebsiella* species can effectively produce 2,3-BD from a cheap substrate. Due to the overexpression of the NADH/NAD^+ regeneration system, *Bacillus amyloliquefaciens* has been shown to produce 2,3-BD at high concentrations of 102.3 g/L, 0.44 g/g, and 1.16 g/Lh of productivity. As a result, there was an increase in 2,3-BD production and no byproduct buildup. Carbon flux to the 2,3-BD branch is increased by the addition of the transcriptional regulator ALsR, which is controlled by the moderate promoter PbdhA (Yang et al., 2015b). By co-overexpressing the *gapA* and *bdH* genes, it was also possible to increase 2,3-BD production and stop the buildup of undesirable byproducts. Because of this, *B. amyloliquefaciens* B10–127 fed-batch fermentations produced more material—up to 132.9 g/L more—than before (Yang et al., 2013). When glucose was used as the substrate in *B. subtilis*, butanediol dehydrogenase (encoded by *K. pneumoniae* budC) was substituted for the (2R, 3R)-BD dehydrogenase gene to produce the new strain BSF9, which produced 103.7 g/L of pure meso-2-BD with a 0.487 g/g yield. Furthermore, it was found that *B. licheniformis* can use hexose and pentose as carbon sources and is a risk-free microorganism for the synthesis of 2,3-BD. Additionally, it has the ability to ferment lignocellulose hydrolysate to create 2,3-BD. For example, two strains, WX-02DbudC and WX-02DgldA, produced optically pure compounds with 32.2 g/L D-2,3-BD and 48.5 g/L meso-2,3-BD, respectively, using feed-batch fermentation. These two strains could also produce D-2,3-BD (10.9 g/L) and meso-2,3-BD (20.2 g/L) using xylose as a carbon source (Fu et al., 2016). In *Serratia marcescens*, the slaC gene was inactivated, which led to a decrease in meso-2,3-butanediol and (2S,3S)-2,3-butanediol accumulation and an increase in (3R)-acetoin accumulation. The bdhA gene from *B. subtilis* 168 was introduced into the slaC strain of *S. marcescens* MG1 to speed up the conversion of (3R)-acetoin to (2R,3R)-2,3-BD. Due to this (2R,3R)-2,3-BD dehydrogenase was produced in excess, and (2R,3R)-2,3-BD accumulated. Because butanediol dehydrogenase was highly expressed, fed-batch fermentation produced the highest amount of 89.81 g/L (2R,3R)-2,3-BD (Bai et al., 2015). Similar outcomes were attained by overexpressing 2,3-BD production genes in *Raoultella ornithinolytica,* which generated 68.27 g/L of 2,3-BD at a pH control of 5.5 and 400 rpm of agitation. The buildup of acetic acid, however, increased to 23.32 g/L. With the overexpression of homologous budABC genes, the production of 2,3-BD increased to 112.19 g/L and 1.35 g/Lh productivity, while the accumulation of acetic acid decreased to 9.71 g/L (Kim et al., 2016b). Yeast has recently received attention as a potential host cell for the synthesis of 2,3-BD, and several techniques have been developed to boost production. Pyruvate is an essential precursor for the synthesis of 2,3-BD in *S. cerevisiae* because it encourages ethanol production at the expense of 2,3-BD. Recent research has demonstrated that *S. cerevisiae* can produce 81.0 g/L 2,3-BD using a fed-batch fermentation technique with glucose as the carbon source. A pyruvate carbon flux tugging strategy was used to increase 2,3-BD production while decreasing dominant ethanol production in order to reach this high concentration (Ishii et al., 2018). In a different study, engineered *S. cerevisiae* that had the xylose metabolic pathway modulated produced a lot of 2,3-BD. In fed-batch fermentation using xylose as a substrate, the engineered strain produced a 2,3-BD titer of 96.8 g/L with 0.58 g/Lh of productivity (Kim et al., 2017b). However, it was discovered that under aerobic conditions and batch fermentation strategies, deletion of the ADH genes ADH1, ADH3, and ADH5, which encode the three major alcohol dehydrogenases (ADH), increased *S. cerevisiae* 2,3-butanediol production. With a yield of 0.113 g/g and a high 2,3-BD titer of 2.29 g/L, the engineered strain, which needed the addition of genes for the 2,3-BD biosynthetic pathway from both *E. coli* and *S. aerogenes*, was used to produce the most. AlsS first converts pyruvate into acetolactate, AlsD then converts acetoin into 2,3-BD, and finally butanediol dehydrogenase completes the process. Glycerol accumulation was stopped so that the desired carbon source could be used effectively by deleting the *GPD1* and *GPD*2 genes. By increasing production, the theoretical yield rose to 84% and 72.9 g/L (Kim and Hahn, 2015). It was also attempted to produce 2,3-BD using the Pichia expression system by introducing the *alsS* and *alsD* genes from *B. subtilis* and the BDH1 gene from *S. cerevisiae* into the genome. This is why, following statistical medium optimization, the strain produced 74.5 g/L (Yang and Zhang, 2018). Similar imports of the 2,3-BD biosynthesis pathway were made by some bacterial strains. For instance, three genes encoding the enzymes butanediol dehydrogenase (Bdh), acetolactate decarboxylase (AldC), and acetolactate synthase (Als) were added to *Z. mobilis* to produce 2,3-BD (Yang et al., 2016). In addition to glucose and xylose, mixed C6 and C5 sugar streams produced by the deacetylation and mechanical refining processes also produced much higher yields of 10 g/L 2,3-BD (Yang et al., 2016). *N. concharum,* on the other hand, makes use of the plentiful and sustainable material acetate to make bulk chemicals. The biosynthesis of acetoin and meso-2,3-BD has been integrated into *N. concharum*, which results in a high titer of 0.38 g/L of meso-2,3-buanediol (Li et al., 2019). Similar to how natural *Clostridium acetobutylicum* species cannot produce the D- and L-stereoisomers of acetoin because they lack acetoin reductase, they can produce 2,3-BD (ACR). *Clostridium acetobutylicum* ATCC 824 overexpressed ACR as a result. Nearly 90% of the naturally occurring acetoin could be converted by the engineered strains into 22 mM D-2,3-BD (Siemerink et al., 2011). In general, 2,3-BD production in microorganisms has been greatly enhanced by genetic engineering. The majority of the research focused on deleting the genes involved in competing pathways, overexpressing the 2, 3-BD pathway genes, or introducing the entire biosynthesis pathway into new organisms. An *in silico* simulation based on a metabolic model at the genomic scale was also tried to increase 2,3-BD yields. Despite all the work, it is still necessary to create top-notch strains that can ferment inexpensive materials while growing and producing more.

TABLE 7.1 Microorganisms Created Specifically for the Purpose of Producing 2, 3-BD

Microorganism	Substrate	Yield (g/g)	Titer (g/L)	Description	References
S. elongatus	CO_2	NA	0.496	Photosynthesis	Oliver et al. (2014)
E. coli	Glucose	0.41	73.8	Fed-batch	Xu et al. (2014)
E. coli	Glucose	0.42	115	Fed-batch	Ji et al. (2015)
B. subtilis	Glucose	0.46	49.29	Shake flask	Fu et al. (2014)
B. subtilis	Glucose	0.487	103.7	Fed-batch	Fu et al. (2016)
B. licheniformis	Glucose	0.4	98	Fed-batch	Qiu et al. (2016)

TABLE 7.1 *(Continued)*

Microorganism	Substrate	Yield (g/g)	Titer (g/L)	Description	References
B. licheniformis	Glucose	NA	123.7	Fed-batch	Ge et al. (2016)
B. amyloliquefaciens	Glucose	0.45	132.9	Fed-batch	Yang et al. (2013a)
B. amyloliquefaciens	Crude glycerol	0.44	102.3	Fed-batch	Yang et al. (2015b)
K. pneumoniae	Glucose	NA	116	Fed-batch	Guo et al. (2014)
K. pneumoniae	Glucose	NA	44	Fed-batch	Park et al. (2017b)
K. oxytoca	Glucose	0.42	142.5	Fed-batch	Cho et al. (2015)
E. aerogens	Sugarcane	NA	140	Fed-batch	Jung et al. (2015)
E. aerogens	Sugarcane molasses	0.395	21	Batch	Um et al. (2017)
E. cloacae	Glucose	0.475	119.4	Fed-batch	Li et al. (2015)
S. marcescens	Sucrose	0.35	89.81	Fed-batch	Bai et al. (2015)
S. cerevisiae	Xylose	NA	96.8	Fed-batch	Kim et al. (2017a)
S. cerevisiae	Glucose	0.404	154.3	Fed-batch	Kim et al. (2016a)
C. glutamicum	Glucose	NA	18.9	Batch	Yang et al. (2015a)
Z. mobilis	Glucose, xylose	0.16	15	Batch	Yang et al. (2016)

7.3.2 MEDIUM OPTIMIZATION

It is effective to increase the production of 2,3-BD by statistically optimizing the medium composition. The Plackett-Burman design is frequently used to pinpoint the essential components of the medium's constituents. For further optimizing level points for important factors, response surface methodologies such as Box-Behnken design and central composite design are used. It was discovered that the three main ingredients in the production of 2,3-BD using *P. polyxyma* ZJ-9 were inulin, potassium phosphate dibasic (K_2HPO_4), and ammonium chloride (NH_4Cl). Using an optimized medium, a 2,3-BD titer of 36.92 g/L was obtained. When cellulosic hydrolysate was used, similar strategies were used to maximize the production of 2,3-BD in *P. polymyxa* ICGEB2008, increasing the yield to 0.33 g/g sugar, which is comparable to that of rich medium (Adlakha and Yazdani, 2015). The highest reported titer of 2,3-BD in *K. pneumoniae* was obtained by adjusting the medium based on a two-level Plackett-Burman design (Ma et al., 2009). *B. amyloliquefaciens* B10–127 was found to produce 62.7 g/L 2,3-BD when grown in an optimized medium based on Plackett-Burman design, central composite design, and response surface analysis, as opposed to 45.7 g/L when grown in an unoptimized medium

(Yang et al., 2012). Notably, a working volume of 20 L could be reached using the optimized medium with a similar titer. Using *B. licheniformis* strain 10-1-A, improved the medium composition (Li et al., 2013b). As a nitrogen source, yeast extract was partially replaced with cheap corn steep liquor powder. With the help of an optimized medium and a productivity of 2.4 g/L/h, the 2,3-BD titer was achieved at 115.7 g/L.

7.3.3 pH OPTIMIZATION

The distribution of metabolites during fermentation for different micro-organisms and substrates is influenced by the pH of the culture medium, which has a significant effect on microbial production of 2,3-BD (Park et al., 2013c). It has been noted that the pH of the cultivation broth and 2,3-BD productivity have a negative relationship. At pH values between 5.0 and 6.5, 2,3-BD is typically the main product, whereas at 7.1 to 8.0, it is lactic acid. Accordingly, maintaining the cultivation broth within the proper pH range may increase the effectiveness of 2,3-BD production. The pH level has the potential to affect the metabolic pathway. Numerous studies have demonstrated that the pH range of 4.5 to 9.0 is where 2,3-BD is produced, with 6.0 being the ideal value (Park et al., 2013a; Wong et al., 2014). The best pH for the fabrication of 2,3-BD is, however, determined by the strains and substrates employed. Numerous studies have shown that a pH of 5.5 affects *Klebsiella oxytoca's* ability to produce 2,3-BD and other organic acids. Enzymes that produce flux (Als), 2,3-BD were more active at pH 5.5 than they were at pH 7.0. However, lactate dehydrogenase and acetate kinase showed higher activities at pH 7.0 compared to pH 5.5 (Park et al., 2013b). These enzymes are in charge of producing other organic acids. The ideal pH of 6.0 for the *Klebsiella* sp. Zmd30 strain may be related to cell growth. 2,3-BD was produced by batch fermentation at a rate of 57.17 g/l and by fed-batch production at a yield of 94% at 110 g/l (Wong et al., 2014). Several studies suggest that fed-batch operations are best suited for profit-oriented 2,3-BD production. The most successful experiment produced 0.86 g/l.h. (PD + BD) and a PD/BD ratio of 7.67 using native *Klebsiella* sp. Ana-WS5 in batch culture at 7.0 pH (Yen et al., 2014). Using glycerol and the forced pH fluctuations method, the highest yield of 2,3-BD was 70 g/l, compared to only 52.5 g/l in the previous study without pH control. The results of the forced pH changes showed that pH has an impact on microbial conversion processes and can increase the production of 2,3-BD (Nie et al., 2011). Similar studies suggest that acetic acid and pH have an impact on the growth of *Enterobacter aerogenes* as well as the production of 2,3-butanediol from glucose. Numerous studies have suggested that many cells can tolerate pH levels between 5.0 and 7.0, which could lead to the maximum (Zeng et al., 1990).

7.3.4 SELECTION OF CULTIVATION TECHNIQUES

Improving 2,3-BD production begins with developing a sound cultivation strategy. Whether batching or fed-batching, it entails managing substrate concentration and analyzing feeding strategies (Cheng et al., 2010). Batch operations typically have low productivity and are easily inhibited by substrates. To address this issue, it is preferable to continuously add substrate, such as glucose or molasses, at the appropriate rates throughout the cultivation using a fed-batch operation (Figure 7.2). Due to the production of large amounts of 2,3-BD, the effects of the initial substrate inhibition are significantly diminished (Wong et al., 2014). For *K. pneumonia*, an aerobic double-fed batch method was also developed. The pulse, constant feed rate, constant residual glucose concentration, and fed-batch exponential strategies are also employed by *K. pneumoniae*. A high titer of 150 g/l 2,3-BD was discovered using a feeding strategy with a constant residual glucose concentration. Continuous culture can significantly boost 2,3-BD productivity, but productivity is typically lower because significant amounts of sugar are still present in the product stream and persistently obstruct 2,3-BD formation (Ji et al., 2011). Bacteria in a fed-batch may use 2,3-BD as a carbon source for cell growth if the feeding is not timed. An improved pH-stat fed-batch culture strategy was developed in response to this situation, in which glucose and sodium hydrate were fed into a glucose level controlled by an automatic pH adjustment. Using this pH-stat fed-batch method, *K. oxytoca* significantly increased the productivity of 2,3-BD, reaching productivity and yield values of 127.9 g/l, 1.78 g/l.h, and 0.48 g/g (2,3-BD/glucose), respectively (Nie et al., 2011). The *E. aerogenes* SUMI014 strain produced 93.75 g/l with 0.49 g/g of yield and 1.74 g/l.h of productivity during batch fermentation. 2,3-BD was produced more abundantly by fed-batch fermentation using an acetate addition strategy, with yields of 126.10 g/l, 0.38 g/g, and productivity of 2.10 g/l.h (Lee et al., 2017). *Paenibacillus polymyxa* was able to produce the most 2,3-BD in a fed-batch fermentation by adjusting the conditions and fermentation medium. In a non-optimized medium, the toxic threshold of 2,3-BD production was found to be 50 g/l, while in an ideal environment, it was 68.54 g/l because fed-batch fermentation produces more because it uses more glucose to support cell growth and maintenance. In fed-batch fermentation, 2,3-BD productivity reached 68.54 g/l of titer, 0.34 g/g of yield, and 0.70 g/l.h under ideal circumstances (Okonkwo et al., 2017). Particularly, it is essential to gauge and monitor the substrate concentration. The experiments have demonstrated that glucose significantly increases 2,3 BD production.

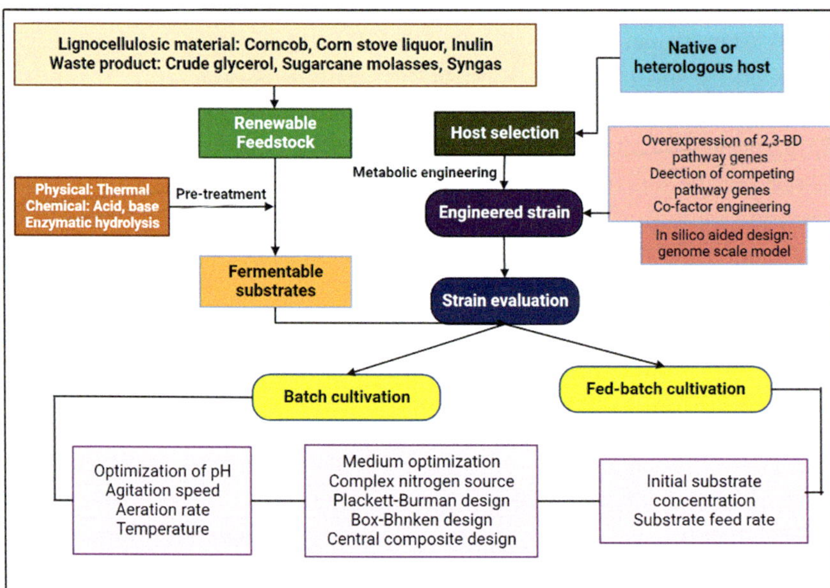

FIGURE 7.2 Approaches to increase 2,3-BD production using renewable feedstocks.

7.3.5 AERATION AND AGITATION

7.3.5.1 AERATION

Controlling the oxygen flow is essential for the fermentation of 2,3-BD. Even though they are facultative anaerobes, *Klebsiella* species produce 2,3-BD in low-oxygen environments to keep the pyridine nucleotide pool in a state of redox equilibrium for biosynthesis and glycolysis (Converti et al., 2003). Butanediol dehydrogenase uses glycolysis in a reversible reaction involving acetoin and 2,3-BD to produce NADH. Thus, the relative formation of acetoin to 2,3-BD may maintain the NAD^+/NADH balance (Blomqvist et al., 1993). Under aerobic conditions, α-acetolactate synthase is rapidly and irreversibly inactivated, according to a study. As a result, a greater supply of O_2 prevents the production of 2,3-BD. Aeration, especially micro aeration, increases its production (Perego et al., 2003). *Klebsiella* species, in particular, can produce energy through respiration and fermentation processes. Both pathways are simultaneously activated when a finite amount of oxygen is present, and as a result, the relative activities of each pathway have an impact on the production of 2,3-BD. However, reducing O_2 supply also lowers cell mass, which lowers 2,3-BD yield (Celinska and Grajek, 2009). When compared to anaerobic culture conditions, *K. oxytoca's* aerobic culture produced the most 2,3-BD and had the highest cell densities. Using the proper agitation and aeration, the concentration of butanediol inhibitors like lactic acid and ethanol was partially suppressed. During prolonged aeration-agitation, butanediol may be lost in favor of acetoin and acetic acid. It has been shown that the suppression of cell growth and a decrease in the rates of diol and acetoin formation occur when there is very little oxygen present. When the conditions for the medium and aeration were optimized, *E. aerogenes* produced 118.05 g/l of 2,3-BD in fed-batch fermentation, as opposed to *Enterobacter* species' 69.5 g/l (Jung et al., 2012). Furthermore, under optimal aeration conditions, *B. licheniformis* ATCC9789 was found to produce a large amount of 2,3-BD. The 2,3-BD yield of 0.44 g/g and productivity of 0.91 g/l/h were found to increase at an optimal oxygen transfer rate of 7 to 15 mmol/l/h (Rebecchi et al., 2018). In order to ensure that 2,3-BD production is carried out as effectively as possible, a strategy for controlling the oxygen supply must be developed because 2,3-BD is typically produced in low-oxygen environments.

7.3.5.2 AGITATION OF THE FERMENTATION MEDIUM

The 2,3-BD fermentation process depends on the agitation speed because it increases the fermentation yield. For example, when agitation speeds were between 200 and 300 rpm, a higher concentration of 2,3-BD was produced, yielding 89.9 g/l, while speeds outside of this range produced lower concentrations, 78.5 g/l at 400 rpm and 79.4 g/l at 100 rpm. The high accumulation of the main byproducts, ethanol and acetoin, may be to blame for this change. It was therefore impossible to achieve high concentration, high yield, and high productivity of 2,3-BD by maintaining a constant agitation speed throughout the entire culture process (Ji et al., 2009). However, a two-stage method of regulating the agitation speed during 2,3-BD fermentation may result in a greater yield of the compound. For instance, the method produced higher concentrations of 2,3-BD (95.5 g/l) and increased productivity by 22.14%. Consequently, a two-stage agitation speed control strategy for improved 2,3-BD production and adequate oxygen supply may be used (Chan et al., 2016; Ji et al., 2009). When agitation speeds were optimized for high 2,3-BD fabrication using *K. oxytoca* M1, it was discovered that increasing agitation speeds increased glucose consumption, dry weight, and 2,3-BD concentrations. The lower agitation speed did, however, result in a lower yield of 2,3-BD. Furthermore, the appropriate agitation speeds resulted in 78.8 g/l, 102.1 g/l, and 105.1 g/l at 200 rpm, 300 rpm, and 400 rpm, respectively (Cho et al., 2015). In a different experiment using *K. oxytoca,* productivity significantly changed from 0.43 to 2.7 g/l.h at 450 rpm, favoring cell growth (Park et al., 2013). This change was caused by increasing the agitation speed from 150 to 450 rpm. Additionally, the 200 rpm to 500 rpm range of agitation speeds examined in *E. coli* showed that up to 400 rpm, both biomass and 2,3-BD production increased and then decreased. At 200 rpm and 400 rpm, respectively, the yield of 2,3 BD increased from 0.38 g/g to 0.46 g/g while the productivity increased from 0.51 g/l.h to 1.48 g/l.h (Xu et al., 2014). An analysis of *Pichia pastoris* at three agitation speeds revealed that the highest productivity of 8.34 g/l was attained at 300 rpm (200 rpm, 300 rpm, and 400 rpm) (Yang and Zhang, 2018). According to the aforementioned findings, the fermentation strain is largely responsible for determining the optimal agitation speed, which is necessary for effective 2,3-BD fabrication.

7.3.6 OPTIMIZATION OF THE TEMPERATURE

Temperature control is crucial during fermentation because it affects enzymatic activity and cellular maintenance. In order to produce the most biomass, between 30 and 37°C is the ideal temperature range for bacterial fermentation (Perego et al., 2003). At these temperatures, the conditions change from aerobic to microaerobic, allowing the production of desired fermentation products rather than biomass. Cells and enzymes are altered above the optimum temperature, which has a minimal impact on metabolism and 2,3-butanediol production. At suboptimal temperatures, both regulation and metabolic rate may fail. It was discovered that lowering *K. pneumoniae* cultures' temperatures from 35°C to 30°C prevented the formation of byproducts and noticeably decreased ethanol production in favor of 2,3-BD synthesis. Around 33°C is believed to be the ideal temperature for *K. pneumoniae* in the given conditions, but temperature variations slightly affect 2,3-BD synthesis (Celinska and Grajek, 2009). *Paenibacillus polymyxa*, on the other hand, was found to grow best at 30°C in batch and fed-batch cultivation, while *E. aerogenes* was found to grow best at a temperature of 39°C (Converti et al., 2003). The ideal temperature is, however, greatly influenced by the strain and the cultivation medium. The optimal value must therefore be established specifically for each strain and substrate used.

7.3.7 SUBSTRATES

Numerous studies indicate that it might not be economically practical to produce 2,3-BD from glucose and sucrose. To cut costs, cheap and renewable substrates are used, in particular low-cost sources like lignocellulosic biomass, corncob molasses, corncob acid hydrolysate, and Jerusalem artichoke tuber hydrolysate (Zeng and Sabra, 2011). One of the strains with a broad range of fermentable substrates is *K. oxytoca,* which can also ferment molasses, enzyme or acid-hydrolyzed cereal mashes, wheat and starch, wood hydrolysates, and sulfate waste liquor. About 90% of the cellulose in the raw material is removed through pretreatment prior to fermentation, and this cellulose is then hydrolyzed and used as a substrate to produce 2,3-BD. Grajek and Celinska (2009) stated that the hydrolysate from biomass materials can have a pentose/glucose ratio of 1:1.5, it is practically considered essential that *K. oxytoca* be able to utilize both pentoses and hexoses. Notable is the ability to convert nearly all of the sugars found in hemicellulose and cellulose hydrolysates into 2,3-BD (Kim et al., 2017b). Succinic acid was increased from 0 g/l to 30 g/l to increase the productivity of 2,3-BD; 10 g/l produced the best outcomes. Additionally, *Enterobacter cloacae* subsp. dissolves SDM could produce 93.9 g/l of 2,3-BD by simultaneously using fermentation and saccharification processes with low-cost materials like crop-biomass cassava powder (Ripoll et al., 2016). To put it briefly, many substrates have been tried to act as a source of carbon. Better strains with higher 2,3-BD yields and the ability to use inexpensive substrates are still desperately needed.

7.4 RECOVERY STRATEGIES FOR 2,3-BD

Steam stripping, solvent extraction, reverse osmosis, and pervaporation are methods for separating 2,3-BD from other substances (Radoš et al., 2016). At a pilot plant, 2,3-BD was previously recovered from whole fermentation broths using a counter-current steam stripping technique. However, the additional labor and energy needed to complete such a process result in higher production costs (Yang et al., 2017). In particular, combining techniques like reverse osmosis and distillation can marginally lower processing costs. The combined method, which employs tributyl phosphate as an extractant, is less expensive than distillation and extraction carried out separately. There has been a lot of interest in liquid-liquid extraction, which includes techniques like solvent extraction and the aqueous two-phase extraction of 2,3-BD in the PEG/dextran system (Um et al., 2017). Additionally, using information from liquid-liquid equilibrium, 2,3-BD was drawn from aqueous streams as it was exposed to tetraoctyl ammonium and 2-methyl-1-naphthoate. The extraction of 2,3-BD was carried out using organic solvents such as alcohols and esters, in particular ethyl acetate, tributyl phosphate, diethyl ether, n-butanol, dodecanol, and oleyl alcohol (Savakis and Hellingwerf, 2015). The research was done to see if 2,3-BD could be recovered using repulsive extraction, potassium chloride (KCl) salting out, or K2CO3 dehydrating. In fact, salting out affects K_2CO_3 by affecting butanol extraction during the acetone-butanol-ethanol (ABE) fermentation.

Additionally, water must be removed before salting out because 2,3-BD, which is present in the fermentation broth in the lowest concentration, can still be salted out even in a saturated KCl or K_2CO_3 solution. Reactive extraction of 1,3-propanediol and 2,3-butanediol under acidic catalysis results in formal, which can react with formaldehyde. The reaction of acid methanol and 2,3-butanediol formal gathered in the top oil phase led to the production of 2,3-BD and methylal. The hydrolysis of methyl alcohol can result in the production of methanol and formaldehyde (Nakashima et al., 2014). Some techniques, such as the anionic extraction method, were demonstrated to be practical for recovering 2,3-BD because they involve the reversible complexation of phenylboronates with an anionic compound.

In addition, 2,3-BD can be recovered by back-extraction into an acidic solution. Conditions for extraction and back-extraction were improved using commercial 2,3-BD before being used with different fermentation broths. Under ideal circumstances, up to 72–93% of the compound 2,3-butanediol may be extracted, and 80–90% of the compound 2,3-BD may also be co-extracted from the broth along with other substances like glycerol or glucose (Dai et al., 2015). Numerous purification technologies have been developed, and various parameters have been optimized to increase the production of 2,3-BD. The development of an effective and financially viable downstream process for the purification of 2,3-BD from its fermentation broth is still a difficult task. This is primarily because it has a low concentration, a high affinity for water molecules, and other dissolved fermentation mash components are present (Bialkowska, 2016).

7.5 A TECHNO-ECONOMIC ANALYSIS OF THE PRODUCTION OF 2,3-BD

A critical stage in the growth of the bioeconomy is the techno-economic analysis of various technologies for the production of biofuels. To determine whether the construction of large-scale biorefineries is feasible, a technoeconomic assessment (TEA) should be conducted. The cost of the feedstock and the technology are important factors influencing the conversion of biomass into biofuels, according to numerous researchers (Bialkowska, 2016). However, there are not many reports on the production of TEA at 2,3-BD. Koutinas et al. (2016) demonstrated that 2,3-BD industrial production is feasible but also emphasized how this has a high minimum selling price (MSP) due to higher raw material costs and insufficient productivity. Glycerol, sucrose, and sugarcane molasses were used in fermentation along with previously reported experimental findings and downstream separation to improve this process (Li et al., 2013). The TEA revealed that the estimated MPS in this experimental test was greater than \$1/kg (Koutinas et al., 2016). In an effort to increase the amount of BD produced in a fed-batch culture using *Enterobacter ludwigii* to ferment very high polarity (VHP) cane sugar, Maina and colleagues (2019) also made an effort to optimize the bioreactor conditions. Higher yields (0.37 g/g) and productivity (3.95 g/L.h) of 2,3-BD values were obtained by keeping the culture at 33.9°C and pH 6.3. However, for a projected 50,000 metric tons of production annually, the estimated costs rise to \$2.67 per kg with such a fermentation system. In both TEA cases, MPS exceeded \$1/kg. The main determining factors are the high cost of the fermentation media, the separation and purification procedure, the fermentation's success, and the particular growth conditions needed by the strain used in the bio-process. There is still a need for high-yielding strains that can utilize easily accessible, inexpensive substrates in order to make 2,3-BD a widely available and reasonably priced platform chemical.

7.6 2,3-BD POTENTIAL APPLICATIONS

It is crucial to use high-price strategies when selling biochemicals. The competition between conventionally produced biochemicals and petrochemicals for the same industrial uses has made it challenging to enter the market. Finding the demand for high-value-added industries that are distinct from petrochemical products is crucial. About 2,3-BD is a promising industrial chemical with numerous potential uses in this field (Figure 7.3). The chemical, cosmetic, agricultural, pharmaceutical, and food industries could all make extensive use of it. As a bulk production

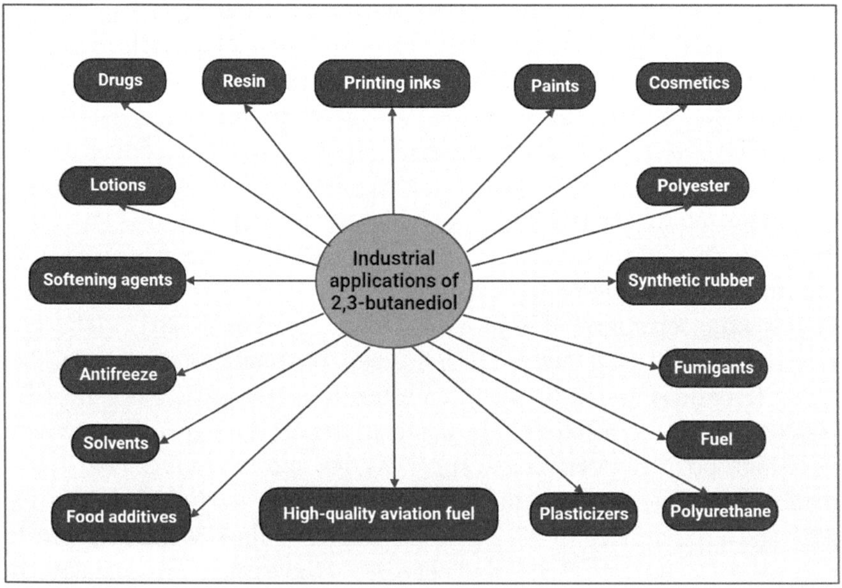

FIGURE 7.3 2,3-BD and their potential applications.

chemical, 2,3-BD has traditionally been used primarily in the chemical and fuel industries. This situation has delayed the commercialization of biological production techniques due to financial limitations (Song et al., 2019). However, new applications for this substance have emerged recently in the food, pharmaceutical, agricultural, and cosmetic sectors. As a result, 2,3-BD is receiving interest once more. These possibilities might contribute to the market and price growth of 2,3-BD.

7.6.1 UTILIZATION OF 2,3-BD IN THE FUEL, BULK CHEMICAL, AND POLYMER INDUSTRIES

The ability to dehydrate 2,3-BD using chemical catalysis to produce 1,3-butadiene, which could then be used to produce synthetic rubber, sparked industrial interest in the compound during World War II (Ji et al., 2011). MEK, a useful fuel additive as well as an industrial solvent for resins and lacquers, is produced by pinacol rearrangement as opposed to 1,2-elimination of a water molecule, which yields 1,3-butadiene. With a hydrogenation reaction, 2,3-BD and MEK can be used to make high-quality octane for aviation fuel (Bialkowska, 2016). Since (2R,3R)-BD has a freezing point below 30°C, it can be used as antifreeze. It can also be used as a monomer to create polymers (Ui et al., 1999). Additionally, by using 2,3-BD as a state-of-the-art chain extender and initiator, polyol and polymeric isocyanate, intermediates of polyurethanes, can be created (Joanna and Bogusław, 2006).

7.6.2 2,3-BD APPLICATION IN THE PERSONAL CARE AND COSMETICS SECTORS

Natural, environmentally friendly, and sustainable cosmetic raw materials are now in demand in the cosmetics industry. As an example, L'Oréal, the top cosmetics brand in the world, announced in 2013 that "Sharing beauty with all" would be its new sustainability goal for 2020 (The L'Oréal Sustainability Commitment, 2013). It has created a framework around four key commitments: creating, sustaining growth, surviving, and innovating. Because of this, 2,3-BD is one of the raw materials that is most well suited for modifications brought about by mega-trends in the cosmetics industry. It has a lot of potential to be used as a cosmetics and personal care product ingredient. For use as a cosmetic raw material, effectiveness and safety are equally crucial. Given that 2,3-BD is readily available in nature, its toxicity is predicted to be minimal. Reliable toxicity test results for 2,3-BD show that it is safe for use in cosmetics and does not result in acute toxicity, skin or eye irritation, or skin sensitization (found in Sigma-Safety Aldrich's data sheets for 2,3-BD and in safety assessment reports of alkane diols; Cosmetic Ingredient Review 2018 Safety Assessment of Alkane Diols as Used in Cosmetics). Regarding potential risks, data show that 2,3-BD can be used in cosmetic products as an emollient, humectant, and antiseptic. (2R,3S)-BD-containing formulations exhibit antibacterial properties as a preservative to enhance the product's storage stability (Baek et al., 2016). These compositions do not have the side effects of chemical preservatives like parabens and phenoxyethanol, which can irritate skin and act as allergens. 2,3-BD has been discovered in a variety of products, including soap, hand sanitizer, body wash, cleansing cream, cleansing gel, cleansing foam, cleansing water, and others (Baek et al., 2016). Additionally, 2,3-BD has already been included in a number of databases, such as the IECIC (Inventory of Existing Cosmetic Ingredients in China), INCI (International Nomenclature of Cosmetic Ingredients), and others. As a result, 2,3-BD is a cosmetic raw material. It is anticipated that 1,3-BD, a chemically derived substance present in many products today, will eventually be replaced by it.

7.6.3 APPLICATION OF 2, 3-BD IN AGRICULTURAL INDUSTRY

A number of uses for 2,3-BD are anticipated in the agricultural sector. It is critical to take into account 2,3-BD's environmental and human safety before using it as an agricultural product. Given that it is a known component of foods like fruits and wines, it is acknowledged to be a safe chemical (Jiang and Zhang, 2010). Additionally, we discovered that 2,3-BD had no phytotoxic effects on agricultural crops like tomato, cucumber, or pepper (unpublished data). Because 2,3-BD is easily biodegradable, there will not be much residual toxicity, a problem with chemical pesticides. Additionally, 2,3-BD has been shown to be effective as a component in the growth and preservation of plant species. Previous studies have demonstrated that the bacterial pathogen *Erwinia carotovora* in Arabidopsis is susceptible to the effects of (2R,3R)-BD, which can significantly promote growth (Ryu et al., 2004). Furthermore, it has been demonstrated that exposure to doses of (2R,3R)-BD ranging from 1 mg to 100 pg/plant can cause tobacco to develop a systemic resistance to *Erwinia carotovora* (Han et al., 2006). According to reports, 2,3-BD is effective in treating bacterial and fungal infections. In comparison to the water control, the soil application of (2R,3R)-BD can reduce the diseased leaf area of *Agrostis stolonifera* by 20–40% for the fungi *Microdochium nivale*, *Rhizoctonia solani*, or *Sclerotinia homoeocarpa*. Furthermore, it can reduce the number of *Colletotrichum orbiculare* lesions on *Nicotiana benthamiana* leaves by 77% (Cortes-Barco et al., 2010). Recent research has revealed some fascinating things, one of which is that 2,3-BD can effectively treat viral diseases. There is no effective antiviral drug at this time. According to a recent study, when compared to water control, 2,3-BD treatment can significantly lower the prevalence of naturally occurring viruses like cucumber mosaic virus and tobacco mosaic virus, increasing the yield of mature pepper fruits (Kong et al., 2018). It has been asserted that 2,3-BD can reduce a number of abiotic stresses. Arabidopsis developed a tolerance to drought after receiving treatments with (2R,3R)-BD. Leaf stomatal closure acted as a mediator for this drought tolerance (Cho et al., 2008). This phenomenon has been shown to be caused by guard cells being stimulated to produce H_2O_2 and NO accumulation (Wu et al., 2018), suggesting the potential use of 2,3-BD as a resistant material to prevent or lessen the effects of a drought disaster. Based on the multifunctional benefits of 2,3-BD, it is expected that agricultural products containing it as an active ingredient will be developed in the near future, potentially having a significant impact on the pesticide market.

7.6.4 ANOTHER APPLICATION FOR 2,3-BD

The pharmaceutical industry may soon become more interested in 2,3-BD because it has been noted that it can activate natural killer cell activity, improve innate immunity, and cause the separation of damaged liver cells (Lai et al., 2012). A previous study showed that 2,3-BD has anti-inflammatory properties. It can reduce the acute lung damage endotoxins cause in rats (Hsieh et al., 2007). These results suggest that 2,3-BD might be useful in human therapeutics. Because it is a natural substance found in naturally occurring or prepared products like wine, beer, fermented foods, soil, and plants (Ryu et al., 2004), 2,3-BD may also be used as a food additive or a health supplement. It is a substance that is safe for the environment and can be made through microbial fermentation.

7.7 CONCLUSION

It is acknowledged that producing 2,3-BD economically through microbial fermentation is a promising way to reduce reliance on fossil fuel-based chemical production methods. Recent improvements in genetic engineering and fermentation methods make microbial production of 2,3-BD very promising. Effective biological production systems could undoubtedly address the lack of synthetic chemicals and fossil fuels. Low yields and high production costs linked to the use of pricey ingredients like glucose continue to impede the production of 2,3-BD. Therefore, there is a need to create and test safe, highly productive microorganisms that can use inexpensive substrates to produce higher yields. Alternately, by stopping the synthesis of byproducts like ethanol, lactate, and acetate, 2,3-BD production could be boosted. The design of effective bioreactors and the optimization of fermentation parameters, such as culture medium, pH, pressure, temperature, and oxygen supply, should receive the most attention. In conclusion, improved fermentation conditions and meticulous work, particularly in the areas of genetic and metabolic engineering, are required for increased 2,3-BD production.

KEYWORDS

- **energy savings**
- **fermentation broth**
- **industrial biotechnology**
- **metabolic engineering**
- **metabolic pathway**
- **renewable substrates**

REFERENCES

Adlakha, N., & Yazdani, S. S. (2015). Efficient production of (R,R)-2,3-butanediol from cellulosic hydrolysate using *Paenibacillus polymyxa* ICGEB2008. *Journal of Industrial Microbiology & Biotechnology, 42*(1), 21–28.

Amraoui, Y., Prabhu, A. A., Narisetty, V., Coulon, F., Chandel, A. K., Willoughby, N., Jacob, S., Koutinas, A., & Kumar, V. (2022). Enhanced 2,3-butanediol production by mutant *Enterobacter ludwigii* using brewers' spent grain hydrolysate: Process optimization for a pragmatic biorefinery loom. *Chemical Engineering Journal, 427*(25), 130851.

Baek, H. S., Woo, B. Y., Yoo, S. J., Joo, Y. H., Shin, S. S., Oh, M. H., Lee, J. H., & Kim, S. Y. (2016). *Composition containing meso-2,3-butanediol.* WO 2016064180 A1.

Bai, F., Dai, L., Fan, J., Truong, N., Rao, B., Zhang, L., et al. (2015). Engineered *Serratia marcescens* for efficient (3R)-acetoin and (2R,3R)-2,3-butanediol production. *Journal of Industrial Microbiology & Biotechnology, 42*, 779–786.

Banoon, S., Ali, Z., & Salih, T. (2020). Antibiotic resistance profile of local thermophilic *Bacillus licheniformis* isolated from Maysan province soil. *Comunicata Scientiae, 11*, e3921.

Białkowska, A. M. (2016). Strategies for efficient and economical 2,3-butanediol production: New trends in this field. *World Journal of Microbiology & Biotechnology, 32*(12), 200.

Blomqvist, K., Nikkola, M., Lehtovaara, P., Suihko, M. L., Airaksinen, U., Stråby, K. B., Knowles, J. K., & Penttilä, M. E. (1993). Characterization of the genes of the 2,3-butanediol operons from *Klebsiella terrigena* and *Enterobacter aerogenes*. *Journal of Bacteriology, 175*(5), 1392–1404.

Celińska, E., & Grajek, W. (2009). Biotechnological production of 2,3-butanediol—Current state and prospects. *Biotechnology Advances, 27*(6), 715–725.

Chan, S., Jantama, S. S., Kanchanatawee, S., & Jantama, K. (2016). Process optimization on micro-aeration supply for high production yield of 2,3-butanediol from maltodextrin by metabolically engineered *Klebsiella oxytoca*. *PLOS ONE, 11*(9), e0161503.

Chen, C., Wei, D., Shi, J., Wang, M., & Hao, J. (2014). Mechanism of 2,3-butanediol stereoisomer formation in *Klebsiella pneumoniae*. *Applied Microbiology and Biotechnology, 98*, 4603–4613.

Chen, J. S., Zidwick, M. J., Rogers, P., et al. (2013). Organic acid and solvent production: Butanol, acetone, and isopropanol; 1,3- and 1,2-propanediol production; and 2,3-butanediol production. In E. Rosenberg (Ed.), *The Prokaryotes: Applied Bacteriology and Biotechnology* (pp. 77–134). Springer, Berlin Heidelberg.

Cheng, K., Liu, Q., Zhang, J., Li, J., Xu, J., & Wang, G. (2010). Improved 2,3-butanediol production from corncob acid hydrolysate by fed-batch fermentation using *Klebsiella oxytoca*. *Process Biochemistry, 45*(4), 613–616.

Cho, S. M., Kang, B. R., Han, S. H., Anderson, A. J., Park, J. Y., Lee, Y. H., Cho, B. H., Yang, K. Y., Ryu, C. M., & Kim, Y. C. (2008). 2R,3R-butanediol, a bacterial volatile produced by *Pseudomonas chlororaphis* O6, is involved in induction of systemic tolerance to drought in *Arabidopsis thaliana*. *Molecular Plant-Microbe Interactions, 21*(8), 1067–1075.

Cho, S., Kim, T., Woo, H. M., Lee, J., Kim, Y., & Um, Y. (2015). Enhanced 2,3-butanediol production by optimizing fermentation conditions and engineering *Klebsiella oxytoca* M1 through overexpression of acetoin reductase. *PLOS ONE, 10*(9), e0138109.

Choi, M. H., Kim, S. J., Kim, J. W., Park, Y. C., & Seo, J. H. (2016). Molecular cloning and expression of *Enterobacter aerogenes* alpha-acetolactate decarboxylase in pyruvate decarboxylase-deficient *Saccharomyces cerevisiae* for efficient 2,3-butanediol production. *Process Biochemistry, 51*, 170–176.

Converti, A., Perego, P., & Del Borghi, M. (2003). Effect of specific oxygen uptake rate on *Enterobacter aerogenes* energetics: Carbon and reduction degree balances in batch cultivations. *Biotechnology and Bioengineering, 82*(3), 370–377.

Cortes-Barco, A. M., Hsiang, T., & Goodwin, P. H. (2010). Induced systemic resistance against three foliar diseases of *Agrostis stolonifera* by (2R,3R)-butanediol or an isoparaffin mixture. *Annals of Applied Biology, 157*, 179–189.

Dai, J. Y., Zhao, P., Cheng, X. L., & Xiu, Z. L. (2015). Enhanced production of 2,3-butanediol from sugarcane molasses. *Applied Biochemistry and Biotechnology, 175*(6), 3014–3024.

Erian, A. M., Gibisch, M., & Pflügl, S. (2018). Engineered *E. coli* W enables efficient 2,3-butanediol production from glucose and sugar beet molasses using defined minimal medium as economic basis 06 biological sciences 0605 microbiology. *Microbial Cell Factories, 17*, 1–17.

Fu, J., Huo, G., Feng, L., Mao, Y., Wang, Z., Ma, H., Chen, T., & Zhao, X. (2016). Metabolic engineering of *Bacillus subtilis* for chiral pure meso-2,3-butanediol production. *Biotechnology for Biofuels, 9*(1).

Fu, J., Wang, Z., Chen, T., Liu, W., Shi, T., Wang, G., Tang, Y. J., & Zhao, X. (2014). NADH plays the vital role for chiral pure D-(−)-2,3-butanediol production in *Bacillus subtilis* under limited oxygen conditions. *Biotechnology and Bioengineering, 111*(10), 2126–2131.

Fulmer, E. I., Christensen, L. M., & Kendall, A. R. (1933). Production of 2,3-butylene glycol by fermentation. *Industrial & Engineering Chemistry Research, 25*, 798–800.

Ge, Y., Li, K., Li, L., Gao, C., Zhang, L., Ma, C., et al. (2016). Contracted but effective: Production of enantiopure 2,3-butanediol by thermophilic and GRAS *Bacillus licheniformis*. *Green Chemistry, 18*(17), 4693–4703.

Guo, X., Cao, C., Wang, Y., Li, C., Wu, M., Chen, Y., Zhang, C., Pei, H., & Xiao, D. (2014). Effect of the inactivation of lactate dehydrogenase, ethanol dehydrogenase, and phosphotransacetylase on 2,3-butanediol production in *Klebsiella pneumoniae* strain. *Biotechnology for Biofuels, 7*(1), 44.

Hakizimana, O., Matabaro, E., & Lee, B. H. (2020). The current strategies and parameters for the enhanced microbial production of 2,3-butanediol. *Biotechnology Reports, 25*, e00397.

Han, S. H., Lee, S. J., Moon, J. H., Park, K. H., Yang, K. Y., Cho, B. H., Kim, K. Y., Kim, Y. W., Lee, M. C., Anderson, A. J., & Kim, Y. C. (2006). GacS-dependent production of 2R,3R-butanediol by *Pseudomonas chlororaphis* O6 is a major determinant for eliciting systemic resistance against *Erwinia carotovora* but not against *Pseudomonas syringae* pv. *tabaci* in tobacco. *Molecular Plant-Microbe Interactions: MPMI, 19*(8), 924–930.

Harden, A., & Walpole, G. (1906). 2,3-Butylene glycol fermentation by *Aerobacter aerogenes*. *Proceedings of the Royal Society of London, 77*, 399–405.

Hazeena, S. H., Sindhu, R., Pandey, A., & Binod, P. (2020). Lignocellulosic bio-refinery approach for microbial 2,3-butanediol production. *Bioresource Technology, 302*, 122873.

Hsieh, S. C., Lu, C. C., Horng, Y. T., Soo, P. C., Chang, Y. L., Tsai, Y. H., Lin, C. S., & Lai, H. C. (2007). The bacterial metabolite 2,3-butanediol ameliorates endotoxin-induced acute lung injury in rats. *Microbes and Infection, 9*(12–13), 1402–1409.

Ioannidou, S. M., Pateraki, C., Ladakis, D., Papapostolou, H., Tsakona, M., Vlysidis, A., Kookos, I. K., & Koutinas, A. (2020). Sustainable production of bio-based chemicals and polymers via integrated biomass refining and bioprocessing in a circular bioeconomy context. *Bioresource Technology, 307*, 123093.

Ishii, J., Morita, K., Ida, K., Kato, H., Kinoshita, S., Hataya, S., Shimizu, H., Kondo, A., & Matsuda, F. (2018). A pyruvate carbon flux tugging strategy for increasing 2,3-butanediol production and reducing ethanol subgeneration in the yeast *Saccharomyces cerevisiae*. *Biotechnology for Biofuels, 11*(1), 180.

Ji, X. J., Huang, H., & Ouyang, P. K. (2011). Microbial 2,3-butanediol production: A state-of-the-art review. *Biotechnology Advances, 29*(3), 351–364.

Ji, X. J., Huang, H., Du, J., Zhu, J., Ren, L., Hu, N., & Li, S. (2009). Enhanced 2,3-butanediol production by *Klebsiella oxytoca* using a two-stage agitation speed control strategy. *Bioresource Technology, 100*(13), 3410–3414.

Ji, X. J., Huang, H., Li, S., Du, J., & Lian, M. (2008). Enhanced 2,3-butanediol production by altering the mixed acid fermentation pathway in *Klebsiella oxytoca*. *Biotechnology Letters, 30*(4), 731–734.

Ji, X. J., Liu, L. G., Shen, M. Q., Nie, Z. K., Tong, Y. J., & Huang, H. (2015). Constructing a synthetic metabolic pathway in *Escherichia coli* to produce the enantiomerically pure (R,R)-2,3-butanediol. *Biotechnology and Bioengineering, 112*(5), 1056–1059.

Jiang, B., & Zhang, Z. (2010). Volatile compounds of young wines from *Cabernet Sauvignon, Cabernet Gernischet* and *Chardonnay* varieties grown in the Loess Plateau region of China. *Molecules (Basel, Switzerland), 15*(12), 9184–9196.

Joanna, P., & Bogusław, C. (2006). New compounds for production of polyurethane foams. *Journal of Applied Polymer Science, 102*(6), 5918–5926.

Jung, M. Y., Jung, H. M., Lee, J., & Oh, M. K. (2015). Alleviation of carbon catabolite repression in *Enterobacter aerogenes* for efficient utilization of sugarcane molasses for 2,3-butanediol production. *Biotechnology for Biofuels, 8*, 106.

Jung, M. Y., Ng, C. Y., Song, H., Lee, J., & Oh, M. K. (2012). Deletion of lactate dehydrogenase in *Enterobacter aerogenes* to enhance 2,3-butanediol production. *Applied Microbiology and Biotechnology, 95*(2), 461–469.

Kanchanatawee, S. (2015). Efficient reduction of the formation of byproducts and improvement of production yield of 2,3-butanediol by a combined deletion of alcohol dehydrogenase, acetate kinase phosphotransacetylase, and lactate dehydrogenase genes in metabolically engineered *Klebsiella oxytoca* in mineral salts medium. *Metabolic Engineering, 30*, 16–26.

Kim, B., Lee, S., Park, J., Lu, M., Oh, M., Kim, Y., & Lee, J. (2012). Enhanced 2,3-butanediol production in recombinant *Klebsiella pneumoniae* via overexpression of synthesis-related genes. *Journal of Microbiology and Biotechnology, 22*(9), 1258–1263.

Kim, D. K., Rathnasingh, C., Song, H., Lee, H. J., Seung, D., & Chang, Y. K. (2013). Metabolic engineering of a novel *Klebsiella oxytoca* strain for enhanced 2,3-butanediol production. *Journal of Bioscience and Bioengineering, 116*(2), 186–192.

Kim, J. W., Kim, J., Seo, S. O., Kim, K. H., Jin, Y. S., & Seo, J. H. (2016a). Enhanced production of 2,3-butanediol by engineered *Saccharomyces cerevisiae* through fine-tuning of pyruvate decarboxylase and NADH oxidase activities. *Biotechnology for Biofuels, 9*, 265.

Kim, S. J., Sim, H. J., Kim, J. W., Lee, Y. G., Park, Y. C., & Seo, J. H. (2017a). Enhanced production of 2,3-butanediol from xylose by combinatorial engineering of xylose metabolic pathway and cofactor regeneration in pyruvate decarboxylase-deficient *Saccharomyces cerevisiae*. *Bioresource Technology, 245*(Pt B), 1551–1557.

Kim, S., & Hahn, J. S. (2015). Efficient production of 2,3-butanediol in *Saccharomyces cerevisiae* by eliminating ethanol and glycerol production and redox rebalancing. *Metabolic Engineering, 31*, 94–101.

Kim, T., Cho, S., Lee, S. M., Woo, H. M., Lee, J., Um, Y., & Seo, J. H. (2016b). High production of 2,3-butanediol (2,3-BD) by *Raoultella ornithinolytica* B6 via optimizing fermentation conditions and overexpressing 2,3-BD synthesis genes. *PLOS ONE, 11*(10), e0165076.

Kim, T., Cho, S., Woo, H. M., Lee, S. M., Lee, J., Um, Y., & Seo, J. H. (2017b). High production of 2,3-butanediol from glycerol without 1,3-propanediol formation by *Raoultella ornithinolytica* B6. *Applied Microbiology and Biotechnology, 101*(7), 2821–2830.

Kong, H. G., Shin, T. S., Kim, T. H., & Ryu, C. M. (2018). Stereoisomers of the bacterial volatile compound 2,3-butanediol differently elicit systemic defense responses of pepper against multiple viruses in the field. *Frontiers in Plant Science, 9*, 90.

Koutinas, A. A., Yepez, B., Kopsahelis, N., Freire, D. M. G., de Castro, A. M., Papanikolaou, S., & Kookos, I. K. (2016). Techno-economic evaluation of a complete bioprocess for 2,3-butanediol production from renewable resources. *Bioresource Technology, 204*, 55–64.

Kuittinen, S., Hietaharju, J., Bhattarai, I., Hassan, M. K., Kupiainen, L., Kangas, J., Tanskanen, J., & Pappinen, A. (2022). Technoeconomic analysis and environmental sustainability estimation of bioalcohol production from barley straw. *Biocatalysis and Agricultural Biotechnology, 43*, 102427.

Lai, H. C., Chang, C. J., Yang, C. H., Hsu, Y. J., Chen, C. C., Lin, C. S., Tsai, Y. H., Huang, T. T., Ojcius, D. M., Tsai, Y. H., & Lu, C. C. (2012). Activation of NK cell cytotoxicity by the natural compound 2,3-butanediol. *Journal of Leukocyte Biology, 92*(4), 807–814.

Lange, J. P. (2007). Lignocellulose conversion: An introduction to chemistry, process and economics. *Biofuels Bioproducts and Biorefining, 1*(1), 39–48.

Lee, S. J., Choi, H. S., Kim, C. K., Thapa, L. P., Park, C., & Kim, S. W. (2017). Process strategy for 2,3-butanediol production in fed-batch culture by acetate addition. *Journal of Industrial and Engineering Chemistry, 56*, 157–162.

Lee, S. M., Oh, B. R., Park, J. M., Yu, A., Heo, S. Y., Hong, W. K., Seo, J. W., & Kim, C. H. (2013). Optimized production of 2,3-butanediol by a lactate dehydrogenase deficient mutant of *Klebsiella pneumoniae*. *Biotechnology and Bioprocess Engineering, 18*(6), 1210–1215.

Lee, Y. G., & Seo, J. H. (2019). Production of 2,3-butanediol from glucose and cassava hydrolysates by metabolically engineered industrial polyploid *Saccharomyces cerevisiae*. *Biotechnology for Biofuels, 12*, 204.

Li, L., Li, K., Wang, Y., Chen, C., Xu, Y., Zhang, L., Han, B., Gao, C., Tao, F., Ma, C., & Xu, P. (2015). Metabolic engineering of *Enterobacter cloacae* for high-yield production of enantiopure (2R,3R)-2,3-butanediol from lignocellulose-derived sugars. *Metabolic Engineering, 28*, 19–27.

Li, L., Zhang, L., Li, K., Wang, Y., Gao, C., Han, B., Ma, C., & Xu, P. (2013b). A newly isolated *Bacillus licheniformis* strain thermophilically produces 2,3-butanediol, a platform and fuel bio-chemical. *Biotechnology for Biofuels, 6*(1), 123.

Li, T., Cheng, J., Huang, R., Zhou, J., & Cen, K. (2015). Conversion of waste cooking oil to jet biofuel with nickel-based mesoporous zeolite Y catalyst. *Bioresource Technology, 197*, 289–294.

Li, W., Pu, N., Liu, C., Yuan, Q., & Li, Z. (2019). Metabolic engineering of the marine bacteria *Neptunomonas concharum* for the production of acetoin and meso-2,3-butanediol from acetate. *Biochemical Engineering Journal, 151*, 107311.

Li, Y., Zhu, J., Wu, Y., et al. (2013a). Reactive-extraction of 2,3-butanediol from fermentation broth by propionaldehyde: Equilibrium and kinetic study. *Korean Journal of Chemical Engineering, 30*(1), 73–81.

Ma, C., Wang, A., Qin, J., Li, L., Ai, X., Jiang, T., Tang, H., & Xu, P. (2009). Enhanced 2,3-butanediol production by *Klebsiella pneumoniae* SDM. *Applied Microbiology and Biotechnology, 82*(1), 49–57.

Mailaram, S., Narisetty, V., Ranade, V. V., Kumar, V., & Maity, S. K. (2022). Techno-economic analysis for the production of 2,3-butanediol from brewers' spent grain using pinch technology. *Industrial & Engineering Chemistry Research, 62*(5), 2195–2205.

Maina, S., Prabhu, A. A., Vivek, N., Vlysidis, A., Koutinas, A., & Kumar, V. (2022). Prospects on bio-based 2,3-butanediol and acetoin production: Recent progress and advances. *Biotechnology Advances, 54*, 107783.

Maina, S., Stylianou, E., Vogiatzi, E., Vlysidis, A., Mallouchos, A., Nychas, G. E., de Castro, A. M., Dheskali, E., Kookos, I. K., & Koutinas, A. (2019). Improvement on bioprocess economics for 2,3-butanediol production from very high polarity cane sugar via optimization of bioreactor operation. *Bioresource Technology, 274*, 343–352.

Nakashima, N., Akita, H., & Hoshino, T. (2014). Establishment of a novel gene expression method, BICES (biomass-inducible chromosome-based expression system), and its application to the production of 2,3-butanediol and acetoin. *Metabolic Engineering, 25*, 204–214.

Nie, Z. K., Ji, X. J., Huang, H., et al. (2011). An effective and simplified fed-batch strategy for improved 2,3-butanediol production by *Klebsiella oxytoca*. *Applied Biochemistry and Biotechnology, 163*(8), 946–953.

Okonkwo, C. C., Ujor, V. C., Mishra, P. K., & Ezeji, T. C. (2017). Process development for enhanced 2,3-butanediol production by *Paenibacillus polymyxa* DSM 365. *Fermentation, 3*(2), 18.

Oliver, J. W., Machado, I. M., Yoneda, H., & Atsumi, S. (2014). Combinatorial optimization of cyanobacterial 2,3-butanediol production. *Metabolic Engineering, 22*, 76–82.

Park, C., Lu, M., Yun, S., Park, K., & Lee, J. (2013b). Effect of pH on the metabolic flux of *Klebsiella oxytoca* producing 2,3-butanediol in continuous cultures at different dilution rates. *Bioprocess and Biosystems Engineering, 36*(6), 845–855.

Park, J. M., Hong, W. K., Lee, S. M., Heo, S. Y., Jung, Y. R., Kang, I. Y., Oh, B. R., Seo, J. W., & Kim, C. H. (2014). Identification and characterization of a short-chain acyl dehydrogenase from *Klebsiella pneumoniae* and its application for high-level production of L-2,3-butanediol. *Journal of Industrial Microbiology and Biotechnology, 41*(9), 1425–1433.

Park, J. M., Rathnasingh, C., & Song, H. (2017). Metabolic engineering of *Klebsiella pneumoniae* based on in silico analysis and its pilot-scale application for 1,3-propanediol and 2,3-butanediol co-production. *Journal of Industrial Microbiology & Biotechnology, 44*(3), 431–441.

Park, J. M., Song, H., Lee, H. J., & Seung, D. (2013a). In silico aided metabolic engineering of *Klebsiella oxytoca* and fermentation optimization for enhanced 2,3-butanediol production. *Journal of Industrial Microbiology and Biotechnology, 40*(9), 1057–1066.

Park, J. M., Song, H., Lee, H. J., & Seung, D. (2013c). Genome-scale reconstruction and in silico analysis of *Klebsiella oxytoca* for 2,3-butanediol production. *Microbial Cell Factories, 12*(1), 20.

Perego, P., Converti, A., & Del Borghi, M. (2003). Effects of temperature, inoculum size and starch hydrolyzate concentration on butanediol production by *Bacillus licheniformis*. *Bioresource Technology, 89*(2), 125–131.

Polyiam, K. P., Khunnonkwao, P., Chan, S., Sangproo, M., Khor, K., Jantama, S. S., & Kanchanatawee, S. (2015). Efficient reduction of the formation of byproducts and improvement of production yield of 2,3-butanediol by a combined deletion of alcohol dehydrogenase, acetate kinase phosphotransacetylase, and lactate dehydrogenase genes in metabolically engineered *Klebsiella oxytoca* in mineral salts medium. *Metabolic Engineering, 30*, 16–26.

Qiu, Y., Zhang, J., Li, L., Wen, Z., Nomura, C. T., Wu, S., & Chen, S. (2016). Engineering *Bacillus licheniformis* for the production of meso-2,3-butanediol. *Biotechnology for Biofuels, 9*, 117.

Radoš, D., Turner, D. L., Catarino, T., Hoffart, E., Neves, A. R., Eikmanns, B. J., Blombach, B., & Santos, H. (2016). Stereospecificity of *Corynebacterium glutamicum* 2,3-butanediol dehydrogenase and implications for the stereochemical purity of bioproduced 2,3-butanediol. *Applied Microbiology and Biotechnology, 100*(24), 10573–10583.

Rebecchi, S., Pinelli, D., Zanaroli, G., et al. (2018). Effect of oxygen mass transfer rate on the production of 2,3-butanediol from glucose and agro-industrial byproducts by *Bacillus licheniformis* ATCC9789. *Biotechnology for Biofuels, 11*(1), 145.

Ripoll, V., de Vicente, G., Morán, B., Rojas, A., Segarra, S., Montesinos, A., Tortajada, M., Ramón, D., Ladero, M., & Santos, V. E. (2016). Novel biocatalysts for glycerol conversion into 2,3-butanediol. *Process Biochemistry, 51*(6), 740–748.

Ryu, C. M., Farag, M. A., Hu, C. H., Reddy, M. S., Kloepper, J. W., & Paré, P. W. (2004). Bacterial volatiles induce systemic resistance in *Arabidopsis*. *Plant Physiology, 134*(3), 1017–1026.

Sabra, W., Groeger, C., & Zeng, A. P. (2016). Microbial cell factories for diol production. *Advances in Biochemical Engineering / Biotechnology, 155*, 165–197.

Savakis, P., & Hellingwerf, K. J. (2015). Engineering cyanobacteria for direct biofuel production from CO_2. *Current Opinion in Biotechnology, 33*, 8–14.

Siemerink, M. A., Kuit, W., López-Contreras, A. M., Eggink, G., van der Oost, J., & Kengen, S. W. (2011). D-2,3-butanediol production due to heterologous expression of an acetoin reductase in *Clostridium acetobutylicum*. *Applied and Environmental Microbiology, 77*(8), 2582–2588.

Song, C. W., Park, J. M., Chung, S. C., Lee, S. Y., & Song, H. (2019). Microbial production of 2,3-butanediol for industrial applications. *Journal of Industrial Microbiology & Biotechnology, 46*(11), 1583–1601.

Tinco, D., Borschiver, S., Coutinho, P. L., & Freire, D. (2021). Technological development of the bio-based 2,3-butanediol process. *Biofuels Bioproducts and Biorefining, 15*(2), 357–376.

Ui, S., Mimura, A., Ohkuma, M., & Kudo, T. (1999). Formation of a chiral acetoinic compound from diacetyl by *Escherichia coli* expressing meso-2,3-butanediol dehydrogenase. *Letters in Applied Microbiology, 28*(6), 457–460.

Um, J., Kim, D. G., Jung, M., Saratale, G. D., & Oh, M. (2017). Metabolic engineering of *Enterobacter aerogenes* for 2,3-butanediol production from sugarcane bagasse hydrolysate. *Bioresource Technology, 245*, 1567–1574.

Van Houdt, R., Aertsen, A., & Michiels, C. W. (2007). Quorum-sensing-dependent switch to butanediol fermentation prevents lethal medium acidification in *Aeromonas hydrophila* AH-1N. *Research in Microbiology, 158*(4), 379–385.

Vivek, N., Hazeena, S. H., Alphy, M. P., Kumar, V., Magdouli, S., Sindhu, R., Pandey, A., & Binod, P. (2021). Recent advances in microbial biosynthesis of C3–C5 diols: Genetics and process engineering approaches. *Bioresource Technology, 322*, 124527.

Wang, D., Oh, B. R., Lee, S., Kim, D. H., & Joe, M. H. (2021). Process optimization for mass production of 2,3-butanediol by *Bacillus subtilis* CS13. *Biotechnology for Biofuels, 14*(1), 15.

Wong, C. L., Yen, H. W., Lin, C. L., & Chang, J. S. (2014). Effects of pH and fermentation strategies on 2,3-butanediol production with an isolated *Klebsiella* sp. Zmd30 strain. *Bioresource Technology, 152*(3), 169–176.

Wu, L., Li, X., Ma, L., Borriss, R., Wu, Z., & Gao, X. (2018). Acetoin and 2,3-butanediol from *Bacillus amyloliquefaciens* induce stomatal closure in *Arabidopsis thaliana* and *Nicotiana benthamiana*. *Journal of Experimental Botany, 69*(22), 5625–5635.

Xie, S., Li, Z., Zhu, G., Song, W., & Yi, C. (2022). Cleaner production and downstream processing of bio-based 2,3-butanediol: A review. *Journal of Cleaner Production, 343*, 131033.

Xu, Y., Chu, H., Gao, C., Tao, F., Zhou, Z., Li, K., Li, L., Ma, C., & Xu, P. (2014). Systematic metabolic engineering of *Escherichia coli* for high-yield production of fuel bio-chemical 2,3-butanediol. *Metabolic Engineering, 23*, 22–33.

Yang, J., Kim, B., Kim, H., Kweon, Y., Lee, S., & Lee, J. (2015a). Industrial production of 2,3-butanediol from the engineered *Corynebacterium glutamicum*. *Applied Biochemistry and Biotechnology, 176*(8), 2303–2313.

Yang, S., Mohagheghi, A., Franden, M. A., Chou, Y. C., Chen, X., Dowe, N., Himmel, M. E., & Zhang, M. (2016). Metabolic engineering of *Zymomonas mobilis* for 2,3-butanediol production from lignocellulosic biomass sugars. *Biotechnology for Biofuels, 9*(1), 189.

Yang, T., Rao, Z., Zhang, X., Xu, M., Xu, Z., & Yang, S. T. (2013). Improved production of 2,3-butanediol in *Bacillus amyloliquefaciens* by over-expression of glyceraldehyde-3-phosphate dehydrogenase and 2,3-butanediol dehydrogenase. *PLOS ONE, 8*(10), e76149.

Yang, T., Rao, Z., Zhang, X., Xu, M., Xu, Z., & Yang, S. T. (2015b). Enhanced 2,3-butanediol production from biodiesel-derived glycerol by engineering of cofactor regeneration and manipulating carbon flux in *Bacillus amyloliquefaciens*. *Microbial Cell Factories, 14*(1), 122.

Yang, T., Rao, Z., Zhang, X., Xu, M., Xu, Z., & Yang, S. T. (2017). Metabolic engineering strategies for acetoin and 2,3-butanediol production: Advances and prospects. *Critical Reviews in Biotechnology, 37*(8), 990–1005.

Yang, T., Zhang, X., Rao, Z., Gu, S., Xia, H., & Xu, Z. (2012). Optimization and scale-up of 2,3-butanediol production by *Bacillus amyloliquefaciens* B10-127. *World Journal of Microbiology & Biotechnology, 28*(4), 1563–1574.

Yang, Z., & Zhang, Z. (2018). Production of (2R,3R)-2,3-butanediol using engineered *Pichia pastoris*: Strain construction, characterization and fermentation. *Biotechnology for Biofuels, 11*, 35.

Yang, Z., & Zhang, Z. (2019). Recent advances on production of 2,3-butanediol using engineered microbes. *Biotechnology Advances, 37*, 569–578.

Yen, H. W., Li, F. T., Wong, C. L., & Chang, J. S. (2014). The pH effects on the distribution of 1,3-propanediol and 2,3-butanediol produced simultaneously by using an isolated indigenous *Klebsiella* sp. Ana-WS5. *Bioprocess and Biosystems Engineering, 78*(3), 425–431.

Zeng, A. P., Biebl, H., & Deckwer, W. D. (1990). Effect of pH and acetic acid on growth and 2,3-butanediol production of *Enterobacter aerogenes* in continuous culture. *Applied Microbiology and Biotechnology, 33*(5), 485–489.

Zeng, A., & Sabra, W. (2011). Microbial production of diols as platform chemicals: Recent progresses. *Current Opinion in Biotechnology, 22*(6), 749–757.

CHAPTER 8

Bacillus-Meditated Fermentation and Its Role in Quality Enhancement of Soybean Fermented Food

SHREETAMA BHATTACHARJEE, ISRA AMAN AZIZ, YASHVARDHAN SINGH, ADITI SINGH,* and MALA TRIVEDI*

Amity Institute of Biotechnology, Amity University Uttar Pradesh, Lucknow Campus, Lucknow, Uttar Pradesh, India

*Corresponding author

ABSTRACT

One of the first biotechnologies for producing food products with desirable qualities, such as a long shelf life and good organoleptic features, is fermentation. The microbiological stability and safety of finished fermented foods are typically increased, and some can even be stored at room temperature. There are many Microorganisms (bacteria, yeast, and fungi) involved in the production of fermented food and beverages. These microbes degrade complex molecules by producing secondary metabolites, alcohols, and acids during the fermentation process. Bacillus, belonging to the family Bacillaceae, is considered one of the most useful genera in microbial fermentations. Many genera like *Bacillus subtilis, Bacillus amyloliquefacians, Bacillus pumilus, Bacillus cereus,* are major participants in the production of soyabean fermentation. *Bacillus* has the ability to synthesize polymers of unique properties like high water holding capacity, fine structured network, low cytotoxicity which supports the production of wide range of commercial products like food additives. This chapter summarizes the important role of genus *Bacillus* in soybean fermentation,

Industrial Microbiology: Current Progress and Novel Technologies.
Prakash Kumar Sarangi, Akhilesh Kumar Singh, Rajesh Kumar Srivastava, and Brijesh Pandey (Eds.)

DOI: 10.1201/9781779642219-8

with a focus on nutritional properties of its fermented food, the fermentation mechanism, many global soybean fermented traditional food using genus *Bacillus*, the health benefits of these fermented foods and profiles of bioactive compounds present in them.

8.1 INTRODUCTION

Microorganisms have long been used to produce food and beverages by the process of fermentation. Moreover, microbial fermentation products are used in food and beverages as additives, preservatives, aroma, or colorants, and so forth. Exploring the physiology and mechanism of the microbial diversity has huge potential due to the role of microbes in nutrient cycle, bioremediation, production of novel pharmaceutical and industrial products, like enzymes, metabolites, fermented food, and beverages, and so forth. But fermentation can be carried out only by a few microorganisms, mainly by yeasts, fungi, and bacteria. Bacteria are the most common group of this diversity, which lives in a variety of ecological niches, including extreme environments in both thc lithosphere and the hydrosphere, where their metabolic activities play an important role in geochemical nutrient cycling and the production of a wide range of industrially significant products (Lyngwi and Joshi, 2014).

In this chapter, we have shed light on the *Bacillus* genus, which contains a variety of industrially significant species that are often employed as starting cultures in the fermentation industries. The Gram-positive endospore forming *Bacillus* genus comprises of both mesophiles and extremophiles; hence, this genus is considered to be one of the most varied genera in the bacilli class. *Bacillus* is generally considered a category of soil dwellers (Elshaghabee et al., 2017). *Bacillus* spp. can be isolated from several sources, including air, water, human, and animal intestines, as well as vegetables and food (Alou et al., 2015). Their cell walls contain a vast range of DNA base and amino acid compositions. Recent findings have discovered that they can be either aerobes or facultative anaerobes, depending on the conditions provided. These microbes are metabolically chemoorganotrophs, which rely on organic chemicals for carbon and energy (Baruzzi et al., 2011). Furthermore, they have the capacity to produce resistant strains which enables them to colonize a wide range of habitats. *Bacillus* strains may be cultured at high temperatures and pH levels to create products that are stable in a variety of harsh settings.

Because of their wide ubiquity in nature and genetic and metabolic diversity, which leads to the production of several antibiotics and enzymes, these bacteria have become increasingly interesting for various biotechnological applications ranging from the production of fermented foods to engineered industrial enzymes used in the food and detergent industries (Priest, 1993). The large divergence in physiological types found in *Bacillus* spp., along with the facts that most members of the genus are non-pathogenic, relatively easy to manipulate by genetics, good secretors of proteins and metabolites, and simple to cultivate, make *Bacillus* one of the preferred hosts for production by fermentation (Arbige et al., 1993).

8.2 FERMENTATIVE MECHANISM OF *BACILLUS*

In simpler terms, fermentation is a chemical process in which complex molecules are broken down into simpler molecules, producing liquid products and gases (Haile and Kang, 2019). Fermented food and beverages were initially produced to prolong the shelf life and yield of raw materials obtained from agriculture and animal husbandry. Nowadays, the scene has changed; the bioprocessing industries aim on employing microorganisms and their enzymes to produce goods with desirable traits such as enhanced shelf life, texture, taste, fragrance, colorants, preservatives, sweeteners, and so forth, through processes like acidification, alkalinization, proteolysis, or amino acid conversions. In addition, it plays a crucial role in producing nutritious foods in a natural and relatively inexpensive and sustainable manner with health-promoting elements. The fermentation process increases the organoleptic qualities of various foods (Dimidi et al., 2019; Sivamaruthi et al., 2018).

Bacillus spp. has been used safely in food and industry for many years (Priest, 1977). The US Food and Drug Administration (FDA) has generally recognized a number of *Bacillus* spp. products as safe for certain usages (Arbige et al., 1993). *Bacillus* species are an excellent source of high-quality bioactive chemicals. Furthermore, a high amount of protein, synthesized by *Bacillus* and its related taxa, goes directly into the growth medium and has been found to be safe; this also makes this species a good choice as a source of heterologous proteins. In fact, *Bacillus* species fermentation produces almost two-thirds of the enzyme market (proteases, amylases, rennet replacements, endonucleases, glucose-dehydrogenase, and pullulanase) for industrial uses (Baruzzi et al., 2011). For example,

B. subtilis has been utilized to produce nucleotides, which are used as food flavor enhancers, as well as amino acids and vitamins like biotin, folic acid, and riboflavin (Queener and Lively, 1989). More than 500 antimicrobial compounds have been identified as bio-preservatives to date (Baruzzi et al., 2011). *Bacillus* spp.'s fastest expanding field of production is the expression of heterologous gene products (Arbige et al., 1993). The *Bacillus* genus has been shown to create about 45 antimicrobial molecules; some of these compounds are clinically useful, some are tested *in vitro* to control food bacteria, and the remainder control plant diseases (Stein, 2005; Urdaci and Pinchuk, 2004).

8.2.1 CHEMICAL/ NUTRITIONAL PROPERTIES OF BACILLUS FERMENTED FOOD

Fermentation causes major chemical changes in fermented foods. Soybean fermentation with *Bacillus* spp. happens in an alkaline environment. Traditional fermentation of *Bacillus* spp. has demonstrated that proteins are hydrolyzed into amines and ammonia, which causes the pH change. The moisture content increases somewhat at the end of the process (Sarkar et al., 1994). Carbohydrates have been shown to decrease as a result of microorganisms using them for energy (Hu et al., 2010).

Bacillus-mediated food fermentations can provide some of the important industrial features of this genus. In Japan, *Bacillus subtilis* is used to make the fermented soybean food named natto. While *B. subtilis* (natto) produces many enzymes, including amylases and cellulases, the most important enzymes in the production of natto are proteases; two proteases having a pH optimum of 8.5 and 10.3–10.8 have been characterized (Yoshimoto et al., 1971). The enzyme, protease, is primarily responsible for the taste in natto, via the hydrolysis of soybeans. D-Ribose is a taste enhancer that is commonly used in food and health industries. Several strains of *B. subtilis* and *Bacillus pumilus* and their mutants are reported to produce significant amounts of D-ribose (De Wulf and Vandamme, 1997). Recent advances in genetic engineering and fermentation technology have helped *Bacillus* fermentations produce more D-ribose (Lyngwi and Joshi, 2014). Furthermore, fermenting black beans increases their antioxidant and anthocyanin content in food supplements.

The majority of soybean fermented products, which are manufactured by traditional or conventional methods in Asia, contain prominent levels of *Bacillus* spp., namely *Bacillus subtilis, Bacillus amyloliquefaciens, Bacillus pumilus, Bacillus licheniformis,* and so forth. Some species of *Bacillus* are listed in Table 8.1. During the cheonggukjang or natto manufacturing process, the fermentation of cooked soybeans by bacteria such

TABLE 8.1 *Bacillus* Species Involved in the Production of Soybean Fermented Foods and Their Health Benefits

Sl. No.	Fermented Soybean Products	*Bacillus* spp.	Health Benefits	References
1.	*Natto* (Japan)	*Bacillus subtilis*	Improved Digestibility, reduces gastric problems, fibrinolytic activity	Sumi et al. (1987)
2.	*Thua Nao* (Thailand)	*Bacillus subtilis; Bacillus megaterium*	Antioxidant; reduces DNA damage; improved glucose control	Inatsu et al. (2006)
3.	*Soy stink tofu* (East Asia)	*Bacillus sphaericus*	Cancer inhibition properties; decrease cholesterol	Gopikrishna et al. (2021)
4.	*Tempeh* (Indonesian)	*Bacillus subtilis*	Reduces duration of Diarrhea; inhibits proliferation of pathogens in gut.	Esaki et al. (1996)
5.	*Cheonggukjang* (Korea)	*Bacillus licheniformis; Bacillus pumilus; Bacillus subtilis*	Bone health, antioxidants, anti-hypertension	Jang et al. (2021); Shin (2015)
6.	*Kinema* (Nepal)	*Bacillus amyloliquefaciens; Bacillus pumilus; Bacillus subtilis; Bacillus licheniformis; Bacillus circulans*	Enhances cognitive functions; removes indigestible oligosaccharides	Yang et al. (2015)
7.	*Miso* (Japan)	*Bacillus subtilis*	Anti- tumor; improved gut health; anti- diabetic; enhance immune system	Da Silva (2011)
8.	*Doenjang* (Korea)	*Bacillus subtilis Bacillus siamensis*	Anti-tumor; anti-inflammatory; anti-obesogenic; cure for liver disease	Gopikrishna et al. (2021)
9.	*Tungrymbai* (Northeast India)	*Bacillus subtilis; Bacillus cereus; Bacillus licheniformis*	ACE Inhibitory activities; anti-microbial properties	Chettri and Tamang (2015)
10.	*Bekang* (Northeast India)	*Bacillus subtilis; Bacillus cereus*	Antioxidant properties; enhances digestibility	Chettri and Tamang (2015)

as *Bacillus subtilis* produces various metabolites, including peptones, peptides, amino acids, sugars, organic acids, nattokinase, levan, and poly-glutamic acid (PGA), which is responsible for the sticky and slimy texture of these foods. Additionally, these secondary metabolites markedly affect the organoleptic and biological properties of the resulting products (Jeon et al., 2018). Several proteinases produced by *Bacillus pumilus* and *Bacillus subtilis* and found in fermented soy products have amyloid-degrading activity, suggesting that they could be turned into anti-aggregation medications (Hsu et al., 2009; Novik and Savich, 2020).

8.2.2 BACILLUS CELLS AS A DESIRABLE HOST

Each strain of microbe produces a large number of enzymes that perform functions such as hydrolysis, oxidation or reduction, and metabolism. Amylase, lipase, protease, and cellulase are particularly important components of microbial enzymes used in the food, beverage, pharmaceutical, and other sectors. The global market for industrial enzymes is expected to reach more than 1.5 billion dollars, with food enzymes accounting for 29% of that, feed enzymes for 15%, and general technical enzymes making a market share of 56%. *Bacillus* spp. enzymes are predicted to account for approximately half of the global enzyme market (Schallmey et al., 2004). *Bacillus* species are widely known for secretion of a vast number of proteolytic enzymes, such as amylases, cellulases, proteases, and so forth. A variety of novel commercial enzymes have been created through extensive genetic engineering and selection of *Bacillus* species. It is clear that *Bacillus* spp. in general is capable of expressing and secreting individual gene products at very high levels (Arbige et al., 1993). Due to the extensive research of these species, these microorganisms can be considered as model organisms for fermentation of food and beverages.

Bacillus spp. found in fermented foods hydrolyze the substrate and create enzymes such as natto kinase, phytase, amylase, protease, cellulase, and lipase (Gopikrishna et al., 2021). However, consistent, and appropriate fermentation conditions are required for *Bacillus* spp. growth to create such enzymes (Nguyen and Nguyen, 2020). These enzymes help break down complicated compounds into simpler ones. For example, the enzyme amylase converts soybean starch into sugar (Gopikrishna et al., 2021). Similarly, protease turns proteins into amino acids and ammonia.

These hydrolytic bacteria are responsible for the utilization and reduction of indigestible oligosaccharides and polysaccharides (Shrestha et al., 2013). *Bacillus* species are extremely heat stable and have the capacity to produce resistant strains which enables them to colonize a wide range of habitats. *Bacillus* strains can be cultivated at high temperatures and pH levels to create products that are stable in a variety of adverse settings, which is why they are considered suitable hosts for food fermentation.

8.2.3 SOYBEAN FERMENTATION

One of the most significant industrial plants in the world is soybean, a plant belonging to the Leguminosae family (Beuchat, 1984). Soybeans provide two thirds of the sources of protein and one third of the edible oils. Soybeans in their natural form are rarely used as direct foods because of low digestion, beany taste, and flavor (Beuchat, 1984; Dimidi et al., 2019). However, fermentation can enhance the nutritional value and eating qualities of soybeans (Beuchat, 1984). Oriental diets have historically included a variety of fermented soybean dishes (Beuchat, 1984). Traditional fermented soy foods with a *Bacillus* subtilis predominance offer a distinctive flavor, texture, and scent that are well-liked in Asian and African nations (Beuchat, 1984). Soaking and cooking the soybeans results in a number of physio-chemical and sensory modifications that greatly increase their digestibility and nutritional value (Gopikrishna et al., 2021). The most prevalent naturally occurring bacterium in raw soybeans belongs to the genus *B. subtilis*. This species grows naturally in a variety of meals made from fermented soybeans without the use of inoculum. Although there are many other microbial species present in the traditionally fermented soy foods, *Bacillus* is the only one responsible for the fermentation. Because the traditional method of preparing soybeans still relies significantly on spontaneous fermentation, the final products will differ depending on which microbial species predominate during fermentation (Shrestha et al., 2013). Fermenting processes will differ by region and country (Shrestha et al., 2013). *Bacillus* spp., *B. amyloliquefaciens*, *B. cereus*, *B. circulans*, *B. licheniformis*, *B. sphaericus*, *B. subtilis*, and *B. thuringiensis* were all found in fermented soybean meals. However, *B. subtilis* is the main functional bacteria involved in soybean fermentation.

Since soy proteins are heat stable, they can be fermented without losing any of their chemical integrity. An alkaline environment is created when

Bacillus spp. ferment soybean. According to a study, *Bacillus siamensis*, which is found in fermented soybean meals, enhances nutritional characteristics by breaking the original structure of soybean protein and lowering-sheet structure by 43.2%, thus promoting protein digestibility (Sarkar et al., 1993).

8.3 METHODS OF SOYBEAN FERMENTATION

The majority of *Bacillus*-fermented soy products are still made using age-old techniques. Therefore, the preparation process varies depending on the country or locale, the convenience of the family, and the supplies accessible (Gopikrishna et al., 2021). Broadly, soybean fermentation calls for soaking and boiling the bean before splitting the cotyledons to hasten the *Bacillus* spp. fermentation process. To maintain an ambient temperature of 20 to 35°C, soybean grit is wrapped in banana, bamboo, or ginger leaves, and roughly 1% firewood ash can be added to keep the fermentation alkaline (Shrestha et al., 2013). Natto, aakhone, bekang, axone, and peruyan are only a few of the foods that have been processed with *Callicarpa arborea* or *Phrynium* spp. Foods such as tungrymbai, hawaijar, and axone are prepared and preserved in bamboo baskets or jute bags lined with banana or fern leaves. After 1–2 days, a white sticky coating and a subtle ammonia aroma emerge, indicating fermentation (Gopikrishna et al., 2021). This fermentation technique extends the shelf life to two to three days and allows it to be stored for up to a month by drying it in the sun for two to three days (Figure 8.1).

8.4 HEALTH BENEFITS OF *BACILLUS* FERMENTED SOYBEAN FOODS

Fermented soybean products high in protein, when consumed instead of animal protein, can lower total cholesterol, low-density lipoproteins (LDLs), and triglyceride levels in the blood. Fermented soy products have been shown to be useful in reducing the effects of diabetes, high blood pressure, cardiac illnesses, and cancer-related problems in addition to their effects on cholesterol levels. Due to their higher levels as compared to non-fermented soy products, the nutritional value of fermented soy products attracts a lot of attention (Gopikrishna et al., 2021; Shrestha et al., 2013).

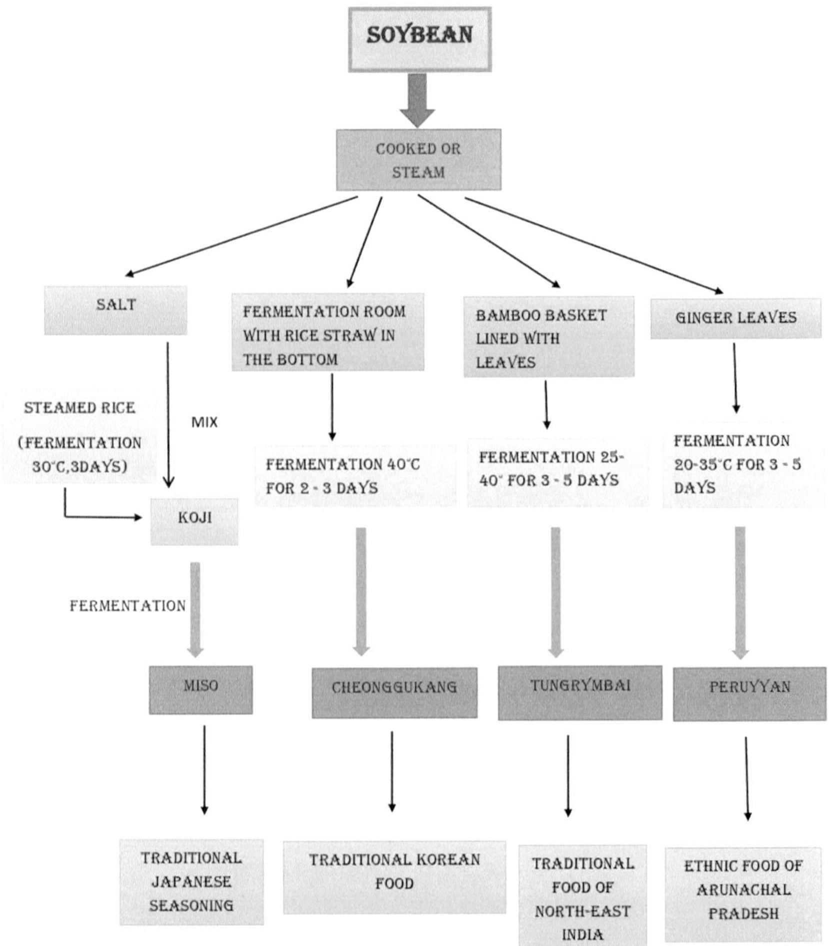

FIGURE 8.1 The production method of various traditional soybean fermented foods.

When soybeans are fermented, especially with *B. subtilis*, all of the macromolecules, including proteins, starch, and lipids, are extensively hydrolyzed into simpler sugars and fatty acids as well as smaller peptides, amino acids, mucilage, and sulfur and nitrogen compounds. Many research studies have suggested that they may provide a variety of health advantages (Figure 8.2) (Cao, 2019). The bioactive chemicals produced by the microorganisms that support soybean fermentation have advantageous benefits too (Cao, 2019).

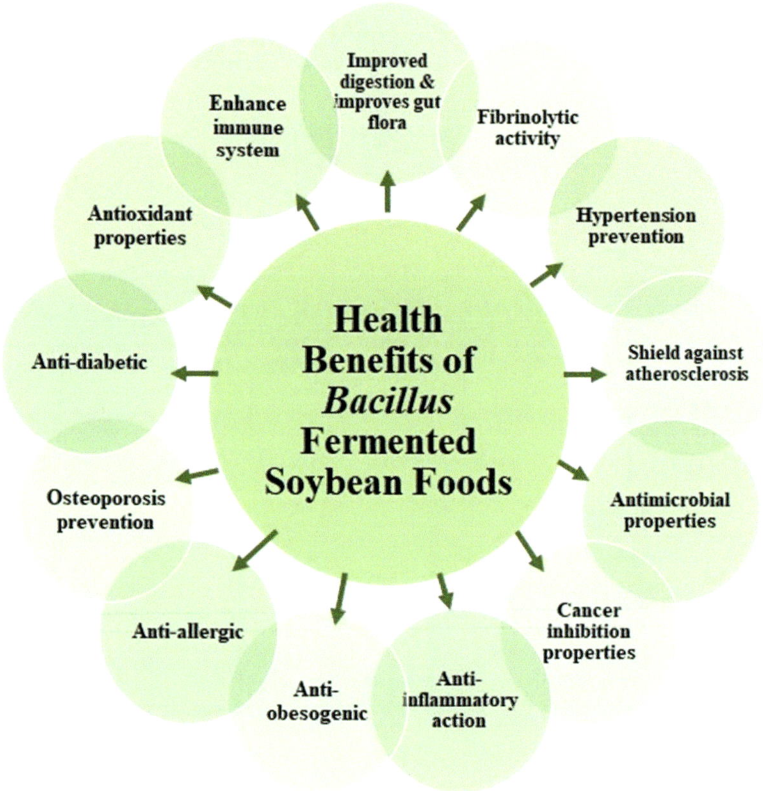

FIGURE 8.2 Health benefits of *Bacillus* fermented soybean foods

8.4.1 IMPROVES DIGESTIBILITY

The amount of highly digestible soluble matter was observed to increase by nearly three times (22 to 65%), while at the same time, nearly 40% of the total defatted dry matter became dialyzable due to the breakdown of high molecular weight compounds (Dimidi et al., 2019). Elimination of almost all of the indigestion-causing oligosaccharides, including sucrose, raffinose, stachyose, and another tri- and tetrasaccharides (Gopikrishna et al., 2021).

8.4.2 FIBRINOLYTIC ACTIVITY

Thrombosis can occur when there is an excessive build-up of fibrin in the blood vessels, which seriously impairs blood flow. Natto and other fermented soybean products have been demonstrated to increase tissue plasminogen activator production and plasma fibrinolytic activity (Hoi, 2014).

8.4.3 HYPERTENSION PREVENTION

Angiotensin I-converting enzyme (ACE)-inhibitory peptide is one of the physiologically active peptides found in fermented soybean foods that has an anti-hypertensive action (Gopikrishna et al., 2021). The proteolytic breakdown of soybean protein fractions yields the ACE inhibitory peptides. From soybean meal fermented by *B. subtilis* natto, a new ACE-inhibitory peptide with activity of 84.1% was identified (Gopikrishna et al., 2021).

8.4.4 SHIELD AGAINST ATHEROSCLEROSIS

Atherosclerosis causes artery constriction due to plaque accumulation. An enzyme called nattokinase (NK), which is generated by the bacteria *B. subtilis* in fermented soybeans, has a potent anti-hyperlipidemic effect that can stop atherosclerosis (Gopikrishna et al., 2021; Güzeler, 2016).

8.4.5 ANTIMICROBIAL PROPERTIES

Bacillus species that are isolated from numerous traditional fermented soybean meals like cheonggukjang, doenjang, and meju have been found to have antibacterial activity. These foods produce antimicrobial substances like proteins, enzymes, lipopeptides, and bacteriocins. *Cheonggukjang* contains phenylacetic acid, an organic compound obtained from the bacteria *Bacillus* licheniformis that inhibits the growth of intestinal infections such as *Staphylococcus aureus* and *E. coli*. Lipopeptides have been identified as the key molecules responsible for fermented soybean foods' antibacterial and antifungal effects.

8.4.6 CANCER INHIBITION PROPERTIES

Consuming traditional soy fermented foods like tofu, tempeh, edamame, and miso may reduce your chance of developing breast cancer, according to a growing body of research (Sanjukta et al., 2015). The biosurfactant found in natto, which is fermented by *B. natto* TK-1, was found to have antitumor action (Sanjukta et al., 2015). *B. subtilis* CSY191 isolated from doenjang produces surfactin, which inhibits the development of human breast cancer cells in a dose-dependent manner (Sanjukta et al., 2015). It has been demonstrated to downregulate both the transforming growth factor (TGF) pathway and inflammation-associated genes (Gopikrishna et al., 2021).

8.4.7 ANTI-INFLAMMATORY ACTION

Doenjang and *cheonggukjang*, two fermented soybean meals, have anti-inflammatory effects by regulating the production of COX-2 and nitric oxide synthase (iNOS) (Muthukumaran and Xu, 2019).

8.4.8 ANTI-OBESOGENIC

Obesity and overweight are risk factors for coronary heart disease, diabetes, hypertension, dyslipidemia, asthma, and arthritis. Doenjang, which is largely fermented by *B. subtilis*, has been found to have anti-obesogenic effects (Gopikrishna et al., 2021).

8.4.9. ANTI-ALLERGIC

Glycinin and β-conglycinin, two food allergens that make up around 30% of soybean meal, are known to make animals hypersensitive as well as to produce abnormalities such as small bowel and diarrhea in newborn animals. According to one study, these major allergens may be destroyed and the allergenicity of soybean meal lowered through solid-state fermentation by microorganisms such as *B. subtilis* (Chi, 2016).

8.4.10 OSTEOPOROSIS PREVENTION

Vitamin K2 activates proteins that regulate calcium transport, including osteocalcin (OC) and matrix-Gla protein (MGP) (Gopikrishna et al., 2021). Vitamin K2 is found in soybean fermented foods like natto (Gopikrishna et al., 2021). Therefore, research suggests that the vitamin

K2 included in traditional soy fermented foods like natto may play a part in bone metabolism (Gopikrishna et al., 2021). Additionally, it might help to preserve bone mineral density and prevent osteoporosis.

8.4.11 ANTI-DIABETIC

In a non-obese T2D animal model, it has been proven that *cheonggukjang* synthesized with *B. amyloliquefaciens* may have anti-diabetic properties that could potentially enhance insulin sensitivity and insulin secretion capacity (Chi, 2016).

8.4.12 ANTIOXIDANT PROPERTIES

B. subtilis MTCC 5480-fermented yellow and black soybeans have been found to have increased antioxidant activity and to aid in digestive function (Gopikrishna et al., 2021). According to studies, the presence of peptides and polyphenols in fermented soy meals from Sikkim increases the antioxidant effects of proteolytic bacteria like *B. subtilis*. The results show that compared to unfermented soybean, fermentation of soybean improves scavenging activity (3.1–24 times) and overall antioxidant activity (Sanjukta et al., 2015).

8.4.13 ENHANCE IMMUNE SYSTEM AND GUT MICROFLORA

Miso, natto, and other fermented soybean products play a crucial role in boosting immune system performance by maintaining a healthy balance of gut flora (Sanchez, 2017). These meals contain advantageous elements including probiotic bacteria, high protein content, folic acid, antioxidants, vitamins, and minerals that strengthen the immune system and aid in the battle against illnesses (Sanchez, 2017).

8.5 TRADITIONAL SOYBEAN FERMENTED FOODS MADE USING *BACILLUS* STRAIN

Soybean processing has produced a wide range of products, including soymilk, tofu, sprouts, and fermented goods. Korean *cheonggukjang* and Japanese *natto* are created in two steps: cooking and fermentation. Other fermented soybean products are *doenjang* (soybean paste), *ganjang* (soy sauce), *Korean gochujang*, *Indonesian tempeh*, *sieng* (Cambodia, Laos), *pepok* (Myanmar), *thua nao* (Thailand), and *kinema* (India, Nepal). *B. subtilis* is the most commonly used strain for soybean fermented foods. Here are some classic soybean fermented meals created with *Bacillus* strains (Figure 8.3).

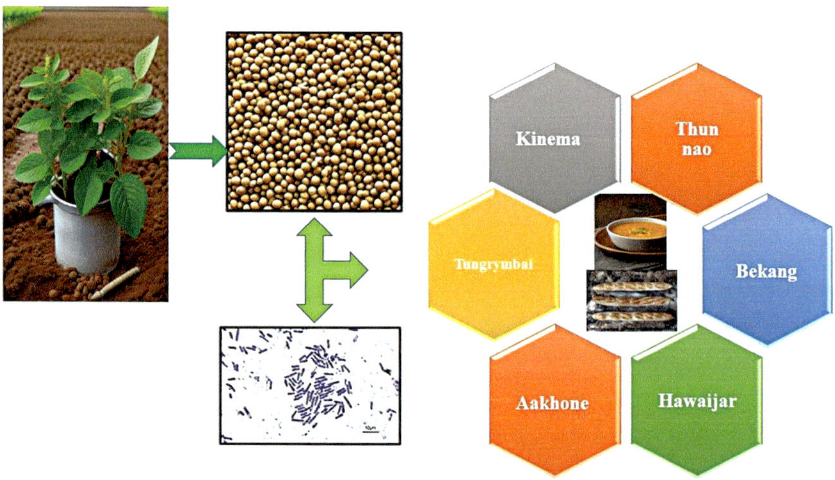

FIGURE 8.3 Traditional soyabean fermented foods made using *Bacillus* strain.

8.5.1 *KINEMA*

The Indian and Nepali community in the Eastern Himalayas consumes a traditional fermented soybean dish called *kinema*, which is a sticky, mildly alkaline snack with a faintly ammoniacal flavor. It is a fermented snack made from whole soybeans with a sticky texture, tan color, and a delicious scent. During the traditional *kinema* production process, the small (6 mm) "yellow cultivar" soybean dry seeds are selected, cleaned, and soaked in water for 8–10 hours (Tamang, 2001).

Freshly made *kinema* has a shelf life without refrigeration of 2–3 days in the summer and up to one week in the winter (Nikkuni, 1997). Drying in the sun for two to three days may help to extend it. For several months, dried *kinema* is kept at room temperature. *Kinema* preparation differs from location to location and is still only permitted within the home. It is interesting to observe that mountain women create *kinema* using their traditional expertise of food production. Among the Limbo, this specific expertise in *kinema*-making is preserved as an inherited right and passed down from mother to daughter.

8.5.2 THUA NAO

Thua nao is a traditional fermented soybean dish from northern Thailand. *Thua nao* is commonly made by cooking and mashing soybeans before fermenting them in banana leaves for two to three days at room temperature. Alternatively, mashed, boiling soybeans can be sun-dried outside. Thua nao that has been sun-dried can be stored at room temperature for months. Sun-dried fermented soybean cuisines are also produced in Nepal, China's Yunnan province, northern Laos, and Myanmar (Tamang, 2015).

Thua nao and other naturally fermented soybean meals may contain *B. subtilis* strains, which have a high potential for creating health-promoting chemicals such as PGA and protease NAT, as well as enzymes such as amylase and protease. However, there is a dearth of information on foods made from fermented soybeans by the *Bacillus* genus except Japanese *natto*.

8.5.3 HAWAIJAR

Manipur's traditional fermented soybean alkaline dish is known as *hawaijar*. It is made from a locally developed, small-seeded soybean type that is grown in the Manipur hillsides. Small-sized soybean seeds are chosen, cleaned, and boiled for two to three hours in an open cooker. The whole soybean seeds are then placed in a small, lidded bamboo basket after any extra water has been drained out and chilled to about 40°C (Tamang, 2015). Fresh fig plant (*Ficus hispida*) leaves or banana leaves, locally referred to as "*assee heibong*" in Meitei, are used to line the basket (Tamang, 2015).

Thua nao and other naturally fermented soybean meals may contain *B. subtilis* strains, which have the ability to produce health-promoting compounds like PGA and protease NAT, as well as enzymes like amylase and protease. The *Meitei* classify fried soybean seeds as good quality *hawaijar* if they emit the distinctive ammonia odor and have a sticky look. Without refrigeration, *hawaijar* has a maximum shelf life of 7 days. It is occasionally dried in the sun for two to three days before being preserved for a few weeks for eating.

8.5.4 TUNGRYMBAI

In Meghalaya, the Khasi and Garo people make a fermented soybean dish known as *tungrymbai*. It resembles *kinema*. It takes around 4–6 hours to gather, clean, wash, and soak soybean seeds in water (Agrahar-Murugkar and Subbulakshmi, 2006). Before cooking, the soybean seed coat is often removed by gently rubbing the wet seeds. Soaked soybeans are cooked for one to two hours, or until the water is fully evaporated. Cooked beans are placed in a bamboo basket, covered with a thick cloth, and left to cool before being filled with fresh *Clinogyne dichotoma* leaves (locally known as "*lamet*"). To produce *tungrymbai*, lay a covered basket over a fire and allow it to ferment organically for three to five days.

Tungrymbai is mashed and added to a kettle of water. The pot is then brought to a boil, stirring frequently. It is seasoned with salt, fried onion, garlic, ginger, chili, and ground black sesame, also known as "*til*," in the region. In Meghalaya, Khasi cook and serve a thick curry as a side dish with steamed rice. *Tungrymbai* is commonly used to create pickles.

8.5.5 BEKANG

Bekang is a fermented soybean dish popular among Mizo people in Mizoram. Kinema and this are comparable. Small, dry soybean seeds are gathered, cleaned, and steeped in water for 10–12 hours to make *bekang* using the traditional method. After removing the extra water, the beans are cooked in an open cooker for two to three hours until mushy. The excess water from cooked beans is removed and then it is wrapped with fresh leaves of either "*nuhlhan*" (local name for *Calliparpa aroria* of Family: Verbanaceae) or "*hnahthial*," (*Phrynium* sp., Family: Merantaceae). Tamang (2015) reports that the wrapped beans are kept in a small bamboo basket. The basket is then allowed to organically ferment for 3–4 days, either near the earthen oven or in a warm place. Customers in the area appreciate the sticky soybeans produced, which have an ammonia-like aroma.

8.5.6 AAKHONE

Aakhone, also known as *axone*, is a traditional Nagaland dish composed of fermented sticky soybeans. The preparation procedure is similar to that of other north-east Indian cuisines based on fermented soybeans. Soybean seeds are soaked and boiled before being placed above the fireplace to ferment for 5–7 days (Mao, 2007). The beans are then wrapped with fresh banana, *Phrynium pubinerve* Blume (Marantaceae), or *Macaranga indica* Wight (Euphorbiaceae) leaves. Freshly fermented *aakhone* has a storage life of one week. Fresh *aakhone* is formed into cakes, dried over an earthen furnace, and baked. On occasion, each fermented bean is manually separated and sun-dried for two to three days. Containers are used to store dried *aakhone* for future consumption. Green chili, tomato, salt, and freshly fermented *aakhone* are combined to make pickle.

8.5.7 PERUYYAN

The Apatani tribes of Arunachal Pradesh consume a fermented soybean dish known as *peruyyan*. The Apatani dialect is where the word peruyyan originated; "*perun*" is the word for beans and "*yannii*" is the word for packaging in leaves (Tamang, 2009). To make peruyyan traditionally, soybean seeds are gathered, cleaned, and boiled for two to three hours until the beans are tender. The excess water is removed and chilled for a bit. The cooked soybeans, known locally as "*taki yannii,*" are stored in a bamboo basket (vessel) lined with fresh ginger leaves.

The basket is placed on a wooden rack over the hearth, lightly covered in ginger leaves, and left to ferment for three to five days. The product is tested for stickiness, and if it is proven to be sufficiently sticky, it is safe to consume. The Apatani tribes of Arunachal Pradesh primarily eat *peruyyan* as a side dish with steamed rice.

8.6 BIOACTIVE COMPOUNDS IN SOYBEAN FERMENTED FOODS

8.6.1 FATTY ACIDS AND AMINO ACIDS

While soybeans are high in all nine essential amino acids, they are poor in those that contain sulfur. Major soybean proteins were broken down through proteolysis as a result of *B. subtilis* fermentation. Soybeans that had not been fermented contained low amounts of free amino acids, compared to 60 times more in the fermented product. In addition, they had a notable rise in fatty acid chain length during fermentation, a finding that was previously made by Kimchi in Natto. However, as previously noted, one disadvantage of Kinema is that an increased amount of free fatty acids (FFA) may be accumulated, and this may lead to inactivation of glycolytic and glucogenic enzymes, hydrolysis of trypsin, and fatty acid synthesis.

8.6.2 MINERALS AND VITAMINS

Microorganisms can create a variety of micronutrients, including vitamins, and can alter mineral profiles. Heating and soaking Kinema decreases its thiamine (Vitamin B1) and riboflavin (Vitamin B2) content. When *B. subtilis* ferments cooked soybeans, quantities of thiamine (5.8 to 8.4 mg/kg), riboflavin (6.8 to 11.6 mg/kg), and niacin (36.4 to 40.8 mg/kg) increase (Murata, 1985). They also stated that the presence of *Enterococcus faecium* caused a significant decrease in all three vitamin levels in Kinema. There is limited investigation into how *B. subtilis* fermentation impacts other vitamin levels in soybeans. Thiamine content in fermented tempeh (mold) was found to be reduced by nearly half, but other B-vitamin levels increased (Murata, 1985).

8.6.3 ANTHOCYANINS, PHENOLIC COMPOUNDS, AND OTHER SIMILAR SUBSTANCES

Due to the presence of anthocyanin, chlorophyll, and diverse combinations of breakdown products from these pigments, soybeans have a variety of seed coat pigmentations, including yellow, green, brown, and black. According to reports, black soybeans in particular contain a variety of bioactive chemicals that inhibit LDL oxidation, scavenge free radicals, and lessen the likelihood that cyclophosphamide would cause DNA damage. The majority of antioxidant nutrients are polyphenolic compounds, which act as chelators, reducing agents, and singlet oxygen quenchers. Traditional fermented foods, such as Natto, have been found to have high levels of these compounds. B Moktan (2008) investigated the antioxidant properties of Kinema, a *B. subtilis* fermented product of soybean and cooked non-fermented (CNF) soybeans using different *in vitro* techniques. They reported 140% more phenolic content in Kinema as compared to CNF soybean, and for free radical scavenging capacity too, the fermented *Bacillus* product, Kinema, was found to be better than its non-fermented counterpart.

8.6.4 ISOFLAVONES AND RELATED CHEMICALS

There are several phytochemical substances of high biological value connected with soybeans, and these chemicals tend to increase during fermentation. Isoflavones are a group of chemical substances linked to flavonoids that serve as phytoestrogen and have a number of positive health effects.

Under acidic or alkaline conditions, β-glucosidase converts isoflavones into their bioavailable form, isoflavone aglycone. In their native form, isoflavones occur as isoflavone glucose (glucoside conjugate). *B. subtilis* contains an enzyme termed β-glucosidase, which converts glucoside isoflavones into aglycone forms (Shrestha et al., 2013).

8.6.5 FLAVORING AGENTS AND AROMA CHEMICALS

Soybeans that have undergone microbial fermentation, in particular *B. subtilis*, taste distinctive (Shrestha et al., 2013). Amylase, proteases, lipases, and sucrose are examples of extracellular hydrolase enzymes that are associated to microbial activity during soybean fermentation and contribute to the shift in volatile chemicals that give soybeans their distinctive flavor (Shrestha et al., 2013).

8.7 CONCLUSION

A highly significant class of legume that has been consumed for a very long time in many nations is the soybean. Due to its tremendous nutritional value, particularly in terms of protein content, it is referred to as the "wonder plant" in many areas. The linoleic and linolenic acids it contains are also a significant source. Fermented soybean foods are among the numerous foods based on soybeans that have been developed over the years. The traditional preparations of these include *Bacillus*-fermented soy foods such as *Kinema, Natto, Thua-Nao, Dawadawa, Miso, Cheong-gukjang, Tungrymbai, Peruyyan, Hawaijar, Bekang, Aakhone,* and others.

Traditional soybean fermented foods provide a cheap source of plant protein which serves as a staple diet for people in South-east Asian countries. In addition to aiding in the fermentation process, the *Bacillus* species found in these fermented soybean foods also creates a number of advantageous substances, including isoflavones, lipopeptides, protein hydrolysates, and enzymes. The bioactive compounds exhibit antioxidant, antitumor, anti-diabetic, antibacterial, antihypertensive, anti-allergic, and immunoregulatory properties. Foods made from fermented soybeans contain *Bacillus* species that have been found to provide a number of health benefits; nevertheless, these foods will not be a viable option for human consumption unless effective food safety and quality procedures are implemented.

8.8 FUTURE PROSPECTS

Recent scientific research on bioactive chemicals derived from FSF, such as isoflavones, lipopeptides, protein hydrolysates, and enzymes, has revealed numerous applications. More study is needed to commercialize FSF as useful products, specifically on the mechanism and efficiency of the bioactive chemicals produced by *Bacillus* spp. against dangerous bacteria. *Bacillus* species isolated from FSF are expected to have effective probiotic function; hence, in a world where antibiotic resistance is on the rise, these microbes may be used as a viable alternative to antibiotics.

Products like *Kinema*, *Thua-Nao*, and *Dawadawa*, and so forth, are still made in a rudimentary manner, much as they were done hundreds of years ago. Eating these items poses a serious health risk because they are frequently prepared in an unclean manner. In order to strengthen their quality, standardization, and year-round availability, fermented soybean products are to be produced more scientifically. It is pertinent to have a complete understanding of the type of microorganisms that dominate the fermented product and then optimizing the procedure for using them as a specific inoculum for fermentation. The future studies must be focused on these directions, which will greatly enhance the finished product quality and safety.

KEYWORDS

- **bioactive compounds**
- **fermentation**
- **nutritional benefits**
- **organoleptic**
- **soybean**
- **traditional fermented food**

REFERENCES

Agrahar-Murugkar, D., & Subbulakshmi, G. (2006). Preparation techniques and nutritive value of fermented foods from the Khasi tribes of Meghalaya. *Ecology of Food and Nutrition, 45*(1), 27–38. https://doi.org/10.1080/03670240500408336 (accessed on 15th November 2025).

Alou, M. T., Rathored, J., Khelaifia, S., Michelle, C., Brah, S., Diallo, B. A., Raoult, D., & Lagier, J. C. (2015). *Bacillus rubiinfantis* sp. Nov. strain mt2T, a new bacterial species isolated from human gut. *New Microbes and New Infections, 8*, 51–60. https://doi.org/10.1016/j.nmni.2015.09.008 (accessed on 15th November 2025).

Arbige, M. V., Bulthuis, B. A., Schultz, J., & Crabb, D. (1993). Fermentation of *Bacillus*. In M. V. Arbige (Ed.), *Bacillus subtilis and other gram-positive bacteria: Biochemistry, physiology, and molecular genetics* (pp. 869–895). Wiley Online Library.

Baruzzi, F., Quintieri, L., Morea, M., & Caputo, L. (2011). Antimicrobial compounds produced by *Bacillus* spp. and applications in food. In A. Méndez-Vilas (Ed.), *Science against microbial pathogens: Communicating current research and technological advances* (pp. 1102–1111).

Beuchat, L. R. (1984). Fermented soybeans foods. *Food Technology*, 64–70.

Cao, Z. H., Green-Johnson, J. M., Buckley, N. D., & Lin, Q. Y. (2019). Bioactivity of soy-based fermented foods: A review. *Biotechnology Advances, 37*(1), 223–238. https://doi.org/10.1016/j.biotechadv.2018.12.001 (accessed on 15th November 2025).

Chen, K. I., Erh, M. H., Su, N. W., Liu, W. H., Chou, C. C., & Cheng, K. C. (2012). Soyfoods and soybean products: From traditional use to modern applications. *Applied Microbiology and Biotechnology, 96*(1), 9–22. https://doi.org/10.1007/s00253-012-4330-7 (accessed on 15th November 2025).

Chettri, R., & Tamang, J. P. (2015). *Bacillus* species isolated from tungrymbai and bekang, naturally fermented soybean foods of India. *International Journal of Food Microbiology, 197*, 72–76.

Chettri, R., Bhutia, M. O., & Tamang, J. P. (2016). Poly-γ-glutamic acid (PGA)-producing *Bacillus* species isolated from kinema, Indian fermented soybean food. *Frontiers in Microbiology, 7*, 971. https://doi.org/10.3389/fmicb.2016.00971 (accessed on 15th November 2025).

Chi, C. H., & Cho, S. J. (2016). Improvement of bioactivity of soybean meal by solid-state fermentation with *Bacillus amyloliquefaciens* versus *Lactobacillus* species and *Saccharomyces cerevisiae*. *LWT – Food Science and Technology, 68*, 619–625. https://doi.org/10.1016/j.lwt.2015.12.002 (accessed on 15th November 2025).

Da Silva, L. H. (2011). Effect of the fermentation of whole soybean flour on the conversion of isoflavones from glycosides to aglycones. *Food Chemistry, 128*, 640–644.

De Wulf, P., & Vandamme, E. J. (1997). Production of D-ribose by fermentation. *Applied Microbial Biotechnology, 48*, 141–148.

Devi, S. L., Khaund, P., & Joshi, S. R. (2010). Thermostable α-amylase from natural variants of *Bacillus* spp. prevalent in eastern Himalayan range. *African Journal of Microbiology Research, 4*(23), 2534–2542.

Dimidi, E., Cox, S. R., Rossi, M., & Whelan, K. (2019). Fermented foods: Definitions and characteristics, impact on the gut microbiota and effects on gastrointestinal health and disease. *Nutrients, 11*(8), 1806.

Elshaghabee, F. M., Rokana, N., Gulhane, R. D., Sharma, C., & Panwar, H. (2017). *Bacillus* as potential probiotics: Status, concerns, and future perspectives. *Frontiers in Microbiology, 8*, 1490. https://doi.org/10.3389/fmicb.2017.01490 (accessed on 15th November 2025).

Esaki, H., Onozaki, H., Kawakishi, S., & Osawa, T. (1996). New antioxidant isolated from tempeh. *Journal of Agricultural and Food Chemistry, 44*, 696–700.

Gopikrishna, T., Suresh, K. H. K., Perumal, K., et al. (2021). Impact of *Bacillus* in fermented soybean foods on human health. *Annals of Microbiology, 71*, 30. https://doi.org/10.1186/s13213-021-01641-9 (accessed on 15th November 2025).

Güzeler, N., & Çağla, Y. Ö. (2016). The utilization and processing of soybean and soybean products. *Journal of Agricultural Faculty of Uludağ University, 30*, 546–555.

Haile, M., & Kang, W. H. (2019). The role of microbes in coffee fermentation and their impact on coffee quality. *Journal of Food Quality, 2019*, 4836709–483715. https://doi.org/10.1155/2019/4836709 (accessed on 15th November 2025).

Hoi, J. H., Chung, M. J., Jeong, D. Y., & Oh, D. H. (2014). Immunostimulatory activity of isoflavone-glycosides and ethanol extract from a fermented soybean product in human primary immune cells. *Journal of Medicinal Food, 17*(10), 1113–1121. https://doi.org/10.1089/jmf.2013.3040 (accessed on 15th November 2025).

Hsu, R. L., Lee, K. T., Wang, J. H., Lee, L. Y., & Chen, R. P. (2009). Amyloid-degrading ability of nattokinase from *Bacillus subtilis* natto. *Journal of Agricultural and Food Chemistry, 57*(2), 503–508. https://doi.org/10.1021/jf803072r (accessed on 15th November 2025).

Hu, Y., Ge, C., Yuan, W., Zhu, R., Zhang, W., Due, L., & Xue, J. (2010). Characterization of fermented black soybeans natto inoculated with *Bacillus natto* during fermentation. *Journal of the Science of Food and Agriculture, 90*, 1194–1202.

Inatsu, Y., Nakamura, N., Yuriko, Y., Fushimi, T., Watanasiritum, L., & Kawamoto, S. (2006). Characterization of *Bacillus subtilis* strains in Thua nao, a traditional fermented soybean food in northern Thailand. *Letters in Applied Microbiology, 43*(3), 237–242. https://doi.org/10.1111/j.1472-765X.2006.01966.x (accessed on 15th November 2025).

Jang, C. H., Oh, J., Lim, J. S., Kim, H. J., & Kim, J. S. (2021). Fermented soy products: Beneficial potential in neurodegenerative diseases. *Foods, 10*(3), 636–648. https://doi.org/10.3390/foods10030636 (accessed on 15th November 2025).

Jeon, H. L., Yang, S. J., Son, S. H., Kim, W. S., Lee, N. K., & Paik, H. D. (2018). Evaluation of probiotic *Bacillus subtilis* P229 isolated from cheonggukjang and its application in soybean fermentation. *LWT – Food Science and Technology, 97*, 94–99. https://doi.org/10.1016/j.lwt.2018.06.054 (accessed on 15th November 2025).

Lyngwi, N. A., & Joshi, S. R. (2014). Economically important *Bacillus* and related genera: A mini review. In S. R. Joshi (Ed.), *Biology of useful plants and microbes* (pp. 33–34). Narosa Publishing House.

Mao, A. A., & Odyuo, N. (2007). Traditional fermented foods of the Naga tribes of Northeastern India. *Indian Journal of Traditional Knowledge, 6*(1), 37–41.

Moktan, B., Saha, J., & Sarkar, P. K. (2008). Antioxidant activities of soybeans as affected by *Bacillus* fermentation to kinema. *Food Research International, 41*(6), 586–593. https://doi.org/10.1016/j.foodres.2008.04.003 (accessed on 15th November 2025).

Murata, K. (1985). Formation of antioxidants and nutrients in tempeh. In *Proceedings from Asian Symposium on Non-Salted Soybeans Fermentation* (pp. 186–198). National Food Research Institute, Tsukuba Science City, Japan.

Muthukumaran, J., & Xu, B. J. (2019). An insight into the health benefits of fermented soy products. *Food Chemistry*, 362–371.

Nguyen, T., & Nguyen, C. H. (2020). Determination of factors affecting the protease content generated in fermented soybean by *Bacillus subtilis* 1423. *Energy Reports, 6*(1), 831–836. https://doi.org/10.1016/j.egyr.2019.11.011 (accessed on 15th November 2025).

Nikkuni, S. (1997). Natto, kinema and Thua-nao: Traditional non-salted fermented soybean foods in Asia. *Farming Japan, 31*, 27–36.

Novik, G., & Savich, V. (2020). Beneficial microbiota. Probiotics and pharmaceutical products in functional nutrition and medicine. *Microbes and Infection, 22*(1), 8–18. https://doi.org/10.1016/j.micinf.2019.06.004 (accessed on 15th November 2025).

Priest, F. G. (1977). Extracellular enzyme synthesis in the genus *Bacillus*. *Bacteriological Reviews, 41*(3), 711–753. https://doi.org/10.1128/br.41.3.711-753.1977 (accessed on 15th November 2025).

Priest, F. G. (1993). Systematics and ecology of *Bacillus*. In A. L. Sonenshein, J. A. Hoch, & R. Losick (Eds.), *Bacillus subtilis and other Gram-Positive Bacteria: Biochemistry, Physiology, and Molecular Genetics* (pp. 3–16). ASM Press.

Queener, S. W., & Lively, D. H. (1989). Screening and selection for strain improvement. In A. L. Demain & N. A. Solomon (Eds.), *Manual of Industrial Microbiology and Biotechnology* (pp. 155–169). ASM Press.

Sánchez, B., Delgado, S., Blanco-Míguez, A., Lourenço, A., Gueimonde, M., & Margolles, A. (2017). Probiotics, gut microbiota, and their influence on host health and disease. *Molecular Nutrition & Food Research, 61*(1). https://doi.org/10.1002/mnfr.201600240 (accessed on 15th November 2025).

Sanjukta, S., Rai, A. K., Muhammed, A., Jeyaram, K., & Talukdar, N. C. (2015). Enhancement of antioxidant properties of two soybean varieties of Sikkim Himalayan region by proteolytic *Bacillus subtilis* fermentation. *Journal of Functional Foods, 14*, 650–658. https://doi.org/10.1016/j.jff.2015.02.033.

Sarkar, P. K., Cook, P. E., & Owens, J. D. (1993). *Bacillus* fermentation of soybeans. *World Journal of Microbiology & Biotechnology, 9*, 295–299. https://doi.org/10.1007/BF00383066 (accessed on 15th November 2025).

Sarkar, P. K., Tamang, J. P., Cook, P. E., & Owens, J. D. (1994). Kinema – A traditional soybean fermented food: Proximate composition and microflora. *Food Microbiology, 11*, 47–55.

Schallmey, M., Singh, A., & Ward, O. P. (2004). Developments in the use of *Bacillus* species for industrial production. *Canadian Journal of Microbiology, 50*(1), 1–17. https://doi.org/10.1139/w03-076 (accessed on 15th November 2025).

Shin, D., & Jeong, D. (2015). Korean traditional fermented soybean products: Jang. *Journal of Ethnic Foods, 2*, 2–7. https://doi.org/10.1016/j.jef.2015.02.002 (accessed on 15th November 2025).

Shrestha, A. K., Dahal, N. R., & Ndungutse, V. (2013). *Bacillus* fermentation of soybean: A review. *Journal of Food Science and Technology Nepal, 6*, 1–9. https://doi.org/10.3126/jfstn.v6i0.8252 (accessed on 15th November 2025).

Sivamaruthi, B. S., Kesika, P., & Chaiyasut, C. (2018). Thai fermented foods as a versatile source of bioactive microorganisms—A comprehensive review. *Scientia Pharmaceutica, 86*(3), 37. https://doi.org/10.3390/scipharm86030037 (accessed on 15th November 2025).

Stein, T. (2005). *Bacillus subtilis* antibiotics: Structures, syntheses and specific functions. *Molecular Microbiology, 56*(4), 845–857. https://doi.org/10.1111/j.1365-2958.2005.04587.x (accessed on 15th November 2025).

Sumi, H., Hamada, H., Tsushima, H. J., Muraki, H., & Mihara, H. (1987). A novel fibrinolytic enzyme (nattokinase) in the vegetable cheese natto; a typical and popular soybean food in the Japanese diet. *Experientia, 43*, 1110–1111.

Tamang, J. P. (2001). Kinema. *Food Culture, 3*, 11 14.

Tamang, J. P. (2015). Naturally fermented ethnic soybean foods of India. *Journal of Ethnic Foods, 2*(1), 8–17. https://doi.org/10.1016/j.jef.2015.02.003 (accessed on 15th November 2025).

Tamang, J. P., Chettri, R., & Sharma, R. M. (2009). Indigenous knowledge of north east women on production of ethnic fermented soybean foods. *Indian Journal of Traditional Knowledge, 8*(1), 122–126.

Urdaci, M. C., & Pinchuk, I. (2004). Antimicrobial activity of *Bacillus* probiotics. In E. Ricca, A. O. Henriques, & S. M. Cutting (Eds.), *Bacterial Spore Formers: Probiotics and Emerging Applications* (pp. 171–182). Horizon Bioscience.

Yang, H. J., Kwon, D. Y., Hyun, J. K., Min, J. K., Kang, H. J., & Kim, D. S. (2015). Fermenting soybeans with *Bacillus licheniformis* potentiates their capacity to improve cognitive function and glucose homeostasis in diabetic rats with experimental Alzheimer's-type dementia. *European Journal of Nutrition, 54*(1), 77–88. https://doi.org/10.1007/s00394-014-0687-y (accessed on 15th November 2025).

Yang, J., Wu, X. B., Chen, H. L., Sun-Waterhouse, D., Zhong, H. B., & Cui, C. (2019). A value-added approach to improve the nutritional quality of soybean meal byproduct: Enhancing its antioxidant activity through fermentation by *Bacillus amyloliquefaciens* SWJS22. *Food Chemistry, 272*, 396–403. https://doi.org/10.1016/j.foodchem.2018.08.037.

Yoshimoto, T., Fukumoto, J., & Tsimut, D. (1971). Studies on bacterial proteases. *International Journal of Protein Research, 3*, 285–295. https://doi.org/10.1111/j.1399-3011.1971.tb01722.x (accessed on 15th November 2025).

CHAPTER 9

Bacterial Cell Factories for Industrially Important Protein Production

OM PRAKASH,[1] PRITI PAL,[2] BABLI KUMARI,[1] ANAND PRAKASH,[1] SAURABH SINGH RATHORE,[1,3] BRIJESH PANDEY,[1] PRAKASH KUMAR SARANGI,[4] and AKHILESH KUMAR SINGH[1,*]

[1]*Department of Biotechnology, School of Life Sciences, Mahatma Gandhi Central University, Motihari, Bihar, India*

[2]*Amity Institute of Biotechnology, Amity University Uttar Pradesh, Lucknow Campus, Lucknow, Uttar Pradesh, India*

[3]*Department of Life Sciences, Inter University Centre for Teacher Education, Bananas Hindu University, Varanasi, Uttar Pradesh, India*

[4]*College of Agriculture, Central Agricultural University, Imphal, Manipur, India*

*Corresponding author

ABSTRACT

Large-scale production of single-cell protein (SCP) and recommended protein (RCP) has the potential not only to solve some of the world's food insecurity and water scarcity crises, but also to reduce the costs associated with the treatment of industrial and agricultural wastewater. However, a class of phototrophic bacteria known as purple non-sulfur bacteria (PNSB) has recently emerged as a promising option for both wastewater treatment and resource recovery. PNSB are metabolically versatile and tolerant to a wide range of conditions, allowing them to thrive in a variety of waste streams. This chapter investigates the comparative advantages of

Industrial Microbiology: Current Progress and Novel Technologies.
Prakash Kumar Sarangi, Akhilesh Kumar Singh, Rajesh Kumar Srivastava, and Brijesh Pandey (Eds.)

DOI: 10.1201/9781779642219-9

using PNSB for SCP production, key components of PNSB-based SCP processing, and potential environmental and economic benefits associated with the process. There is no doubt that recombinant protein production in microbial systems has transformed biochemistry. The ability to express and purify the desired recombinant protein in large quantities enables biochemical characterization, industrial process use, and commercial product development. Nonetheless, advancements in recombinant protein expression and purification are ongoing. In this chapter, we discuss recent advances in recombinant protein and single cell protein. Current challenges with PNSB-based SCP production are also discussed, as are future prospects for recombinant protein.

9.1 INTRODUCTION

Bacteria are microscopic, single-celled organisms that live in almost every environment, including soil, water, organic matter, and multicellular animal bodies. Bacteria are distinguished in part by morphological and genetic characteristics, such as spherical, rod-like, or spiral shapes. They are also classified as gram-positive or gram-negative based on the structure of their cell wall and their reaction to the Gram stain. Many bacteria swim using flagella (see flagellum). Most bacteria have a single circular chromosome that is distributed throughout the cytoplasm rather than being enclosed within a membrane-enclosed nucleus. While some bacteria can cause food poisoning and infectious diseases in humans, the vast majority are harmless and even beneficial. They are used in a variety of industrial processes, particularly in the food industry (e.g., the production of yogurt, cheeses, and pickles). Bacteria and archaea are genetically distinct. They are also distinct from eukaryotes as prokaryotic organisms (no membrane-bound nucleus). Sulfur bacteria, budding bacteria, coliform bacteria, cyanobacteria, denitrifying bacteria, nitrifying bacteria, sheathed bacteria (Kadner et al., 2023). Bacteria have historically been extremely important to both humans and the environment. Apart from their role in geochemical cycling and environmental restoration, bacteria are widely used in industries to synthesize products that are beneficial to humans. The growing utility of bacteria in industrial applications has propelled the microbial industries into more diverse and valuable horizons. Microbial cells act as mini bio factories, producing novel, economically valuable products such as enzymes, organic acids, vitamins, antibiotics, antibodies, hormones, carotenoids, steroids, alkaloids, alcoholic beverages, interferons, and vaccines. Bacteria are also involved in wastewater treatment, bioremediation, biomineralization, bio recovery of precious metals and metalloids, biocontrol agents, the production of prebiotics and probiotics, biogas, and the production of eco-friendly nanomaterials. This chapter provides an overview of the diverse industrial applications of bacteria, ranging from beneficial products to environmental rehabilitation (Akoijam et al., 2022).

Lactic acid bacteria (LAB) produce several inhibitory factors. This factor is frequently derived from metabolic by-products that act as antibiotics, and they are known as bacteriocins because they are bactericidal proteins with antimicrobial activity (Goa et al., 2022; Oliveira et al., 2008). Many subtilisin's have been characterized and found to be potent, with activity comparable to antibiotics (Chandrakasan et al., 2019). There is growing concern about the rise of antibiotic-resistant microorganisms and the likelihood of current therapies becoming ineffective in the near future. These issues highlight the importance of developing alternative strategies. However, due to the negative effects of chemical preservatives on human health, interest in bio preservation of food has grown significantly. Bacteriocins are well-characterized bacterial peptides with antimicrobial activity that have been used in a variety of industries, including the food industry and medicine (Rahmdel et al., 2019).

9.2 BACTERIA'S ROLE IN INDUSTRY

In the dairy industry, LAB play an important role in the dairy industry's fermentation process. *Streptococcus lactis*, *S. thermophilus*, *Lactococcus lactis*, *Lactobacillus plantarum*, *Lactobacillus casei*, *Lactobacillus acidophilus*, *Lactobacillus helveticus*, *Lactobacillus bulgaricus*, and other LAB are used to make butter, cheese, curd, and other dairy products. These bacteria convert lactose in milk into lactic acid, which aids in curd coagulation and texture formation during cheese production. In the vinegar industry, acetic acid bacteria (AAB) are used in the production of certain foods and vinegar. These are Gram-negative bacteria from the Acetobacteraceae family. They can produce acetic acid during oxidative fermentation by performing an oxidation reaction that produces vinegar as a by-product, for example, *Acetobacter aceti*, *A. cerevisiae*, *A. malorum*,

A. oeni, *A. pasteurianus*, *A. pomorum*, *Gluconacetobacter entanii*, *G. liquefaciens*, *G. oxydans*, *Komagataeibacter europaeus*, K (Tanamool et al., 2020) (Table 9.1).

TABLE 9.1 Bacterial Substrate

Bacteria	Substrate	Reference
Strepyomyces species	Methanol	Turnbull et al. (1992)
Rhodopsedomonas gelatinous	Wheat bran	Mahasneh et al. (1997)
Bacillus cereus	Ram horn	Lang et al. (1999)
Cellulomonas species	Fructose and glucose	Ugalde and Castrillo (2002)
Rastonia species	Natural gas	Ravinder et al. (2003)
Rhodopseudomonas palustris	Rubber waste	Queiroz et al. (2007)
Methylomonas species	Methane broth	Queiroz et al. (2007)
Corynebacterium glutamicum	Ram horn	Zepka et al. (2008)
Methanomonas methamtropus	Agricultural waste	Sousa et al. (2008)
Bacillus lechnifertis	Potato waste	Sousa et al. (2008)
Bacillus plenitis	Potato waste	Raja et al. (2008)
Methylococcaceae family	C1 compounds	Kadim et al. (2015)

9.3 PROCESS FOR MAKING BACTERIAL PROTEIN

Transcription and translation are the two stages of protein production in cells, according to the central dogma of molecular biology. In other words, the recombinant protein (RCP) expression steps include transcription and translation. The gene is isolated and cloned into an expression vector to produce RCPs. The protein expression system, protein purification system, and protein identification system are all required for the production of RCPs (Yilmaz et al., 2019).

The following are the basic steps to obtain RCP:

1. Gene amplification of interest.
2. Incorporate into the cloning vector.
3. Subcloning into the expression vector.
4. Infection with a protein-expressing host (bacteria (*E. coli*), yeast, mammalian cells, or baculovirus-insect cell system).
5. Perform a RCP identification test (fluorescence or Western blot).
6. Large-scale production (fermentation on a large scale).
7. Purification and isolation.

9.4 ADVANCED PROTEIN PRODUCTION TECHNIQUES

9.4.1 DEVELOPMENT OF DOWNSTREAM PROCESSES

The downstream purification procedure is crucial because it helps to get medicinal goods approved for human use (Owczarek et al., 2019). RCPs are typically purified by centrifugation, chromatography, and membrane filtering processes. In addition to these procedures, recombinant biopharmaceuticals made from mammalian cells utilize viral inactivation (Zydney et al., 2016). Recovery of the desired proteins, expressed as intracellular IBs, requires cell disruption. High-pressure homogenizers and bead mills are two frequently employed techniques for cell disruption on a big scale (Mevada et al., 2019). Figure 9.1 displays a flow chart for the purification of RCP that has been expressed by the IB. To obtain pure and physiologically active protein, this involves cell harvesting utilizing centrifugation, cell lysis, IB solubilization, refolding, diafiltration, and chromatographic purification procedure. The three methods for cell harvesting and cell separation that are most frequently employed are centrifugation, depth filtration, and tangential flow microfiltration (TFF-MF) (Besnard et al., 2016; Khoo et al., 2020; Mancini et al., 2020). There have been other descriptions of the various clarifying technologies utilized for the downstream processing of antibodies (Singh and Chollangi, 2017; Singh et al., 2016).

9.4.2 BACTERIA IS A GOOD SOURCE OF PROTEIN

Due to their high growth rates, bacteria, in particular, are relevant as a source of microbial protein (doubling times of 20–30 minutes). Yeast only needs around 2 to 3 hours to multiply, while yeast, algae, and fungi all need about 16 hours. Protein is abundant in bacteria. Their greater levels of lysine and sulfur-containing amino acids contribute to their excellent amino acid profile and high protein quality. Some species of *actinomycetes* are of interest for protein synthesis because they grow at speeds similar to bacteria. However, bacteria have a few drawbacks, including a high nucleic acid concentration that renders them unfit for use as human food and an exceedingly tiny cell size that makes separation energy costly. Many different kinds of bacteria have been considered as potential hosts for protein production. Growth rates, cell mass yield on substrate, pH and temperature tolerance, oxygen needs, and genetic stability are all crucial characteristics (Mordes et al., 2018). There is a direct correlation between

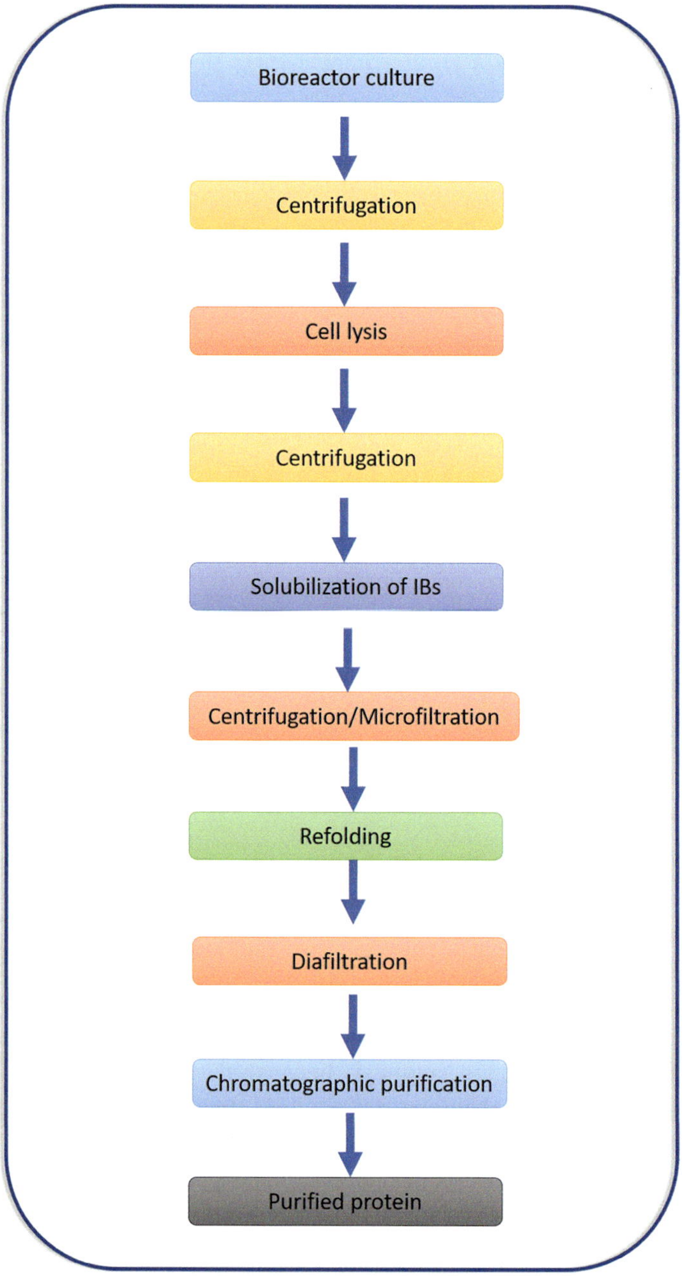

FIGURE 9.1 Flow chart of the purification of recombinant proteins expressed as IBs using *E. coli*.

the availability of nitrogen and the growth of bacteria, and the same holds true for carbon. Therefore, the ratio of carbon to nitrogen in growth media should be kept at 10:1 or below. The best nitrogen sources are ammonia or ammonium salts. Mineral salts are typically present in sufficient proportions in natural water sources, although occasionally supplements are needed. Such additions must be selected with care to prevent corrosion issues (Garbey et al., 2014). pH levels during growth can be managed between 5 and 7. Another crucial quality of bacterial strains can be their ability to tolerate different temperatures. Transfer of oxygen to developing cells is particularly crucial for aerobic strains. All microbes must be mass-cultivated; therefore, contamination must be avoided. Compared to other, slower-growing microorganisms, fast-growing bacteria find it easier to meet this requirement (Zhou et al., 2018).

9.5 PROTEIN PRODUCTION

Protein synthesis is a biotechnological method for making a particular protein. It is frequently performed by changing an object's gene expression to make it express a recombinant gene in massive volumes. In this procedure, mRNA is read and translated into polypeptide chains that may later be folded into useful proteins and chosen for intracellular or extracellular distribution (Hallberg et al., 2014). Expression systems, used in biotechnology, medicine, and the life sciences, are methods for creating proteins. Many expressions system-derived proteins and enzymes are employed in molecular biology studies, such as cloning with restriction endonucleases, polymerase chain reaction (PCR) with DNA polymerase, RNA analysis with reverse transcriptase, and protein expression vectors for testing drug targets. Expression systems have important uses in industrial fermentation, particularly in the synthesis of enzymes and biopharmaceuticals such as human insulin for the treatment of diabetes. According to the Census Bureau of the United States, the number of people on Earth has expanded by up to 250% in the past six decades, rising from 2.6 to 7 billion. If the current rate is maintained, the population may reach 9 billion by 2042 (Gabriel et al., 2014). As the global population rises, so does the strain on food production. Agriculture, livestock raising, and fishing cannot stand alone as sources of nutrition. However, the majority of affluent nations have seen growth in their agricultural sectors. Nevertheless, a lot of people experience problems including hunger, malnutrition, food insecurity, and dietary abnormalities

(Gabriel et al., 2014). Up to 12 million people every year in developing nations succumb to hunger, malnutrition, and related causes, according to the World Health Organization (WHO) (Gabriel et al., 2014; Israelidis et al., 2008). For the future's indiscriminately growing population, it is necessary to investigate alternative food sources. SCP is getting popular as a healthy new food option, and microbes are becoming a valuable resource for generating large food sources. *Spirullina* (blue green algae) has been found to be a rich source of SCP among the microorganisms. By giving all of the nutrients it needs, this microalga can be mass-cultivated in water reservoirs outdoors and in laboratories. The function of bacteria in supplying future generations with nutritious food is discussed in this chapter.

9.6 BACTERIAL CELLULOSE (BC)

Bacterial cellulose (BC) is a byproduct of exocellular bacteria such as *Achromobacter, Alcaligenes, Aerobacter, Agrobacterium, Azotobacter, Komagataeibacter* (previously Gluconacetobacter), *Pseudomonas, Rhizobium, Sarcina, Dickeya*, and *Rhodobacter*. According to Ul-Islam et al. (2016), this word was first used by R. M. Brown in 1886, who defined it as "a kind of moist skin (formerly Gluconacetobacter), *Pseudomonas, Rhizobium, Sarcina, Dickeya*, and *Rhodobacter*." Cell-free systems are also employed to produce BC in addition to microbial synthesis (Ullah et al., 2016). While BC's molecular formula is the same as that of cellulose obtained from plants, Hibbert and Barsha discovered in 1931 that BC had different qualities. They did this by analyzing the chemical composition and structural characteristics of BC. Since its exceptional qualities have been explored, it is receiving more and more attention in the domains of medical engineering, biosensors, papermaking, cosmetics, and food (Ul-Islam et al., 2019).

9.7 TYPES OF BACTERIA RESPONSIBLE FOR PROTEIN PRODUCTION.

9.7.1 *ESCHERICHIA COLI*

Escherichia coli is often used as a recombinant host because (i) its genome can be modified rapidly and accurately; (ii) it grows rapidly to high cell densities; (iii) it can be cultured in inexpensive media; and (iv) its protease activity can be readily reduced. Heterologous proteins may account for up to half of *E. coli* dry cell weight (Vilkhovoy et al., 2018). The formation of inclusion bodies composed of inactive, aggregated, and inaccessible heterologous proteins is another drawback. Other constraints include a lack of PTMs, the presence of toxic pyrogens in cell walls, and the inability to modify proteins at the genetic level. In 1993, recombinant E. coli methods produced products with an estimated total market value of $5 billion. These products include insulin, HGH, α-, β-, and γ-interferons, and granulocyte colony-stimulating factor (G-CSF).

Plasmid expression vectors are commonly used to insert DNA into *E. coli*, which is then utilized as an expression host. Well-established methods for overexpression in *E. coli* include increasing the number of gene copies or the binding strength of the promoter to boost transcription. The transformation of *E. coli* may occur when a DNA sequence encoding a target protein is cloned or subcloned into a plasmid with a high copy number and the lac (typically LacUV5) promoter. IPTG, a lactose analogue, activates the lac promoter in bacteria, causing them to make protein. A couple of the most common *E. coli* strains for this are BL21 and BL21 (DE3). Since these organisms are of the B lineage, they lack the proteases lon and OmpT that ordinarily break down the proteins they produce. The T7 RNA polymerase (controlled by the LacUV5 promoter) may be used in lieu of the native promoter in BL21(DE3) thanks to the DE3 prophage (Horesh et al., 2021).

9.7.2 *CORYNEBACTERIUM*

Commensal gram-positive *Corynebacterium* species produce different amino acids in industrial processes. The *C. glutamicum* species is used to produce the amino acids glutamate and lysine, which are often used in the food and pharmaceutical industries (Brinkrolf et al., 2010). *C. glutamicum* has been shown to express human epidermal growth factor, indicating that the organism may be used to produce large quantities of human proteins (Jia et al., 2018). The secretion of proteins that have been produced may be directed by either the twin-arginine translocation mechanism or the general secretory pathway (Sec) (Tat). *Corynebacterium*, a gram-positive bacterium, lacks the lipopolysaccharides that act as antigenic endotoxins in humans, making it immune to these assaults (Prajapati et al., 2021).

9.7.3 PSEUDOMONAS FLUORESCENS

The non-pathogenic gram-negative bacterium *Pseudomonas fluorescens* is used to mass produce RCPs, which are then utilized to make vaccines and bio-therapeutics. Due to *P. fluorescens'* adaptable metabolism, complex proteins can be developed quickly and through high throughput screening. The most notable characteristic of *P. fluorescens* is its capacity to create high titers of active, soluble protein rapidly and effectively (Retallack et al., 2012).

9.7.4 PROBIOTIC BACTERIA

By inhibiting the growth of enteric and food-borne microbial pathogens, probiotic bacteria have been found to improve immunity and ward against gastrointestinal and urinary tract infections (UTI) such as diarrhea, gastritis, and yeast infections. The industry for functional foods that boast ingredients like probiotic bacteria and probiotic fiber is growing, and experts predict that demand will only increase in the future (Rathore et al., 2012). Numerous dairy products with extra nutritional content have recently emerged in response to the rapidly increasing customer demand for nutrient-dense diets (Shiby et al., 2013). The broad availability of RTD protein drinks in supermarkets has contributed to a recent uptick in their appeal. It is well known that proteins are valuable as a supply of vital amino acids, but only recently has it become clear that dietary proteins also serve a variety of additional functions in the body, thanks to physiologically active peptides. Digestive processes produce inactive peptides within the original protein's sequence during fermentation, ripening, and gastrointestinal transit (Korhonen et al., 2009). Amino acids are essential for cell growth, maintenance, and energy production, and milk proteins are thought to be a good supply of these compounds (Unal et al., 2012). The powerful nutritional and functional qualities of whey protein have led to its increased acceptance as a functional food component. Generally Recognized as Safe (GRAS) certification applies to food items containing commercial whey proteins (Dima et al., 2015). Sweet whey powder (SWP), whey protein concentrate (WPC), whey protein isolate (WPI), and specialty WPCs are some of the whey components that might be used in the creation of lactic acid fermentation products like yogurt and fermented beverages (Krolczyk et al., 2016). Due to the high concentration of branched-chain amino acids found in whey proteins, they have a greater biological value than proteins derived from other sources such as eggs, soy, and milk casein (Pescuma et al., 2010).

9.8 PROTEIN FORMED BY DIFFERENT BACTERIA

9.8.1 RECOMBINANT PROTEIN

Making a decision on the host system is the initial step in RCP manufacturing. Analyze the chemical properties of the desired RCP before selecting a host; if the protein is heavily commented or glycosylated, a eukaryotic expression system is often used, because bacteria cannot adapt such modifications, which are normally required for effective protein function. However, disulfide bonds (DSB) can occur in bacteria (Landeta et al., 2006). Recent research on the formation of PTM similar to that of eukaryotic cells in bacterial hosts was reviewed by (Nothaft et al., 2013). Lower medium costs and quicker turnaround times mean that bacterial operations are often less costly than eukaryotic cell processes. *Escherichia coli*, an enteric bacterium with a long tradition of safe use in labs and industry, is the most popular organism used to create RCPs (Huang et al., 2012). *E. coli* is an excellent host organism since its genome was one of the first to be sequenced in its whole (Forde et al., 2014) and because a variety of molecular biology tools are available for changing its DNA sequences to produce novel abilities. Because of its dependability and affordable production costs, RCP synthesis using *E. coli* has earned the endorsement of drug regulatory bodies. Somatostatin was the first human protein to be generated in *E. coli* (Rosano et al., 2014). The first recombinant medication to be sold commercially was recombinant human insulin, which was created in *E. coli* and given FDA approval in 1982. Since then, several recombinant drugs that were first created in bacteria and then transferred to *E. coli* have been authorized for use in humans. Antibody subunits, hormones, growth factors, interferons, and thrombolytics are all examples. According to Ferrer et al. about 30% of all recombinant biologic products available today are made in *E. coli* (Ferrer et al., 2009). RCP therapeutics made in *E. coli* include the multiple sclerosis treatment interferon b-1b Extavia1 and the gout therapy Krystexxa1. In individuals with compromised renal function, Voraxaze1 is used to treat delayed methotrexate clearance. Since its genome was one of the first to be sequenced in its whole (Browng et al., 2023), and since there are a variety of molecular biology tools accessible for modifying its DNA sequences to develop unique abilities, *E. coli* is an excellent host organism. RCP manufacture utilizing *E. coli* has been approved by drug regulatory bodies due to its dependability and inexpensive production costs. Human somatostatin was the first protein to be artificially produced in *E. coli* (Samazan et al., 2015). There are no recombinant biopharmaceuticals authorized for use in humans that are produced using bacteria other than *E. coli*. This basic review will go through the steps involved in producing RCPs in *E. coli* and the choices that must be taken when formulating plans for their production. I proceed by troubleshooting the production processes that are involved in producing them, emphasizing many solutions and strategies to improving RCP output and function. Many RCPs are difficult to produce; therefore, I move on to do that. Finally, I go through a few potential future routes for bacterial RCP manufacturing.

9.9 BACTERIA CELL FACTORIES FOR RECOMBINANT PROTEIN PRODUCTION: EXPANDING THE CATALOGUE

Since the initial methods for creating recombinant DNA were created for *E. coli* genetic material and infecting bacteriophages, *E. coli* has been the most often used host for the production of RCPs. *E. coli* is the best host when trying to develop a new protein since more and more is being known about the genetics and physiology of these bacteria and because there are more and more genetic engineering methods designed specifically for this bacterium. Since simple culture processes and quick growth are crucial components, it is still the favored approach for manufacturing a large quantity of proteins in the laboratory and on a commercial scale. Finding the best system for protein synthesis may be challenging, and it is clear that this bacterial species does not provide particularly good circumstances for protein production. The ability of *E. coli* to produce proteins is constrained by a number of elements, including plasmid loss and antibiotic-based maintenance, unfavorable chemical gene-inducing chemicals, plasmid/ protein-mediated physiologic load and stress responses, a lack of post-translational changes (such as the failure to develop disulfide bonds), little to no secretion, protein aggregation and proteolytic digestion, endotoxin infection, and complex pathways. Pharmaceutical companies typically need to properly glycosylate proteins, and when looking for cost-effective bioproduction, the simplest purification methods become more appealing. In this scenario, whole plants and animals (as transgenic systems), insect cells, and the yeast *Saccharomyces cerevisiae* are all used to produce proteins. For use as protein treatments, several of these medicines have previously been granted licenses (Ferrer et al., 2009). The flexibility and promise of such platforms as protein producers or, more broadly, as cell factories producing value-added goods like chemicals, amino acids, or vitamins have recently been highlighted in tests and research (Navarrete et al., 2020). The biological intricacy of some of these systems, however, necessitates that large-scale manufacturing procedures be occasionally expensively adjusted (Figure 9.2).

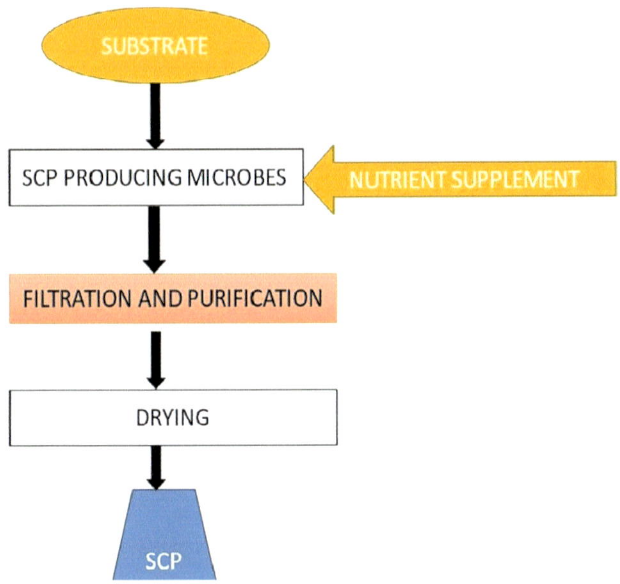

FIGURE 9.2 Single cell protein production.

9.10 SINGLE CELL PROTEIN PRODUCTION

Despite the fact that man has known about microbes for a very long time as food and feed, they have been used as a significant part of diets from the start of time. Bread using fermented yeast (*Saccharomyces* sp.) as a leavener has been found as far back as 2500 B.C. (Barranco et al., 2018). In Egypt and Greece between 50 and 100 B.C., LAB (*Lactobacillus and Streptococcus*) were used to make fermented milk and cheese (Barranco et al., 2018). The protein that may be obtained by removing cells from bacteria, yeast, and mold is known as SCP (Jones et al., 2020). The words "mixed culture" and "source of mixed protein" in this context denote, respectively, pure, and mixed cultures of bacteria, fungi, algae, and yeast, as well as the protein produced from these species. Microbes are a prominent source of single cell protein that may be utilized to supplement the diets of animals and humans due to their high protein content. It is also a novel protein, a petro crop, and a micro meal. Prof. Scrimshow of the Massachusetts Institute of Technology presented it as a more appealing alternative to the terms "microbial protein" and "bacterial protein." Carol L. Wilson first used the term single cell protein in 1966, and it may be more suitable given that most microorganisms grow as single or filamentous individuals (Adedayo et al., 2011). Industry has taken advantage of the potential of microorganisms to transform inferior organic material into highly nutritious proteaceous meals. Mass production of SCP food would not commence until the 20th century, despite the fact that the technology for doing so had been developed over the course of the previous century. During World War I (1914–1915), the necessity to mass-produce microorganisms for human consumption was first realized in Germany. Therefore, molasses medium with additional oxygen and ammonium salts was used to produce *Saccharomyces cerevisiae* (baker's yeast) (Gabriel et al., 2014; Simon et al., 2014) (Table 9.2).

TABLE 9.2 Showing Bacterial Protein as a Single Cell Protein on Specific Substrate

Bacteria	Substrate	Protein Content (%)	References
Bacillus subtilis	Ram horn	71	Kurbanoglu et al. (2002)
Rhodopseudomonas palustris	Latex rubber sheet	66	Kurbanoglu et al. (2002)
Escherichia coli	Ram horn	68	Kurbanoglu et al. (2002)
Methylomonas sp.	Methane salt broth	69	Yazdian et al. (2005)
Corynebacterium ammoniagenes	Glucose + Fructose	61	Wang et al. (2013)
Cupriavidus necator	synthetic growth medium	40 to 46	Kunasundari et al. (2013)
Bacillus pumilus	Potato starch processing waste	46	Liu et al. (2013)
Bacillus licheniformis	Potato starch processing waste	38	Liu et al. (2014)
Bacillus subtilis	SOYBEAN HALL	12	Wongputtisin et al. (2014)
Haloarcula sp. *IRU1*	Petrochemical wastewater	76	Taran et al. (2014)
Rhizospheric diazotrophs	Brewery wastewater	55	lee et al. (2015)

However, Germany manufactured aerobic yeasts as a food and feed supplement during the Second World War (1939–1945). Additionally, since World War II, a lot of work has been put into developing methods for the mass production of SCP by creating various growth media and improving microorganism culture (Gabriel et al., 2014). For instance, after World War II, Germany relied on imported SCP to make up for a 60% drop in food production, all thanks to the industrialization of the *Saccharomyces cerevisiae* yeast (Alves et al., 2015). British Petroleum scientists devised the "proteins-from-oil approach" in the 1960s to cultivate yeast on waxy n-paraffins, a by-product of oil refining, in order to create SCP. Numerous oil firms in the 50s were curious about the effects of microbial development on alkanes (Agitos et al., 2011; Gour et al., 2015). The first areas where microorganisms colonized were uncultivated surfaces and industrial waste. To reduce microbial protein synthesis, grape juice, cashew apple juice, cellulose, and hemi-cellulose were all employed as carbon sources (Kam et al., 2012). A few microbes have undergone modifications to become suitable sources of animal food. Numerous new companies have joined the SCP industry, including BP (UK), *Kanegafuchi* (Japan), and *Liquichimica* (Italy). Additional substrates for SCP synthesis include bagasse, citrus waste, sulphatic waste liquor, molasses, animal wastes, whey, starch, sewage, and others (Gour et al., 2015). Despite India being the birthplace of the modern mushroom industry in the early 1950s, there has been surprisingly less attention being paid to the manufacture of single cell proteins. However, efforts to grow mushrooms have been successful in Solan, Himachal Pradesh, since 1970. The Central Food Technologies Research Institute (CFTRI) located in Mysore, and the National Botanical Research Institute (NBRI), located in Lucknow, have recently built facilities for the mass synthesis of single cell protein from *cyanobacteria*. Single cell protein is created at the NBRI from sewage, and this protein is then used as animal feed.

9.11 ADVANTAGES OF INDUSTRIALLY FORMED PROTEIN

Global food security is under threat from the expanding human population and the anticipated fall in agricultural output brought on by climate change (Godfray et al., 2010). To keep up with population increase, Tilman et al. (2011) estimate that by 2050, agricultural protein production would need to be 110% greater than in 2005. However, there is a finite quantity of land and water available for agriculture. Poore et al. (2018) report that 43% of all agricultural land and all land that is not covered by ice or desert is used for agriculture. Urbanization, the forestry sector, the development of biofuels, and the protection of carbon stocks through trees are other competing demands for the use of land and freshwater. About a quarter of all anthropogenic greenhouse gas emissions and the great majority of nutrient emissions originate from agricultural, forestry, and other land uses, which is bad enough (Csilik et al., 2019). Most of these resources are used to produce animal proteins, which only provide about a third to world nutrition (Campbell et al., 2016; Rojas-Downing et al., 2017). In view of this, new approaches are needed to facilitate the production of protein that is climate-resilient, environmentally benign, less reliant on land and water, and adaptable to a variety of environmental conditions (Campbell et al., 2016; Rosegrant et al., 2009). Protein's availability, quality, and environmental effect are all important indications of food safety because of its central role in human nutrition (Coles et al., 2016). When high-quality protein is scarce, malnutrition may occur, and this is especially truc in developing countries. Millions of children, for instance, have insufficient access to essential amino acids in their diets (Semba et al., 2016). High-quality protein will be considerably harder to come by in the future as a result of rising demand brought on by climate change, population increase, and rising income. Especially those based on bacteria, microbial proteins (MPs) are thought to be practical replacements for conventional animal and plant-based proteins for the production of high-quality proteins to maintain food security with the least amount of environmental harm (Matassa et al., 2016). It has been shown that MPs made from the gram-negative soil bacterium *Cupriavidus necator* provide the whole range of amino acids needed for animal feed development, making them functionally like basic protein sources like soymeal and fishmeal (Volova et al., 2010). There are a number of ways in which bacterial proteins excel above other kinds of protein. Bacterial MPs have high protein concentrations (up to 83% of the dry biomass) and develop quickly in comparison to other protein sources (Jach et al., 2022; Matassa et al., 2016; Upadhaya et al., 2016). When compared to alternative MP sources like yeasts and fungi, as well as traditional protein suppliers, bacteria are far more efficient in transforming nutrients into protein-rich biomass. MP production frequently makes use of waste carbon substrates, which are both cheap and readily available

(Matassa et al., 2016; Upadhaya et al., 2016). In addition, MP production may be mapped out as a closed system that is relatively immune to seasonal changes, allowing for efficient fertilizer utilization with minimal environmental runoff and no need for pesticides or herbicides (Israelidis et al., 2003; Pikaar et al., 2017). Although the efficiency of solar-to-biomass conversion for MP production applications is likely to be greater than for crops, this efficiency varies greatly depending on the application. The efficiency of converting solar energy into biomass is typically less than 1% per year for perennial plants that may be harvested yearly. C3 plants, which include most common grains and vegetables, and C4 plants, which include maize (corn), sorghum, and sugarcane, may achieve an energy consumption efficiency during the growing season of 3.5% and 4.3%, respectively. Theoretical maximum efficiency is just 6% (Blankenship et al., 2011). *Arthrospira platensis*, a cyanobacterium, has just a 2% success rate, whereas H_2-oxidizing bacteria may have a 10% rate of success (Jiang et al., 2010). It has long been recognized that H_2-oxidizing bacteria might be used to produce MP. With the aid of CO_2 and O_2, H_2-oxidizing bacteria can use H_2 as an energy source for microbial development. In a bioreactor, electrolyzing water can yield hydrogen (H_2) and oxygen (O_2), and carbon dioxide (CO_2) can be collected from the air by use of adsorbents (Leung et al., 2022). Despite the high energy requirements of CO_2 separation and water electrolysis, there is a growing interest in discovering techniques to manufacture MP since it can be made using sustainable energy (Liu et al., 2016; Matassa et al., 2016). A rise in utility instability and periodic power shortages may be attributed to the growing capacity of wind and solar energy. In particular, solar energy is becoming more widely available as a result of the sharp drop in production prices. As a consequence, in the near future, it will be important to have flexible applications that can use electrical loads for demand response. From H_2-oxidizing bacteria, MP can be produced using the excess electricity. When the necessary materials cannot be supplied to the bioreactor from renewable energy sources during the night, *C. necator* survives the cycle and resumes growth during the day (Liu et al., 2016). When there is a surplus or when it is necessary, MP production may be carried out using solar or wind energy generation in areas with a lack of access to food and large amounts of non-arable land. Africa has a theoretical yearly energy production potential of 660 PWh for photovoltaic (PV) energy and 460 PWh for wind energy, respectively, according to the International Renewable Energy Agency. The world has far more potential for energy production than the 25 x 570 TWh needed to meet global power consumption in 2017. This is valid even while arable land is not taken into consideration (IEA, 2018). Despite MP production's increasing popularity, there are not many studies that evaluate how sustainable it is for the environment. Water electrolysis utilizes less space and produces less emissions from the burning of fossil fuels than other methods of MP production (Pikaar et al., 2018). We are aware of no studies that have evaluated the energy needs of cutting-edge technologies like direct air capture (DAC), bioreactors with in-built water electrolyzes and separating the microbial biomass from the growing medium for MP synthesis by H_2-oxidizing bacteria. MP is produced by binding CO_2 to biomass, which lowers ambient CO_2 levels overall and makes the product far more transportable than processes that use point sources of CO_2 (Matassa et al., 2016; Pikaar et al., 2018). The main advantage of this technique is that it uses far less water and land for production than traditional protein sources, with the energy source utilized having the most influence on the climate. We considered a renewable energy source that increases land usage in order to offer a conservative assessment of the potential to minimize land and water consumption. The purpose of this study is to investigate whether or whether the production of bacterial MPs utilizing direct CO_2 air capture, a water electrolyze, and renewable energy sources may use less water and land than soybean farming and other MPs already on the market with known requirements for these inputs. Because of their high protein content and large yields, soybeans are a popular source of plant-based protein, and their protein serves as a great example of a plant-based protein (Liu et al., 2018; Mekonnen et al., 2013). We calculated the overall amount of land and freshwater utilized by the system based on the energy and raw material requirements of the system's components.

9.12 DRAWBACK OF INDUSTRIAL BASED PROTEIN

In most cases, the complexity of bacterial gene regulation networks precludes their direct use in the development of synthetic circuits. Standardized components for use in synthetic biology may be gathered, produced, and enhanced by utilizing the desirable qualities of well-characterized, isolated cis- and trans-acting essential gene expression and regulatory components. This chapter focuses on the varied approaches taken by various regulatory regimes in addressing these SynBio applications. The development of specific gene-silencing antibiotics, bioremediation, diagnostics, treatments, and well-characterized artificial gene expression systems for microbial manufacture are just a few of the many possible applications in human and animal health care that could benefit from a greater knowledge of transcriptional and post-transcriptional gene control mechanisms. Compared to the conventional, petrochemical production methods, the latter is gaining popularity as an ecologically preferable choice (Jullesson et al., 2015). To avoid having a significant impact on the metabolism of the host organism, synthetic biology for microbial production must have stable, predictable, and controlled orthogonal gene expression systems that demonstrate maximal independence from and minimal interaction with the host metabolism (Mutalik et al., 2013; Singh et al., 2014). The use of elements that are functionally distinct from the operations of the endogenous gene expression and regulatory machineries and, ideally, only recognize their corresponding elements is referred to as orthogonal gene expression and orthogonal gene function and regulatory elements. Therefore, when introduced into a host, heterologous orthogonal sigma factors will only bind to the homologous promoters of these genes and will not interact with the promoters of the indigenous genes. Orthogonal ribosomes are also subject to the same constraints, since they can only interact with ribosome-binding sites that are identical to their own (RBS). Sequence homology gene expression often still depends on some of the host's basic gene-expression machinery, or at the very least on the host's metabolic energy-providing system and other cellular resources.

However, by using orthogonal gene regulatory elements, we may alter and optimize the expression of a subset of genes without substantially altering the overwhelming majority of the endogenous genes, hence reducing metabolic demand. Such an expression system needs a toolbox with well-characterized "components," such as regulatory elements and promoter and ribosome-binding sites (libraries) of various strengths that may be assembled in a combinatorial fashion (Bradley et al., 2016; Davis et al., 2011). This article goes into detail on the functions of several regulatory elements, including short regulatory RNAs, riboswitches, alternative sigma factors, DNA- and RNA-binding regulatory proteins, and DNA- and RNA-binding regulatory proteins, all of which have a lot of promise in this field. However, only a small portion of synthetic circuits requires optimization of gene expression. There may be additional significant bottlenecks in the production of heterologous materials as a result of post-translational modifications (PTM), protein folding and stability, feedback suppression of enzyme activity, or allosteric modulation of enzyme activity, though they are not discussed in this review.

9.13 FUTURE RESEARCH REQUIREMENTS AND THE VALIDITY OF ANALYSIS

When compared to standard agricultural production methods, the suggested protein manufacturing method requires a lot more energy (Wang et al., 2010). Since the electrolysis process provides around 90% of the energy required for the application, electrolysis efficiency, and the mass transfer of H_2 and O_2 are critical elements. The procedure, which was trialed in a lab-scale bioreactor, has to be scaled up for higher production levels. It has been shown that scaling up designs for producing biomass based on H_2-oxidizing bacteria is not difficult, even if testing a facility with a greater production capacity is necessary to confirm the findings of this study (Leung et al., 2022). Only the cost of energy is necessary at the current price of electricity. In the future, things might change if electricity costs go down, soybean prices go up, or demand-response services for the grid are provided by electro bioreactors. Other technologies besides those mentioned in this method can also be used to reduce energy costs. Therefore, before the proposed application is commercialized, techno-economic study is preferred. Although no electricity is needed to give the nutrients for biomass development at night, H_2-oxidizing bacteria continue to grow during the day when renewable electricity is available. There is possibility for reduced energy expenditures and/or incentive payouts through the use of demand response to coordinate bacterial biomass production with the grid's variable output and load (Abdullatif et al., 2017). Additional research is needed to see whether the program can be used as a demand-response instrument. It is intriguing to think about how long it would take to reach peak MP productivity. Since this application traps CO_2 in the air, it would be interesting to see how much of a greenhouse effect it has on the whole manufacturing process. The real energy and material flows of the manufacturing processes, as well as the suitable production sizes, may be evaluated using a pilot-scale application. The material flows of a product, from raw materials to finished goods, are all included into a cradle-to-grave impact study. Research is necessary to identify the extent to which MPs can serve as a replacement for plant-based feeds, as well as the amount of renewable energy that can be used to generate MPs.

9.14 CONCLUSION

Microbial protein, has recently gained popularity as a component in food. Microorganisms are used to develop this protein because they have several benefits over traditional protein sources, including a quicker time to double, an easier culture process, the flexibility to employ a variety of inexpensive/widely accessible substrates as energy sources, a low area need for growth, and the ability to adapt to climatic variations. Even though SCP could be used as a source of protein, it cannot compete because of the issues it faces. It also confronts challenges that prevent it from currently competing with conventional proteins. The possibility for disease transmission, the low digestibility, and the high cost of production due to substrate costs, utility costs, capital loads, and product-specific variables are some of these challenges. Another is the high nucleic acid content when made using bacteria. These problems can be lessened by developing cheaper downstream processes, using genetically modified bacteria that use more substrate, and improving suitable fed-batch fermentation. Its acceptance will also rise and improve with the implementation of efficient toxicological testing, raising its cost in comparison to other proteins and making it more affordable for the consumer.

KEYWORDS

- **bacterial protein production**
- ***Escherichia coli***
- **functional foods**
- **metabolic engineering**
- **recombinant protein**
- **single cell protein**

REFERENCES

Abdellatif, L., Ben-Mahmoud, O. M., Yang, C., Hanson, K. G., Gan, Y., & Hamel, C. (2017). The H_2-oxidizing rhizobacteria associated with field-grown lentil promote the growth of lentil inoculated with Hup^+ rhizobium through multiple modes of action. *Journal of Plant Growth Regulation, 36*(2), 348–361.

Adedayo, M. R., Ajiboye, E. A., Akintunde, J., & Odaibo, A. B. (2011). Single cell proteins: As nutritional enhancer. *Advances in Applied Science Research, 2*, 396–409.

Ageitos, J. M., Vallejo, J. A., Veiga-Crespo, P., & Villa, T. G. (2011). Oily yeasts as oleaginous cell factories. *Applied Microbiology and Biotechnology, 90*, 1219–1227.

Akoijam, N., Kalita, D., & Joshi, S. R. (2022). Bacteria and their industrial importance. In P. Verma (Ed.), *Industrial Microbiology and Biotechnology*. Springer.

Alves, Z., Melo, A., Figueiredo, A. R., Coimbra, M. A., Gomes, A. C., & Rocha, S. M. (2015). Exploring the *Saccharomyces cerevisiae* volatile metabolome: Indigenous versus commercial strains. *PLOS ONE, 10*(11). https://doi.org/10.1371/journal.pone.0143641 (accessed on 15th November 2025).

Barranco, C. R., Pérez, L. A., Alfaro, N. P., & Escorcia, J. Y. (2018). *Evaluation of natural culture media for the maintenance of Colletotrichum sp., and Saccharomyces cerevisiae.*

Besnard, L., Fabre, V., Fettig, M. P., Gousseinov, E., Kawakami, Y., Laroudie, N., Scanlan, C., & Pattnaik, P. (2016). Clarification of vaccines: An overview of filter-based technology trends and best practices. *Biotechnology Advances, 34*(1), 1–13.

Bradley, R. W., Buck, M., & Wang, B. (2016). Tools and principles for microbial gene circuit engineering. *Journal of Molecular Biology, 428*(5), 862–888.

Brinkrolf, K., Schröder, J., Pühler, A., & Tauch, A. (2010). The transcriptional regulatory repertoire of *Corynebacterium glutamicum*: Reconstruction of the network controlling pathways involved in lysine and glutamate production. *Journal of Biotechnology, 149*(3), 173–182.

Browning, D. F., Hobman, J. L., & Busby, S. J. W. (2017). Laboratory strains of *Escherichia coli* K-12: Things are seldom what they seem. *Microbial Genomics, 9*(2).

Campbell, B. M., Vermeulen, S. J., Aggarwal, P. K., Corner-Dolloff, C., Girvetz, E. H., Loboguerrero, A. M., Ramirez-Villegas, J., Rosenstock, T. S., Sebastian, L. S., Thornton, P. K., & Wollenberg, E. K. (2016). Reducing risks to food security from climate change. *Global Food Security, 11*, 34–43.

Chandrakasan, G., Rodríguez-Hernández, A. I., del Rocío López-Cuellar, M., Palma-Rodríguez, H. M., & Chavarría-Hernández, N. (2019). Bacteriocin encapsulation for food and pharmaceutical applications: Advances in the past 20 years. *Biotechnology Letters, 41*, 453–469.

Coles, G. D., Wratten, S. D., & Porter, J. R. (2016). Food and nutritional security requires adequate protein as well as energy, delivered from whole-year crop production. *PeerJ, 4*, e1522.

Csillik, O., Kumar, P., Mascaro, J., et al. (2019). Monitoring tropical forest carbon stocks and emissions using Planet satellite data. *Scientific Reports, 9*, 17831.

Davis, J. H., Rubin, A. J., & Sauer, R. T. (2010). Design, construction and characterization of a set of insulated bacterial promoters. *Nucleic Acids Research, 39*, 1131–1141.

Dima, C., & Dima, S. (2015). Essential oils in foods: Extraction, stabilization, and toxicity. *Current Opinion in Food Science, 5*, 29–35.

Ferrer-Miralles, N., Domingo-Espín, J., Corchero, J. L., Vázquez, E., & Villaverde, A. (2009). Microbial factories for recombinant pharmaceuticals. *Microbial Cell Factories, 8*, 1–8.

Forde, B. M., Ben Zakour, N. L., Stanton-Cook, M., Phan, M. D., Totsika, M., Peters, K. M., et al. (2014). The complete genome sequence of *Escherichia coli* EC958: A high-quality reference sequence for the globally disseminated multidrug-resistant *E. coli* O25b:H4-ST131 clone. *PLOS ONE, 9*(8), e104400.

Gabriel, A. Y., Victor, N., & Preez, J. C. (2014). Cactus pear biomass, a potential lignocellulose raw material for single cell protein production (SCP): A review. *International Journal of Current Microbiology and Applied Sciences, 3*, 171–197.

García-Garibay, M., Gómez-Ruiz, L., & Cruz-Guerrero, A. E. (2014). Single cell protein: Yeasts and bacteria. In C. A. Batt & M. L. Tortorello (Eds.), *Encyclopedia of Food Microbiology* (2nd ed., pp. 431–438). Academic Press.

Goa, T., Beyene, G., Mekonnen, M., & Gorems, K. (2022). Isolation and characterization of lactic acid bacteria from fermented milk produced in Jimma Town, Southwest Ethiopia, and evaluation of their antimicrobial activity against selected pathogenic bacteria. *International Journal of Food Science*, 2076021.

Godfray, H. C., Beddington, J. R., Crute, I. R., Haddad, L. J., Lawrence, D., Muir, J. F., Pretty, J., Robinson, S., Thomas, S. M., & Toulmin, C. (2010). Food security: The challenge of feeding 9 billion people. *Science, 327*, 812–818.

Gour, S., Mathur, N., Singh, A., & Pradeep, B. (2015). Single cell protein production: A review. *International Journal of Current Microbiology and Applied Sciences, 4*(9), 251–262.

Hallberg, A. R. (2014). Aberrant CpG methylation of the *TFAP2A* gene constitutes a mechanism for loss of *TFAP2A* expression in human metastatic melanoma. *Epigenetics, 9*(12), 1641–1647.

Horesh, G., Blackwell, G. A., Tonkin-Hill, G., Corander, J., Heinz, E., & Thomson, N. R. (2021). A comprehensive and high-quality collection of *Escherichia coli* genomes and their genes. *Microbial Genomics, 7*(2), 000499.

Huang, C., Lin, H., & Yang, X. (2012). Industrial production of recombinant therapeutics in *Escherichia coli* and its recent advancements. *Journal of Industrial Microbiology & Biotechnology, 39*, 383–399.

Israelidis, C. J. (2003). Nutrition–Single cell protein, twenty years later. In *Proceedings from First Biointernational Conference*.

Jach, M. E., Serefko, A., Ziaja, M., & Kieliszek, M. (2022). Yeast protein as an easily accessible food source. *Metabolites, 12*(1), 63.

Jia, H., Li, H., Zhang, L., & Xu, D. (2018). Development of a novel gene expression system for secretory production of heterologous proteins via the general secretory (Sec) pathway in *Corynebacterium glutamicum*. *Iranian Journal of Biotechnology, 16*(1).

Jiang, T., Li, L., Ma, C., & Xu, P. (2010). Strategies for regulating multiple genes in microbial cell factories. *Sheng Wu Gong Cheng Xue Bao*, 1419–1425.

Jones, S. W., Karpol, A., Friedman, S., Maru, B. T., & Tracy, B. P. (2020). Recent advances in single cell protein use as a feed ingredient in aquaculture. *Current Opinion in Biotechnology, 61*, 189–197.

Jullesson, D., David, F., Pfleger, B. F., & Nielsen, J. B. (2015). Impact of synthetic biology and metabolic engineering on industrial production of fine chemicals. *Biotechnology Advances, 33*(7), 1395–1402.

Kadim, I. T., Mahgoub, O., Baqir, S., Faye, B., & Purchas, R. (2015). Cultured meat from muscle stem cells: A review of challenges and prospects. *Journal of Integrative Agriculture, 14*(2), 222–233.

Kadner, R. J., & Rogers, K. (2023). Bacteria. *Encyclopedia Britannica.*

Kam, S., Abedian, K. A., Kenari, A., & Younesi, H. (2012). Production of single cell protein in stickwater by *Lactobacillus acidophilus* and *Aspergillus Niger*. *Journal of Aquatic Food Product, 21.*

Khoo, K. S., Chew, K. W., Yew, G. Y., Leong, W. H., Chai, Y. H., Show, P. L., & Chen, W. H. (2020). Recent advances in downstream processing of microalgae lipid recovery for biofuel production. *Bioresource Technology, 304*, 122996.

Korhonen, H. (2009). Milk-derived bioactive peptides: From science to applications. *Journal of Functional Foods, 1*(2), 177–187.

Królczyk, J., Dawidziuk, T., Janiszewska-Turak, E., & Sołowiej, B. (2016). Use of whey and whey preparations in the food industry – A review. *Polish Journal of Food and Nutrition Sciences, 66*, 157–165.

Kurbanoğlu, E. B., & Algur, Ö. F. (2002). Use of ram horn hydrolysate as peptone for bacterial growth. *Turkish Journal of Biology, 26*, 115–123.

Landeta, C., Boyd, D., & Beckwith, J. (2018). Disulfide bond formation in prokaryotes. *Nature Microbiology, 3*(3), 270–280.

Lang, V., Bellisle, F., Alamowitch, C., Craplet, C., Bornet, F. R., Slama, G., & Guy-Grand, B. (1999). Varying the protein source in mixed meal modifies glucose, insulin and glucagon kinetics in healthy men, has weak effects on subjective satiety and fails to affect food intake. *European Journal of Clinical Nutrition, 53*(12), 959–965.

Lee, J. Z., Logan, A., Terry, S., & Spear, J. R. (2015). Microbial response to single-cell protein production and brewery wastewater treatment. *Microbial Biotechnology, 8*(1), 65–76.

Leung, P. M., Daebeler, A., Chiri, E., Hanchapola, I., Gillett, D. L., Schittenhelm, R. B., & Greening, C. (2022). A nitrite-oxidizing bacterium constitutively consumes atmospheric hydrogen. *The ISME Journal, 16*(9), 2213–2219.

Liu, C., Colón, B. C., Ziesack, M., Silver, P. A., & Nocera, D. G. (2016). Water splitting–biosynthetic system with CO_2 reduction efficiencies exceeding photosynthesis. *Science, 352*, 1210–1213.

Liu, L., Liu, Y., Shin, H. D., Chen, R. R., Wang, N. S., Li, J., Du, G., & Chen, J. (2013). Developing *Bacillus* spp. as a cell factory for production of microbial enzymes and industrially important biochemicals in the context of systems and synthetic biology. *Applied Microbiology and Biotechnology, 97*(14), 6113–6127.

Liu, X., Yu, L., Li, W., Peng, D., Zhong, L., Li, L., & Gong, P. (2018). Comparison of country-level cropland areas between ESA-CCI land cover maps and FAOSTAT data. *International Journal of Remote Sensing, 39*(20), 6631–6645.

Mahasneh, I. A. (1997). Production of single cell protein from five strains of the microalga *Chlorella* spp. (*Chlorophyta*). *Cytobios, 90*(362, 363), 153–161.

Mancini, E., Mansouri, S. S., Gernaey, K. V., Luo, J., & Pinelo, M. (2020). From second generation feedstocks to innovative fermentation and downstream techniques for succinic acid production. *Critical Reviews in Environmental Science and Technology, 50*(18), 1829–1873.

Matassa, S., Boon, N., Pikaar, I., & Verstraete, W. (2016). Microbial protein: Future sustainable food supply route with low environmental footprint. *Microbial Biotechnology, 9*(5), 568–575.

Mekonnen, M. M., & Hoekstra, A. Y. (2014). Water footprint benchmarks for crop production: A first global assessment. *Ecological Indicators, 46*, 214–223.

Mevada, J., Devi, S., & Pandit, A. B. (2019). Large scale microbial cell disruption using hydrodynamic cavitation: Energy saving options. *Biochemical Engineering Journal.*

Morales, M., Sánchez, L., & Revah, S. (2018). The impact of environmental factors on carbon dioxide fixation by microalgae. *FEMS Microbiology Letters, 365*(3).

Mutalik, V. K., Guimaraes, J. C., Cambray, G., Lam, C., Christoffersen, M. J., Mai, Q., Tran, A. B., Paull, M. L., Keasling, J. D., Arkin, A. P., & Endy, D. (2013). Precise and reliable gene expression via standard transcription and translation initiation elements. *Nature Methods, 10*, 354–360.

Navarrete, C., Jacobsen, I., Martínez, R. J., & Procentese, A. (2020). Cell factories for industrial production processes: Current issues and emerging solutions. *Processes, 8*, 768.

Nothaft, H., & Szymanski, C. M. (2013). Bacterial protein N-glycosylation: New perspectives and applications. *The Journal of Biological Chemistry, 288*, 6912–6920.

Oliveira, R. B., de Oliveira, A. L., & Glória, M. B. (2008). Screening of lactic acid bacteria from vacuum packaged beef for antimicrobial activity. *Brazilian Journal of Microbiology, 39*, 368–374.

Owczarek, B., Gerszberg, A., & Hnatuszko-Konka, K. (2019). A brief reminder of systems of production and chromatography-based recovery of recombinant protein biopharmaceuticals. *BioMed Research International.*

Pescuma, M., Hebert, E. M., Mozzi, F., & de Valdez, G. F. (2010). Functional fermented whey-based beverage using lactic acid bacteria. *International Journal of Food Microbiology, 141*(1, 2), 73–81.

Pikaar, I., Matassa, S., Rabaey, K., Bodirsky, B. L., Popp, A., Herrero, M., & Verstraete, W. (2017). Microbes and the next nitrogen revolution. *Environmental Science & Technology, 51*(14), 7297–7303.

Poore, J., & Nemecek, T. (2018). Reducing food's environmental impacts through producers and consumers. *Science, 360*(6392), 987–992.

Prajapati, B., Bernal-Cabas, M., López-Álvarez, M., Schaffer, M., Bartel, J., Rath, H., Steil, L., Becher, D., Völker, U., Mäder, U., & van Dijl, J. M. (2021). Double trouble: *Bacillus* depends on a functional Tat machinery to avoid severe oxidative stress and starvation upon entry into a NaCl-depleted environment. *Biochimica et Biophysica Acta. Molecular Cell Research, 1868*(2), 118914.

Queiroz, M. I., Lopes, E. J., Zepka, L. Q., Bastos, R. G., & Goldbeck, R. (2007). The kinetics of the removal of nitrogen and organic matter from parboiled rice effluent by cyanobacteria in a stirred batch reactor. *Bioresource Technology, 98*(11), 2163–2169.

Rahmdel, S., Shekarforoush, S. S., Hosseinzadeh, S., Torriani, S., & Gatto, V. (2019). Antimicrobial spectrum activity of bacteriocinogenic *Staphylococcus* strains isolated from goat and sheep milk. *Journal of Dairy Science, 102*(4), 2928–2940.

Raja, R., Hemaiswarya, S., Kumar, N. A., Sridhar, S., & Rengasamy, R. (2008). A perspective on the biotechnological potential of microalgae. *Critical Reviews in Microbiology, 34*(2), 77–88.

Rathore, S., Salmerón, I., & Pandiella, S. S. (2012). Production of potentially probiotic beverages using single and mixed cereal substrates fermented with lactic acid bacteria cultures. *Food Microbiology, 30*(1), 239–244.

Ravinder, R., Linga, V., & Ravindra, P. (2013). Studies on *Aspergillus oryzae* mutants for the production of single cell proteins from deoiled rice bran. *RNA, 41*, 243–246.

Retallack, D. M., Jin, H., & Chew, L. (2012). Reliable protein production in a *Pseudomonas fluorescens* expression system. *Protein Expression and Purification, 81*(2), 157–165.

Richardson, A., & Walker, J. (2018). Continuous solids-discharging centrifugation: A solution to the challenges of clarifying high–cell-density mammalian cell cultures. *Bioprocess International*.

Rojas-Downing, M. M., Nejadhashemi, A. P., Harrigan, T. M., & Woznicki, S. A. (2017). Climate change and livestock: Impacts, adaptation, and mitigation. *Climate Risk Management, 16*, 145–163.

Rosano, G. L., & Ceccarelli, E. A. (2014). Recombinant protein expression in *Escherichia coli*: Advances and challenges. *Frontiers in Microbiology, 5*, 172.

Samazan, F., Rokbi, B., Seguin, D., Telles, F., Gautier, V., Richarme, G., Chevret, D., Varela, P. F., Velours, C., & Poquet, I. (2015). Production, secretion and purification of a correctly folded staphylococcal antigen in *Lactococcus lactis*. *Microbial Cell Factories, 14*, 104.

Semba, R. D. (2016). The rise and fall of protein malnutrition in global health. *Annals of Nutrition and Metabolism, 69*, 79–88.

Shiby, V. K., Radhakrishna, K., & Bawa, A. S. (2013). Development of whey-fruit-based energy drink mixes using D-optimal mixture design. *International Journal of Food Science and Technology, 48*, 742–748.

Simion, A., & Grigoras, C. (2014). New possibilities of fodder yeast production. *Scientific Papers. Series D, Animal Science, 23*, 71–78.

Singh, N., & Chollangi, S. (2017). Next-generation clarification technologies for the downstream processing of antibodies. In *Process Scale Purification of Antibodies* (pp. 81–112).

Singh, N., Arunkumar, A., Chollangi, S., Tan, Z., Borys, M. C., & Li, Z. J. (2016). Clarification technologies for monoclonal antibody manufacturing processes: Current state and future perspectives. *Biotechnology and Bioengineering, 113*, 698–716.

Sousa, I., Gouveia, L., Batista, A., Raymundo, A., & Bandarra, N. (2008). Microalgae in novel food products. In *Algae: Nutrition, Pollution Control and Energy Sources*.

Tanamool, V., Chantarangsee, M., & Soemphol, W. (2020). Simultaneous vinegar fermentation from a pineapple by-product using the co-inoculation of yeast and thermotolerant acetic acid bacteria and their physiochemical properties. *3 Biotech, 10*(3), 115.

Taran, M., Rad, M., & Alavi, M. (2017). Antibacterial activity of copper oxide (CuO) nanoparticles biosynthesized by *Bacillus* sp. FU4: Optimization of experiment design. *Pharmaceutical Sciences, 23*(3), 198–206.

Tilman, D., Balzer, C., Hill, J. D., & Befort, B. L. (2011). Global food demand and the sustainable intensification of agriculture. *Proceedings of the National Academy of Sciences, 108*, 20260–20264.

Turnbull, W. H., Leeds, A. R., & Edwards, D. G. (1992). Mycoprotein reduces blood lipids in free-living subjects. *The American Journal of Clinical Nutrition, 55*(2), 415–419.

Ugalde, U. O., & Castrillo, J. I. (2002). Single cell proteins from fungi and yeasts. In G. G. Khachatourians & D. K. Arora (Eds.), *Agriculture and Food Production* (pp. 123–149). Elsevier.

Ul-Islam, M., Subhan, F., Islam, S. U., Khan, S. S., Shah, N., Manan, S., Ullah, M. W., & Yang, G. (2019). Development of three-dimensional bacterial cellulose/chitosan scaffolds: Analysis of cell-scaffold interaction for potential application in the diagnosis of ovarian cancer. *International Journal of Biological Macromolecules, 137*, 1050–1059.

Ul-Islam, M., Ullah, M. W., Khan, S. S., Kamal, T., Ul-Islam, S., Shah, N., & Park, J. (2016). Recent advancement in cellulose-based nanocomposite for addressing environmental challenges. *Recent Patents on Nanotechnology, 10*(3), 169–180.

Ullah, M. W., Ul-Islam, M., Khan, S. S., Kim, Y., & Park, J. (2016). Structural and physico-mechanical characterization of bio-cellulose produced by a cell-free system. *Carbohydrate Polymers, 136*, 908–916.

Unal, G., & Akalın, A. (2012). Antioxidant and angiotensin-converting enzyme inhibitory activity of yogurt fortified with sodium calcium caseinate or whey protein concentrate. *Dairy Science and Technology, 92*, 627–639.

Upadhyaya, S., Tiwari, S. H. A. S. H. A. N. K., Arora, N., & Singh, D. P. (2016). Microbial protein: A valuable component for future food security. In *Microbes and Environmental Management, 8*.

Vilkhovoy, M., Horvath, N., Shih, C. H., Wayman, J. A., Calhoun, K., Swartz, J., & Varner, J. D. (2018). Sequence specific modeling of *E. coli* cell-free protein synthesis. *ACS Synthetic Biology, 7*(8), 1844–1857.

Volova, T. G., & Barashkov, V. A. (2010). Characteristics of proteins synthesized by hydrogen-oxidizing microorganisms. *Applied Biochemistry and Microbiology, 46*, 574–579.

Wang, X., Kim, Y., Ma, Q., Hong, S. H., Pokusaeva, K., Sturino, J. M., & Wood, T. K. (2010). Cryptic prophages help bacteria cope with adverse environments. *Nature Communications, 1*(1), 147.

Wongputtisin, P., Khanongnuch, C., Kongbuntad, W., Niamsup, P., Lumyong, S., & Sarkar, P. K. (2014). Use of *Bacillus subtilis* isolates from Tua-nao towards nutritional improvement of soybean hull for monogastric feed application. *Letters in Applied Microbiology, 59*(3), 328–333.

Yazdian, F., Hajizadeh, S., Jahanshahi, M., Alsadati, S. A. S., Khalilzadeh, R., & Nosrati, M. (2005). Production of single cell protein from natural gas: Parameter optimization and RNA evaluation. *Iranian Journal of Biotechnology, 3*, 235–242.

Yilmaz, H., Faburay, B., Turan, N., Cotton-Caballero, M., Cetinkaya, B., Gurel, A., Yilmaz, A., Cizmecigil, U. Y., Aydin, O., Tarakci, E. A., Bayraktar, E., & Richt, J. A. (2019). Production of recombinant N protein of infectious bronchitis virus using the baculovirus expression system and its assessment as a diagnostic antigen. *Applied Biochemistry and Biotechnology, 187*(2), 506–517.

Zepka, L., Jacob-Lopes, E., & Goldbeck, R., Queiroz, M. (2008). Production and biochemical profile of the microalgae *Aphanothece microscopica Nägeli* submitted to different drying conditions. *Chemical Engineering and Processing, 47*, 1305–1310.

Zhou, Y., Han, L., He, H., Sang, B., Yu, D., Feng, J., & Zhang, X. (2018). Effects of agitation, aeration and temperature on production of a novel glycoprotein GP-1 by

Streptomyces kanasenisi ZX01 and scale-up based on volumetric oxygen transfer coefficient. *Molecules, 23*.

Zydney, A. L. (2016). Continuous downstream processing for high value biological products: A review. *Biotechnology and Bioengineering, 113*, 465–475.

CHAPTER 10

Effective Methods and Approaches for Utilizing Microbial Biopesticides

PANKAJ KUMAR SINGH,[1] PAWAN KUMAR,[2] NISHAT FATIMA,[1] ANJALI SINGH,[1] BRIJESH PANDEY,[2,*] and KESHAV SINGH[1,*]

[1]*Department of Zoology, Vermibiotechnology Laboratory, D.D.U. Gorakhpur University, Gorakhpur, Uttar Pradesh, India*

[2]*Department of Biotechnology, Mahatma Gandhi Central University, Motihari, Bihar, India*

*Corresponding author

ABSTRACT

Microbial biopesticides have become a promising alternative to conventional chemical pesticides (synthetic pesticides) due to their eco-friendly nature, environmental safety, cost-effectiveness, efficacy, biodegradability, and target-specific action against pests. Living microbes or their metabolites are utilized as microbial biopesticides in public health, forestry, agriculture, and other fields to manage pests and diseases. This book chapter covers the application techniques and strategies essential for maximizing the efficacy of microbial biopesticides in the management of pests. It addresses formulation development, microbial strain selection, and a range of delivery techniques, including seed treatments, soil drenching, and foliar sprays. The book chapter explores integration with Integrated Pest Management (IPM) strategies, emphasizing how crucial it is to combine microbial biopesticides with other pest management techniques for increased efficacy and sustainability. The book chapter also discusses essential factors, such as regulatory considerations and environmental

Industrial Microbiology: Current Progress and Novel Technologies.
Prakash Kumar Sarangi, Akhilesh Kumar Singh, Rajesh Kumar Srivastava, and Brijesh Pandey (Eds.)

DOI: 10.1201/9781779642219-10

conditions, that affect the effectiveness and persistence of microbial biopesticides in agricultural ecosystems. In order to provide insights into overcoming challenges and maximizing the potential of microbial biopesticides in modern agriculture, case examples demonstrating effective uses and future prospects in microbial biopesticide research and implementation are studied.

10.1 INTRODUCTION

Synthetic (chemical) pesticides have been used for agricultural pest control since earlier times, with the goal of improving crop productivity (Anani et al., 2020). Chemicals and carriers, such as polymers, that are specific to certain pests are used to produce synthetic insecticides (Rakhimol et al., 2020). According to their classification, they range from those used to control algae (algicides), bacteria (bactericides), weeds (herbicides), fungi (fungicides), insects (insecticides), mollusks (molluscicides), nematodes (nematicides), ticks or mites (acaricides/miticides), rodents (rodenticides), and termites (termiticides) (Anakwue, 2019). Pesticides can also be categorized based on the active ingredients they contain, such as dichlorvos, organochlorines, diazinon, diamide, and chlorpyrifos. Synthetic pesticides can increase agricultural productivity and yield, but they can also have detrimental effects on animals, aquatic life, soil biodiversity, and humans (Farooq et al., 2019). According to Pelosi et al. (2021) and Pertile et al. (2020), synthetic pesticides typically cause the soil to become brittle, reduce soil respiration, and minimize the activity of some soil microorganisms, such as earthworms. Additionally, they maintain an animal's vitality, lower its resistance to disease, and alter the characteristics of its progeny (Syromyatnikov et al., 2020). According to Kumar and Kumar (2019), they have an adverse effect on soil microorganisms by reducing their biological services in the synthesis of numerous substances that support plant growth, notably siderophores, nitrogen, and indole-3-acetic acid. There are numerous ways that synthetic pesticides can find their way into the environment, such as leaching, erosion, droplet drift, vapor movements, and improper dumping. They might therefore come into contact with some non-targeted plants, which could lessen their capacity to compete for seed production and photosynthetic capacity (Hashimi et al., 2020).

Pesticides may contaminate water and kill aquatic organisms when they find their way into bodies of water through runoff. Furthermore, pesticides have the potential to spread from aquatic life to terrestrial animals and humans, whereupon their biomagnification may cause the emergence of deadly diseases like diabetes, cancer, renal disease, and skin rashes (Jayaraj et al., 2017; Manfo et al., 2020; Sabarwal et al., 2018). However, biopesticides have been developed and have proven to be very effective in controlling different types of pests. According to Borges et al. (2021), biopesticides are affordable, environmentally friendly, selective in their mode of action, residue-free, sustainable, and not linked to the release of greenhouse gases. In comparison to their synthetic counterparts, biopesticides are safer, more effective, and have a lesser environmental impact. They additionally prevent bioaccumulation because they only target particular pests (Kumar et al., 2021; Saberi et al., 2020). These biopesticides can be divided into three categories: microbial pesticides (of microbial origin; Harish et al., 2021), Phyto pesticides (of plant origin; Idris et al., 2022), and nano biopesticides (nanoparticles produced by biological agents; Abdollahdokht et al., 2022; Pan et al., 2023). Microbial biopesticides function specifically, do not need expensive chemicals to be sourced, and have no long-term detrimental impact on the environment, in contrast to chemical pesticides (Harish et al., 2021; Hummadi et al., 2021). In addition to exhibiting a variety of mechanisms of action due to the broad spectrum of phytochemical compounds, Phyto pesticides pose a lower risk to human health than synthetic pesticides on the market and do not release greenhouse gases (Idris et al., 2022; Malahlela et al., 2021). Nano biopesticides provide greater pesticidal activity, targeted or controlled release, and improved biocompatibility and biodegradability when compared to synthetic pesticides (Abdollahdokht et al., 2022; Pan et al., 2023).

Biopesticides act in a number of methods, including inhibiting and destroying the pathogens or pests' ability to translate proteins into plasma membranes. Though they have some drawbacks that have decreased their acceptability and commercial utilization, biopesticides are less persistent in the soil environment, have a short shelf life, are more targeted than synthetic pesticides, and originate from sustainable raw materials (Kumar et al., 2021). Some of the positive effects of biopesticides that were previously mentioned could also be regarded as disadvantages. For instance, the specialization of their pest-management strategy could work against them if the goal is to control several pests at once. They also break down more quickly and are less persistent in the environment due to their short shelf life, but this becomes a drawback if the goal is to eradicate all of the existing pests and stop future pest growth after the application of biopesticides (Figure 10.1).

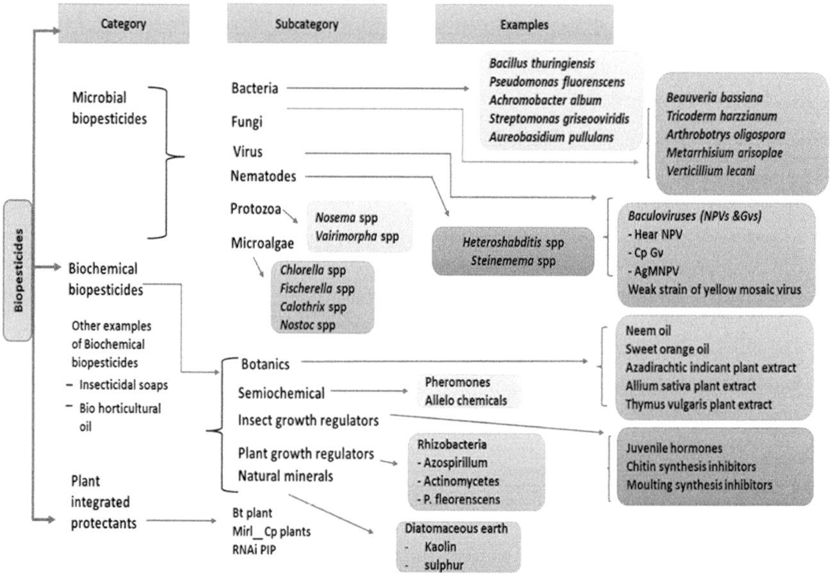

FIGURE 10.1 Three different biopesticides categories with selected examples.

TABLE 10.1 Registered Biopesticides Lists in India Under Insecticides Act, 1968

Sl.No.	Biopesticides Name
1.	*Bacillus thuringiensis* subsp. *israelensis*
2.	*B. thuringiensis* subsp. *kurstaki*
3.	*B. thuringiensis* subsp. *galleriae*
4.	*B. sphaericus*
5.	*Trichoderma harzianum*
6.	*Trichoderma viride*
7.	*Pseudomonas fluorescens*
8.	Nuclear polyhedrosis virus (NPV) of *Helicoverpa armigera*
9.	*Beauveria bassiana*
10.	Nuclear polyhedrosis virus (NPV) of *Spodoptera litura*
11.	Neem based pesticides
12.	*Cymbopogon*

10.2 MICROBIAL BIOPESTICIDES

Genetically engineered microbial biopesticides are a recent development. It is common practice to employ organisms like bacteria, viruses, fungi, algae, and protozoans. Compared to traditional chemical pesticides, they develop a toxin unique to a particular insect that causes disease and inhibits the growth of other microbes through antagonistic interactions or other non-toxic mechanisms of action (Bellinger, 2007). Normally, live microorganisms that are harmful to the target pest are employed as microbial biopesticides. *Phytophthora*, *Pseudomonas*, *Trichoderma*, and *Bacillus* are examples of bio fungicides, as are bioinsecticides (Quarles, 2011). Bacterial biopesticides account for 74% of the market for the majority of crops, with the remaining 10% going to fungal, 10% to viral, 8% to predator, and 3% to other biopesticides (Thakore, 2006).

Microbial biopesticides, which are used to manage pests, are substances derived from microorganisms such as algae, bacteria, fungi, viruses, and protozoa (Adeleke et al., 2022a). In order to eliminate and prevent the growth of pests, microorganisms produce toxic metabolites. Multidisciplinary approaches such as emulsion, fluidized bed, electro spraying system, spray drying, extrusion, lyophilization, spray chilling, and coacervation are employed to apply microbial pesticides to environmental surfaces (De Oliveira et al., 2021). The main categories of microorganisms used as biopesticides are the following: bacteria genera *Chromobacterium*, *Pseudomonas*, and *Yersinia*; nematodes of the genera *Steinernema* and *Heterorhabditis*; and fungal genera *Beauveria*, *Paecilomyces*, *Verticillium*, *Hirsutella*, *Metarhizium*, and *Yersinia* (Adeleke et al., 2022b; Chang et al., 2003; Kumar et al., 2021). According to reports, some other fungi in the soil that cause black gram, green gram, and groundnut and chickpea root rot are inhibited by a species of fungus known as *Trichoderma* spp. (Samada and Tambunan, 2020). *Beauveria bassiana* and *M. brunneum* have been reported to be effective in suppressing thrips, aphids, beetles, weevil, whiteflies, and mite infestation in ornamental crops, fruits, and vegetables (Arthurs and Dara, 2019; Dara, 2017). Tables 10.1 and 10.2 contain more lists of microbiological pesticides. The most well-known bacterial insecticide is *Bacillus thuringiensis* (Bt), which has been formulated into products available for commercial purpose (Ruiu, 2018; Ujvary, 2010). The Gram-positive bacterium *Bacillus thuringiensis* produces endospores and deadly parasporin crystals as exudates, which are used as an insecticide.

The endospores are dissolved by the alkaline environment in the midgut of insects that consume these materials, generating delta-endotoxin, a protein that kills insects (Xiao and Wu, 2019). *Bacillus thuringiensis* reduces pest infestation in plants, such as potatoes and cabbage, by controlling lepidopterans in a range of plant species (Berini et al., 2018; Samada and Tambunan, 2020).

In addition to their beneficial advantages, commercially available *Bt* sounds have drawbacks such as limited environmental awareness and sensitivity, short activity duration, rapid deactivation when exposed to light, and slow lethal rate (Xiao and Wu, 2019). The main drawbacks to using microbial pesticides are their short shelf life and sensitivity to the environment, which decrease user awareness and usage (Pathak et al., 2017). For example, baculoviruses can only exist within their host and are unable to reproduce outside of it; as a result, they cannot be exploited in other environments (Borges et al., 2021). Their use could be hazardous, and their host could have a negative environmental impact. Plant pests can also be controlled with the help of fungi. Mycoinsecticide is a microbial pesticide that contains a living fungus that has an antagonistic impact on insects and other arthropod pests. Some fungus strains cause the pest to release metabolites that can either kill or harm the insect (Zaki et al., 2020). As there are so few uncommon fungal strains that can be successfully transformed into mycoinsecticides, the technique is currently in its early stages of development. According to Zaki et al. (2020), mycoinsecticides have six general modes of action: attachments, germination, penetration, invasion, reproduction, and host death. Spores can land on the target host's cuticle and stick to it when the designed product is applied diluted and according to label directions.

The main mechanism of adhesion is hydrophobic interactions between the spore and its cuticle. Their effectiveness is based on the spore's quantity adhered to the host's body. In response to chemical cues on the cuticle, the spore germinates and then produces an aspersorium, or piercing structure. According to Zaki et al. (2020), the fungus penetrates through the cuticle's layers by breaking down enzymes and applying mechanical pressure.

Due to their safe and target-specific ingredients, microbial pesticides typically have no negative environmental effect, consumers of agricultural products or farmers (Guven et al., 2021). Furthermore, its application results in reduced greenhouse gas emissions as compared to chemical pesticides (Llamas et al., 2021). In addition, a vast variety of organisms can be used to produce microbial biopesticides, which guarantee sustainability and address the issue of resistance. It may be difficult for suppliers, manufacturers, marketers, and final consumers to handle the storage and shipping of various microorganisms employed as biopesticides since they may require varying storage conditions (Figure 10.2).

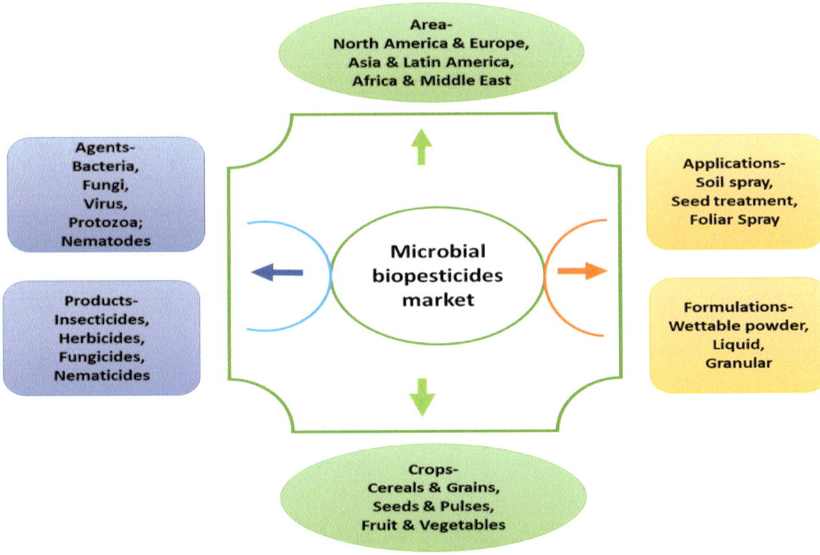

FIGURE 10.2 Potential market for application of microbial biopesticides.

The use of biopesticides is being encouraged as an alternative to synthetic pesticides due to their disadvantages (Ayilara et al., 2022; Ojuederie et al., 2021). In comparison to their synthetic counterparts, biopesticides are safer, more effective, and have less environmental impact. Additionally, because they target specific pests, they prevent bioaccumulation (Kumar et al., 2021; Saberi et al., 2020). Microbial biopesticides are a sustainable kind of pest management because they are derived from natural materials such as biologically derived nanoparticles, microorganisms, and plants (Kumar et al., 2021). There have been some reported achievements when using biopesticides to manage pests that are typically treated with chemical pesticides. The various modes of action of biopesticides, such as those that control gut disruption, growth of insects, and pest metabolism, contribute to their efficacy in pest management.

TABLE 10.2 Different Microbial Biopesticides and the Target Insects and Pest's Types They Control

Source	Organisms	Target Insects and Pest's Types	References
Viruses	Imported cabbageworm (PiraGV) NPV (AucaMNPV)	*Arteria (Pieris) rapae*	Singh et al. (2019)
	Spodoptera exigua NPV, *Helicoverpa zea* NPV	*Lepidoptera*	Elvira et al. (2013); Fan et al. (2022); Lacey et al. (2008); Lasa et al. (2007); Luna-Espino et al. (2018);
	Potato tuber moth GV (PhopGV)	*Phthorimaea operculella*	Singh et al. (2019)
	Cydia pomonella GV	*Codling moth*	Kadoi´c Balaško et al. (2020)
	Nucleopolyhedro (NPV) viruses	*Lepidoptera*	Harish et al. (2021)
Fungi	*Beauveria bassiana*	Whitefly	McGuire and Northfield (2020)
		Thysanoptera, *Hemiptera*, *Coleoptera*	Dannon et al. (2020)
	Isaria fumosorosea	Beetles, caterpillars, grasshoppers, and termites	Gautam (2020)
		Hemiptera	Kumar et al. (2017); Luo et al. (2022); Stauderman et al. (2012)
	Paecilomyces lilacinus	Plant-parasitic Soil nematodes	Giri et al. (2022); Khan et al. (2006); Kiewnick and Sikora (2004); Moreno-Gavira et al. (2020)
	P. fumosoroseus	Insects and mealy bugs	Abbas (2020)
	Hirsutella thompsonii	Spider mites and whitefly	Saranya et al. (2021)
		Coleoptera, Diptera, Hemiptera	Sinha et al. (2016)
	Lagenidium giganteum	Pest mosquito species	Kaczmarek and Boguś (2021)
	Lecanicillium lecanii	*Orthoptera*, *Thysanoptera*	Goettel et al. (2005)
	L. muscarium	*Hemiptera*, *Thysanoptera*	Marshall et al. (2003); North et al. (2006)
	Metarhizium brunneum	Nematodes (pathogens)	Hummadi et al. (2021)
		Thysanoptera, *Hemiptera*, *Coleoptera*	Bischoff et al. (2009)

TABLE 10.2 *(Continued)*

Source	Organisms	Target Insects and Pest's Types	References
	Metarhizium anisopliae	*Coleoptera*, *Orthoptera*	Adatia et al. (2010); Erler and Ates (2015); Magalh˜aes et al. (2001)
	Myrothecium verrucaria	Nematodes	Hagag (2021)
	Verticillium lecanii	Mites and thrips, mealy bugs, nematodes, scale insects, etc.	Pathania et al. (2022)
Bacteria	*B. thuringiensis* Subsp. *kurstaki*	Caterpillars, fungi (*Botrytis*)	Bravo et al. (2007); Koul (2011)
		Moths	Cuyk et al. (2011); Devi et al. (2020)
	Bacillus thuringiensis Subsp. *tenebrionis*	Alfalfa weevil, Elm Leaf Beetle	Saberi et al. (2020)
		Colorado potato beetle	Nault et al. (2000)
	B. thuringiensis subsp. *aizawai*	Wax moths	Gounina-Allouane et al. (2022)
	B. thuringiensis Subsp. *israelensis*	Fungus gnats, black flies, larvae of mosquitoes	Lee et al. (2021)
		Mosquitoes, black flies	Boisvert and Boisvert (2000)
	Paenibacillus popilliae	Japanese beetle	Jenson (2014); Khachatourians (2019)
	Beauveria bassiana	Whitefly	McGuire and Northfield (2020)
	B. sphaericus, *B. lentimorbus*, and *B. popilliae*	*Larvae of Aedes* spp., *Culiseta*, *Psorophora*, and *Culex mosquitoes*	Falqueto et al. (2021)
Nematodes	*Heterorhabdits taysearae*	*Bactrocera dorsalis*	Godjo et al. (2018)
	Heterorhabditis bacteriophora Steinernema feltiae	soilborne pests	de Doucet et al. (1999)
	Steinernema carpocapsae, *Heterorhabditis bacteriophora*	Larvae of cabbage white butterfly	Aioub et al. (2021)
	Steinernema riobrave,	Armyworm	Gozel and Gozel (2021)
	S. carpocapsae	soilborne pests	Taylor (2019)
	S. bicornutum	Leaf miners	Abbas (2022)

According to Dar et al. (2021), Fenibo et al. (2021), and Sparks and Nauen (2015), and other studies, biopesticides function by denaturing protein, inducing metabolic disturbance and paralysis, triggering target-poisoning mechanisms, displaying multisite inhibitory activities, and releasing neuromuscular toxins and bioactive substances. Compared to chemical pesticides, biopesticides have the ability to change the trajectory of pest resistance because of these various effects. Research has shown that biopesticides have low toxicity, are biodegradable, ecologically safe, and have a targeted impact with negligible or no adverse effects on organisms that are not their intended targets (Deravel et al., 2014; Kalpana and Anil, 2021a). Contrary to biopesticides, conventional pesticides contribute significantly to environmental pollution, which increases the likelihood of pest resistance in food crops due to bioaccumulation and high post-harvest contamination (Fenibo et al., 2021).

10.3 SELECTION OF MICROBIAL BIOPESTICIDES

In agriculture, microbial biopesticides provide efficient pest control with minimal environmental damage, making them a sustainable substitute for chemical pesticides. When selecting suitable microbes for the production of biopesticides, several crucial considerations need to be made. Microbes ought to show pest-specificity while avoiding harming non-target organisms. This guarantees accurate pest management while protecting beneficial species (Glare et al., 2012). In order to effectively induce disease and inhibit population growth, the chosen microorganism must have great pathogenicity towards the target pest (Bailey and Lazarovits, 2003). Microbial biopesticides should be safe for the environment, with minimal threat to non-target organisms, soil, water, or air quality.

Ecological sustainability is guaranteed by this. The most important thing is safety. To ensure consumer and environmental safety, microbes used for biopesticides must not be harmful to humans, animals, or beneficial organisms (Glare and O'Callaghan, 2000). To provide long-term efficacy and reliable pest control, microbes should exhibit stability and persistence in field conditions. To ensure effectiveness across a variety of climates and cropping systems, microbes in fields should be able to adapt to a wide range of environmental variables (Berg, 2009). To ensure practicality and accessibility for farmers, microbes should be simple to synthesize into stable products and compatible with current application procedures. In order to ensure the safety and product efficacy, market approval and commercialization of microbial biopesticides require adherence to regulatory standards and recommendations (Glare and O'Callaghan, 2000). Based on production, formulation, and application costs, microbial biopesticides should be a more affordable option than chemical pesticides (Bonmatin et al., 2015). The identification of promising microbial species for the development of biopesticides is largely dependent on screening approaches. By using these methods, researchers can assess the safety, specificity, and effectiveness of microorganisms against specific pests while maintaining environmental compatibility. In this process, several screening techniques are used. Microbial isolates are tested in controlled laboratory conditions using *in vitro* assays (Benkova et al., 2020). These tests evaluate a number of characteristics, including enzymatic activity, antagonistic behavior towards the target pest, and pathogenicity. Growth inhibition tests, enzyme activity tests, and toxin generation tests are examples of common assays (Lahlali et al., 2022). Microbial isolates are tested *in vivo* experiments, which simulate or mimic natural field conditions. These tests assess the ability of microorganisms to suppress specific pests in their natural habitat. Field tests, greenhouse experiments, and bioassays carried out on pest targets are a few examples (Borges, 2021).

Molecular techniques are used in screening procedures for microbial biopesticides in order to find and classify microorganisms that may have pesticidal activities (Aguilar-Marcelino et al., 2020). These techniques provide quick, precise, and sensitive tools to assess the efficacy of biopesticides. A popular method for amplifying certain DNA sequences from microbial samples is PCR. PCR can be used in biopesticide screening to find genes linked to pesticidal activity, like those that encode antifungal or insecticidal toxins. This makes it possible to quickly identify possible microorganisms that produce biopesticides (Porcar and Juárez-Pérez, 2003). A PCR variant called qPCR makes it possible to quantify DNA molecules. Finding the prevalence of pesticidal genes in microbial populations can be helpful in assessing the possible efficacy of biopesticide candidates. Microbial genomes can be sequenced at high throughput using NGS technologies, such as those found in Oxford Nanopore and Illumina sequencing. Through entire genetic analysis of microorganisms, Next-Generation Sequencing (NGS) enables the identification of new biopesticide candidates and their functional genes (Gupta and Verma, 2019). Direct sequencing of DNA taken from environmental samples is known as metagenomics, because it bypasses the need to isolate and culture individual microbes. This approach broadens the pool of candidates for further research by enabling the identification of diverse microbial communities with potential for biopesticide action. The functions of particular genes in conferring pesticidal action can be clarified with the aid of functional genomics approaches like gene knockdown or gene expression analysis. Through the manipulation of microbial genomes and subsequent evaluation of phenotypic alterations, researchers can identify crucial genetic factors that influence the effectiveness of biopesticides (Bonaterra et al., 2022). Generally, the screening of microbial biopesticides using molecular techniques provides effective means of identifying, characterizing, and optimizing eco-friendly substitutes for chemical pesticides. Researchers can effectively assess microbial diversity and utilize the biopesticide capabilities of various microbial communities with the help of these techniques. Microbial biopesticide screening is greatly aided by bioinformatics analysis (Rastegari et al., 2020). Molecular data from screening experiments is analyzed computationally on a large scale. Finding potential biopesticide candidates is made easier by the application of bioinformatics techniques, which include gene annotation, sequence alignment, and protein function prediction. In addition to predicting the safety and environmental impact of microbial strains, researchers can additionally infer their potential pesticidal efficacy by analyzing genetic and protein sequences (Iquebal et al., 2022). Comparing the genomes of different microbial genomes to find conserved genes linked to pesticidal properties is another way that bioinformatics is helpful. The overall effectiveness and efficiency of microbial biopesticide screening techniques are improved by bioinformatics analysis, which helps in identifying and development of ecologically acceptable pest control solutions. An essential component of screening techniques for microbial biopesticides is ecological compatibility (Santos et al., 2021). It involves assessing whether biopesticides affect ecosystems as an entire and non-target organism. Laboratory analysis and field investigations are commonly used in the ecological compatibility screening process to assess how microbial biopesticides affect beneficial insects, plants, animals, and microbes (Borges et al., 2021). These evaluations aim to ensure that non-target species are not harmed by biopesticides or that the ecological balance is not disrupted. Ecological compatibility screening also takes persistence, bioaccumulation, and the possibility of off-target impacts into account. Researchers can develop sustainable pest management strategies that minimize harm to beneficial organisms and ecosystems by thoroughly assessing the ecological impact of microbial biopesticides (Thakur et al., 2020).

10.4 FORMULATION TECHNIQUES

High-quality formulation techniques are needed to produce biopesticides. A formulation's most important requirements are to be safe, efficient, simple to use, and have a prolonged shelf life (Bejarano and Puopolo, 2020). Unfortunately, it is extremely uncommon to find a biocontrol solution that satisfies all of these requirements. A few key quality criteria, such as safe and effective delivery with an extended shelf life, are emphasized in the CIBRC registration requirements for biopesticide formulations before they are registered and put on the market (Ram et al., 2022). Few biocontrol products on the market today adhere completely with CIBRC criteria. Aqueous suspension (AS), suspension concentrations (SC), wettable powder (WP), and wettable granules (WG) are among the formulations used in India. At present, clay, solid carriers such as talc, grinded corn cob, peat, lignite, wheat husk, rice bran, fly ash, and sawdust are used to produce biocontrol products. Products made from *Bacillus* are available for purchase as fly ash, charcoal, WP and granules, dust, and aqueous solution (Tikar and Prakash, 2017). AS, which contain 0.5–5% active component, are a common kind of liquid formulations made by many companies these days. There are relatively few fungal (*Beauveria bassiana*) AS; instead, most are *Bt* products and viral biopesticides.

WG and WP are the first choices for the remaining biopesticides, such as those based on Trichoderma and Pseudomonas. The main obstacle facing microbe-based biopesticides is increasing the microbial cell's activity and survival rate in various carrier materials (Figure 10.3).

Based on their physical state, biopesticide formulations can be classified into two categories: dry formulations and liquid formulations. Dry formulations are typically used for applying direct treatments against insect pests. These dry formulations or products come in a variety of forms. The biologically active substance that is present on translucently ground as a solid of mineral fine particles (clay and talc) is absorbed to

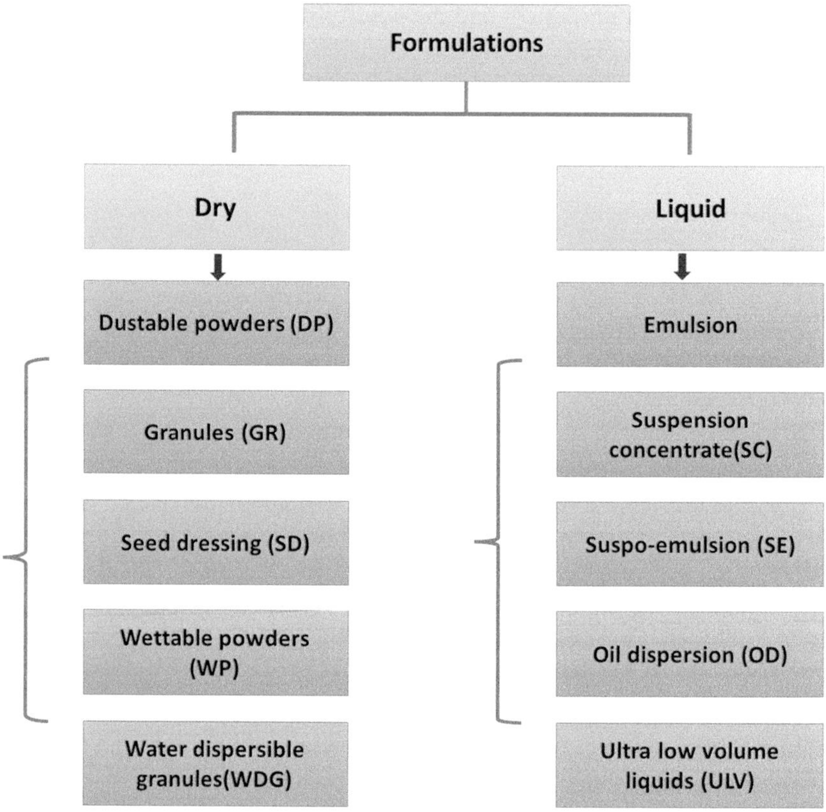

FIGURE 10.3 Different types of formulations.

provide the 10% active ingredient concentration used in dust formulations. According to Gašic and Tanovic (2013), the inert (lifeless) ingredients included in dust formulations are typically anticaking mediators, sticky materials (such as stickers), and UV protectants that increase absorption. The granules' active ingredient concentration ranges from 2% to 20%. Additionally, the particle's exterior is coated with the active ingredients, or they are wrapped in granules. Granules may be coated in polymers or resins to control the pace of effectiveness of active ingredients following appliance. Granules are mostly helpful in controlling weeds, nematodes, and insect pests that are present in that specific soil. Groundnut plant residue, kaolin, starch, silica, dry fertilizers, and polymers are used to make granules with a coarse size of 100–600 mm (Gašic and Tanovic, 2013; Tadros, 2005). Several of the granules release their active ingredients when the soil is properly moist. Formulations for biopesticides are typically used when the active ingredients are incorporated with a dust carrier, which makes the product inert for seed coverings. By dipping seeds into a product made to stick to them, powders are applied to seeds for seed dressing. Additionally, coloring mediators are used to notify the red pigment in the treated seed to act as a security marker. Usually, WPs can be extracted through the melting and blending of active ingredients like dispersing agents, synergists, surfactants, and inert fillers. Safe procedures are typically used since they are less dusty, which can lead to health problems whenever applied. According to Knowles (2008), WPs are administered using conventional spraying techniques and have good miscibility in water and long stability in storage. WDGs are designed in such a way for suspension in water to overcome the drawbacks associated with WPs, with dust-free and high-quality storage stability (Gašic and Tanovic, 2013).

There are various forms of liquid formulations, including capsule suspension (CS), emulsion, oil dispersion (OD), suspension concentrate (SC), suspo-emulsion (SE), and ultralow volume liquids (ULV). Designed to be applied with water to make an inert emulsion that is either water in oil or oil in water. Significantly, stability cannot be achieved without appropriate emulsifier selection. Water in oil mixture because oil is present at the formulation's periphery, evaporation losses, and consequent spray drift are minimal (Gašic and Tanovic, 2013). Prepared by combining an active ingredient that is translucently solid and mostly dispersed in the liquid phase, typically in water. As the solid particles are difficult to dissolve in liquid, a demonstration is usually necessary prior to application in order to ensure the element's equal distribution. With a particle size of 1–10 mm, active substances can enter plant tissue more easily and exhibit improved bio efficiency. This developed formulation is popular due to its environmentally friendly nature. The development of a homogenous emulsion ingredient with the particle suspension element is necessary for the mixture of SCs and emulsion, a highly desired formulation, to make the final product stable (Knowles, 2008). The procedure used to make the product of this formulation follows the same process as that used to produce SC. The appropriate selection of inert ingredients helps to prevent problems with instability (Vernner and Bauer, 2007). The active ingredients in this are generally formulated by a stable liquid suspension in microencapsulation, which is diluted with water prior to use. The bioagents are protected from the adverse environment by being encapsulated in capsules made of cellulose, starch, gelation, and polymers. Frequently, interfacial polymerization is the encapsulation technique that produces smaller-sized, highly effective formulations (Winder et al., 2003). Formulations are not diluted with water and have a high concentration of active ingredients. They have an active ingredient that is a suspended biological control agent and are easy to transport (Tijjani et al., 2016) (Table 10.3).

TABLE 10.3 Main Formulations of Microbial Pesticides (Lin et al., 2023)*

Microbial Pesticide Type	Microbial Pesticide Name	Main Formulations
Bacteria	*Bacillus amyloliquefaciens*	Suspension Concentrate (SC), Wettable Powder (WP), and Water Dispersible Granules (WG)
	Bacillus cereus	Wettable Powder (WP) and Suspension Concentrate (SC)
	Bacillus firmus	Wettable Powder (WP)
	Bacillus licheniformis	Soluble Liquid (SL)
	Bacillus marine	Wettable Powder (WP)
	Bacillus sphaericus	Suspension Concentrate (SC)
	Bacillus subtilis	Flowable Concentrate for Seed Treatment (FS) and Wettable Powder (WP)
	Bacillus thuringiensis	Suspension Concentrate (SC), Wettable Powder (WP), and Water Dispersible Granules (WG)
	Bacillus velezensis	Technical Concentrate (TK) and Water Dispersible Granules (WG)
	Brevibacterium brevis	Suspension Concentrate (SC)
	Paenibacillus polymyxa	Suspension Concentrate (SC) and Wettable Powder (WP)
	Pseudomonas fluorescens	Wettable Powder (WP) and Water Dispersible Granules (WG)
Fungus	*Beauveria bassiana*	Wettable Powder (WP)
	Locust Microsporidia	Suspension Concentrate (SC)
	Metarhizium anisopliae	Wettable Powder (WP)
	Paecilomyces lilacinus	Dustable Powder (DP) and Granules (GR)
	Trichoderma	Granules (GR) and Wettable Powder (WP)
	Verticillium pachyspora	Wettable Powder (WP)

TABLE 10.3 *(Continued)*

Microbial Pesticide Type	Microbial Pesticide Name	Main Formulations
Virus	*Autographa californica Nucleopolyhedrovirus* (NPV)	Suspension Emulsion (SE)
	Ectropis obliqua Nucleopolyhedrovirus (NPV)	Technical Concentrate (TK)
	Helicoverpa armigera Nucleopolyhedrovirus (NPV)	Suspension Concentrate (SC) and Wettable Powder (WP)
	Mamestra brassicae Nucleopolyhedrovirus (NPV)	Granules (GR), Suspension Concentrate (SC), Technical Concentrate (TK), and Wettable Powder (WP)
	Pieris rapae Granulosis Virus	Technical Concentrate (TK)
	Plutella xylostella Granulosis Virus	Wettable Powder (WP)
	Spodoptera exigua Nucleopolyhedrovirus (NPV)	Suspension Concentrate (SC), Technical Concentrate (TK), and Water Dispersible Granules (WG)
	Spodoptera litura Nucleopolyhedrovirus (NPV)	Suspension Concentrate (SC), Water Dispersible Granules (WG), and Wettable Powder (WP)

*No permission is needed being open access source." https://www.mdpi.com/2076-2607/11/2/364.

10.5 FACTORS AFFECTING THE STABILITY AND EFFICACY OF FORMULATIONS

Microbial biopesticide formulations are susceptible to several variables that may affect their stability and effectiveness. These variables may be inherent in the microbial organism, the way it is formulated, the environmental conditions, or how it interacts with other elements.

The selection of a microbial strain is essential because different strains may differ in their ability to withstand environmental conditions and in how effective they are at eliminating target pests. The selection of formulation ingredients such as adjuvants, stabilizers, surfactants, and carriers can affect the microbial biopesticides' stability and effectiveness (Lin et al., 2023). Additives used in formulations should not interfere with the microbiological organism's viability or effectiveness. The environmental factors that can impact the stability of microbial biopesticides include temperature, humidity, pH, and UV radiation. Certain pH and temperature ranges may be necessary for microbial organisms to remain stable and active (Šunjka and Mechora, 2022). Maintaining the viability and efficacy of microbial biopesticides requires appropriate storage conditions. Degradation of the formulation can be avoided by controlling variables such temperature, humidity, light exposure, and oxygen levels. Microbial biopesticide formulations' stability and effectiveness are significantly impacted by their compatibility with chemical pesticides. It is possible to combine different pesticides without adversely affecting their stability or effectiveness that is known as compatibility. There may be chemical interactions between chemical pesticides and microbial biopesticides that result in reduced stability or efficacy (Santos et al., 2021). Chemical reactions, the breakdown of active ingredients, or changes in microbial viability may all be a part of these interactions. For example, the survivability of microbiological agents may be impacted by acidic or alkaline chemical pesticide formulations, which could reduce the pesticides' effectiveness against the intended targets. Testing for compatibility is crucial before formulation. In this study, various concentrations of chemical pesticides are mixed with microbial biopesticides, and any adverse effects on stability or efficacy are observed. Formulators can make sure that the final product retains the appropriate level of efficacy against pests by doing such tests. Formulations can be made more compatible with chemical pesticides by adding adjuvants or stabilizers (Lin et al., 2023). In mixed formulations, these additions can help mitigate adverse interactions and increase the stability of microbial biopesticides. Compatibility might also be influenced by the application technique. The effectiveness of the microbial agents may be impacted, for example, if chemical pesticides are applied prior to microbial biopesticides. Such adverse effects can be reduced by applying pesticides in a proper sequence and at the right time. Moreover, compatibility testing is often required by regulatory agencies as a requirement for pesticide product registration. The efficacy and stability of the formed product must be demonstrated by formulators under realistic field circumstances, including situations in which chemical pesticides are combined with microbial biopesticides. Maintaining the stability and effectiveness of microbial biopesticide formulations requires compatibility with chemical pesticides (Bharti and Ibrahim, 2020). Through careful formulation design, compatibility testing, and adherence to regulatory requirements, manufacturers can develop integrated pest management (IPM) strategies that optimize the advantages of both chemical and microbial pesticides, while simultaneously reducing potential risks and adverse effects.

Microbial biopesticide formulation stability and effectiveness are significantly influenced by persistence in the environment. The longevity and efficacy of microbial agents for controlling specific pests are contingent upon their capacity to persist in their environment (Ayilara et al., 2023). Microbial biopesticides' persistence can be affected by a number of variables, including temperature, humidity, soil composition, and exposure to ultraviolet (UV) light. Furthermore, the survival and activity of microbial agents can be influenced over time by interactions with natural predators and soil microorganisms. Encapsulation and incorporation into matrices are two formulation techniques that increase the stability and persistence of biopesticides, which can increase their efficacy by extending their duration of action in the field. For formulation techniques to be optimized and for microbial biopesticides to be effective over the long term in pest management programs, it is crucial to understand the environmental conditions that impact persistence. The processes of formulation development and registration may be impacted by regulations and guidelines pertaining to microbial biopesticides. Regulations pertaining to safety, effectiveness, and environmental impact must be met by formulations. To ensure consistency and effectiveness, quality control procedures are crucial during the formulation production process. Critical aspects of quality control are monitoring the viability of microorganisms, the stability of formulations, and the shelf life of products.

Applications of biopesticides have proven to be effective in a variety of agricultural and environmental contexts, demonstrating their potential for ecological sustainability and pest management. Here are a few notable examples:

1. ***Bacillus thuringiensis* (Bt):** *Bacillus thuringiensis*, or *Bt*, is a popular microbial biopesticide which acts effectively against insect pests like beetles, caterpillars, and mosquitoes. Controlling the European corn borer (*Ostrinia nubilalis*) in corn crops is one prominent usage for it. Applying *Bt* formulations to maize plants has proven to be successful in killing maize borer larvae while causing the least amount of harm to beneficial insects and the environment.
2. **Entomopathogenic Nematodes (EPNs):** Biopesticides that target soil-dwelling insects, such as *Heterorhabditis bacteriophora* and

Steinernema feltiae, are effective. EPNs have been utilized in agricultural settings to control pests such as cutworms, root weevils, and white grubs. Application techniques that result in focused pest reduction with minimal impact on non-target organisms include soil drenching, injection, or incorporation into irrigation systems.

3. ***Trichoderma* spp.:** Trichoderma species are well-researched biocontrol agents that exhibit antagonistic activity towards plant pathogens. Trichoderma-based biopesticides have been used in horticulture and agriculture to inhibit diseases caused by fungus including *Fusarium, Rhizoctonia*, and *Pythium*. These pesticides are applied to the soil, seeds, or plant surfaces. They are useful tools for managing diseases because of their capacity to colonize plant roots and stimulate plant defenses.
4. **Microbial Consortia:** In pest management, the use of microbial consortia—groups of several helpful microorganisms—has demonstrated promising results. For example, a mixture of Pseudomonas species, *Bacillus* species, and mycorrhizal fungi has been used as a biopesticide in agricultural systems to prevent soilborne diseases and encourage plant development. The general health of crops and pest suppression are improved by the synergistic interactions between various microorganisms.
5. **Virus-based Biopesticides:** *Nucleopolyhedroviruses* (NPVs) and *granuloviruses* (GVs) are two examples of viruses that have been developed as biopesticides to manage insect pests. In apple and pear orchards, for instance, the application of *Cydia pomonella* granulovirus (CpGV) has been beneficial in controlling infestations of the codling moth (*Cydia pomonella*). These viral biopesticides provide focused pest management with minimal environmental impact by specifically targeting and infecting particular insect species.

10.6 APPLICATION METHODS

Microbial biopesticides' effectiveness may vary depending on how they are applied. The distribution and efficacy of the microbial organism can be affected by variables such as spray volume, droplet size, and timing of application. The effective use of microbial biopesticides in agricultural settings is greatly aided by spray application techniques. By distributing the biopesticide precisely and uniformly over the target area, these strategies maximize its efficacy while minimizing its impact on the environment. When using spray application to apply microbial biopesticides, there are a few things to keep in mind. Spray coverage and droplet size are significantly determined by the spray nozzle selection. The uniform distribution and penetration of microbial biopesticides into plant canopies are contingent upon the use of nozzles which produce fine droplets, as these methods frequently need a thorough coverage of plant surfaces. Determining an appropriate spray volume is essential to achieving sufficient coverage of the intended area with the least amount of waste and runoff (He et al., 2022). The optimal spray volume required to properly administer microbial biopesticides is influenced by various factors, including insect distribution, canopy structure, and crop density. Consistent spray pressure must be maintained in order to attain homogeneous droplet size and distribution. Sprays with high pressure can result in drift and excessive atomization, while those with low pressure can provide inadequate coverage. The effective distribution of microbial biopesticides to target pests and diseases is ensured by optimizing spray pressure. For the biopesticide to be as effective as possible and to coincide with the life cycle of the insect or pathogen, spray application timing is essential. The best control is ensured when microbial biopesticides, which target certain phases of pest or disease development, are applied at the right time to coincide with peak susceptibility. The efficiency of spray application and the performance of biopesticides are greatly impacted by environmental conditions such as temperature, humidity, and wind speed (Tudi et al., 2021). To minimize drift and maximize retention on target surfaces, the ideal spray application conditions are usually mild temperatures, moderate humidity, and minimal wind. To ensure efficiency and prevent negative impacts on microbial viability and performance, compatibility testing is required when tank-mixing microbial biopesticides with other agricultural inputs like fertilizers or pesticides. Testing for compatibility helps in preventing chemical reactions that may compromise the efficacy of biopesticides (Kopparthi, 2020). Adjuvants, such as spreader-stickers or surfactants, can improve microbial biopesticides' adherence to plant surfaces and increase their penetration into insect populations. Adjuvant compatibility with microbial formulations can enhance the effectiveness of biopesticides and increase spray coverage (Gaskin and Steele, 2009). Accurate distribution of microbial biopesticides and avoidance of under- or over-application are ensured by regular calibration of spray apparatus. Proper calibration guarantees uniform application rates throughout the target area and takes into consideration differences in equipment performance. Growers may maximize crop protection, minimize environmental impact, and promote sustainable pest control methods by carefully considering these spray application strategies and considerations. This will maximize the efficiency of microbial biopesticides. Effective management of pests and diseases with minimal impact on the environment depends on determining when and how often to apply microbial biopesticides (Ayilara et al., 2023). Understanding the life cycle and behavior of the target pest, considering humidity and temperature into account, and determining crop growth stage compatibility are important elements that influence this decision. While incorporating microbial biopesticides into IPM programs optimizes application timing and frequency alongside other control methods, monitoring pest pressure and disease incidence levels enables proactive intervention. By preventing the emergence of pest resistance, rotation strategies and resistance management procedures ensure long-term effectiveness. By taking these factors into account, growers may boost sustainable pest management techniques, reduce dependency on synthetic pesticides, and optimize the efficiency of microbial biopesticides. Growers may maximize the efficiency of microbial biopesticides in sustainable pest management programs, eliminating the need for synthetic chemicals and fostering ecological balance in agricultural systems, by taking these factors into account and adjusting application timing and frequency to specific pest and crop requirements. Microbial consortiums—groups of various microbe species—have drawn interest as effective tools for biological control in agriculture. These consortia protect plants from pests and diseases through utilizing the strength of many microorganisms to support plant growth and health. Microbial consortiums provide a sustainable substitute for chemical pesticides by effectively suppressing pest populations and mitigating disease outbreaks through the utilization of complex interactions such mutualism, competition, and predation (Glare et al., 2012). Moreover, microbial consortiums demonstrate adaptability in response to diverse environmental circumstances and agricultural practices, ensuring their effectiveness across a range of cropping systems (Berg et al., 2014). The significance of microbial consortiums in enhancing plant development and disease resistance via synergistic interactions among beneficial microorganisms is highlighted by studies conducted by Berendsen et al. (2012).

10.7 INTEGRATED PEST MANAGEMENT (IPM) STRATEGIES

In IPM strategies, the combination of chemicals and microbial biopesticides provides a multifaceted method to pest control that maximizes efficacy while minimizing environmental impact (Fahad et al., 2021). Growers can target pests through multiple strategies, lowering the risk of resistance development and improving overall control efficacy, by combining chemical pesticides with microbial biopesticides. Because chemical treatments may be strategically used to suppress pest populations, and microbial biopesticides contribute to maintaining ecological balance and long-term pest suppression, this integrated method enables more accurate and sustainable pest management. It is ensured that combined treatments are applied responsibly and effectively by ensuring careful consideration to compatibility, mode of action, application timing, and environmental conditions. Ultimately, IPM that combines chemical and microbial biopesticides decreases dependency on synthetic pesticides while fostering resilient and ecologically sustainable agricultural systems. Sustainable agriculture driven by biopesticides is more socially acceptable, produces more products economically, and reduces environmental hazards. The ternate approach to sustainable development is represented by these features. The application of biopesticides has a number of benefits. Various mechanisms of action, including metabolic poisoning, neuromuscular toxins, gastrointestinal disruptors, non-specific multisite inhibitors, and growth regulators, can cause their inhibition (Sparks and Nauen, 2015). These numerous modes of action against the targeted pests absolutely eliminate the possibility of resistance developing. This explains the increasing demand for biopesticides and the declining usage of classical pesticides. Deravel et al. (2014) and Kalpana and Anil (2021b) list the benefits as reduced pest resistance, reciprocal use of synthetic pesticides, eco-friendliness, low toxicity, specificity (does not affect humans or non-target organisms), biodegradability, stability against abiotic stress, and compatibility with IPM. Similar to other natural resources like biostimulators and biofertilizers, the use of biopesticides in contemporary farming achieves a balance between the preservation of the environment, economic productivity, and socio-cultural relevance. This is why biopesticide application is essential to sustainable agriculture. Accordingly, the foundation of sustainable development would be the incorporation of public policy into all four sections (including technology) (Marteel-Parrish et al., 2017; Singh et al., 2016).

10.8 ENHANCING EFFICACY AND PERSISTENCE

Enhancing the efficiency of microbial biopesticides requires a number of key strategies. First, effective pest management is ensured by choosing microbial strains with high virulence and specificity. Adjuvants increase adhesion and penetration on plant surfaces, whereas formulation optimization increases microbial survival and dissemination (Lin et al., 2023). Advanced application methods, like electrostatic spraying, provide uniform coverage and time synchronization with pest life cycles. By incorporating biopesticides into IPM programs, control techniques are synchronized and resistance is managed through rotational approaches. The use of bio stimulants increases plant resilience, and ongoing research encourages the development of novel microbial biopesticides with enhanced environmental safety and efficacy. When combined, these strategies maximize pest control while lowering environmental impact (Fadiji et al., 2022). The environmental durability of microbial biopesticides plays a crucial role in determining their effectiveness in controlling pests along with potential effects on non-target organisms. These variables include environmental factors that can either promote or inhibit microbial viability, such as temperature, humidity, and UV radiation. Microbial strain characteristics and formulation techniques are important factors, but so are soil characteristics like pH and organic matter concentration. Persistence is additionally influenced by interactions with native microbial communities and exposure to environmental stressors such as pesticides or heavy metals. The application techniques and rates that determine the environmental fate of microbial biopesticides are determined by regulatory regulations. In order to ensure sustainable pest management strategies, it is imperative to comprehend and address these factors in order to maximize their effectiveness and minimize associated hazards.

10.9 REGULATORY AND SAFETY CONSIDERATIONS

An eco-friendly and promising substitute for chemical pesticides is microbial biopesticides, which use naturally existing microorganisms to manage pests. However, regulation is essential to ensuring their effectiveness and safety. Microbial biopesticide regulations differ from nation to nation, but they usually entail extensive evaluations of the products' effectiveness, environmental impact, and safety for humans and non-target species.

To monitor, register, and control the quality of these biopesticides, a number of rules have been established internationally (Neelam et al., 2020). Various regulatory bodies, such as committees, boards, and special organizations, are established by governments that support regulation and manage its operation. According to Desai et al. (2016), these regulatory bodies draft the dossier specifications for biopesticides and update them on a periodic schedule to reflect changes in the market and local requirements. Three federal agencies are primarily responsible for regulating biopesticides in the United States. The agencies in question are the Animal and Plant Health Inspection Service (USDAAPHIS), the Food and Drug Administration (FDA), and the Environmental Protection Agency (EPA). USDAAPHIS is in charge of making sure that these biopesticides are available for field and laboratory testing.

The EPA supports their use if the studies show that there is no obvious negative impact on the environment or human health. Furthermore, it ensures protection against pesticide residue in food and feed under the Federal Food, Drug, and Cosmetic Act (FFDCA Act) and permits the sale of pesticides under the Federal Insecticide, Fungicide, and Rodenticide Act (FIFRA Act). Moreover, the FDA assesses these biopesticides to determine whether or not they pose a threat food, feed, or animals (Desai et al., 2016). Biopesticide registration and regulation in China are controlled by the Ministry of Agriculture's Institute for the Control of Agrochemicals. Based on Chinese data requirements for pesticide registration, they are essentially divided into six categories: botanical, biochemical, biological, microbiological, genetically modified organisms (GMOs), and agro-antibiotics. Six country representatives from the West African countries of Kenya, Uganda, Ethiopia, Tanzania, Nigeria, and Ghana carried out a regional inventory of the regulatory environments as part of the Commercial Products (COMPRO II) project, which is managed by the International Institute of Tropical Agriculture (IITA) (Arora et al., 2016).

In the EU (European Union), biopesticide registration seems to be more difficult than in other nations since efficacy evaluation is required in addition to the toxicological and environmental testing that is part of the dossier. Directive 91/414/EEC (EU, 1991), which was initially designed for chemical pesticides, governed microbes, plants, and pheromones in the EU. In order to incorporate particular requirements for microorganisms, Directive 91/414 was modified in 2001/36/EC (EC, 2001) and 2005/25/EC (EC, 2005). In addition, the EU added a new plant protection regulation in 2009 (Arora et al., 2016). The Insecticides Act of 1968 and the Insecticides Rules of 1971, which regulate the use of biopesticides, are enforced by the Central Insecticides Board and Registration Committee (CIBRC) in India (Kumar et al., 2019). CIB, as the highest-level advisory body, keeps a strong network of eminent scholars from all relevant fields. Aspiring producers of biopesticides must apply for registrations and licenses from the RC. The entire procedure follows a set protocol. A multitude of quality control approaches are employed to properly analyze the innovative biopesticide formulation, and any potential dangers to human health or the environment are suitably assessed (Chetan et al., 2016). In order for the safety of people and animals, this board provides the federal and state governments with expert guidance on issues pertaining to the production, promotion, distribution, and use of all insecticides, including biopesticides. The CIBRC registration committee licenses private and public companies to produce, distribute, and retail biopesticides to relevant parties after carefully examining their formulations and confirming the toxicity, efficacy, and packaging information provided by the importer or manufacturer. Sections 9(3) of the Insecticides Act (regular registration) and 9(3B) of the Act (provisional registration for a unique active component used in India) allow manufacturers to register new products (Kumar et al., 2019).

Commercial producers of microbial pesticides considered generally safe can use this procedure to get provisional registration and keep up product development while waiting for full registration, therefore lowering commercial hurdles to product development. Comparing 9(3)B to 9(3), there are less details needed for registration. When it comes to specific crops, for instance, 9(3) B demands efficacy evidence from two locations over two seasons, while 9(3) wants the same from three. The content must meet the CIB's specified quality requirements in terms of moisture content, secondary non-pathogenic microbial load, shelf life, and virulence of the organism as determined by LC50. Chetan et al. (2016) established protocols for evaluating these quality aspects. The Department of Biotechnology (DBT) and the Ministry of Agriculture and Farmers Welfare are in charge of marketing biopesticides to farmers. Other regulatory bodies in India include the Directorate of Biological Control, the National Centre for IPM (NCPM), located in Faridabad, the Central Integrated Pest Management Centre (CIPMC), and the Indian Agricultural Research Council. The Department of Biotechnology (DBT) provides funding for studies aimed at developing biopesticides, in addition to the regulatory bodies previously listed. The National Agricultural Research System (NARS) and the National Accreditation Board (NBA) both instruct state agriculture departments in quality control procedures and perform biopesticide quality control tests. According to Chetan et al.

(2016), the latter is crucial in this regulatory chain to guarantee farmers obtain biopesticides at a sustainable rate.

Regulation, which is often seen as a development hurdle, is thus obviously necessary for the commercialization of biopesticides. Regulations vary from country to country, but the primary concerns are that (i) they do not differentiate between synthetic molecules and living organisms; (ii) they require costly testing regardless of market size, which makes niche biopesticide markets unprofitable; and (iii) the registration process can take an extended period of time (Travis et al., 2012). Therefore, by organizing conferences, workshops, and meetings to discuss the state of biopesticides and bioformulations, governments can establish policies at the global level. Internationally recognized regulatory standards need to be established by governments. There are currently issues with import, export, registration, and use because different nations have different laws and regulations in place. As a result, a single policy governing the use of biopesticides is required, and the regulation may establish consistent practices or laws that could be accepted internationally (Chandler et al., 2011; Keswani et al., 2022; Singh et al., 2016). Regulations reviews that aim to ensure the sustainable use of microbial biopesticides while reducing risks must include assessments of the safety of these products for the environment and human health. Comprehensive studies of the effects of microbial biopesticides on ecosystems, non-target organisms, and human health are included in these evaluations. Environmental risk assessments focus on analyzing characteristics such as persistence, dispersal, and ecological interactions in order to determine potential consequences on biodiversity and environmental processes. In order to safeguard users, consumers, and observers, human health risk assessments look into the toxicological properties, exposure routes, and potential health effects associated with the use of biopesticides. International guidelines and standards, such those offered by the U.S. EPA and the European Food Safety Authority, serve as frameworks for conducting these assessments and ensure that microbial biopesticides are used in a manner that minimizes adverse effects on the environment and human health.

10.10 FUTURE DIRECTIONS AND EMERGING TRENDS

Microbial biopesticides are shaping the future of sustainable pest management in agriculture. Since increasing numbers of consumers are demanding ecologically friendly and sustainable farming methods, there is a growing focus on biologically based pest management solutions. Minimizing risks to the environment and decreasing dependency on conventional pesticides, microbial biopesticides are a viable substitute for synthetic insecticides. Increasingly, the management of pests and diseases is being achieved through the use of microbial consortia, which are groups of helpful microorganisms. More comprehensive and durable pest management solutions are provided by the synergistic interactions between several microbial species, which improve disease prevention, pest suppression, and plant growth promotion. Microbial biopesticides can now be applied precisely and site-specifically through developments in precision agricultural technologies. Growers can maximize efficacy, minimize waste, and optimize biopesticide distribution with the help of innovations like automated application systems, GPS-guided machinery, and remote sensing. To increase plant health and resistance to pests and diseases, there is growing interest in combining microbial biopesticides with bio stimulants and microbial enhancers. In addition to augmenting the effects of biopesticides, bio stimulants also increase plant defenses and encourage beneficial microbial activity in the rhizosphere. Research and development on microbial biopesticides are being revolutionized by advances in omics technologies such as transcriptomics, metagenomics, and genomics. These technologies make it possible to quickly identify and characterize new microbial strains, optimize microbial formulations, and fully understand the interactions between different microorganisms in complex agricultural ecosystems. It is becoming more and more common to combine microbial biopesticides with other biological control agents including nematodes, parasitoids, and predatory insects. Combining the use of various biological control agents provides pest management solutions that are complimentary and synergistic, increasing overall efficacy and lowering the need for chemical pesticides. Beyond conventional pest control, microbial biopesticides are being used for bioremediation of contaminated soils and water bodies. Some types of microorganisms are able to break down and remove toxins from the environment, providing environmentally friendly ways to clean up the environment. Global regulatory support for microbial biopesticides is growing, which makes it easier to register products and get them approved for sale. Microbial biopesticides are in high demand due to growing public awareness of the health and environmental risks associated with synthetic pesticides. This is encouraging investment and expansion in the business.

A variety of limitations and challenges accompany the use of microbial inoculants as biopesticides, despite the fact that they present a viable substitute for chemical fertilizers. According to Qiu et al. (2019), they possess a limited shelf life and may lose their effectiveness if not used promptly or maintained appropriately. Formulations with a longer shelf life should be made possible by the development of new biotechnological methods. Furthermore, there exist specific limitations concerning the application of microbial inoculants. These include the requirement for monitoring uniformity, quality consistency, application techniques, compatibility with chemical fertilizers, and environmental factors like pH, moisture, and temperature (Shahwar et al., 2023). The most important requirement for a successful product is the capacity to protect or improve yield. This is absolutely the most important component for the overall success of the product, even more so than the others. But success in the lab or greenhouse does not guarantee success outside of it. An important factor in deciding an inoculant or biocontrol product's success is how well it is used by both the manufacturer and the end user. According to Parnell et al. (2016), the product should ideally have a low adoption barrier and work with the farmer's equipment and farming methods.

For microbial biopesticides to be as successful as possible, there are obstacles that must be overcome in agricultural settings. Growers' lack of awareness regarding biopesticides is one of these difficulties, but it can be addressed by outreach and education initiatives. Field tests and the sharing of data are necessary to instill confidence in growers who might have doubts about the efficacy of biopesticides. Adoption may be hampered by financial concerns, but it is important to highlight long-term advantages including decreased pesticide resistance and environmental impact. By working with regulatory bodies and supplying scientific data, regulatory hurdles can be overcome. Maintaining the effectiveness of biopesticides requires careful handling, application, and storage methods requiring equipment that is easy to use and training. It may be difficult to incorporate biopesticides into current IPM programs, but assistance and direction can help. Infrastructure investment and stakeholder participation are necessary for increasing production in order to meet demand. Through collaboration, support, and education, these challenges can be overcome, allowing microbial biopesticides to reach their full potential for environmentally friendly pest control in agriculture.

10.11 CONCLUSION

In conclusion, microbial biopesticide application techniques and strategies offer a promising direction for modern agricultural methods. It is clear from the investigation of different application techniques and strategic approaches that microbial biopesticides provide environmentally friendly and sustainable pest management solutions with minimal negative effects on human health. We have explored a wide variety of application methods in this chapter, including soil drenches, foliar sprays, and seed treatments, all of which are customized to fit particular pest species and environmental circumstances. The significance of formulation technologies in enhancing the effectiveness and durability of microbial biopesticides, ensuring their uniform performance in the field, has also been discussed. Furthermore, a promising strategy to improve pest control while lowering dependency on conventional chemical pesticides is the incorporation of microbial biopesticides into IPM programs. Farmers may develop comprehensive strategies that are environmentally friendly and effective by integrating several pest management techniques, such as host plant resistance, biological control agents, and cultural practices. It is crucial to understand, nevertheless, that additional research and development are needed to ensure the widespread use of microbial biopesticides. To fully realize the potential of microbial-based solutions in agriculture, challenges like formulation optimization, shelf-life extension, and regulatory barriers must be resolved.

Additionally, in order to raise awareness and encourage farmers and other stakeholders to embrace microbial biopesticides, outreach and education initiatives are essential. Collaboration between researchers, industry partners, lawmakers, and growers is essential to promoting the development and application of innovative solutions to the evolving pest management challenges. Integrating biopesticides into traditional agriculture will require a better understanding of the mechanisms by which they enhance their activity spectra against pests, better field performance, advancements in biopesticide delivery systems, longer shelf lives, low production costs, accessibility, farmer awareness, and straightforward registration and regulation policies (Figure 10.4).

In conclusion, microbial biopesticide application techniques and strategies present a viable route forward for sustainable agriculture, biodiversity conservation, improved food security, and ecosystem health preservation.

We can grow better crops, reduce pest pressure, and build a more resilient agricultural system for coming generations by utilizing the potential of beneficial microbes.

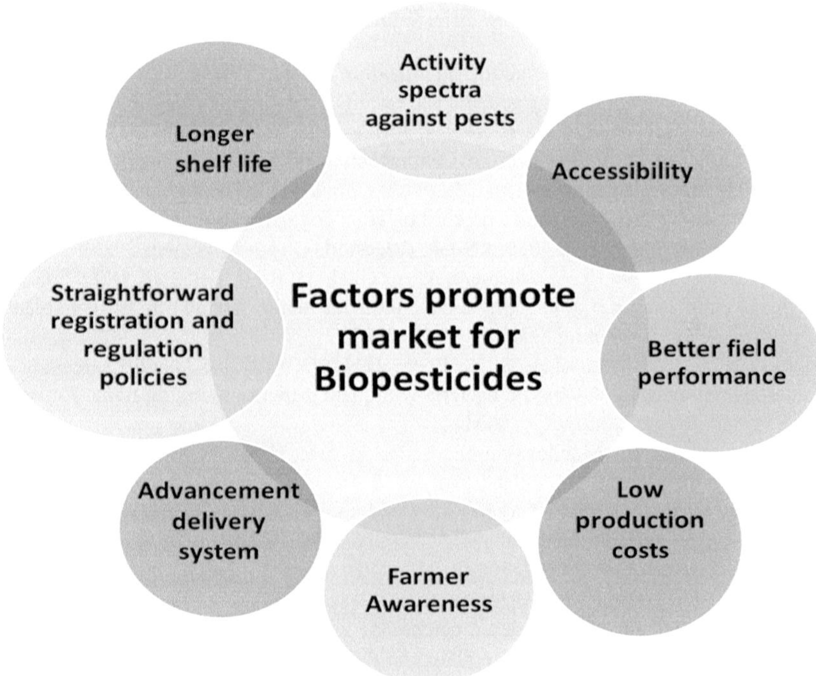

FIGURE 10.4 Factors that promote the biopesticide market.

KEYWORDS

- **application techniques**
- **formulation techniques**
- **integrated pest management**
- **microbial biopesticides**
- **seed treatment**
- **sustainability**

REFERENCES

Abbas, M. S. T. (2020). Interactions between entomopathogenic fungi and entomophagous insects. *Advances in Entomology, 8*, 130–146.

Abbas, M. S. T. (2022). Pathogenicity of entomopathogenic nematodes to dipteran leaf miners, house flies and mushroom flies. *Egyptian Journal of Biological Pest Control, 32*, 1–9.

Abdollahdokht, D., Gao, Y., Faramarz, S., Poustforoosh, A., Abbasi, M., Asadikaram, G., et al. (2022). Conventional agrochemicals towards nano-biopesticides: An overview on recent advances. *Chemical and Biological Technologies in Agriculture, 9*, 13.

Adatia, A., Johnson, D., & Entz, S. (2010). Pathogenicity of two new isolates of *Metarhizium anisopliae* from Canadian soil to *Melanoplus bivittatus* (Orthoptera: Acrididae) and *Tenebrio molitor* (Coleoptera: Tenebrionidae). *The Canadian Entomologist, 142*, 128–134. https://doi.org/10.4039/n09-041 (accessed on 15th November 2025).

Adeleke, B. S., Akinola, S. A., Adedayo, A. A., Glick, B. R., & Babalola, O. O. (2022a). Synergistic relationship of endophyte-nanomaterials to alleviate abiotic stress in plants. *Frontiers in Environmental Science, 10*, 1015897.

Adeleke, B. S., Ayangbenro, A. S., & Babalola, O. O. (2022b). *In vitro* screening of sunflower associated endophytic bacteria with plant growth-promoting traits. *Frontiers in Sustainable Food Systems, 6*, 903114.

Aguilar-Marcelino, L., Mendoza-de-Gives, P., Al-Ani, L. K. T., López-Arellano, M. E., Gómez-Rodríguez, O., Villar-Luna, E., & Reyes-Guerrero, D. E. (2020). Using molecular techniques applied to beneficial microorganisms as biotechnological tools for controlling agricultural plant pathogens and pest. In *Molecular Aspects of Plant Beneficial Microbes in Agriculture* (pp. 333–349). Academic Press.

Aioub, A. A., El-Ashry, R. M., Hashem, A. S., Elesawy, A. E., & Elsobki, A. E. (2021). Compatibility of entomopathogenic nematodes with insecticides against the cabbage white butterfly, *Pieris rapae* L. (Lepidoptera: Pieridae). *Egyptian Journal of Biological Pest Control, 31*, 1–12.

Anakwue, R. (2019). Cardiotoxicity of pesticides: Are Africans at risk? *Cardiovascular Toxicology, 19*, 95–104.

Anani, O. A., Mishra, R. R., Mishra, P., Enuneku, A. A., Anani, G. A., & Adetunji, C. O. (2020). Effects of toxicant from pesticides on food security: Current developments. In P. Mishra, R. R. Mishra, & C. O. Adetunji (Eds.), *Innovations in Food Technology* (pp. 313–321). Springer.

Arora, N. K., Verma, M., Prakash, J., & Mishra, J. (2016). Regulation of biopesticides: Global concerns and policies. In *Bioformulations: For Sustainable Agriculture* (pp. 283–299). Springer.

Arthurs, S., & Dara, S. K. (2019). Microbial biopesticides for invertebrate pests and their markets in the United States. *Journal of Invertebrate Pathology, 165*, 13–21.

Ayilara, M. S., Adeleke, B. S., & Babalola, O. O. (2022). Bioprospecting and challenges of plant microbiome research for sustainable agriculture: A review on soybean endophytic bacteria. *Microbial Ecology*, 1–23.

Ayilara, M. S., Adeleke, B. S., Akinola, S. A., Fayose, C. A., Adeyemi, U. T., Gbadegesin, L. A., Omole, R. K., Johnson, R. M., Uthman, Q. O., & Babalola, O. O. (2023).

Biopesticides as a promising alternative to synthetic pesticides: A case for microbial pesticides, phytopesticides, and nanobiopesticides. *Frontiers in Microbiology, 14*, 1040901.

Bailey, K. L., & Lazarovits, G. (2003). Suppressing soil-borne diseases with residue management and organic amendments. *Soil and Tillage Research, 72*(2), 169–180.

Bejarano, A., & Puopolo, G. (2020). Bioformulation of microbial biocontrol agents for a sustainable agriculture. In *How Research Can Stimulate the Development of Commercial Biological Control Against Plant Diseases* (pp. 275–293).

Bellinger, R. G. (2007). *Organic pesticides and biopesticides*. Clemson Extension, Home and Garden Information Center (HGIC). Available online: http://www.clemson.edu/extension/hgic (accessed on 15 May 2012).

Benkova, M., Soukup, O., & Marek, J. (2020). Antimicrobial susceptibility testing: Currently used methods and devices and the near future in clinical practice. *Journal of Applied Microbiology, 129*(4), 806–822.

Berendsen, R. L., Pieterse, C. M. J., & Bakker, P. A. H. M. (2012). The rhizosphere microbiome and plant health. *Trends in Plant Science, 17*(8), 478–486.

Berg, G. (2009). Plant-microbe interactions promoting plant growth and health: Perspectives for controlled use of microorganisms in agriculture. *Applied Microbiology and Biotechnology, 84*(1), 11–18.

Berg, G., Grube, M., Schloter, M., & Smalla, K. (2014). Unraveling the plant microbiome: Looking back and future perspectives. *Frontiers in Microbiology, 5*, 148.

Berini, F., Katz, C., Gruzdev, N., Casartelli, M., Tettamanti, G., & Marinelli, F. (2018). Microbial and viral chitinases: Attractive biopesticides for integrated pest management. *Biotechnology Advances, 36*, 818–838.

Bharti, V., & Ibrahim, S. (2020). Biopesticides: Production, formulation and application systems. *International Journal of Current Microbiology and Applied Sciences, 9*, 3931–3946. https://doi.org/10.20546/ijcmas.2020.910.453.

Bischoff, J. F., Rehner, S. A., & Humber, R. A. (2009). A multilocus phylogeny of the *Metarhizium anisopliae* lineage. *Mycologia, 101*, 512–530. https://doi.org/10.3852/07-202.

Boisvert, M., & Boisvert, J. (2000). Effects of *Bacillus thuringiensis* var. *israelensis* on target and nontarget organisms: A review of laboratory and field experiments. *Biocontrol Science and Technology, 10*, 517–561. https://doi.org/10.1080/09583150750016361.

Bonaterra, A., Badosa, E., Daranas, N., Francés, J., Roselló, G., & Montesinos, E. (2022). Bacteria as biological control agents of plant diseases. *Microorganisms, 10*(9), 1759. https://doi.org/10.3390/microorganisms10091759 (accessed on 15th November 2025).

Bonmatin, J. M., Giorio, C., Girolami, V., Goulson, D., Kreutzweiser, D. P., Krupke, C., Liess, M., Long, E., Marzaro, M., Mitchell, E. A. D., Noome, D. A., Simon-Delso, N., & Tapparo, A. (2015). Environmental fate and exposure: Neonicotinoids and fipronil. *Environmental Science and Pollution Research, 22*(1), 35–67.

Borges, S., Alkassab, A. T., Collison, E., et al. (2021). Overview of the testing and assessment of effects of microbial pesticides on bees: Strengths, challenges and perspectives. *Apidologie, 52*, 1256–1277. https://doi.org/10.1007/s13592-021-00900-7 (accessed on 15th November 2025).

Bravo, A., Gill, S. S., & Soberón, M. (2007). Mode of action of *Bacillus thuringiensis* Cry and Cyt toxins and their potential for insect control. *Toxicon, 49*, 423–435.

Chandler, D., Bailey, A. S., Tatchell, G. M., Davidson, G., Greaves, J., & Grant, W. P. (2011). The development, regulation and use of biopesticides for integrated pest management. *Philosophical Transactions of the Royal Society B: Biological Sciences, 366*(1573), 1987–1998. https://doi.org/10.1098/rstb.2010.0390.

Chang, J. H., Choi, J. Y., Jin, B. R., Roh, J. Y., Olszewski, J. A., Seo, S. J., et al. (2003). An improved baculovirus insecticide producing occlusion bodies that contain *Bacillus thuringiensis* insect toxin. *Journal of Invertebrate Pathology, 84*, 30–37.

Chetan, K., Sarma, B. K., & Singh, H. B. (2016). Synthesis of policy support, quality control, and regulatory management of biopesticides in sustainable agriculture. In *Agriculturally Important Microorganisms* (pp. 3–12). Springer.

Cuyk, S. V., Deshpande, A., Hollander, A., Duval, N., Ticknor, L., Layshock, J., Gallegos-Graves, L., & Omberg, K. M. (2011). Persistence of *Bacillus thuringiensis* subsp. *kurstaki* in urban environments following spraying. *Applied and Environmental Microbiology, 77*, 7954–7961. https://doi.org/10.1128/AEM.05207-11 (accessed on 15th November 2025).

Dannon, H. F., Dannon, A. E., Douro-Kpindou, O. K., Zinsou, A. V., Houndete, A. T., Toffa-Mehinto, J., Elegbede, I. A. T. M., Olou, B. D., & Tamò, M. (2020). Toward the efficient use of *Beauveria bassiana* in integrated cotton insect pest management. *Journal of Cotton Research, 3*, 24. https://doi.org/10.1186/s42397-020-00061-5 (accessed on 15th November 2025).

Dar, S. A., Wani, S. H., Mir, M., Showkat, A., Dolkar, T., & Dawa, T. (2021). Biopesticides: Mode of action, efficacy and scope in pest management. *Journal of Advanced Research in Biochemistry and Pharmacology, 4*, 1–8.

Dara, S. (2017). Microbial control of arthropod pests in small fruits and vegetables in temperate climate. In L. A. Lacey (Ed.), *Microbial Control of Insect and Mite Pests* (pp. 209–221). Elsevier.

De Doucet, M. M., Bertolotti, M. A., Giayetto, A. L., & Miranda, M. B. (1999). Host range, specificity, and virulence of *Steinernema feltiae*, *Steinernema rarum*, and *Heterorhabditis bacteriophora* from Argentina. *Journal of Invertebrate Pathology, 73*, 237–242. https://doi.org/10.1006/jipa.1998.4831.

De Oliveira, J. L., Fraceto, L. F., Bravo, A., & Polanczyk, R. A. (2021). Encapsulation strategies for *Bacillus thuringiensis*: From now to the future. *Journal of Agricultural and Food Chemistry, 69*, 4564–4577.

Deravel, J., Krier, F., & Jacques, P. (2014). Biopesticides, a complementary and alternative approach to the use of agrochemicals: A review. *Biotechnologie, Agronomie, Société et Environnement, 18*, 220–232.

Desai, S., Kumar, G. P., Amalraj, E. L. D., Talluri, V. R., & Peter, A. J. (2016). Challenges in regulation and registration of biopesticides: An overview. In *Microbial Inoculants in Sustainable Agricultural Productivity: Vol. 2: Functional Applications* (pp. 301–308). Springer.

Devi, P. S. V., Duraimurugan, P., Chandrika, K. S. V. P., Vineela, V., & Hari, P. P. (2020). Novel formulations of *Bacillus thuringiensis* var. *kurstaki*: An eco-friendly approach for management of lepidopteran pests. *World Journal of Microbiology and Biotechnology, 36*, 78. https://doi.org/10.1007/s11274-020-02849-8 (accessed on 15th November 2025).

Elvira, S., Ibargutxi, M. A., Gorria, N., Muñoz, D., Caballero, P., & Williams, T. (2013). Insecticidal characteristics of two commercial *Spodoptera exigua* nucleopolyhedrovirus strains produced on different host colonies. *Journal of Economic Entomology, 106*, 50–56. https://doi.org/10.1603/EC12283 (accessed on 15th November 2025).

Erler, F., & Ates, A. O. (2015). Potential of two entomopathogenic fungi, *Beauveria bassiana* and *Metarhizium anisopliae*, as biological control agents against the June beetle. *Journal of Insect Science, 15*(44). https://doi.org/10.1093/jisesa/iev029 (accessed on 15th November 2025).

Fadiji, A. E., Babalola, O. O., Santoyo, G., & Perazzolli, M. (2022). The potential role of microbial biostimulants in the amelioration of climate change-associated abiotic stresses on crops. *Frontiers in Microbiology, 12*, 829099. https://doi.org/10.3389/fmicb.2021.829099 (accessed on 15th November 2025).

Falqueto, S. A., Pitaluga, B. F., de Sousa, J. R., Targanski, S. K., Campos, M. G., De Oliveira Mendes, T. A., et al. (2021). *Bacillus* spp. metabolites are effective in eradicating *Aedes aegypti* larvae with low toxicity to non-target species. *Journal of Invertebrate Pathology, 179*, 107525.

Fan, J., Jehle, J. A., Rucker, A., & Nielsen, A. L. (2022). First evidence of CpGV resistance of codling moth in the USA. *Insects, 13*(6), 533. https://doi.org/10.3390/insects13060533 (accessed on 15th November 2025).

Farooq, M. A., Arif, M. J., Gogi, M. D., Nawaz, A., & Atta, B. (2019). Comparative efficacy of different pesticide residue mitigation modules in mango. *American Journal of Biomedical Science & Research, 4*, 214–219.

Fenibo, E. O., Ijoma, G. N., & Matambo, T. (2021). Biopesticides in sustainable agriculture: A critical sustainable development driver governed by green chemistry principles. *Frontiers in Sustainable Food Systems, 5*, 619058.

Gašić, S., & Tanović, B. (2013). Biopesticide formulations, possibility of application and future trends. *Pesticidi i Fitomedicina, 28*(2), 97–102.

Gaskin, R., & Steele, K. D. (2009). A comparison of sticker adjuvants for their effects on retention and rainfastening of fungicide sprays. *New Zealand Plant Protection, 62*, 339–342. https://doi.org/10.30843/nzpp.2009.62.4809 (accessed on 15th November 2025).

Gautam, U. K. (2020). *Effect of the entomopathogenic fungus Isaria fumosorosea on physiological processes in insects*. Doctoral dissertation, University of South Bohemia, Faculty of Science.

Giri, B., Rawat, R., Saxena, G., Manchanda, P., Wu, Q. S., & Sharma, A. (2022). Effect of *Rhizoglomus fasciculatum* and *Paecilomyces lilacinus* in the biocontrol of root-knot nematode, *Meloidogyne incognita*, in *Capsicum annuum* L. *Communicative & Integrative Biology, 15*(1), 75–87. https://doi.org/10.1080/19420889.2021.2025195 (accessed on 15th November 2025).

Glare, T. R., & O'Callaghan, M. (2000). *Bacillus Thuringiensis: Biology, Ecology and Safety*. John Wiley & Sons.

Glare, T., Caradus, J., Gelernter, W., Jackson, T., Keyhani, N., Köhl, J., Marrone, P., Morin, L., & Stewart, A. (2012). Have biopesticides come of age? *Trends in Biotechnology, 30*(5), 250–258.

Godjo, A., Zadji, L., Decraemer, W., Willems, A., & Afouda, L. (2018). Pathogenicity of indigenous entomopathogenic nematodes from Benin against mango fruit fly (*Bactrocera dorsalis*) under laboratory conditions. *Biological Control, 117*, 68–77.

Goettel, M. S., Eilenberg, J., & Glare, T. (2005). Entomopathogenic fungi and their role in regulation of insect populations. In L. I. Gilbert (Ed.), *Comprehensive Molecular Insect Science* (Vol. 6, pp. 361–405). Elsevier. https://doi.org/10.1016/B0-44-451924-6/00088-0.

Gounina-Allouane, R., Acheuk, F., & Sahir-Halouane, F. (2022). Efficient and cost-effective production of *Bacillus thuringiensis* subsp. *aizawai* spores and delta-endotoxins using agricultural raw materials and agro-industrial wastes under submerged fermentation. *Bioresource Technology Reports, 18*, 101001. https://doi.org/10.1016/j.biteb.2022.101001.

Gozel, U., & Gozel, C. (2021). Entomopathogenic nematodes in pest management. In *Integrated Pest Management—Environmentally Sound Pest Management* (Chapter 55). https://doi.org/10.5772/intechopen.96394.

Gupta, N., & Verma, V. K. (2019). Next-generation sequencing and its application: Empowering in public health beyond reality. In M. K. J. Nasim (Ed.), *Microbial Technology for the Welfare of Society* (pp. 313–341). Springer. https://doi.org/10.1007/978-981-13-8844-6_15.

Guven, O., Aydin, T., Karaca, I., & Butt, T. (2021). Biopesticides offer an environmentally friendly solution for control of pine processionary moth (*Thaumetopoea wilkinsoni* Tams) larvae and pupae in urban areas. *Biocontrol Science and Technology, 31*(1), 35–52.

Hagag, E. S. (2021). Evaluation of metabolites of *Myrothecium verrucaria* as biological nematicide against root-knot nematode, *Meloidogyne incognita*, in vitro and in vivo on sugar beet plants. *Journal of Plant Protection and Pathology, 12*, 47–53.

Harish, S., Murugan, M., Kannan, M., Parthasarathy, S., Prabhukarthikeyan, S., & Elango, K. (2021). Entomopathogenic viruses. In Omkar (Ed.), *Microbial Approaches for Insect Pest Management* (pp. 1–57). Springer.

Hashimi, M. H., Hashimi, R., & Ryan, Q. (2020). Toxic effects of pesticides on humans, plants, animals, pollinators, and beneficial organisms. *Asian Plant Research Journal, 5*, 37–47.

He, Y., Wu, J., Fu, H., Sun, Z., Fang, H., & Wang, W. (2022). Quantitative analysis of droplet size distribution in plant protection spray based on machine learning method. *Water, 14*(2), 175. https://doi.org/10.3390/w14020175 (accessed on 15th November 2025).

Hummadi, E. H., Dearden, A., Generalovic, T., Clunie, B., Harrott, A., Cetin, Y., et al. (2021). Volatile organic compounds of *Metarhizium brunneum* influence the efficacy of entomopathogenic nematodes in insect control. *Biological Control, 155*, 104527.

Idris, H., Suryani, E., Gustia, H., & Ramadhan, A. I. J. R. I. E. (2022). The effect of various essential oil and solvent additives on the botanical pesticide of *Piper aduncum* essential oil on formulation antifungal activity. *Results in Engineering, 16*, 100644.

Iquebal, M. A., Jagannadham, J., Jaiswal, S., Prabha, R., Rai, A., & Kumar, D. (2022). Potential use of microbial community genomes in various dimensions of agriculture productivity and its management: A review. *Frontiers in Microbiology, 13*, 708335. https://doi.org/10.3389/fmicb.2022.708335 (accessed on 15th November 2025).

Jayaraj, R., Megha, P., & Sreedev, P. (2017). Organochlorine pesticides: Their toxic effects on living organisms and their fate in the environment. *Interdisciplinary Toxicology, 9*(3, 4), 90–100.

Jenson, I. (2014). *Bacillus* | Introduction. In C. A. Batt & M. L. Tortorello (Eds.), *Encyclopedia of Food Microbiology* (2nd ed., pp. 111–117). Academic Press. https://doi.org/10.1016/B978-0-12-384730-0.00018-5 (accessed on 15th November 2025).

Kaczmarek, A., & Boguś, M. I. (2021). Fungi of entomopathogenic potential in *Chytridiomycota* and *Blastocladiomycota*, and in fungal allies of the *Oomycota* and *Microsporidia*. *IMA Fungus, 12*(1), 1–13.

Kadoić, B. M., Bažok, R., Mikac, K. M., Lemic, D., & Pajač, Ž. I. (2020). Pest management challenges and control practices in codling moth: A review. *Insects, 11*(1), 38. https://doi.org/10.3390/insects11010038.

Kalpana, T., & Anil, K. (2021a). A review of biopesticides and their plant phytochemicals information. *Annals of the Romanian Society for Cell Biology, 25*, 3576–3588.

Kalpana, T., & Anil, K. (2021b). High performance liquid chromatography (HPLC) and phytochemical screening of three plants: *Malvastrum coromandelianum*, *Medicago lupulina*, and *Parthenium hysterophorus*. *Annals of the Romanian Society for Cell Biology, 25*, 3589–3596.

Keswani, C., Dilnashin, H., Birla, H., Roy, P., Tyagi, R. K., Singh, D., Rajput, V. D., Minkina, T., & Singh, S. P. (2022). Global footprints of organochlorine pesticides: A pan-global survey. *Environmental Geochemistry and Health, 44*, 149–177.

Khachatourians, G. G. (2019). Insecticides, microbial. In *Reference Module in Life Sciences*. Elsevier. https://doi.org/10.1016/B978-0-12-809633-8.13066-3 (accessed on 15th November 2025).

Khan, A., Williams, K. L., & Nevalainen, H. K. M. (2006). Infection of plant-parasitic nematodes by *Paecilomyces lilacinus* and *Monacrosporium lysipagum*. *BioControl, 51*, 659–678. https://doi.org/10.1007/s10526-005-4242-x (accessed on 15th November 2025).

Kiewnick, S., & Sikora, R. A. (2004). Optimizing the efficacy of *Paecilomyces lilacinus* (strain 251) for the control of root-knot nematodes. *Communications in Agricultural and Applied Biological Sciences, 69*, 373–380.

Knowles, A. (2008). Recent developments of safer formulations of agrochemicals. *Environmentalist, 28*, 35–44.

Kopparthi, S. (2020). Compatibility of biopesticides with insecticides in IPM. *Indian Journal of Entomology, 82*, 588. https://doi.org/10.5958/0974-8172.2020.00146.7 (accessed on 15th November 2025).

Koul, O. (2011). Microbial biopesticides: Opportunities and challenges. *CAB Reviews: Perspectives in Agriculture, Veterinary Science, Nutrition and Natural Resources, 6*, 1–26.

Kumar, J., Ramlal, A., Mallick, D., & Mishra, V. (2021). An overview of some biopesticides and their importance in plant protection for commercial acceptance. *Plants, 10*(6), 1185. https://doi.org/10.3390/plants10061185 (accessed on 15th November 2025).

Kumar, K. K., Sridhar, J., Murali-Baskaran, R. K., Senthil-Nathan, S., Kaushal, P., Dara, S. K., & Arthurs, S. (2019). Microbial biopesticides for insect pest management in India: Current status and future prospects. *Journal of Invertebrate Pathology, 165*, 74–81.

Kumar, V., & Kumar, P. (2019). Pesticides in agriculture and environment: Impacts on human health. In V. Kumar, R. Kumar, J. Singh, & P. Kumar (Eds.), *Contaminants in Agriculture Environment: Health Risks Remediation* (Vol. 1, pp. 76–95). Nature and Environmental Science Academy.

Kumar, V., Avery, P. B., Ahmed, J., Cave, R. D., McKenzie, C. L., & Osborne, L. S. (2017). Compatibility and efficacy of *Isaria fumosorosea* with horticultural oils for mitigation of the Asian citrus psyllid, *Diaphorina citri* (Hemiptera: Liviidae). *Insects, 8*(4), 119. https://doi.org/10.3390/insects8040119 (accessed on 15th November 2025).

Lacey, L. A., Thomson, D., Vincent, C., & Arthurs, S. P. (2008). Codling moth granulovirus: A comprehensive review. *Biocontrol Science and Technology, 18*, 639–663. https://doi.org/10.1080/09583150802267046 (accessed on 15th November 2025).

Lahlali, R., Ezrari, S., Radouane, N., Kenfaoui, J., Esmaeel, Q., El Hamss, H., Belabess, Z., & Barka, E. A. (2022). Biological control of plant pathogens: A global perspective. *Microorganisms, 10*(3), 596. https://doi.org/10.3390/microorganisms10030596 (accessed on 15th November 2025).

Lasa, R., Pagola, I., Ibañez, I., Belda, J. E., Williams, T., & Caballero, P. (2007). Efficacy of *Spodoptera exigua* multiple nucleopolyhedrovirus as a biological insecticide for beet armyworm control in greenhouses of southern Spain. *Biocontrol Science and Technology, 17*, 221–232. https://doi.org/10.1080/09583150701211335 (accessed on 15th November 2025).

Lee, Y. K., Jin, N. Y., Kim, H. J., Yu, Y. M., Yasunaga-Aoki, C., & Youn, Y. N. (2021). Investigation and characteristics of novel biological control agents for controlling dipteran pests. *Journal of the Faculty of Agriculture, 66*, 183–190.

Lin, F., Mao, Y., Zhao, F., Idris, A. L., Liu, Q., Zou, S., Guan, X., & Huang, T. (2023). Towards sustainable green adjuvants for microbial pesticides: Recent progress, upcoming challenges, and future perspectives. *Microorganisms, 11*(2), 364. https://doi.org/10.3390/microorganisms11020364.

Llamas, B., Suarez-Rodriguez, M. C., Gonzalez-Lopez, C. V., Mora, P., & Acien, F. G. (2021). Techno-economic analysis of microalgae-related processes for CO_2 bio-fixation. *Algal Research, 57*, 102339.

Luna-Espino, J. C., Castrejón-Gómez, V. R., Pineda, S., Figueroa, J. I., & Martínez, A. M. (2018). Effect of four multiple nucleopolyhedrovirus isolates on the larval mortality and development of *Spodoptera exigua* (Lepidoptera: Noctuidae): Determination of virus production and mean time to death. *Florida Entomologist, 101*, 153–159.

Luo, Y., Wu, S., He, X., Wang, D., He, Y., & Nian, X. (2022). Identification of a *Cordyceps fumosorosea* fungus isolate and its pathogenicity against Asian citrus psyllid, *Diaphorina citri* (Hemiptera: Liviidae). *Insects, 13*(4), 374. https://doi.org/10.3390/insects13040374 (accessed on 15th November 2025).

Magalhães, B. P., de Faria, M. R., Lecoq, M., Schmidt, F. G. V., Silva, J. B. T., Frazão, H. S., Balança, G., & Foucart, A. (2001). The use of *Metarhizium anisopliae* var. *cridum* against the grasshopper *Rhammatocerus schistocercoides* in Brazil. *Journal of Orthoptera Research, 10*, 199–202. https://doi.org/10.1665/1082-6467(2001)010[0199:TUOMAV]2.0.CO;2 (accessed on 15th November 2025).

Malahlela, M., Thibane, V. S., & Mudau, F. N. (2021). Nematocidal activity of fermented extracts from *Lantana camara* plant parts against *Meloidogyne javanica* on tomato. *International Journal of Vegetable Science, 27*(1), 20–28.

Manfo, F. P. T., Mboe, S. A., Nantia, E. A., Ngoula, F., Telefo, P. B., Moundipa, P. F., et al. (2020). Evaluation of the effects of agro-pesticide use on liver and kidney function in farmers from Buea. *Journal of Toxicology, 2020*, 2305764. https://doi.org/10.1155/2020/2305764 (accessed on 15th November 2025).

Marshall, R. K., Lester, M. T., Glare, T. R., & Christeller, J. T. (2003). The fungus, *Lecanicillium muscarium*, is an entomopathogen of passionvine hopper (*Scolypopa australis*). *New Zealand Journal of Crop and Horticultural Science, 31*, 1–7. https://doi.org/10.1080/01140671.2003.9514229 (accessed on 15th November 2025).

Marteel-Parrish, A., Newcity, K. M., & Matthey, J. (2017). Highlights of the impacts of green and sustainable chemistry on industry, academia and society in the USA. *Johnson Matthey Technology Review, 61*, 207–221.

McGuire, A. V., & Northfield, T. D. (2020). Tropical occurrence and agricultural importance of *Beauveria bassiana* and *Metarhizium anisopliae*. *Frontiers in Sustainable Food Systems, 4*, 6.

Moreno-Gavíra, A., Huertas, V., Diáñez, F., Sánchez-Montesinos, B., & Santos, M. (2020). *Paecilomyces* and its importance in the biological control of agricultural pests and diseases. *Plants, 9*, 1746. https://doi.org/10.3390/plants9121746 (accessed on 15th November 2025).

Nault, B. A., Costa, S. D., & Kennedy, G. G. (2000). Colorado potato beetle (Coleoptera: Chrysomelidae) feeding, development, and survival to adulthood after continuous exposure to *Bacillus thuringiensis* subsp. *tenebrionis*-treated potato foliage from the field. *Journal of Economic Entomology, 93*, 149–156. https://doi.org/10.1603/0022-0493-93.1.149 (accessed on 15th November 2025).

Neelam, T., Kaur, S., Tomar, P., Thakur, S., & Yadav, A. N. (2020). Microbial biopesticides: Current status and advancement for sustainable agriculture and environment. In *New and future developments in microbial biotechnology and bioengineering* (pp. 243–282). Elsevier.

North, J. P., Cuthbertson, A. G. S., & Walters, K. F. A. (2006). The efficacy of two entomopathogenic biocontrol agents against adult *Thrips palmi* (Thysanoptera: Thripidae). *Journal of Invertebrate Pathology, 92*, 89–92. https://doi.org/10.1016/j.jip.2006.02.006 (accessed on 15th November 2025).

Ojuederie, O. B., Chukwuneme, C. F., Samuel, O., Olanrewaju, M. A., Adegboyega, T. T., & Babalola, O. O. (2021). Contribution of microbial inoculants in sustainable maintenance of human health, including test methods and evaluation of safety of microbial pesticide microorganisms. *Biopesticides: Botanical and Microorganisms for Improved Agriculture and Human Health, 1*, 207–240.

Pan, X., Guo, X., Zhai, T., Zhang, D., Rao, W., Cao, F., et al. (2023). Nanobiopesticides in sustainable agriculture: Developments, challenges, and perspectives. *Environmental Science: Nano, 10*, 41–61.

Parnell, J. J., Berka, R., Young, H. A., Sturino, J. M., Kang, Y., Barnhart, D. M., & DiLeo, M. V. (2016). From the lab to the farm: An industrial perspective of plant beneficial microorganisms. *Frontiers in Plant Science, 7*, 1110.

Pathak, D., Yadav, R., & Kumar, M. (2017). Microbial pesticides: Development, prospects and popularization in India. In D. Singh, H. Singh, & R. Prabha (Eds.), *Plant-microbe interactions in agro-ecological perspectives* (pp. 455–471). Springer.

Pathania, A., Dutta, J., & Mhatre, P. H. (2022). In-vitro efficacy of *Verticillium lecanii* (Zimm.) Viegas against Estonian cyst nematode, *Cactodera estonica*. *Indian Phytopathology, 75*, 1167–1171.

Pelosi, C., Bertrand, C., Daniele, G., Coeurdassier, M., Benoit, P., Nelieu, S., et al. (2021). Residues of currently used pesticides in soils and earthworms: A silent threat? *Agriculture, Ecosystems & Environment, 305*, 107167.

Pertile, M., Antunes, J. E. L., Araujo, F. F., Mendes, L. W., Van den Brink, P. J., & Araujo, A. S. F. (2020). Responses of soil microbial biomass and enzyme activity to herbicides imazethapyr and flumioxazin. *Scientific Reports, 10*, 1–9.

Porcar, M., & Juárez-Pérez, V. (2003). PCR-based identification of *Bacillus thuringiensis* pesticidal crystal genes. *FEMS Microbiology Reviews, 26*(5), 419–432. https://doi.org/10.1111/j.1574-6976.2003.tb00624.x (accessed on 15th November 2025).

Qiu, Z., Egidi, E., Liu, H., Kaur, S., & Singh, B. K. (2019). New frontiers in agriculture productivity: Optimised microbial inoculants and in situ microbiome engineering. *Biotechnology Advances, 37*(6), 11.

Quarles, W. (2011). New biopesticides for IPM and organic production. *IPM Practitioner, 13*, 7, 8.

Rakhimol, K., Thomas, S., Volova, T., & Jayachandran, K. (2020). *Controlled Release of Pesticides for Sustainable Agriculture*. Springer.

Ram, R. M., Debnath, A., Negi, S., & Singh, H. B. (2022). Use of microbial consortia for broad spectrum protection of plant pathogens: Regulatory hurdles, present status and future prospects. *Biopesticides*, 319–335.

Rastegari, A. A., Yadav, A. N., & Yadav, N. (2020). *New and future developments in microbial biotechnology and bioengineering: Trends of microbial biotechnology for sustainable agriculture and biomedicine systems: Diversity and functional perspectives*. Elsevier.

Ruiu, L. (2018). Microbial biopesticides in agroecosystems. *Agronomy, 8*, 235.

Sabarwal, A., Kumar, K., & Singh, R. P. (2018). Hazardous effects of chemical pesticides on human health–cancer and other associated disorders. *Environmental Toxicology and Pharmacology, 63*, 103–114.

Saberi, F., Marzban, R., Ardjmand, M., Shariati, F. P., & Tavakoli, O. (2020). Optimization of culture media to enhance the ability of local *Bacillus thuringiensis* var. *tenebrionis*. *Journal of the Saudi Society of Agricultural Sciences, 19*, 468–475.

Samada, L. H., & Tambunan, U. S. F. (2020). Biopesticides as promising alternatives to chemical pesticides: A review of their current and future status. *Online Journal of Biological Sciences, 20*, 66–76.

Santos, M. S., Rodrigues, T. F., Nogueira, M. A., & Hungria, M. (2021). The challenge of combining high yields with environmentally friendly bioproducts: A review on the compatibility of pesticides with microbial inoculants. *Agronomy, 11*(5), 870. https://doi.org/10.3390/agronomy11050870 (accessed on 15th November 2025).

Saranya, S., Ramaraju, K., & Jeyarani, S. (2021). Ovicidal action of different fungal pathogens against two spotted spider mite, *Tetranychus urticae* (Koch) under laboratory conditions. *Journal of Biological Control, 35*, 37–40.

Shah, F., Shah, S., Adnan, A., Ali, A. A., Bajwa, H., Shah, H., Battaglia, M., Adnan, M., Wahid, F., Datta, R., Babur, E., Danish, S., Zarei, T., & Irshad, I. (2021). Bio-based integrated pest management in rice: An agro-ecosystems friendly approach for agricultural sustainability. *Journal of the Saudi Society of Agricultural Sciences, 20*(2), 94–102. https://doi.org/10.1016/j.jssas.2020.12.004 (accessed on 15th November 2025).

Shahwar, D., Mushtaq, Z., Mushtaq, H., Alqarawi, A. A., Park, Y., Alshahrani, T. S., & Faizan, S. (2023). Role of microbial inoculants as biofertilizers for improving crop productivity: A review. *Journal of Applied Microbiology, 9*, e16134.

Singh, A., Bhardwaj, R., & Singh, I. K. (2019). Biocontrol agents: Potential of biopesticides for integrated pest management. In B. Giri, R. Prasad, Q. S. Wu, & A. Varma (Eds.), *Biofertilizers for sustainable agriculture and environment* (pp. 413–433). Springer.

Singh, H. B., Sarma, B. K., & Keswani, C. (2016). *Agriculturally important microorganisms: Commercialization and regulatory requirements in Asia*. Springer.

Sinha, K. K., Choudhary, A. K., & Kumari, P. (2016). Entomopathogenic fungi. In Omkar (Ed.), *Ecofriendly pest management for food security* (pp. 475–505). Academic Press. https://doi.org/10.1016/B978-0-12-803265-7.00015-4 (accessed on 15th November 2025).

Sparks, T. C., & Nauen, R. (2015). IRAC: Mode of action classification and insecticide resistance management. *Pesticide Biochemistry and Physiology, 121*, 122–128.

Stauderman, K., Avery, P., Aristizábal, L., & Arthurs, S. (2012). Evaluation of *Isaria fumosorosea* (Hypocreales: Cordycipitaceae) for control of the Asian citrus psyllid, *Diaphorina citri* (Hemiptera: Psyllidae). *Biocontrol Science and Technology, 22*, 747–761. https://doi.org/10.1080/09583157.2012.686599 (accessed on 15th November 2025).

Šunjka, D., & Mechora, Š. (2022). An alternative source of biopesticides and improvement in their formulation: Recent advances. *Plants, 11*(23), 3172. https://doi.org/10.3390/plants11223172 (accessed on 15th November 2025).

Syromyatnikov, M. Y., Isuwa, M. M., Savinkova, O. V., Derevshchikova, M. I., & Popov, V. N. (2020). The effect of pesticides on the microbiome of animals. *Agriculture, 10*(1), 1–14.

Tadros, T. (2005). Surfactants in pharmaceutical formulations. In *Applied SURFACTANTS: PRINCIPLES and Applications* (p. 10). Wiley-VCH Verlag GmbH & Co.

Taylor, R. A. J. (2019). Nematodes and other worms. In R. A. J. Taylor (Ed.), *Taylor's Power Law* (pp. 143–234). Academic Press. https://doi.org/10.1016/B978-0-12-810987-8.00007-0.

Thakore, Y. (2006). The biopesticide market for global agricultural use. *Industrial Biotechnology, 2*, 194–208.

Thakur, N., Kaur, S., Tomar, P., Thakur, S., & Yadav, A. N. (2020). Microbial biopesticides: Current status and advancement for sustainable agriculture and environment. In *New and Future Developments in Microbial Biotechnology and Bioengineering* (pp. 243–282). Elsevier.

Tijjani, A., Bashir, K., Mohammed, I., Muhammad, A., Gambo, A., & Musa, H. (2016). Biopesticides for pest control: A review. *Journal of Biopesticides and Agriculture, 3*, 6–13.

Tikar, S., & Prakash, S. (2017). Fly ash-based *Bacillus thuringiensis israelensis* formulation: An ecofriendly approach. *Indian Journal of Medical Research, 146*(6), 680.

Travis, G., Caradus, J., Gelernter, W., Jackson, T., Keyhani, N., Köhl, J., Marrone, P., Morin, L., & Stewart, A. (2012). Have biopesticides come of age? *Trends in Biotechnology, 30*(5), 250–258.

Tudi, M., Ruan, D., Wang, L., Lyu, J., Sadler, R., Connell, D., Chu, C., & Phung, D. T. (2021). Agriculture development, pesticide application and its impact on the

environment. *International Journal of Environmental Research and Public Health, 18*(3), 1112. https://doi.org/10.3390/ijerph18031112.
Ujváry, I. (2010). Pest control agents from natural products. In B. T. Ikem (Ed.), *Hayes' handbook of pesticide toxicology* (pp. 119–229). Elsevier.
Vernner, R., & Bauer, P. (2007). Q-TEO, a formulation concept that overcomes the incompatibility between water and oil. *Pflanzenschuty-Nachrichten Bayer, 60*, 7–26.
Winder, R., Wheeler, J., Conder, N., Otvos, I., Nevill, R., & Duan, L. (2003). Microencapsulation: A strategy for formulation of inoculum. *Biocontrol Science and Technology, 13*, 155–169.
Xiao, Y., & Wu, K. (2019). Recent progress on the interaction between insects and *Bacillus thuringiensis* crops. *Philosophical Transactions of the Royal Society B, 374*, 20180316.
Zaki, O., Weekers, F., Thonart, P., Tesch, E., Kuenemann, P., & Jacques, P. (2020). Limiting factors of mycopesticide development. *Biological Control, 144*, 104220.

CHAPTER 11

Microbial Synthesis of Biodegradable Polymers: A Focus on Polylactic Acid (PLA)

MAHARUDRA PRATAP SINGH,[1] AKHILESH KUMAR SINGH,[1,*] BRIJESH PANDEY,[1] PRAKASH KUMAR SARANGI,[2] and ANAND PRAKASH[1,*]

[1]*Department of Biotechnology, School of Life Sciences, Mahatma Gandhi Central University, Motihari, Bihar, India*

[2]*College of Agriculture, Central Agricultural University, Imphal, Manipur, India*

*Corresponding author

ABSTRACT

The escalating global apprehension over environmental degradation and plastic pollution has spurred interest in seeking sustainable alternatives to traditional petroleum-based polymers. Biodegradable polymers, synthesized through microbial processes, offer promising solutions to address these ecological challenges. Polylactic acid (PLA) is a leading biodegradable polymer derived from renewable resources. Its production involves fermenting sugars obtained from sources like corn starch or biomass, using lactic acid bacteria. Microorganisms such as *Lactobacillus* and *Streptococcus* are crucial to this fermentation process. PLA's outstanding biodegradability makes it an appealing choice for environmentally conscious applications, undergoing degradation through both physical and biological processes. PLA's versatility is evident in its expanding applications, providing a biodegradable alternative in packaging, medical

Industrial Microbiology: Current Progress and Novel Technologies.
Prakash Kumar Sarangi, Akhilesh Kumar Singh, Rajesh Kumar Srivastava, and Brijesh Pandey (Eds.)

DOI: 10.1201/9781779642219-11

sutures, drug delivery systems, textiles, 3D printing filaments, and disposable cutlery. Despite PLA's potential, there are numerous challenges that hinder its widespread adoption. Ongoing research focuses on addressing these challenges, employing metabolic engineering techniques to alter the physiology of microbial strains for increased lactic acid production and enhancing the overall feasibility of PLA production. As the world strives for a more sustainable future, microbial-synthesized biodegradable polymers like PLA present viable solutions that align with environmental consciousness and ecological responsibility.

11.1 INTRODUCTION

In the modern world, managing plastic trash presents a big environmental challenge. The extensive manufacture and use of plastics in many aspects of daily life have led to the deterioration of the environment. Notably, burning typical polymers that are not biodegradable, such as polyethylene, polypropylene, and polyvinyl chloride, releases carbon dioxide into the atmosphere, hence intensifying worries about global warming (Narancic and O'Connor, 2019). In response, there is an increasing focus on creating biodegradable polymers that can replace non-biodegradable materials in a variety of applications. As they break down over time into easily disposed of byproducts, biodegradable polymers present a possible option. Numerous materials, such as food waste, cellulose, starch, and agricultural leftovers, can be used to create these polymers (Alcântara et al., 2020). Sustainable bioplastics originating from microorganisms are becoming more popular because they are less expensive than petropolitics. By adding biodegradable polymers to a variety of products, it can reduce environmental pollution and carbon emissions associated with global warming. Microorganisms like fungi and bacteria aid in the breakdown of biodegradable polymers, converting them into harmless compounds like carbon dioxide, water, and methane (Tarrahi et al., 2020). Biodegradation is affected by various factors such as the molecular weight, structural configuration, and the compositional makeup of the polymers. Furthermore, biopolymers provide environmental benefits by replenishing raw materials and lowering carbon footprints. Using renewable polymers not only solves environmental issues but also deals with the problem of depleting petroleum supplies. However, the broad use of biodegradable plastics demands comprehensive waste management systems, such as composting and anaerobic digestion, to ensure proper disposal and use (Raza et al., 2018). Biodegradable polymers are used not just in traditional applications like plastic bags and packaging, but also in vital fields like healthcare, where soft tissue engineering and transplantation are applied. When it comes to the environment, natural fibers are more cost-effective than synthetic ones and can be recycled, renewed, and reused. Unfortunately, there are obstacles in the way of obtaining compatibility between hydrophilic natural fibers and hydrophobic polymers, requiring a variety of methods of treatment to enhance adhesion and performance (Alcântara et al., 2020). The suitability of biodegradable polymers, natural fibers, and composites for different applications is largely determined by their mechanical, thermal, and visco-elastic properties. Additionally, progress in materials science has prompted research into biodegradable polymers like PLA, which is produced from renewable resources such as corn, potatoes, and sugarcane (Castro-Aguirre et al., 2016). PLA is suited for a variety of applications, including sports equipment, automobiles, medical science, and agriculture, because of its advantageous properties including low density, high flexibility, and stiffness. Because of its 3D printing compatibility, it is possible to customize and innovate with its design while maintaining optimal performance and sustainability. In conclusion, biodegradable polymers provide a viable substitute for conventional plastics, resolving environmental issues and satisfying industry objectives at the same time. These materials have the potential to completely change how trash is managed and help create a more sustainable future with further study and innovation (Nduko and Taguchi, 2019).

11.2 POLYLACTIC ACID

Poly (lactic acid), or polylactide (PLA), is a biodegradable and renewable aliphatic polyester that has garnered considerable attention in both research and real-world applications. With its potential to supplant conventional petrochemical-based polymers across various industries and its prominence as a biomaterial in medical contexts, PLA shows a significant advancement. In packaging, PLA serves as a successful alternative to petrochemical-based polymers owing to its biodegradability and renewability. It boasts characteristics of a thermoplastic, high-strength, high-modulus polymer, all crafted from annually renewable resources (Balla et al., 2021). PLA is a versatile polymer that can be processed using conventional plastic manufacturing equipment, making it suitable for a broad spectrum of applications, such as industrial packaging and biocompatible medical devices. The stereochemical variations of PLA include poly(L-lactide) (PLLA), poly(D-lactide) (PDLA), and poly(DL-lactide) (PDLLA), which are derived from the enantiomers L-lactic acid and D-lactic acid. Lactic acid, the fundamental building block for PLA, is produced via the fermentation of sugars sourced from renewable materials like sugarcane or corn starch. This production method supports the eco-friendly nature of PLA and ensures its safety for human use (Swetha et al., 2023). PLA has been approved by the United States Food and Drug Administration (FDA) as generally recognized as safe (GRAS) and is commonly utilized in food packaging, posing no health risks. Since receiving FDA approval, interest in lactic acid-based polymers for medical applications has significantly increased, alongside a broad expansion in industrial uses over the past 20 years. While the synthesis of PLA dates back to 1932, when Carothers at DuPont first achieved low molecular weight (Mw) PLA, advancements have led to the production of high-Mw PLA through ring-opening polymerization (ROP) of lactide. Despite various fabrication methods, PLA synthesis remains complex, demanding precise control of conditions, catalysts, and extended polymerization times (Cheng et al., 2009). Compared to other biopolymers, PLA boasts numerous advantages including eco-friendliness, biocompatibility, and ease of processing. Derived from renewable resources, PLA is both biodegradable and recyclable, with production consuming carbon dioxide. Its biocompatibility and non-toxic degradation products makes it ideal for biomedical applications, as approved by the FDA (Table 11.1). Despite its advantages, PLA's low toughness limits its suitability for certain uses, primarily due to its brittleness and low tensile strength. While the tensile strength and modulus of PLA are similar to those of poly (ethylene terephthalate) (PET), its limited flexibility hinders its use in situations requiring plastic deformation. In essence, PLA emerges as a promising candidate in the realm of sustainable polymers, offering a blend of eco-friendliness, biocompatibility, and processability, albeit with considerations for its inherent limitations in toughness (Trivedi et al., 2023). PLA stands out as a biodegradable polymer that has garnered significant interest lately due to its eco-friendliness and compatibility with living organisms. Sourced from renewable origins like corn starch, sugarcane, or potato starch, PLA presents a sustainable alternative to conventional petroleum-derived plastics. Its versatility shines through various applications including packaging, biomedical implants, and drug delivery systems. The properties of PLA are determined by various factors such as the processing temperature, molecular weight, and duration of annealing, and the specific types of isomers used. Crucially, crystallinity, which impacts melting point, rigidity, and tensile strength, is pivotal. PLA crystallizes into three distinct structural forms (α, β, and γ) owing to different thermal and/or mechanical treatments. The rate of crystallinity is key in selecting PLA for specific purposes (Shekhar and Mondal, 2024). PLA synthesis involves different stereoisomer proportions (L-lactide, D-lactide, and meso-lactide), significantly affecting its traits. Manipulating PLA properties is feasible with specialized catalysts of varying isotactic and syndiotactic content. An excess of PLLA content leads to crystallinity, while lower levels yield an amorphous structure. The melting and glass transition temperatures decline as PLLA content decreases, impacting physical attributes such as density and mechanical properties. PLA exhibits solubility in solvents like dioxane, chloroform, and acetonitrile but resists water, alcohol, and unsubstituted hydrocarbons. These solubility traits find utility in coatings, films, fibers, and drug delivery systems (de França et al., 2022). PLA degrades mainly through hydrolysis when exposed to moisture over time. The pace at which degradation occurs depends on various factors, including the properties of the particles, temperature, humidity levels, crystallinity, isomeric composition, residual lactic acid concentration, molecular weight, water diffusion, and the existence of catalyst impurities. Studies, both *in vitro* and *in vivo*, on PLA used in surgical implants, highlight the importance of pH in the degradation process. PLAs with lower molecular weights undergo quicker degradation, resulting in the conversion into lactic acid at a faster rate. This process diminishes the likelihood of PLA accumulation within tissues. PLA's properties can be customized via copolymerization with lactone-type monomers or blending with other materials, significantly altering resultant properties based on component mechanical properties and blend microstructure (Elsawy et al., 2017). Optical and barrier properties are crucial, with studies delving into stereoisomer proportions using ellipsometry measurements and establishing equations for the index of refraction. PLA's superior barrier properties compared to OPS render it suitable for diverse food applications, while water content remains a critical consideration. In essence, PLA emerges as a versatile biodegradable polymer, with properties modifiable through factors like crystallinity, stereoisomer content, and blending. Its distinct characteristics position PLA as a promising material across industries spanning food packaging, textiles, and biomedical implants (Farah et al., 2016).

TABLE 11.1 Properties of Bioplastics and Petroplastic

Bioplastic	Petroplastic
Environment friendly	Pollute the air
Completely biodegradable	Eroded soil
From a wide range of sources	Health impacts
Normally obtained from renewable sources	Non-renewable and degradable
Non-toxic and recyclable	Limited resources for oil production

11.3 LACTIC ACID AS POLYLACTIC ACID PRECURSOR

Lactic acid (LA) is identified as an organic acid, commonly found in the form of a white powder or a colorless to yellow liquid. Its chemical formula is CH_3-CHOHCOOH. LA has two isomeric forms: D (-) and L (+) LA (Figure 11.1). The chemical behavior of LA is influenced by various physicochemical attributes, including its acidity in aqueous solutions, functional properties due to carboxyl and hydroxyl groups, and the asymmetric optical activity of C_2 (Table 11.2). LA is recognized as GRAS by the US FDA for use as a food ingredient. However, it is important to note that D (-) LA may, in certain cases, contribute to acidosis and decalcification, potentially affecting human metabolism adversely (Swetha et al., 2023). In 1780, the Swedish chemist Scheele succeeded in isolating the initial sample of lactic acid from sour milk. Later, in 1857, Louis Pasteur made a pivotal discovery, elucidating that lactic acid was not a constituent of milk but rather a byproduct of fermentation induced by microorganisms. To validate this groundbreaking finding, Edmond Frémy undertook the task of synthesizing lactic acid through fermentation. Subsequently, the first commercially available lactic acid entered the United States market in 1881. At standard conditions, liquid lactic acid exhibits a density of 1.209 grams per milliliter at 25°C, while its solid form boasts a density of 1.33 grams per milliliter at 20°C. The melting points of various lactic acid configurations differ: racemic DL at 53.0°C, D-LA at 52.7°C, and

L-LA at 16.8°C. Correspondingly, their boiling points under reduced pressure (1.87 kilopascals) are 82.0°C, 103°C, and 122°C, respectively. Additionally, the dissociation constants for these configurations stand at 3.79 kilopascals, 3.83 kilopascals, and 3.73 kilopascals, respectively (Arshad et al., 2022). LA is gaining popularity because of its potential as a monomer for making PLA, a biodegradable and environmentally friendly biopolymer. Projections suggest that the annual production of LA is poised to surge by 11.5% by 2026, reaching an estimated volume of 270,000 tons. In addition to fermentative production methods like sugarcane juice and starch fermentation, chemical synthesis is another method used to make LA. The environmental impact, low production cost, and decreased reliance on fossil fuel-based feedstock make biological LA production noteworthy. Biological LA synthesis has several advantages, including reduced reliance on fossil feedstock, lower CO_2 emissions, the utilization of biocatalysts, and high product specificity (Figure 11.2). The efficiency of LA fermentation is determined by several factors, including the specific strains responsible for LA production, the substrates used in fermentation, and the methodologies employed, such as the activity of enzymes like L-lactate dehydrogenase (L-LDH) or D-lactate dehydrogenase (D-LDH) (Table 11.3) (Yu et al., 2023). Most commercial LA is produced by specific organisms belonging to the *Lactobacillus* genus through the bacterial fermentation of carbohydrates. These organisms, which thrive in anaerobic conditions, utilize pyruvic acid and enzymatically convert it into lactate. The increasing interest in LA stems from its potential as a monomer in the manufacturing of PLA, a biodegradable and environmentally friendly biopolymer. In recent years, low-cost biomass sources such as starchy and cellulosic agricultural waste or food industry byproducts have replaced pure sugars and food crops. Starchy substrates can be aerobically converted into LA by amylolytic fungal species such as *Rhizopus oryzae*. In solid-state fermentation, the combination of *R. oryzae* and *Lactobacillus* has been shown to synthesize LA from sugarcane bagasse as a substrate. Bacterial strains like *Bacillus* sp. and *Lactobacillus pentosus* possess the capability to metabolize pentose and hexose carbohydrates, thereby facilitating efficient synthesis of LA. In order to produce LA efficiently, researchers are creating sustainable fermentation techniques (Abedi and Hashemi, 2020). An affordable method is to produce LA in an acidic environment using microbial fermentation. It solves the main problem of gypsum generation in conventional fermentation methods by using strains of bacteria resistant to acid for the manufacture of LA. As a result, the LA recovery process is more straightforward, economical, and effective. Fermenting lactic acid bacteria (LAB) that can ferment in acidic environments is a prerequisite for biological processes that produce LA in an economically feasible manner. This lowers total fermentation costs, streamlines downstream processing, and does away with the need for neutralizers. The use of biocatalysts, the manufacture of optically pure isomers, reduced carbon dioxide emissions, cost-effectiveness, and environmental benefits have all contributed to the growing interest in the biological approach to LA production (Singhvi et al., 2018). Around 90% of the global LA output is generated through fermentation processes, while the remaining 10% is synthesized chemically via lactonitrile hydrolysis. Before fermentation became a more effective option, petrochemicals were the main method used to produce LA. Depending on the bacteria used, fermentation techniques can be divided into two categories: homo-fermentative and hetero-fermentative. In contrast, hetero-fermentation yields a range of substances including acetic acid, lactic acid, carbon dioxide, and ethanol, whereas homo-fermentation specifically generates lactic acid through the metabolism of hexose (Pau et al., 2022).

TABLE 11.2 Properties of PLA and Its Stereo Configurations

Properties / Polymers	Density (g/cm^3)	Melting Point (°C)	Molecular Weight (g/mol)	Tensile Strength (MPa)	References
PLA	1.25	120–170	66,000	21–60	Farah et al. (2016)
PLLA	1.29	173–178	<350,000	15.5–150	Korhonen et al. (2001)
PDLA	1.248	120–150	21,000–67,000	15.5–150	Södergård and Stolt (2002)
PDLLA	1.25	230–240	<350,000	27.6–50	Shao et al. (2015)

11.4 MICROORGANISMS INVOLVED IN LA PRODUCTION

11.4.1 BACTERIA

LAB, a group of gram-positive bacteria, play a crucial role in industrial-scale LA production due to their inherent properties. Various factors such as pH levels, temperature, nutrient availability, and the specific strains of LAB employed significantly influence the efficiency of LA production.

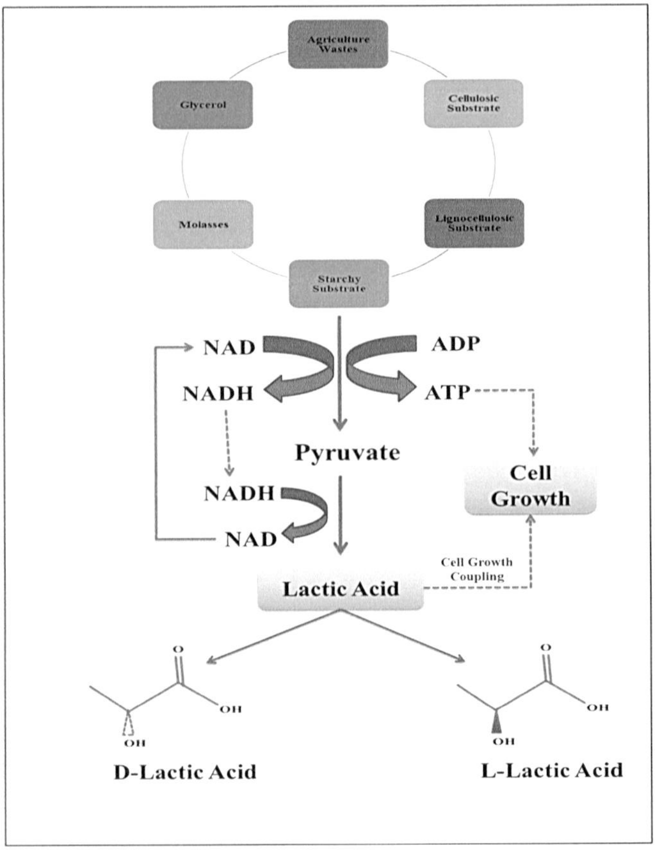

FIGURE 11.1 Substrate for lactic acid production in its two-enantiomer form.

Among the LAB species utilized as starter cultures in industrial fermentation processes are *Lactobacillus, Lactococcus, Leuconostoc, Streptococcus,* and *Pediococcus* (Reddy et al., 2008). *Lactobacillus strains* hold significant commercial promise owing to their remarkable acid tolerance, high yield, and productivity. Nonetheless, employing LAB for industrial LA production faces hurdles, primarily stemming from the requirement for intricate nutrient sources and the necessity of low fermentation temperatures. Yet, certain alkaliphilic LAB strains such as *Marinilactibacillus, Halolactibacillus, Alkalibacterium* spp., and others within the LAB genera exhibit the capability to produce LA at elevated pH levels. This characteristic diminishes the risk of contamination during fermentation, thereby offering a potential solution to industrial challenges associated with LA production (Singh et al., 2006). Genome editing and plasmid transformation are two genetic engineering approaches that have been investigated to increase the efficiency of LA fermentation. Enzymes like D-LDH have the potential to be synthesized within microorganisms such as *Corynebacterium glutamicum*, *Escherichia coli*, and various *yeasts*. This can be achieved through the incorporation of genes sourced from LAB, bovines, and fungi into these microorganisms lacking the necessary enzymes. Consequently, this method facilitates the efficient production of D-LA with high optical purity (Zheng et al., 2020). A pH that is slightly acidic to neutral is usually required for the fermentation of glucose by homofermentative LAB. However, too much acidity can prevent the formation of LA by impeding cellular metabolism. In order to prevent this, lime is frequently added to fermenters to maintain a neutral pH, which causes calcium lactate to form as the primary product. Lactic acid is produced by acidifying the calcium lactate-containing broth with sulfuric acid. However, there are drawbacks to excessive sulfuric acid usage, including increased production costs, corrosion, and waste management (John et al., 2009). In order to increase *Lactobacillus* sp. strains' ability to withstand acidity and flourish in the acidified fermentation medium, metabolic engineering techniques have been performed. Exposing strains to UV radiation and nitrosoguanidine treatments has resulted in the development of enhanced variants capable of producing lactic acid under neutral pH conditions. These strains exhibit rates and yields similar to those achieved through conventional methodologies (Abdel-Rahman et al., 2013). Moreover, *Lb. lactis's* survivability at pH 3.0 has been markedly increased by the production of enzymes such as Propionibacterium Fredericia's trehalose 6-phosphate phosphatase. LAB pathways include both homofermentative and heterofermentative processes. At least two molecules of lactic acid are produced for every mole of glucose consumed by homofermentative LAB. On the other hand, glucose is metabolized by heterofermentative LAB into carbon dioxide, ethanol, format, acetic acid, lactic acid, and diacetyl. The use of heterofermentative LAB as dairy culture starters is limited due to difficulties such as CO_2 release and the synthesis of various organic acids, which can cause packaging issues and degrade product quality (Abdel-Rahman et al., 2011).

Regulatory bodies have officially recognized several strains of *Bacillus* spp. as safe for the production of LA. *Corynebacterium glutamicum*, a

Gram-positive aerobic bacterium, has been demonstrated to produce amino acids such as L-lysine and L-glutamate, along with small quantities of various organic acids including LA, succinic acid (SA), and acetic acid (AA) in industrial settings. This synthesis of organic acids occurs under anaerobic conditions, prompted by cell growth inhibition and stimulated by various sugars like D-glucose, L-arabinose, D-xylose, D-cellobiose, and their combinations in a mineral salt medium (Song et al., 2012). Engineered strains of *C. glutamicum* have demonstrated significant potential in efficiently producing LA without the need for complex nutritional requirements. For instance, *C. vitaeruminis* MTCC 5488 yielded 38.5 g/l of LA through fed-batch fermentation. Nevertheless, both *C. glutamicum* and *E. coli* exhibit limited tolerance to acidic conditions, mandating LA synthesis to occur at nearly neutral pH levels, typically around 7.0. The simultaneous synthesis of organic and LA has presented difficulties for the best possible LA production. Many strategies, such as modifying the medium and applying engineering approaches, have been investigated to improve the synthesis of LA by *C. glutamicum* fermentation (Kawaguchi et al., 2006). A new method using *C. glutamicum* high-density cells (HDC) has significantly raised cell concentrations, increasing the volumetric productivity of LA. When compared to the original strain, *C. glutamicum* has shown promise for increased productivity and purity in the manufacture of D-LA through genetic manipulation. This was accomplished by inactivating the L-LDH gene and upregulating the D-LDH gene (Okino et al., 2005). A genetically modified variant of *C. glutamicum* has been developed, exhibiting the capability to enzymatically convert glucose into D-lactyl-CoA and 3-hydroxybutyryl-CoA. Through the activity of lactate polymerizing enzymes, these compounds can co-polymerize, resulting in the synthesis of poly (LA-co-3HB) with an exceptionally high LA content of 96.8 mol%. Investigations indicate that this engineered strain of *C. glutamicum* is proficient in utilizing not only hexose sugars such as glucose and galactose but also pentose carbohydrates like xylose and arabinose (Yukawa et al., 2007). In a study by Kawaguchi et al. (2008), genes xylA and xylB sourced from *E. coli* were introduced into the *C. glutamicum* R strain. These genes, responsible for encoding xylose isomerase and xylulokinase, respectively, played a crucial role in altering the production profile and enhancing the growth rate of organic acids. Notably, there was a significant increase in the yields and productivity rates of L-LA and SA, making this genetic modification particularly intriguing. Pyruvate kinase

(Pyk) is an essential component of the glycolysis pathway involved in the synthesis of pyruvate and ATP, which is necessary for controlling the distribution of carbon flux. Chai et al. (2016) discovered NCgl2809 as a new Pyk2 in *C. glutamicum*, indicating its involvement in increasing L-LA synthesis, particularly under oxygen-deprived circumstances.

Escherichia coli possesses the natural ability to ferment carbohydrates like hexoses and pentoses, producing L-LA alongside a combination of ethanol and organic acids. Engineered strains of *Escherichia coli* exhibit enhanced efficacy in L-LA fermentation compared to their wild counterparts. Through the substitution of L-LDH enzymes sourced from various organisms, such as LAB and cattle, for D-LDH, these modifications impede aerobic L-LDH activity, thereby preventing the formation of racemic mixtures of D- and L-lactates (Mazumdar et al., 2010). Engineered strains of *E. coli* exhibit versatility in utilizing a range of sugars, encompassing disaccharides like sucrose and monosaccharides such as glucose, xylose, and glycerol. However, they confront constraints such as diminished productivity (1.04g/L/h) and inadequate tolerance to acidic environments, attributed to the generation of L-LA, in contrast to LAB (Mazumdar et al., 2013).

11.4.2 FILAMENTOUS FUNGI

Filamentous fungi like *Rhizopus oryzae* and *R. arrhizus* have demonstrated the ability to synthesize ethanol, fumaric acid, and lactic acid from diverse carbon sources. Specifically, for L-LA production, these fungi undergo aerobic digestion of glucose. *Rhizopus* strains exhibit versatility in L-LA synthesis, utilizing various renewable carbon sources such as corncob hydrolysate, xylose, glucose, molasses, paper pulp sulfite liquor, potato hydrolysate, chicken feather protein hydrolysate, wheat straw, cassava pulp hydrolysis, and glycerol. Additionally, supplementation with Lucerne green juice and inorganic nutrients enhances L-LA production efficiency (Meussen et al., 2012). Nitrogen nutrients play a crucial role in facilitating rapid growth and encouraging the synthesis of chitin instead of LA. However, prolonged nitrogen deficiency can lead to reduced cellular function and output. To mitigate this challenge, two strategies can be employed: altering cell morphology to enhance LA productivity and yield, and adjusting the medium composition by incorporating high carbon content alongside reduced nitrogen concentrations (Bai et al., 2008). Urea serves as a crucial nitrogen source utilized by *Rhizopus* species in their metabolic processes. Its presence aids in controlling biofilm proliferation, retarding sporulation, and sustaining optimal cell viability and lactic acid production rates. In comparison to LAB, *Rhizopus strains* have the following advantages: they are amylolytic, they can grow in simple medium formulations, they can grow filamentous or pelletized in fermentation media, and they can produce useful fermentation byproducts. However, *R. oryzae* has a unique growth shape during fermentation, which means it requires a lot of aeration and can lower LA output (Zhang et al., 2007). Alternative byproducts like ethanol and fumaric acid are formed as a result of insufficient oxygen delivery and mass transfer, which causes unsatisfactory LA productivity in fungal fermentation. Small pellets are preferred in industrial fermentations because of their excellent rheological qualities, improved mass transfer, and long-term utilization via multiple batch fermentation cycles (Liu et al., 2006a). It has been determined that *R. oryzae's* cultivation of mycelia flocks with a morphology resembling cotton can increase LA productivity. To produce pure L-LA with higher cell density and high yields and productivity, a variety of fungal fermentation techniques have been investigated (Liu et al., 2006b). Strategies to enhance the production of LA involve various approaches. These include immobilization techniques to stabilize the microorganisms, managing unwanted byproducts such as fumaric acid and ethanol, incorporating inhibitors of alcohol dehydrogenase (ADH) to minimize ethanol formation, and boosting the activity of LDH to enhance cell biomass and LA yield. Additionally, to optimize LA output, metabolic engineering techniques have been employed, which entail the removal of genes responsible for malate dehydrogenase and ADH within the strain (Maneeboon et al., 2010).

11.4.3 YEASTS

LABs are frequently employed in the production of LA, but their vulnerability to acidic conditions necessitates the use of substantial quantities of neutralizing agents, resulting in gypsum as a byproduct. Conversely, yeasts can tolerate low pH environments, thereby diminishing the requirement for neutralizers and lowering downstream processing expenses. However, wild-type yeasts have a lower output of LA. To overcome this limitation, genetically modified yeasts offer an effective solution (van Maris et al., 2004). Yeast engineering has emerged as a powerful tool for boosting LA production through the inhibition of pyruvate decarboxylase (PDC) and/or pyruvate dehydrogenase activities. This innovative approach enables a shift away from ethanol production towards increased LA generation, resulting in improved productivity and yield. Incorporating the gene responsible for L-LDH from external sources can further augment lactic acid output and bolster natural acid resistance. Successful integration of this gene has been demonstrated in organisms such as *Saccharomyces cerevisiae* and *Candida utilis*. Researchers are exploring various yeast taxa including *Kluyveromyces* spp., *Saccharomyces cerevisiae, Zygosaccharomyces bailii, Pichia stipites,* and *Candida* spp. for the production of LA (Bianchi et al., 2001).

Saccharomyces *cerevisiae* excels at producing D-lactic acid because it can withstand low pH levels effectively. When supplied with essential anaerobic growth elements such as oleic acid, nicotinic acid, and ergosterol, this yeast can proliferate aerobically using glucose. Additionally, *Zygosaccharomyces bailii* is a suitable candidate for LA production due to its resilience to high sugar levels, acidic conditions, and elevated temperatures. This yeast's rapid growth and high biomass yield could enhance LA production (Abdel-Rahman et al., 2011). Engineered strains of *Zygosaccharomyces bailii* have been developed to enhance LA production by incorporating a foreign LDH gene, which redirects the glycolytic pathway towards lactate synthesis. Additionally, *Pichia stipitis* is capable of fermenting both pentose and hexose sugars present in lignocellulose hydrolysates to produce ethanol. By knocking out the ADH1 gene and inserting the L-LDH gene under the control of the ADH1 promoter, the genetically modified strain was able to produce 41 g/l of lactic acid from 100 g/l of glucose and 100 g/l of xylose. This modification also resulted in a 15–30% reduction in ethanol production compared to the wild-type strain (Abdel-Rahman et al., 2013).

Candida sonorensis is capable of fermenting both pentose and hexose sugars into ethanol. This yeast thrives in acidic conditions and can also grow effectively on a basic growth medium. Scientists have engineered mutants of this organism to produce both ethanol and LA by introducing L-LDH genes from *Bacillus megaterium, Rhizopus oryzae,* and *Lactobacillus helveticus.* To halt ethanol production, they deleted two PDC genes. The resulting *C. sonorensis* strain no longer produced ethanol, while still growing well and accumulating LA up to 92 g/l with a yield of 0.94 g/g of glucose. Another engineered strain of *C. sonorensis* was able to produce 31 g/l of LA from 50 g/l of D-xylose, without generating any ethanol (Koivuranta et al., 2014). Through genetic engineering, researchers developed a Crabtree-negative, methylotrophic diploid strain of *Candida boidinii* capable of producing substantial quantities of L-LA. In contrast to the wild-type strain, the knockout of the PDC1 gene resulted in a 17% reduction in ethanol production. Furthermore, the introduction of the bovine L-LDH gene enabled the engineered strain to produce 85.9 g/l of lactic acid with a productivity rate of 1.79 grams per liter per hour (g/L/h) (Osawa et al., 2009). *Candida utilis*, a Crabtree-negative yeast, is proficient in producing valuable compounds like RNA, glutathione, and SCP. Its ability to utilize cost-effective substrates, such as pulping-waste liquors, is advantageous for L-LA production. Research indicates that a genetically modified strain of *C. utilis*, created by integrating two copies of the bovine L-LDH gene into the genome of a CuPdc1-null strain and knocking out the pyruvate decarboxylase gene (CuPDC1), efficiently produced L-LA. This engineered strain yielded 103.3 g/L of L-LA with a purity of 99.9% and a conversion rate of 95.1% (Ikushima et al., 2009).

Kluyveromyces lactis, a Crabtree-negative yeast, is utilized for LA synthesis after undergoing genetic modification. Unlike other yeast species, *Kluyveromyces lactis* possesses a solitary gene, KlPDC1, responsible for encoding PDC activity. By deleting the KlPDC1 gene, strains are developed that lack PDC activity. This genetic alteration enhances LA production without generating ethanol. The absence of KlPDC1 ensures that ethanol is not produced, even amidst intense competition for pyruvate between the native PDC and introduced LDH activities, leading to improved LA yields (Bianchi et al., 2001). In experiments where both KlPDC1 and pyruvate dehydrogenase genes were simultaneously removed, *Kluyveromyces marxianus* demonstrated a notable surge in LA production. This yeast species holds significant economic promise in both commercial and industrial domains due to its rapid proliferation in elevated temperatures and its capacity to thrive on inexpensive carbon substrates. The deletion of these genes prompted a redirection of pyruvate flux towards homolactic fermentation, consequently amplifying LA synthesis (Pecota et al., 2007).

11.4.4 MICROALGAE AND CYANOBACTERIA

Photosynthetic microorganisms, including algae and cyanobacteria, are capable of thriving in diverse environments and have rapid harvesting cycles ranging from 1 to 10 days. These organisms are not only adaptable but also valuable for producing various useful compounds, such as biofuels like ethanol, hydrogen (H_2), lactic acid (LA), acetic acid (AA), and formic acid (FA). One notable application is their potential to serve as an alternative source of lactic acid, which can help avoid the high costs associated with traditional carbohydrate feed media (Hirayama and Ueda, 2004). Different strains of microalgae, such as *Scenedesmus obliquus strain D3* and *Nannochlorum* sp. *26A4,* were examined for their capacity to generate LA. Under dark, anaerobic conditions with starch as the carbon source (constituting 40% of dry weight), *Nannochlorum* sp. *26A4* displayed remarkable LA production, yielding 26 g/L with an impressive yield of 70% and an optical purity of 99.8%. Analysis of *Nannochloropsis salina* biomass revealed a composition of 40% lipids, 20% carbohydrates, and 40% proteins. Upon further investigation, the lipid-free residue was found to contain significant sugar content, with 64.3% of the residue comprising glucose and xylose after neutralization and concentration. Utilizing this biomass, co-fermentation of *N. salina* and *Lactobacillus pentosus* under anaerobic conditions yielded 10.1 g/L of LA, achieving a remarkable conversion rate of 92.8% (Talukder et al., 2012).

11.5 FERMENTATION PROCESSES FOR LA PRODUCTION

11.5.1 CONVENTIONAL FERMENTATION APPROACHES

Batch fermentation operates as a self-contained procedure where both substrates and microorganisms are introduced into the system at the outset, with no significant components being extracted until fermentation concludes. This approach mitigates the risk of contamination and strain variation, optimizes substrate conversion rates, and yields elevated concentrations of LA (Wang et al., 2021). Nonetheless, the build-up of LA can lead to acidification, impeding microbial proliferation and LA synthesis. Batch fermentation stands out as the most prevalent and straightforward model for elucidating novel processes. Investigations have delved into *Lactobacillus pentosus*, sugar metabolism, and LA generation by *E. faecalis* under batch conditions. Moreover, batch fermentation serves as a tool for assessing LA production from sustainable substrates and refining processes (Anagnostopoulon et al., 2022). Continuous fermentation represents a strategic approach in bioprocessing where nutrients are continuously

TABLE 11.3 A Collection of Organisms that Have Been Researched for the Production of LA

Organism		Lactic Acid (g/L)	Source	References
Homo and Heteroferme-ntative LAB	*Lb. delbruckii* NCIMB 8130	90.0	Molasses	Kotzamanidis et al. (2002)
	Lb. delbrueckii sp. delbrueckii ATCC 9649	58	Glucose	Abdel-Rahman et al. (2013)
	Lb. delbrueckii sp. bulgaricus AU	20	Whey permeates	Abdel-Rahman et al. (2013)
	Lb. delbrueckii NCIM 2025	81.9	Cassava bagasse	John et al. (2006)
	Lb. delbrueckii subsp. delbrueckii IFO 3202	28	Defatted rice bran	Abdel-Rahman et al. (2013)
	Lb. delbrueckii mutant Uc-3	67	Sugarcane bagasse waste	Adsul et al. (2007)
	Streptococcus sp.	66.5	Highly viscous food waste	Peinemann et al. (2019)
	Streptococcus sp. (indigenous consortium)	50–69	Mixed food waste	Demichelis et al. (2017)
	Engineered *Pediococcus acidilactici*	130.8	Wheat straw	Qiu et al. (2018)
Bacillus strains	*B. coagulans*	20.1	Sucrose	Payot et al. (1999)
	B. coagulans 36D1	80	Cellulose	Ou et al. (2011)
	B. coagulans strains 36D1	92	Paper sludge	Budhavaram and Zhang (2009)
	B. coagulans strains P4–102B	91.7	Paper sludge	Budhavaram and Zhang (2009)
	B. coagulans DSM2314	58.7–70.4	Sugarcane bagasse	van der Pol et al. (2016)
	B. coagulans	79.4–93.7	Glucose, xylose and cellobiose	Glaser and Venus (2018)
	B. coagulans	79.1	Lignocellulosic corncob residue	Jiang et al. (2019)
E. coli	*Engineered E. coli*	60–62.2	Glucose	Li et al. (2004)
	Engineered E. coli	45.5–51.8	Glucose	Zhou et al. (2003)
	Engineered E. coli	40	Xylose	Zhao et al. (2013)
	Engineered E. coli	56.8	Glycerol	Hong et al. (2009)
	E. coli AC-521	85	Sucrose	Wang et al. (2012)
	E. coli K12 strain	32	Glycerol	Mazumdar et al. (2010)
	E. coli	75	Molasses	Wang et al. (2013)

TABLE 11.3 *(Continued)*

Organism		Lactic Acid (g/L)	Source	References
Corynebacteria glutamicum	*C. glutamicum*	120	Glucose	Okino et al. (2005)
	C. glutamicum	60.27	D-ribose	Chai et al. (2016)
Fungi	*R. oryzae* ATCC 52311	83	Glucose	Taherzadeh et al. (2003)
	R. oryzae NRRL 395	49.1	Water paper	Ikeda et al. (2006)
	R. oryzae GY18	68.5	Xylose	Guo et al. (2010)
	R. oryzae ATCC 9363	113	Glucose	Yu et al. (2007)
	R. oryzae NRRL 395	91	Corn starch	Abdel-Rahman et al. (2013)
	Monascus ruber	129–190	Glucose	Weusthuis et al. (2017)
	Engineered *Aspergillus brasiliensis* from *Rhizopus oryzae*	12.1–32.2	Glucose	Laud et al. (2015)
	Aspergillus niger	7.7	Glucose	Dave and Punekar (2015)
Yeast	Engineered *P. stipitis*: LDH from *L. helveticus* (integrated, 1 copy)	15–58	Glucose	Ilmén et al. (2007)
	Engineered *S. cerevisiae* LDH from *L. casei* (multicopy vector)	12	Glucose	Abdel-Rahman et al. (2013)
	Engineered *S. cerevisiae* LDH from *L. plantarum* (integrated, 1 copy)	58	Glucose	Colombie et al. (2003)
	Engineered *S. cerevisiae* LDH from *B. taurus* (integrated, 2 copies)	50.6	Glucose	Ishida et al. (2005)
	Engineered *S. cerevisiae* LDH from *B. taurus* (integrated, 6 copies)	120	Glucose	Ishida et al. (2005)
Candida	Engineered *Candida utilis*: LDH	93	Xylose	Tamakawa et al. (2012)
	Engineered *Candida utilis*: LDH from *B. taurus*–optimized (integrated, 2 copies)	103.3	-	Ikushima et al. (2009)

TABLE 11.3 *(Continued)*

Organism		Lactic Acid (g/L)	Source	References
	Engineered *Candida boidinii*: LDH from *B. taurus*–optimized (integrated, 1 copy)	85.9	Glucose	Osawa et al. (2009)
	Candida sonorensis	92	Glucose	Ilmén et al. (2007)
Kluyveromyces	Engineered *K. marxianus* from *Actobacillus plantarum*	122–130	Jerusalem artichoke tuber powder	Bae et al. (2018)
	Engineered *K. marxianus* from *Homo sapiens* (HsLDH), *Bacillus subtilis* (BsLDH), *Bacillus* megaterium (BmLDH), *Lactococcus lactis* (LlLDH), *Rhizopus oryzae* (RoLDH), and *Plasmodium falciparum* (PfLDH)	25–105	Alkali-pretreated corncob	Kong et al. (2019)
	Engineered *K. marxianus* LDH from *L. helveticus* (integrated into PDC1 locus)	99	Glucose	Hause et al. (2009)
	Engineered *K. marxianus* LDH from *L. helveticus* (integrated into PDC1 locus)	9.1	Glucose	Hause et al. (2009)
	Engineered *K. lactis* LDH from *B. taurus* (low copy number plasmid, 5 copies)	109	Glucose	Abdel-Rahman et al. (2013)
Microalgae and cyanobacteria	Engineering of *Schizosaccharomyces pombe*	24.4–25.2	Glucose and cellobiose	Ozaki et al. (2017)
	Naturally inhabiting bacteria in garbage	64	Kitchen refuse	Zhang et al. (2008)
	Naturally inhabiting bacteria in garbage	37.7	Garbage	Akao et al. (2007)
	Microbial *consortium* CEE-DL15 *Clostridium sensustricto* (57.29%), *Escherichia* (34.22%), and *Enterococcus* (5.32%)	112.3	Sugarcane molasses	Sun et al. (2019)
	Anaerobic activated sludge as *inocula*	28.4	Methanogenic sludge and fresh food waste	Tang et al. (2017)

infused into a fermenter while simultaneously extracting culture broth. This method is prized for its ability to achieve heightened productivity while minimizing production hindrances. Nonetheless, maintaining a consistent operational state poses challenges, chiefly stemming from the heightened risk of microbial mutations and the perpetual need to thwart contamination. The design of bioreactors tailored for continuous fermentation introduces added layers of complexity, necessitated by the imperative to avert contamination and sustain optimal growth conditions (Garcia et al., 2021). A research endeavor conducted by Veeravalli and Mathews (2018) underscored the impact of manipulating the dilution rate within the continuous fermentation process. Their findings revealed that elevating the dilution rate led to reductions in concentrations of LA and AA, albeit with minimal impact on the ratio between these two acids. However, cell washout occurred when the dilution rate matched the maximum specific cell growth rate. Determining more about the dynamics, it was found that LA yield was positively impacted by using a low pH environment during the continuous LA fermentation of dairy effluent. Significant differences were found when food waste was analyzed in two different ways: batch fermentation and continuous mode, while the continuous strategy resulted in a 38% rise in LA concentration and a 36% increase in LA yield (Peinemann et al., 2019). A development of batch fermentation techniques is fed-batch fermentation, in which nutrient-rich media is gradually added to the vessel over time to allow for more control over the growth of the cells (Pejin et al., 2017). This strategic approach allows for the sustained synthesis of metabolites. According to research, it has a significant impact on specific metabolite yields, as seen by increased lactate dehydrogenase concentration, yield, and volumetric productivity (de Oliveira et al., 2021). Fed-batch fermentation enhanced the yield of *L. casei* by 11%, leading to a notable 69.24% increase in LA production. For *L. pentosus*, switching to fed-batch mode raised the LA concentration from 106.7 g/L to 157.0 g/L. However, this change resulted in a decrease in overall productivity, from 3.09 (g/L/h) to 2.0 (g/L/h). Additionally, tailored feeding strategies positively influence the metabolic activity of microorganisms (Lobeda et al., 2022). Using softwood hydrolysate and inhibitors in a model medium at a low feed rate, for example, made it easier for *P. acidilactici* to adapt to the inhibitors and turn furfural and softwood hydrolysate into LA while also detoxifying the media. This highlights the flexibility of feeding plans to meet the unique metabolic needs of microbes and maximize the synthesis of metabolites. Basically, fed-batch fermentation shows promise as a useful technique to maintain high metabolite production for longer periods of time (Campos et al., 2023).

11.5.2 CELL IMMOBILIZATION

Cell immobilization is the transfer of cells onto a water-insoluble carrier, allowing them to carry out key life functions such as growth, reproduction, and metabolism inside a limited space. This technology allows for the recurrent use of cells in fermentation processes, which results in strong fermentation due to high cell densities (Lu et al., 2020). Cell immobilization can be accomplished using several methods, such as entrapment, encapsulation, adsorption, and containment within synthetic polymers. A study by Wang et al. (2020) showed that encapsulating *Lactobacillus pentosus* in a composite carrier made of sodium alginate (SA) and polyvinyl alcohol (PVA) led to stable and efficient LA production. When compared to free cells, the productivity and yield of immobilized *L. pentosus* were significantly greater under ideal fermentation and immobilization conditions, by 13.6% and 67.3%, respectively. In a similar vein, *L. pentosus* immobilized by SA-PVA showed increased productivity and LA yield when film-coated with chitosan. Additionally, cost-effective and eco-friendly immobilization supports can be created using natural materials such as sugar beet pulp, brewers' spent grain, sunflower seed hulls, and loofah sponge. Among these, sugar beet pulp has proven to be the most efficient, achieving an LA productivity of 1.48 (g/L/h) and an LA concentration of 80.10 g/L (Radosavljevi´c et al., 2021).

11.5.3 MEMBRANE-BASED CELL RETENTION

In continuous LA fermentation, there are challenges like the potential for cell culture washout in a continuous flow system. To address this issue, membrane-based cell recovery systems have emerged as effective solutions for continuous fermentation. These systems minimize cell loss and support optimal cell growth, ultimately enhancing the production of LA (López-Gómez et al., 2019). Research indicates significant productivity gains when utilizing hollow-fiber filters with bakery waste hydrolysates and corn stover hydrolysate. Integrating membrane-based cell recycling with B vitamin supplementation has yielded an impressive 86% reduction in nutrient sourcing costs. Continuous fermentations of agricultural residues exhibit notable LA productivity, particularly when employing molasses (Ma et al., 2022). Optimizing the dilution rate via membrane microfiltration cell recovery has shown promising results in boosting LA output from sugarcane molasses to 27.6 (g/L/h), achieving yields exceeding 0.95 g/g. However, challenges remain regarding the cost implications linked to the establishment and upkeep of large-scale fermentation systems tailored for industrial usage, despite the notable advancements in productivity (Alexandri et al., 2020).

11.5.4 SIMULTANEOUS SACCHARIFICATION AND CO-FERMENTATION

Simultaneous saccharification and co-fermentation (SSCF) represents an integrated approach where cellulose and hemicellulose undergo enzymatic degradation, yielding fermentable sugars that are promptly converted into desired products within a singular container. This method increases the productivity and yield of LA by avoiding substrate inhibition, which is frequently found in separate hydrolysis and fermentation (SHF) settings (Zhou et al., 2016). In addition, it significantly reduces the costs related to large-scale industrial manufacturing. When comparing the production of LA using SHF and SSCF, a comparative study showed that the concentration of LA increased significantly by 43.73%, while processing time decreased by 25.00% and the dosage of fungal cellulase decreased by 33.3%. Saccharification was conducted on both the starch and cellulose fractions of cassava bagasse, yielding a notable productivity rate of 2.74 (g/L/h) of LA. In a distinctive experiment, *L. paracasei* was cultivated directly on bakery trash using amylase and Amyloglucosidase, resulting in over 26.4 g/L of LA. However, it is noteworthy that a sizable amount of the total expenses in SSCF are attributed to enzyme costs (Tian et al., 2021). Li et al. (2021) developed a method for synthesizing enzymes directly at the site using paper mill waste. These enzymes were then used in conjunction with *Bacillus coagulans* in a SSCF process to produce optically pure L-LA. Meanwhile, Zhang et al. (2022) achieved high levels of productivity and LA concentration by utilizing a continuous SSCF process on a substrate made from dry acid-pretreated and bio detoxified wheat straw.

11.5.5 CO-CULTURE

In a fermentation process, microbial co-culture refers to the use of several different organisms at the same time. Co-cultivation offers the chance to capitalize on the qualities and potential of several strains, simulating organic synergies, in contrast to conventional microbial fermentation, which usually uses a single strain or culture. Compared to single-strain fermentation, this method improves production efficiency and makes complex synthetic routes easier to execute. Compared to pure sugar media, agricultural waste is a problem because of its various contaminants, inhibitors, and possibly hazardous components. Additionally, cellulose and hemicellulose provide a sufficient amount of pentose and hexose sugars. Low LA yields and reduced fermentation efficiency are the results of inefficient conversion of both sugar types into LA (Ricci et al., 2019). Klongklaew et al. (2023), explored co-culture fermentation with *Enterococcus mundtii*, known for its tolerance to furfural and its ability to utilize xylose, and *Lactobacillus rhamnosus*, which utilizes glucose, to produce LA from corn stover hydrolysate without prior detoxification. This combination achieved an impressive 31.4 g/L of highly pure (99.9%) L-LA, with a yield of 0.90 g/g and a productivity rate of 1.73 (g/L/h). Additionally, co-culturing *Lactobacillus plantarum* and *Lactobacillus paracasei* demonstrated superior LA fermentation efficiency from orange peel compared to their individual cultures. On the other hand, although *Lactobacillus brevis* can ferment both xylose and glucose, its monoculture in a xylose-glucose mixture only achieved a yield of 0.52 g/g and resulted in the formation of significant by-products (Ge et al., 2009). In the co-culture system, *Lactobacillus brevis* was primarily engaged in metabolizing xylose, whereas *Lactobacillus plantarum* played a crucial role in boosting yield and reducing ethanol production. When fermenting polar hydrolysate and alkaline-treated maize biomass in co-culture, the resulting LA yields were 0.78 g/g and 0.80 g/g, respectively. In another approach involving mixed cultures, one bacterium is responsible for saccharifying the biomass, and ferments the hydrolysate into LA. To directly utilize *Jerusalem artichoke* tubers, which are rich in inulin, a mixed culture of *Aspergillus niger* and *Lactobacillus* species was employed. In this setup, *Aspergillus niger* produced inulin's and invertase to hydrolyze inulin, while the *Lactobacillus* species converted the resulting sugars into LA (Zhang and Vadlani, 2015).

11.5.6 GENETIC AND METABOLIC ENGINEERING

Agricultural waste hydrolysates are rich in several sugars, such as arabinose, xylose, and glucose. But some strains of LAB are the only ones that can ferment pentose carbohydrates into LA. Carbon catabolite suppression, which essentially prevents the intake of less advantageous sugars, presents a barrier for a large number of LAB (Jiang et al., 2018). One strategy is to introduce the genes that are in charge of consuming these less desirable sugars in order to overcome this restriction and improve the utilization of all fermentable carbohydrates. By changing carbon metabolism routes, it is feasible to reduce the generation of unwanted by-products, increasing LA yields while also lowering downstream purifying costs (Ou et al., 2016). Ensuring the production of LA in acidic environments is essential. To enhance the optical purity of the resulting LA, it might be necessary to disrupt and substitute the dehydrogenase gene responsible for producing unwanted racemic mixtures. By integrating both metabolic and genetic engineering methods into a single strain, a significant improvement in productivity can be achieved (Aso et al., 2019). Sahoo and Jayaraman (2019) conducted a study wherein they removed all L-lactate genes from

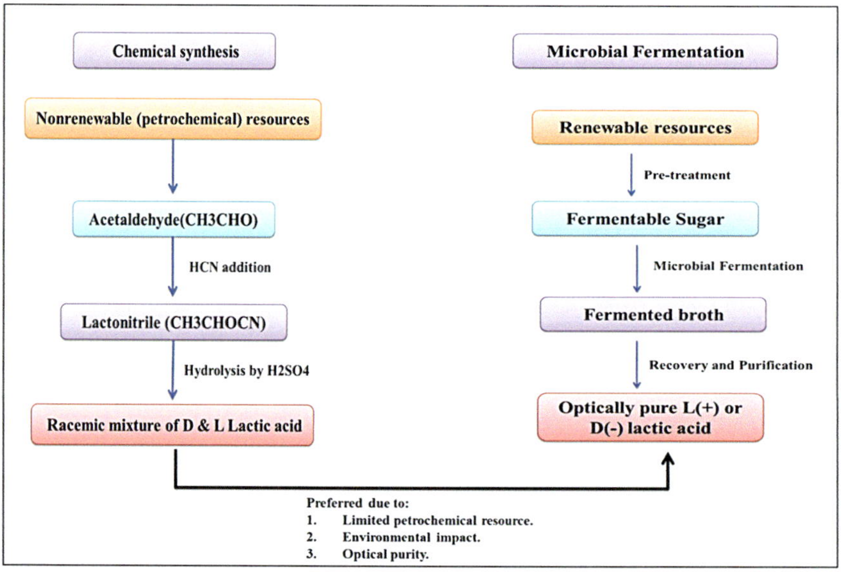

FIGURE 11.2 Chemical vs. microbial method of LA production.

Lactococcus lactis and simultaneously overexpressed a *D-LDH* gene from *Lactococcus bulgaricus*. As a result of this modification, optically pure D-LA was obtained. Subsequently, a co-culture batch process using a recombinant *L. lactis* strain with whey permeate was implemented, leading to notable increases in both the concentration of D-LA (45 g/L) and the yield of LA (0.90 g/g).

11.6 DOWNSTREAM PROCESSING AND RECOVERY OF LA

The optical purity of LA plays a vital role in the production of PLA. To achieve high crystallinity in PLA, it is essential to use optically pure forms of the D (−) or L (+) LA isomers. Racemic mixtures of isomers are unsuitable for this purpose. Microbial fermentation is the only method for producing optically pure LA, influenced by bacterial strains and fermentation medium composition. The purity of LA depends on its intended use, with pharmaceutical grade requiring over 90% purity and food grade requiring 80% to 90% purity. Various downstream processing techniques have been developed to obtain significant amounts of LA, such as microbial fermentation, precipitation, solvent extraction, membrane separation, electrodialysis, adsorption, molecular distillation, and esterification (Albuquerque et al., 2021).

11.6.1 PRECIPITATION

Compounds containing calcium are added to aid in the precipitation of LA, which is then either crystallized or separated via the calcium lactate crystallization-acidolysis process. The formation of LA during fermentation might cause a reduction in the pH of the solution, influencing the metabolic pathways of the microorganisms involved. Ensuring that the pH level of the solution remains higher than the acid dissociation constant (pKa) of LA, which is 3.86, is essential for maximizing the synthesis of LA (Komesu et al., 2017). While maintaining a pH range of 5–7, the addition of calcium hydroxide or calcium carbonate helps to convert LA into calcium lactate. The calcium lactate solution is then filtered to remove impurities, and then activated carbon is used in concentration, crystallization, and decolorization procedures to produce calcium lactate. $CaOH_2$ is added before filtration to guarantee that all free LA molecules are converted to calcium lactate. Moreover, heating the fermentation solution helps any unreacted proteins present precipitate and get denatured (De Oliveira et al., 2020).

11.6.2 SOLVENT EXTRACTION

Solvent extraction entails the separation of components from a solution by employing solvents. This method is employed to isolate pure LA from the fermentation broth. The choice of solvents depends on the solubility of the solute. Extraction agents such as high molecular weight aliphatic amines, phosphorus-bonded oxygen-bearing compounds, and oxygen-bearing hydrocarbons are frequently used. Various techniques for solvent extraction are available for extracting LA (Kumar et al., 2019a). An inorganic salt and an organic solvent are added to the broth in a predetermined quantity using the straightforward salting-out extraction procedure. The extraction medium's pH affects the amount of LA recovered; at pH 2.5, the most LA is recovered. Reactive extraction, which has 70% recovery efficiency, is essential for obtaining LA from fermentation broth. Because the extract has a stronger affinity than the water phase, it dissolves more quickly when combined with an extractant in an aqueous solution (Kumar et al., 2018). LA can additionally be procured from fermentation broth through a method involving emulsion liquid membranes (ELM). This process necessitates the emulsification of two phases that are immiscible, which are then continuously stirred into a feed phase. This facilitates the extraction of solute molecules even at low concentrations (Kumar et al., 2019b). In ideal settings, the green emulsion ionic liquid membrane (GEILM) demonstrates an extraction efficiency of around 90%. To enhance the efficiency of LA extraction while reducing chemical consumption, a liquid membrane in Taylor flow (LMTF) has been devised. Yet, its insufficient stability hampers its suitability for industrial purposes. To address these challenges and retain the benefits of ELM, a hybrid model integrating LMTF with fermentation has been conceived. This presents a more cost-effective and efficient approach to LA production (P´erez and Fontalvo, 2019).

11.6.3 MEMBRANE SEPARATION

Membrane separation has emerged as a vital technique in the realm of separating, enriching, and purifying LA from various components present in fermentation broth. Its integration into fermentation processes plays a crucial role in sustaining elevated cell densities and facilitating efficient cell recycling (Taleghani et al., 2018). When integrated with bipolar electrodialysis and standard fermentation units, the nitrogen trifluoride (NF3) composite polyamide membrane was able to retain 94% of the sugar, yielding L-LA with an optical purity of 85.6%. Membrane separation is an uncomplicated and energy-efficient method that maintains both efficacy and product quality. However, challenges such as high costs, reduced flux, membrane fouling, and other technical issues limit its widespread adoption in industrial applications (Alexandri et al., 2018).

11.6.4 ELECTRODIALYSIS

Electrodialysis works by using ion-exchange membranes to enable the transport of chemicals across solutions caused by electric potential differences. This approach uses separate cation and anion exchange membranes to selectively extract and concentrate LA from fermentation broth while keeping the pH stable (Chen et al., 2016). The process of recovering sodium hydroxide (NaOH) and hydrochloric acid (HCl) for subsequent fermentation and downstream applications is facilitated by its demonstrated effectiveness, achieving up to a 90% yield of LA (Wang et al., 2006). Combining bipolar membrane electrodialysis with *in situ* fermentation provides a highly effective approach for LA extraction, achieving an impressive recovery rate of 86.05%. This combination lowers the initial cost of membrane material and expedites the process timeline while simultaneously managing electrodialysis and fermentation (Cheng et al., 2008). However, issues arise when LAB adhere to the anionic membrane. This adhesion heightens resistance and diminishes the efficiency of the electrodialysis process. Even with benefits like lower reagent consumption, less waste production, and streamlined purification processes, electrodialysis has drawbacks such membrane fouling, concentration polarization, and expensive material costs that prevent it from being widely used. To improve this technique's scalability and optimization, more research is needed to overcome these obstacles (Min-Tian et al., 2005).

11.6.5 ADSORPTION

Adsorption is an essential process that employs solid adsorbents to eliminate impurities from liquids, gases, and extracts, including LA.

This approach emerges as a promising route for LA separation due to its strong affinity for the target element, ease of regeneration, and low energy consumption. In adsorption, a variety of materials are used, such as polyvinyl pyridine resins, activated carbon, and ion exchange resins (Shi et al., 2020). Delgado et al. (2018) research demonstrates the effectiveness of the simulated moving bed (SMB) approach, producing concentrated LA with a notable 95% recovery rate for the LA/water combination (Pradhan et al., 2017). Othman et al. (2018) found that weakly basic resin beats strong base resin in LA adsorption, with Amberlite IRA 67 resin displaying greater LA removal and reusability. The LA fermentation process stands out for its effective workflow, low energy consumption, and high-quality product outcomes, showcasing a strong compatibility with adsorption techniques.

11.6.6 MOLECULAR DISTILLATION

Molecular distillation, also known as short-path distillation, stands as a sophisticated method for liquid separation under high vacuum conditions, espccially beneficial for processing heat-sensitive compounds. Its applications span various sectors such as food processing, chemical manufacturing, pharmaceuticals, and petrochemical industries (Othman et al., 2018). Because it can function at lower temperatures and with faster reaction times, molecular distillation offers a number of advantages when it comes to purifying LA. These benefits include minimizing degradation and guaranteeing superior yield and purity (Dantas et al., 2020). In a study by Yu et al. (2015), they employed various techniques such as molecular distillation, solvent extraction, and calcium precipitation to isolate and purify LA from fermentation broth. Response surface methodology (RSM) was utilized to identify the appropriate molecular distillation parameters, which led to a 92.39% purity level for LA and a 74.09% yield under optimal conditions. Molecular distillation is mostly used in small-scale LA synthesis, despite its high cost and strict vacuum requirements.

11.6.7 ESTERIFICATION

Reactive distillation, also referred to as the esterification hydrolysis process, stands out as an exceptionally efficient technique in separating LA due to its remarkable capacity for achieving both high yield and purity.

This process makes it easier for alcohols and LA to combine to form lactic acid ester, which then makes it possible for LA to be produced through hydrolysis (Chandane et al., 2020). Interestingly, this method can also be used to remove other organic acids from LA fermentation broth, like acetic acid. Komesu et al. (2015) found that the reactive distillation system is extremely effective in creating considerable amounts of LA, with a production rate of up to 99.94%. Additionally, a range of alcohol variants can serve as reactants in esterification reactions, enhancing the flexibility of this method. Despite the numerous advantages of reactive distillation, including its low capital expenditure and the attainment of exceptionally pure end products, challenges related to production persist. These difficulties include high energy use, azeotrope separation, and water extraction (Zhao et al., 2009). To address these challenges, forthcoming research endeavors should prioritize process intensification methodologies, such as advancements in equipment, process integration, and thermal optimization. Implementing these strategies could enhance the sustainability and efficiency of reactive distillation techniques for separating LA (Zhang and Vadlani, 2015).

11.7 FACTORS AFFECTING LA PRODUCTION

A number of parameters have a substantial impact on the generation of LA, and one of the most important ones that affect microbial development is temperature. Research has indicated that a temperature range of 32 to 42°C is ideal for microbial development. The highest output of *Lactobacillus delbrueckii* is usually obtained at approximately 39°C when using agricultural waste. Deviations from this range give unsatisfactory results (Teleghani et al., 2016). Research shows that LA synthesis gradually increases between 35 and 42°C, with different substrates and strains providing different results. Another factor influencing LA synthesis is pH, with the optimal pH range for increased microbial growth and yield being 6.5 to 8.5. For example, the maximum output of *Lactobacillus delbrueckii* ssp. *delbrueckii* CECT 286 is observed at pH 5.8, whereas higher LA synthesis with *Lactobacillus pentosus* and xylose substrates is promoted at pH 7 (Beitel et al., 2020). Furthermore, a high-quality LA output depends on an ideal ratio between the carbon and nitrogen sources, which has a substantial impact on microbial performance. Research has indicated that the optimal yield of LA was achieved by maintaining a carbon-to-nitrogen ratio of 10:1, although yield is also impacted by changes in nitrogen concentration. Purifying the LA broth after fermentation is critical for producing pure LA for subsequent applications, with downstream procedures accounting for a significant amount of manufacturing expenses (Cheng et al., 2019). Purification can be achieved by a number of methods, including chromatography, filtration, evaporation, precipitation, crystallization, distillation, and electrodialysis. More advanced techniques for purification have been developed recently, which has increased LA output and purity. Despite these advances, certain restrictions remain, demanding further research and development in this area (Yang et al., 2018).

11.8 CONVERSION OF LA TO PLA

PLA, an exciting family of polymers sourced from renewable materials, undergoes a complex production sequence starting from the production of LA and culminating in polymerization. Various methods can be employed to synthesize PLA from LA, such as azeotropic dehydration, enzymatic polymerization, direct polycondensation, and ROP (Farah et al., 2016).

11.8.1 AZEOTROPIC DEHYDRATION

In the azeotropic condensation polymerization technique, LA undergoes conversion into high molecular weight PLA utilizing organic solvents like xylene, toluene, and diphenyl ether. This approach effectively eliminates excess water through distillation, streamlining the polymerization process by minimizing challenges associated with water removal. Consequently, LA can readily undergo polycondensation, yielding a polymer with high molecular weight. However, this method often results in high molecular weight polymers containing substantial amounts of catalyst impurities. These impurities contribute to undesired degradation, unstable hydrolysis rates, and concerns regarding catalyst toxicity, posing challenges to the quality and safety of the final product (Desai et al., 2023). In this process, low-boiling organic solvent and potent catalysts like boric acid and sulfuric acid are employed. Water is generated as a byproduct and must be removed through azeotropic distillation to enable the catalyst to dry and regenerate. The standard procedure involves distilling LA for a period of two to three hours at 130°C under reduced pressure. Subsequently, the organic solvent and catalyst are introduced, and the reaction proceeds for 30–40 hours at 130°C. Despite purification efforts on the resulting PLA, significant residues of catalyst persist due to the high concentration necessary to optimize reaction kinetics (Swetha et al., 2023).

11.8.2 ENZYMATIC POLYMERIZATION

Enzymatic polymerization is a crucial method in the production of PLA, employing either ROP of LA or direct polycondensation of LA. Among these methods, enzymatic ring-opening polymerization (eROP) stands out as the preferred approach due to its ability to yield water as a byproduct. It is noteworthy that polycondensation mostly produces low molecular weight PLA, whereas conventional PLA synthesis usually yields large molecular weights. Hydrolases and lipases have the ability to speed up the enzymatic polymerization of aliphatic polyesters. To synthesize PLA, for example, lipase derived from Candida rugosa played a crucial role in polymerizing L-LA (Swetha et al., 2023). PLA was synthesized with remarkable efficiency, boasting a high molecular weight of 5428 g/mol and an average molecular weight of 2854 g/mol, achieving an impressive 93% yield. The resulting PLA exhibited a melting point of 120°C and a crystallinity of 31%. Additionally, enzymatic catalysis offers an alternative route for polyester production from lactic acid. Lipase-catalyzed polymerization of cyclic LA facilitates PLA production, with lipases demonstrating a strong preference for converting LA enantiomers into PLLA and PDLA. By selecting an appropriate lipase, polymerization of D and L-LA can yield PDLLA (Yu et al., 2023).

11.8.3 DIRECT POLYCONDENSATION

A cost-effective approach involves direct polycondensation, where LA is polymerized at reduced pressure using a catalyst. However, achieving high molecular weight polymers exceeding 100,000 Da poses challenges due to the laborious task of removing water from the thick reaction mixture. Consequently, the process yields low molecular weight polymers (ranging from 1000 to 5000 Da) with inferior mechanical properties, lacking the required modulus for plastic packaging. To address this issue, acidic catalysts like boric acid and sulfuric acid are combined with high molecular weight coupling agents or additives to expedite esterification and trans-esterification processes (Yu et al., 2023). However, when temperatures exceed 120°C, these catalysts inadvertently stimulate secondary reactions. DL-LA polycondensation in an aqueous solution consistently fails to achieve average molecular weights exceeding 3000. An alternative method involves first dehydrating pre-condensates at temperatures between 120°C and 180°C, followed by heating them in the presence of non-acidic trans-esterification catalysts like lead (II) oxide (PbO). This approach proves effective when using azeotropic solvents, which form oligomers with average molecular weights hundreds to thousands of times smaller than those produced through single-step polycondensation of LA (Desai et al., 2023).

11.8.4 RING-OPENING POLYMERIZATION (ROP)

Over the past decade, the industrial manufacturing sector has extensively utilized the ROP process as the exclusive method to produce high molecular weight PLA, reaching up to 100,000 in molecular weight. Numerous factors play pivotal roles in determining the properties of the polymer. These include the specific initiator system employed, the co-initiator which functions as a chain control agent, the concentration of the catalyst, the ratio of monomer to initiator, as well as the polymerization temperature and duration. Additionally, these variables significantly impact the molecular weight, temperature sensitivity, and the physical and mechanical characteristics of both PLA and its copolymers. The enantiomeric purity and chain microstructure are contingent upon levels of racemization and transesterification occurring during the polymerization and copolymerization processes (Castro-Aguirre et al., 2016). The catalyst stannous octoate ($Sn(Oct)_2$) has gained approval and recognition from the US FDA for its frequent utilization in contemporary PLA polymerization methods. Transition metal compounds, including aluminum, tin, iron, and zinc, alongside less reactive metal alkoxides, oxides, and carboxylates, facilitate the synthesis of high molecular weight PLA. The polymerization process, termed "coordination-insertion," involves two primary stages: firstly, covalent bonds undergo rearrangement subsequent to the formation of a complex between the catalyst and the monomer; secondly, integration of the metal into the elongating chain transpires via an alkoxide bond. The efficacy of polymerization, as well as its transesterification impact, is subject to various factors such as temperature, duration, monomer-to-catalyst ratio, and catalyst type. Crucially, temperature and duration play pivotal roles in regulating chain length and impeding degradation processes (Rao et al., 2021).

11.9 APPLICATION OF PLA

PLA stands out from its aliphatic polyester rivals due to its remarkable properties such as biodegradability, strong mechanical strength, biocompatibility, and ease of manufacturing. As PLA becomes more widely used, its properties are being enhanced by methods such as copolymerization, blending changes, and heat resistance augmentation. As a result, PLA is becoming more well-known in a variety of industries. PLA, a thermoplastic and biodegradable aliphatic polyester, serves as a versatile material in food packaging. The ratio of LA's L (+) and D (−) optical isomers influences its properties. PLA is used as packaging film because of its high melting and crystallinity points, whereas 90%-10% D/L copolymers meet bulk packing requirements (Song et al., 2022). PLA stands out as a highly promising substitute for conventional plastics like polystyrene (PS), polyethylene terephthalate (PET), low-density polyethylene (LDPE), and high-density polyethylene (HDPE). Its superiority lies in several key attributes: a substantial molecular weight, exceptional processability, resilience against water solubility, and innate biodegradability. Furthermore, PLA exhibits comparable tensile strength modulus, texture, and odor barrier capabilities akin to polyethylene, PET, and flexible polyvinyl chloride (PVC). It also shares similar traits in terms of processability, temperature resilience, and resistance to grease (Taib et al., 2022). PLA, among the pioneering bio-based polymers embraced by the commercial sector, offers versatility in applications including injection molding for various goods, coatings, and film production. Particularly advantageous is its efficacy in preserving perishable items with high respiration rates, such as vegetables, fruits, and baked goods. However, despite its utilization in the production of juice and water bottles, PLA encounters limitations in commercial packaging due to drawbacks such as inadequate carbon dioxide barrier properties and insufficient resistance to creep (Parente et al., 2022).

The automobile industry has witnessed a notable surge in the utilization of natural-based materials. This trend is driven by the potential advantages it offers, such as enhanced fuel efficiency and decreased greenhouse gas emissions attributable to the lighter weight of vehicles. Currently, automobiles are responsible for 80% of environmental deterioration and 23% of worldwide carbon emissions. For every kilogram of weight lost, a vehicle's fuel efficiency can be increased by up to 7%, which translates to a 20 kg reduction in CO_2 emissions. Bio composite materials have two benefits: they reduce total vehicle weight and increase fuel economy. They also reduce carbon dioxide emissions, which is in line with environmental goals (Andrew and Dhakal, 2022). Similarly, nanocomposite materials show promise in reducing vehicle body weight by 40–55%. To meet the demands of the imminent surge in electric vehicles, it is imperative to reduce the weight of automotive components, thereby optimizing the weight-to-battery capacity ratio. Among the available options, bioplastics like PLA and its variants stand out as viable choices for automotive use. These materials boast remarkable mechanical and physical attributes alongside their environmentally friendly traits such as biodegradability and recyclability. Concerns regarding heat stability and stiffness can be mitigated effectively by incorporating suitable modifiers and additives into the bioplastic formulations (Mehmood et al., 2023).

PLA stands as a groundbreaking, eco-friendly fiber, finding its applications across diverse industries such as electronics, textiles, and medicine. PLA fibers behave similarly to conventional nylon and polyester fibers since they come from renewable resources. In the medical field, PLA bioplastics are popular because of unique characteristics including spontaneous hydrolysis, which makes it simple to remove implanted equipment without the need for additional treatments. This reduces the risk of immunological reactions and lowers healthcare costs (Trivedi et al., 2023). PLA-based nanocomposites have emerged as the material of choice in various biomedical fields including tissue engineering, drug delivery, and synthetic bone replacement. Meanwhile, Poly (lactic-glycolic acid) (PLGA) has garnered significant attention within the medical community for its utilization in dental, orthopedic, and craniofacial surgeries, where it serves as a fundamental component in the production of polymer-based fixtures like screws, plates, and pins. A great deal of study has been focused on using PLA and PLGA to create porous scaffolds and aid in bone mending. PLA materials' excellent breakdown qualities reduce surgical risks and patient suffering by eliminating the need for additional surgeries and implant extraction (Ranakoti et al., 2023). Furthermore, PLA helps to avoid stress blocks and lower operating dangers. Nevertheless, pure PLA materials have poorer regeneration potential and degradation behavior, and they lack the bone-bonding efficacy of PLA-based nanocomposites. Tissue engineering is a rapidly developing subject in medical therapy, with scaffolds playing critical roles in bone regeneration, brain repair, and vascularization. Because of their remarkable biocompatibility and ability to regenerate and degrade, PLA materials are widely used as scaffolds (DeStefano et al., 2020). Plasticulture, a technology developed in the 1950s, has been used in agriculture for a variety of objectives, including greenhouse passage protection and soil preservation. Although conventional plastics were formerly the norm, sustainable substitutes such as biodegradable polymers like PLA, have been developed. However, pure PLA is inadequate due to its weak mechanical and thermal characteristics; therefore it must be taken into account. PLA-based mulch films are commonly produced by blending PLA with biodegradable polyesters like LA derivatives or oligomers. The subject of plasticulture is still evolving, necessitating breakthroughs in cost-effective polymers and the further integration of bioplastics into agricultural operations (Andrew and Dhakal, 2022).

11.10 CONCLUSION

The microbial manufacture of biodegradable polymers, notably polylactic acid (PLA), is a potential route in the search for environmentally friendly materials. PLA has enormous potential as an environmentally friendly alternative to conventional plastics manufactured from fossil fuels. This process eliminates dependency on nonrenewable resources while minimizing the environmental impact of traditional plastic production methods, so helping to the transition to a circular economy. PLA stands out as a crucial biomaterial due to its inherent biodegradability, offering immense versatility across various industries. Its applications span from packaging and biomedical engineering to agriculture and textiles, promising sustainable solutions while minimizing environmental impact. Its capacity to decompose into non-toxic molecules under the right conditions reduces the long-term environmental impact of plastic waste accumulation. Breakthroughs in bioprocess engineering, strain enhancement, and fermentation techniques have enabled the large-scale, cost-effective production of PLA. This progress has opened avenues for integrating PLA-derived materials into everyday consumer goods, fueling the demand for eco-friendly plastic substitutes in the market. On the other hand, there are still a number of opportunities and problems in the field of PLA microbial synthesis. More research is needed to improve microbial strain productivity, substrate utilization efficiency, and PLA yield, in order to assure economic viability and competition with petrochemical-based polymers. Optimization of fermentation operations, downstream purification procedures, and polymer processing technologies will be crucial in overcoming technological constraints and satisfying quality criteria for PLA-based products. Investigating new feedstock sources, such as lignocellulosic biomass and agricultural leftovers, could broaden the raw materials utilized in PLA synthesis while reducing competition from food crops. The development of novel ideas and the quick adoption of microbial PLA production technologies will require interdisciplinary cooperation between microbiologists, bioprocess engineers, material scientists, and industry stakeholders. To sum up, the production of PLA through microbial synthesis constitutes a paradigm shift in the direction of environmentally friendly products. Through the utilization of microbial fermentation processes, scientists have discovered a scalable and sustainable method for generating PLA using sustainable resources.

KEYWORDS

- **biodegradable polymers**
- **fermentation**
- **industrial microbiology**
- **lactic acid bacteria**
- **microbial production**
- **polylactic acid**
- **sustainable plastics**

REFERENCES

Abdel-Rahman, M. A., Tashiro, Y., & Sonomoto, K. (2011). Lactic acid production from lignocellulose-derived sugars using lactic acid bacteria: Overview and limits. *Journal of Biotechnology, 156*, 286–301.

Abdel-Rahman, M. A., Tashiro, Y., & Sonomoto, K. (2013). Recent advances in lactic acid production by microbial fermentation processes. *Biotechnology Advances, 31*(6), 877–902.

Abedi, E., & Hashemi, S. M. B. (2020). Lactic acid production – producing microorganisms and substrates sources-state of art. *Heliyon, 6*(10), e04974.

Adsul, M. G., Varma, A. J., & Gokhale, D. V. (2007). Lactic acid production from waste sugarcane bagasse derived cellulose. *Green Chemistry, 9*, 58–62.

Akao, S., Tsuno, H., & Cheon, J. (2007). Semi-continuous L-lactate fermentation of garbage without sterile condition and analysis of the microbial structure. *Water Research, 41*, 1774–1780.

Albuquerque, T. L. D., Júnior, J. E. M., Queiroz, L. P. D., Ricardo, A. D. S., & Rocha, M. V. P. (2021). Polylactic acid production from biotechnological routes: A review. *International Journal of Biological Macromolecules, 186*, 933–951.

Alcântara, J. M. C., Distante, F., Storti, G., Moscatelli, D., Morbidelli, M., & Sponchioni, M. (2020). Current trends in the production of biodegradable bioplastics: The case of polyhydroxyalkanoates. *Biotechnology Advances, 42*, 107582.

Alexandri, M., Blanco-Catalá, J., Schneider, R., Turon, X., & Venus, J. (2020). High L(+)-lactic acid productivity in continuous fermentations using bakery waste and lucerne green juice as renewable substrates. *Bioresource Technology, 316*, 123949.

Alexandri, M., Schneider, R., & Venus, J. (2018). Membrane technologies for lactic acid separation from fermentation broths derived from renewable resources. *Membranes (Basel), 8*.

Alves De, O. R., Alexandri, M., Komesu, A., Venus, J., Vaz Rossell, C. E., & Maciel, F. R. (2020). Current advances in separation and purification of second-generation lactic acid. *Separation and Purification Reviews, 49*, 159–175.

Anagnostopoulou, C., Kontogiannopoulos, K. N., Gaspari, M., Morlino, M. S., Assimopoulou, A. N., & Kougias, P. G. (2022). Valorization of household food wastes to lactic acid production: A response surface methodology approach to optimize fermentation process. *Chemosphere, 296*, 133871.

Andrew, J. J., & Dhakal, H. N. (2022). Sustainable biobased composites for advanced applications: Recent trends and future opportunities – a critical review. *Composites Part C: Open Access, 7*, 100220.

Arshad, H., Irfan, M., Shakir, H. A., Khan, M. T., Ali, S., Saeed, S., & Franco, M. (2022). Microbial production of lactic acid – A review. *Current Biotechnology, 11*(2), 107–116.

Aso, Y., Hashimoto, A., & Ohara, H. (2019). Engineering *Lactococcus lactis* for D-lactic acid production from starch. *Current Microbiology, 76*, 1186–1192.

Bae, J. H., Kim, H. J., Kim, M. J., Sung, B. H., Jeon, J. H., Kim, H. S., Jin, Y. S., Kweon, D. H., & Sohn, J. H. (2018). Direct fermentation of Jerusalem artichoke tuber powder for production of L-lactic acid and D-lactic acid by metabolically engineered *Kluyveromyces marxianus*. *Journal of Biotechnology, 266*, 27–33.

Bai, D. M., Li, S. Z., Liu, Z. L., & Cui, Z. F. (2008). Enhanced L(+)-lactic acid production by an adapted strain of *Rhizopus oryzae* using corncob hydrolysate. *Applied Biochemistry and Biotechnology, 144*, 79–85.

Balla, E., Daniilidis, V., Karlioti, G., Kalamas, T., Stefanidou, M., Bikiaris, N. D., Vlachopoulos, A., Koumentakou, I., & Bikiaris, D. N. (2021). Poly(lactic acid): A versatile biobased polymer for the future with multifunctional properties – From monomer synthesis, polymerization techniques and molecular weight increase to PLA applications. *Polymers, 13*(11), 1822.

Beitel, S. M., Coelho, L. F., & Contiero, J. (2020). Efficient conversion of agroindustrial waste into D(-) lactic acid by *Lactobacillus delbrueckii* using fed-batch fermentation. *BioMed Research International, 2020*, 4194052.

Bianchi, M., Brambilla, L., Protani, F., Liu, C. L., Lievense, J., & Porro, D. (2001). Efficient homolactic fermentation by *Kluyveromyces lactis* strains defective in pyruvate utilization and transformed with the heterologous *LDH* gene. *Applied and Environmental Microbiology, 67*, 5621–5625.

Budhavaram, N. K., & Zhang, F. (2009). Production of lactic acid from paper sludge using acid-tolerant, thermophilic *Bacillus coagulans* strains. *Bioresource Technology, 100*(23), 5966–5972.

Campos, J., Tejada, L. G., Bao, J., & Lidén, G. (2023). Fed-batch strategies for biodetoxification in production of optically pure lactic acid from softwood hydrolysate using *Pediococcus acidilactici*. *Process Biochemistry, 125*, 162–170.

Castro-Aguirre, E., Iñiguez-Franco, F., Samsudin, H., Fang, X., & Auras, R. (2016). Poly(lactic acid) – Mass production, processing, industrial applications, and end of life. *Advanced Drug Delivery Reviews, 107*, 333–366.

Chai, X., Shang, X., Zhang, Y., Liu, S., Liang, Y., Zhang, Y., & Wen, T. (2016). A novel pyruvate kinase and its application in lactic acid production under oxygen deprivation in *Corynebacterium glutamicum*. *BMC Biotechnology, 16*, 79.

Chandane, V. S., Rathod, A. P., & Wasewar, K. L. (2020). Pervaporation reactor for enhanced esterification of lactic acid and isobutyl alcohol. *Chemical Engineering & Technology, 43*, 282–288.

Chen, G. Q., Eschbach, F. I. I., Weeks, M., Gras, S. L., & Kentish, S. E. (2016). Removal of lactic acid from acid whey using electrodialysis. *Separation and Purification Technology, 158*, 230–237.

Cheng, G., Wang, Q., Sun, X., Meng, H., & Li, J. (2008). Experimental study on concentration of ammonium lactate solution from kitchen garbage fermentation broth by two-compartment electrodialysis. *Separation and Purification Technology, 62*, 205–211.

Cheng, K. K., Zeng, J., Jian, J. H., Zhu, J. F., Zhang, G. X., & Liu, D. H. (2019). Model-based temperature control for improving lactic acid production from glycerol. *RSC Advances, 9*(21), 11614–11620.

Cheng, Y., Deng, S., Chen, P., & Ruan, R. (2009). Polylactic acid (PLA) synthesis and modifications: A review. *Frontiers of Chemistry in China, 4*(3), 259–264.

Colombie, S., Dequin, S., & Sablayrolles, J. (2003). Control of lactate production by *Saccharomyces cerevisiae* expressing a bacterial *LDH* gene. *Enzyme and Microbial Technology, 33*, 38–46.

Dantas, T. N. C., Cabral, T. J. O., Neto, A. A. D., & Moura, M. C. P. A. (2020). Enrichment of patchoulol extracted from patchouli (*Pogostemon cablin*) oil by molecular distillation

using response surface and artificial neural network models. *Journal of Industrial and Engineering Chemistry, 81*, 219–227.
Dave, K. K., & Punekar, N. S. (2015). Expression of lactate dehydrogenase in *Aspergillus Niger* for L-lactic acid production. *PLoS One, 10*, e0145459.
de França, J. O. C., da Silva, V. D., Paiva, M. F., Dias, S. C. L., & Dias, J. A. (2022). Polymers based on PLA from synthesis using D,L-lactic acid (or racemic lactide) and some biomedical applications: A short review. *Polymers, 14*(12), 2317.
de Oliveira, P. M., Santos, L. P., Coelho, L. F., Avila, N. P. M., Sass, D. C., & Contiero, J. (2021). Production of L(+) lactic acid by *Lactobacillus casei* Ke11: Fed-batch fermentation strategies. *Fermentation, 7*, 151.
Delgado, J. A., Agueda, V. I., Uguina, M. A., García, A., Matarredona, J., & Moral, R. (2018). Modeling of the separation of lactic acid from an aqueous mixture by adsorption on polyvinylpyridine resin and desorption with methanol. *Separation and Purification Technology, 200*, 307–317.
Demichelis, F., Pleissner, D., Fiore, S., Mariano, S., Navarro, G. I. M., Schneider, R., & Venus, J. (2017). Investigation of food waste valorization through sequential lactic acid fermentative production and anaerobic digestion of fermentation residues. *Bioresource Technology, 241*, 508–516.
Desai, H., Mehta, T., & Shah, N. (2023). Azeotropic dehydrative (solution) polycondensation of lactic acid to polylactic acid (PLA): An in-depth review of an overlooked method for manufacturing PLA. *Polymer-Plastics Technology and Materials, 62*(11), 1394–1402.
DeStefano, V. F., Khan, S. A., & Tabada, A. (2020). Applications of PLA in modern medicine. *Engineering Regeneration, 1*, 76–87.
Elsawy, M. A., Kim, K., Park, J., & Deep, A. (2017). Hydrolytic degradation of polylactic acid (PLA) and its composites. *Renewable and Sustainable Energy Reviews, 79*, 1346–1352.
Farah, S., Anderson, D. G., & Langer, R. (2016). Physical and mechanical properties of PLA, and their functions in widespread applications—A comprehensive review. *Advanced Drug Delivery Reviews, 107*, 367–392.
Garcia, B. E., Rodriguez, E., Salazar, Y., Valle, P. A., Flores-Gallegos, A. C., Rutiaga-Quiñones, M., & Rodriguez-Herrera, R. (2021). Primary model for biomass growth prediction in batch fermentation. *Symmetry, 13*, 1468.
Ge, X. Y., Qian, H., & Zhang, W. G. (2009). Improvement of L-lactic acid production from Jerusalem artichoke tubers by mixed culture of *Aspergillus Niger* and *Lactobacillus* sp. *Bioresource Technology, 100*, 1872–1874.
Glaser, R., & Venus, J. (2018). Co-fermentation of the main sugar types from a beechwood organosolv hydrolysate by several strains of *Bacillus coagulans* results in effective lactic acid production. *Biotechnology Reports, 18*, e00245.
Guo, Y., Yan, Q., Jiang, Z., Teng, C., & Wang, X. (2010). Efficient production of lactic acid from sucrose and corncob hydrolysate by a newly isolated *Rhizopus oryzae* GY18. *Journal of Industrial Microbiology & Biotechnology, 37*, 1137–1143.
Hause, B., Rajgarhia, V., & Suominen, P. (2009). Methods and materials for the production of L-lactic acid in yeast. *Google Patents*.
Hirayama, S., & Ueda, R. (2004). Production of optically pure D-lactic acid by *Nannochlorum* sp. 26A4. *Applied Biochemistry and Biotechnology, 119*, 71–77.

Hong, A. A., Cheng, K. K., Peng, F., Zhou, S., Sun, Y., Liu, C. M., & Liu, D. H. (2009). Strain isolation and optimization of process parameters for bioconversion of glycerol to lactic acid. *Journal of Chemical Technology & Biotechnology, 84*, 1576–1581.

Ikeda, Y., Park, E. Y., & Okuda, N. (2006). Bioconversion of waste office paper to gluconic acid in a turbine blade reactor by the filamentous fungus *Aspergillus Niger*. *Bioresource Technology, 97*(8), 1030–1035.

Ikushima, S., Fujii, T., Kobayashi, O., Yoshida, S., & Yoshida, A. (2009). Genetic engineering of *Candida utilis* yeast for efficient production of L-lactic acid. *Bioscience, Biotechnology, and Biochemistry, 73*(8), 1818–1824.

Ilmén, M., Koivuranta, K., Ruohonen, L., Suominen, P., & Penttilä, M. (2007). Efficient production of L-lactic acid from xylose by *Pichia stipitis*. *Applied and Environmental Microbiology, 73*(1), 117–123.

Ishida, N., Saitoh, S., Tokuhiro, K., Nagamori, E., Matsuyama, T., Kitamoto, K., & Takahashi, H. (2005). Efficient production of L-lactic acid by metabolically engineered *Saccharomyces cerevisiae* with a genome-integrated L-lactate dehydrogenase gene. *Applied and Environmental Microbiology, 71*, 1964–1970.

Jiang, S., Xu, P., & Tao, F. (2019). L-lactic acid production by *Bacillus coagulans* through simultaneous saccharification and fermentation of lignocellulosic corncob residue. *Bioresource Technology Reports, 6*, 131–137.

Jiang, T., Zhang, C., He, Q., Zheng, Z., & Ouyang, J. (2018). Metabolic engineering of *Escherichia coli* K12 for homofermentative production of L-lactate from xylose. *Applied Biochemistry and Biotechnology, 184*, 703–715.

John, R. P., Anisha, G. S., Nampoothiri, K. M., & Pandey, A. (2009). Direct lactic acid fermentation: Focus on simultaneous saccharification and lactic acid production. *Biotechnology Advances, 27*, 145–152.

John, R. P., Nampoothiri, K. M., & Pandey, A. (2006). Simultaneous saccharification and fermentation of cassava bagasse for L(+)-lactic acid production using lactobacilli. *Applied Biochemistry and Biotechnology, 134*, 263–272.

Kawaguchi, H., Sasaki, M., Vertes, A. A., Inui, M., & Yukawa, H. (2008). Engineering of an L-arabinose metabolic pathway in *Corynebacterium glutamicum*. *Applied Microbiology and Biotechnology, 77*, 1053–1062.

Kawaguchi, H., Vertes, A. A., Okino, S., Inui, M., & Yukawa, H. (2006). Engineering of a xylose metabolic pathway in *Corynebacterium glutamicum*. *Applied and Environmental Microbiology, 72*, 3418–3428.

Klongklaew, A., Unban, K., Kalaimurugan, D., Kanpiengjai, A., Azaizeh, H., Schroedter, L., Schneider, R., Venus, J., & Khanongnuch, C. (2023). Bioconversion of dilute acid pretreated corn stover to L-lactic acid using co-culture of furfural tolerant *Enterococcus mundtii* WX1 and *Lactobacillus rhamnosus* SCJ9. *Fermentation, 9*, 112.

Koivuranta, K. T., Ilmén, M., Wiebe, M. G., Ruohonen, L., Suominen, P., & Penttilä, M. (2014). L-lactic acid production from D-xylose with *Candida sonorensis* expressing a heterologous lactate dehydrogenase encoding gene. *Microbial Cell Factories, 13*, 107.

Komesu, A., Maciel, R. M., de Oliveira, R., Helena, L., & Filho, R. M. (2017). Purification of lactic acid produced by fermentation: Focus on non-traditional distillation processes. *Separation and Purification Reviews, 46*, 241–254.

Komesu, A., Martinez, P. F. M., Lunelli, B. H., Filho, R. M., & Maciel, M. R. W. (2015). Lactic acid purification by reactive distillation system using design of experiments. *Chemical Engineering and Processing - Process Intensification, 95*, 26–30.

Kong, X., Zhang, B., Hua, Y., Zhu, Y., Li, W., Wang, D., & Hong, J. (2019). Efficient L-lactic acid production from corncob residue using metabolically engineered thermo-tolerant yeast. *Bioresource Technology, 273*, 220–230.

Korhonen, H., Helminen, A., & Seppälä, J. V. (2001). Synthesis of polylactides in the presence of coinitiators with different numbers of hydroxyl groups. *Polymer, 42*, 7541–7549.

Kotzamanidis, C., Roukas, T., & Skaracis, G. N. (2002). *World Journal of Microbiology and Biotechnology, 18*(5), 441–448.

Kumar, A., Thakur, A., & Panesar, P. S. (2018). Lactic acid extraction using environmentally benign green emulsion ionic liquid membrane. *Journal of Cleaner Production, 181*, 574–583.

Kumar, A., Thakur, A., & Panesar, P. S. (2019a). Recent developments on sustainable solvents for emulsion liquid membrane processes. *Journal of Cleaner Production, 240*, 118250.

Kumar, A., Thakur, A., & Panesar, P. S. (2019b). A review on emulsion liquid membrane (ELM) for the treatment of various industrial effluent streams. *Reviews in Environmental Science and Biotechnology, 18*, 153–182.

Laud, N., Rosso, M. N., Fabre, N., Crapart, S., Herpoël-Gimbert, I., Sigoillot, J. C., Raouche, S., & Levasseur, A. (2015). L-lactic acid production by *Aspergillus brasiliensis* overexpressing the heterologous *ldhA* gene from *Rhizopus oryzae*. *Microbial Cell Factories, 14*, 66.

Li, J., Shi, S., Wang, Y., & Jiang, Z. (2021). Integrated production of optically pure L-lactic acid from paper mill sludge by simultaneous saccharification and co-fermentation (SSCF). *Waste Management, 129*, 35–46.

Li, Y., Ruan, R., Chen, P. L., Liu, Z., Pan, X., Lin, X., Liu, Y., Mok, C., & Yang, T. (2004). Enzymatic hydrolysis of corn stover pretreated by combined dilute alkaline treatment and homogenization. *Transactions of the ASAE, 47*, 821.

Liu, T., Miura, S., Yaguchi, M., Arimura, T., Park, E. Y., & Okabe, M. (2006a). Scale-up of L-lactic acid production by mutant strain *Rhizopus* sp. MK-96-1196 from 0.003 m^3 to 5 m^3 in airlift bioreactors. *Journal of Bioscience and Bioengineering, 101*, 9–12.

Liu, Y., Liao, W., Liu, C., & Chen, S. (2006b). Optimization of L(+)-lactic acid production using pelletized filamentous *Rhizopus oryzae* NRRL 395. In *Twenty-Seventh Symposium on Biotechnology for Fuels and Chemicals* (pp. 844–853). Springer.

Lobeda, K., Jin, Q., Wu, J., Zhang, W., & Huang, H. (2022). Lactic acid production from food waste hydrolysate by *Lactobacillus pentosus*: Focus on nitrogen supplementation, initial sugar concentration, pH, and fed-batch fermentation. *Journal of Food Science, 87*, 3071–3083.

López-Gómez, J. P., Alexandri, M., Schneider, R., & Venus, J. (2019). A review on the current developments in continuous lactic acid fermentations and case studies utilizing inexpensive raw materials. *Process Biochemistry, 79*, 1–10.

Lu, J., Peng, W., Lv, Y., Jiang, Y., Xu, B., Zhang, W., Zhou, J., Dong, W., Xin, F., & Jiang, M. (2020). Application of cell immobilization technology in microbial cocultivation systems for biochemicals production. *Industrial & Engineering Chemistry Research, 59*, 17026–17034.

Ma, K., Cui, Y., Zhao, K., Yang, Y., Wang, Y., Hu, G., & He, M. (2022). D-lactic acid production from agricultural residues by membrane integrated continuous fermentation coupled with B vitamin supplementation. *Biotechnology for Biofuels and Bioproducts, 15*, 1–14.

Maneeboon, T., Vanichsriratana, W., Pomchaitaward, C., & Kitpreechavanich, V. (2010). Optimization of lactic acid production by pellet-form *Rhizopus oryzae* in 3-L airlift bioreactor using response surface methodology. *Applied Biochemistry and Biotechnology, 161*, 137–146.

Mazumdar, S., Blankschien, M. D., Clomburg, J. M., & Gonzalez, R. (2013). Efficient synthesis of L-lactic acid from glycerol by metabolically engineered *Escherichia coli. Microbial Cell Factories, 12*, 7.

Mazumdar, S., Clomburg, J. M., & Gonzalez, R. (2010). *Escherichia coli* strains engineered for homofermentative production of D-lactic acid from glycerol. *Applied and Environmental Microbiology, 76*, 4327–4336.

Mehmood, A., Raina, N., Phakeenuya, V., Wonganu, B., & Cheenkachorn, K. (2023). The current status and market trend of polylactic acid as biopolymer: Awareness and needs for sustainable development. *Materials Today: Proceedings, 72*, 3049–3055.

Meussen, B. J., de Graaff, L. H., Sanders, J. P., & Weusthuis, R. A. (2012). Metabolic engineering of *Rhizopus oryzae* for the production of platform chemicals. *Applied Microbiology and Biotechnology, 94*, 875–886.

Min-Tian, G., Koide, M., Gotou, R., Takanashi, H., Hirata, M., & Hano, T. (2005). Development of a continuous electrodialysis fermentation system for production of lactic acid by *Lactobacillus rhamnosus*. *Process Biochemistry, 40*, 1033–1036.

Narancic, T., & O'Connor, K. E. (2019). Plastic waste as a global challenge: Are biodegradable plastics the answer to the plastic waste problem? *Microbiology, 165*, 129–137.

Nduko, J. M., & Taguchi, S. (2019). Microbial production and properties of LA-based polymers and oligomers from renewable feedstock. In Z. Fang, R. Smith, & X. Tian (Eds.), *Production of materials from sustainable biomass resources* (Vol. 9). Singapore: Springer.

Okino, S., Inui, M., & Yukawa, H. (2005). Production of organic acids by *Corynebacterium glutamicum* under oxygen deprivation. *Applied Microbiology and Biotechnology, 68*, 475–480.

Osawa, F., Fujii, T., Nishida, T., Tada, N., Ohnishi, T., Kobayashi, O., Komeda, T., & Yoshida, S. (2009). Efficient production of L-lactic acid by Crabtree-negative yeast *Candida boidinii. Yeast, 26*(9), 485–496.

Othman, M., Ariff, A. B., Kapri, M. R., Rios-Solis, L., & Halim, M. (2018). Growth enhancement of probiotic *Pediococcus acidilactici* by extractive fermentation of lactic acid exploiting anion-exchange resin. *Frontiers in Microbiology, 9*, 1–11.

Ou, M. S., Awasthi, D., Nieves, I., Wang, L., Erickson, J., Vermerris, W., Ingram, L. O., & Shanmugam, K. T. (2016). Sweet sorghum juice and bagasse as feedstocks for the production of optically pure lactic acid by native and engineered *Bacillus coagulans* strains. *BioEnergy Research, 9*, 123–131.

Ou, M. S., Ingram, L. O., & Shanmugam, K. (2011). L(+)-Lactic acid production from non-food carbohydrates by thermotolerant *Bacillus coagulans*. *Journal of Industrial Microbiology and Biotechnology, 38*, 599–605.

Ozaki, A., Konishi, R., Otomo, C., Kishid, M., Takayama, S., Matsumoto, T., Tanaka, T., & Kondo, A. (2017). Metabolic engineering of *Schizosaccharomyces pombe* via CRISPR-Cas9 genome editing for lactic acid production from glucose and cellobiose. *Metabolic Engineering Communications, 5*, 60–67.

Parente, A. G., Cabrera, M. P., & Neri, D. F. (2022). Bio-based polymer films with potential for packaging applications: A systematic review of the main types tested on food. *Polymer Bulletin, 80*(5), 4689–4717.

Pau, S., Tan, L. C., Arriaga, S., & Lens, P. (2022). Lactic acid fermentation of food waste at acidic conditions in a semicontinuous system: Effect of HRT and OLR changes. *Biomass Conversion and Biorefinery, 14*(10), 10979–10994.

Payot, T., Chemaly, Z., & Fick, M. (1999). Lactic acid production by *Bacillus coagulans*—Kinetic studies and optimization of culture medium for batch and continuous fermentations. *Enzyme and Microbial Technology, 24*(3–4), 191–199.

Pecota, D. C., Rajgarhia, V., & Da Silva, N. A. (2007). Sequential gene integration for the engineering of *Kluyveromyces marxianus*. *Journal of Biotechnology, 127*, 408–416.

Peinemann, J. C., Demichelis, F., Fiore, S., & Pleissner, D. (2019). Techno-economic assessment of non-sterile batch and continuous production of lactic acid from food waste. *Bioresource Technology, 289*, 121631.

Pejin, J., Radosavljević, M., Kočić-Tanackov, S., Mladenović, D., Djukic-Vukovic, A., & Mojovic, L. (2017). Fed-batch L-(+)-lactic acid fermentation of brewer's spent grain hydrolysate. *Journal of the Institute of Brewing, 123*, 537–543.

Pérez, A. D., & Fontalvo, J. (2019). A new concept of liquid membranes in Taylor flow: Performance for lactic acid removal. *Chemical Engineering and Processing: Process Intensification, 139*, 95–102.

Pradhan, N., Rene, E. R., Lens, P. N. L., Dipasquale, L., D'Ippolito, G., Fontana, A., Panico, A., & Esposito, G. (2017). Adsorption behavior of lactic acid on granular activated carbon and anionic resins: Thermodynamics, isotherms and kinetic studies. *Energies, 10*, 1–16.

Qiu, Z., Gao, Q., & Bao, J. (2018). Engineering *Pediococcus acidilactici* with xylose assimilation pathway for high titer cellulosic L-lactic acid fermentation. *Bioresource Technology, 249*, 9–15.

Radosavljević, M., Lević, S., Belović, M., Pejin, J., Djukic-Vukovic, A., Mojovic, L., & Nedović, V. (2021). Encapsulation of *Lactobacillus rhamnosus* in polyvinyl alcohol for the production of L-(+)-lactic acid. *Process Biochemistry, 100*, 149–160.

Ranakoti, L., Gangil, B., Bhandari, P., Singh, T., Sharma, S., Singh, J., & Singh, S. (2023). Promising role of polylactic acid as an ingenious biomaterial in scaffolds, drug delivery, tissue engineering, and medical implants: Research developments and prospective applications. *Molecules, 28*(2), 485.

Rao, W., Cai, C., Tang, J., Wei, Y., Gao, C., Yu, L., & Ding, J. (2021). Coordination insertion mechanism of ring-opening polymerization of lactide catalyzed by stannous octoate. *Chinese Journal of Chemistry, 39*(7), 1965–1974.

Raza, Z. A., Abid, S., & Banat, I. M. (2018). Polyhydroxyalkanoates: Characteristics, production, recent developments and applications. *International Biodeterioration & Biodegradation, 126*, 45–56.

Reddy, G., Altaf, M., Naveena, B., Venkateshwar, M., & Kumar, E. V. (2008). Amylolytic bacterial lactic acid fermentation—A review. *Biotechnology Advances, 26*, 22–34.

Ricci, A., Diaz, A. B., Caro, I., Bernini, V., Galaverna, G., Lazzi, C., & Blandino, A. (2019). Orange peels: From by-product to resource through lactic acid fermentation. *Journal of the Science of Food and Agriculture, 99*, 6761–6767.

Sahoo, T. K., & Jayaraman, G. (2019). Co-culture of *Lactobacillus delbrueckii* and engineered *Lactococcus lactis* enhances stoichiometric yield of D-lactic acid from whey permeate. *Applied Microbiology and Biotechnology, 103*, 5653–5662.

Shao, J., Liu, Y. L., Xiang, S., Bian, X. C., Sun, J. R., Li, G., Chen, X. S., & Hou, H. Q. (2015). The stereocomplex formation and phase separation of PLLA/PDLA blends with different optical purities and molecular weights. *Chinese Journal of Polymer Science, 33*, 1713–1720.

Shekhar, N., & Mondal, A. (2024). Synthesis, properties, environmental degradation, processing, and applications of polylactic acid (PLA): An overview. *Polymer Bulletin*. https://doi.org/10.1007/s00289-023-05020-w (accessed on 15th November 2025).

Shi, H., Ni, J., Zheng, T., Wang, X., Wu, C., & Wang, Q. (2020). Remediation of wastewater contaminated by antibiotics: A review. *Environmental Chemistry Letters, 18*, 345–360.

Singh, S. K., Ahmed, S. U., & Pandey, A. (2006). Metabolic engineering approaches for lactic acid production. *Process Biochemistry, 41*, 991–1000.

Singhvi, M., Zendo, T., & Sonomoto, K. (2018). Free lactic acid production under acidic conditions by lactic acid bacteria strains: Challenges and future prospects. *Applied Microbiology and Biotechnology, 102*(14), 5911–5924.

Södergård, A., & Stolt, M. (2002). Properties of lactic acid-based polymers and their correlation with composition. *Progress in Polymer Science, 27*, 1123–1163.

Song, L., Yang, D., Liu, R., Liu, S., Dai, L., & Dai, X. (2022). Microbial production of lactic acid from food waste: Latest advances, limits, and perspectives. *Bioresource Technology, 345*, 126052.

Song, Y., Matsumoto, K. I., Yamada, M., Gohda, A., Brigham, C. J., Sinskey, A. J., & Taguchi, S. (2012). Engineered *Corynebacterium glutamicum* as an endotoxin-free platform strain for lactate-based polyester production. *Applied Microbiology and Biotechnology, 93*, 1917–1925.

Sun, Y., Xu, Z., Zheng, Y., Zhou, J., & Xiu, Z. (2019). Efficient production of lactic acid from sugarcane molasses by a newly microbial consortium CEE-DL15. *Process Biochemistry, 81*, 132–138.

Swetha, T. K., Bora, A., Mohanrasu, K., Balaji, P., Raja, R., Ponnuchamy, K., Muthusamy, G., & Arun, A. P. (2023). A comprehensive review on polylactic acid (PLA) – Synthesis, processing and application in food packaging. *International Journal of Biological Macromolecules, 234*, 123715.

Taherzadeh, M. J., Fox, M. B., Hjorth, H., & Edebo, L. (2003). Production of mycelium biomass and ethanol from paper pulp sulfite liquor by *Rhizopus oryzae*. *Bioresource Technology, 88*(3), 167–177.

Taib, N. A. B., Rahman, M. R., Huda, D., Kuok, K. K., Hamdan, S., Bakri, M. K. B., Julaihi, M. R. M., & Khan, A. A. (2022). A review on poly(lactic acid) (PLA) as a biodegradable polymer. *Polymer Bulletin, 80*(2), 1179–1213.

Taleghani, H. G., Ghoreyshi, A. A., & Najafpour, G. D. (2018). Thin film composite nanofiltration membrane for lactic acid production in membrane bioreactor. *Biochemical Engineering Journal, 132*, 152–160.

Taleghani, H. G., Najafpour, G. D., & Ghoreyshi, A. A. (2016). A study on the effect of parameters on lactic acid production from whey. *Polish Journal of Chemical Technology, 18*(1), 58–63.

Talukder, M. M. R., Das, P., & Wu, J. C. (2012). Microalgae (*Nannochloropsis salina*) biomass to lactic acid and lipid. *Biochemical Engineering Journal, 68*, 109–113.

Tamakawa, H., Ikushima, S., & Yoshida, S. (2012). Efficient production of L-lactic acid from xylose by a recombinant *Candida utilis* strain. *Journal of Bioscience and Bioengineering, 113*, 73–75.

Tang, J., Wang, X. C., Hu, Y., Zhang, Y., & Li, Y. (2017). Effect of pH on lactic acid production from acidogenic fermentation of food waste with different types of inocula. *Bioresource Technology, 224*, 544–552.

Tarrahi, R., Fathi, Z., Seydibeyoglu, M. Ö., Doustkhah, E., & Khataee, A. (2020). Polyhydroxyalkanoates (PHA): From production to nanoarchitecture. *International Journal of Biological Macromolecules, 146*, 596–619.

Tian, X., Liu, X., Zhang, Y., Chen, Y., Hang, H., Chu, J., & Zhuang, Y. (2021). Metabolic engineering coupled with adaptive evolution strategies for the efficient production of high-quality L-lactic acid by *Lactobacillus paracasei*. *Bioresource Technology, 323*, 124549.

Trivedi, A. K., Gupta, M. L., & Singh, H. (2023). PLA-based biocomposites for sustainable products: A review. *Advances in Industrial and Engineering Polymer Research, 6*(4), 382–395.

van der Pol, E. C., Eggink, G., & Weusthuis, R. A. (2016). Production of L(+)-lactic acid from acid pretreated sugarcane bagasse using *Bacillus coagulans* DSM2314 in a simultaneous saccharification and fermentation strategy. *Biotechnology for Biofuels, 9*, 248.

van Maris, A. J., Winkler, A. A., Porro, D., van Dijken, J. P., & Pronk, J. T. (2004). Homofermentative lactate production cannot sustain anaerobic growth of engineered *Saccharomyces cerevisiae*: Possible consequence of energy-dependent lactate export. *Applied and Environmental Microbiology, 70*, 2898–2905.

Veeravalli, S. S., & Mathews, A. P. (2018). Continuous fermentation of xylose to short chain fatty acids by *Lactobacillus buchneri* under low pH conditions. *Chemical Engineering Journal, 337*, 764–771.

Wang, J., Huang, J., Guo, H., Jiang, S., Zhang, J., Ning, Y., Fang, M., & Liu, S. (2020). Optimization of immobilization conditions for *Lactobacillus pentosus* cells. *Bioprocess and Biosystems Engineering, 43*, 1071–1079.

Wang, J., Jiang, S., Huang, J., Guo, H., Bi, X., Hou, M., Chen, X., Hou, S., Lin, H., YuMing, L., Lv, H., Qiao, J., Yang, R., & Liu, S. (2021). Optimization of initial cation concentrations for L-lactic acid production from fructose by *Lactobacillus pentosus* cells. *Applied Biochemistry and Biotechnology, 193*, 1496–1512.

Wang, Q., Cheng, G., Sun, X., & Jin, B. (2006). Recovery of lactic acid from kitchen garbage fermentation broth by four-compartment configuration electrodialyzer. *Process Biochemistry, 41*, 152–158.

Wang, Y. Z., Li, K. P., Huang, F., Wang, J. H., Zhao, J. F., Zhao, X., Garza, E., Manow, R., Grayburn, S., & Zhou, S. D. (2013). Engineering and adaptive evolution of *Escherichia coli* W for L-lactic acid fermentation from molasses and corn steep liquor without additional nutrients. *Bioresource Technology, 148*, 394–400. https://doi.org/10.1016/j.biortech.2013.08.163.

Wang, Y., Tian, T., Zhao, J., Wang, J., Yan, T., Xu, L., Liu, Z., Garza, E., Iverson, A., Manow, R., Finan, C., & Zhou, S. (2012). Homofermentative production of D-lactic acid from sucrose by a metabolically engineered *Escherichia coli. Biotechnology Letters, 34*(11), 2069–2075. https://doi.org/10.1007/s10529-012-1007-0 (accessed on 15th November 2025).

Weusthuis, R. A., Mars, A. E., Springer, J., Wolbert, E. J. H., van der Wal, H., de Vrije, T. G., Levisson, M., Leprince, A., Houweling-Tan, G. B., Moers, A. P. H. A., Hendriks, S. N. A., Mendes, O., Griekspoor, Y., Werten, M. W. T., Schaap, P. J., van der Oost, J., & Eggink, G. (2017). *Monascus ruber* as cell factory for lactic acid production at low pH. *Metabolic Engineering, 42*, 66–73. https://doi.org/10.1016/j.ymben.2017.05.006 (accessed on 15th November 2025).

Yang, E., Fan, L., Yan, J., Jiang, Y., Doucette, C., Fillmore, S., & Walker, B. (2018). Influence of culture media, pH and temperature on growth and bacteriocin production of bacteriocinogenic lactic acid bacteria. *AMB Express, 8*(1), 10. https://doi.org/10.1186/s13568-017-0536-x (accessed on 15th November 2025).

Yu, J., Xu, S., Liu, B., Wang, H., Qiao, F., Ren, X., & Wei, Q. (2023). PLA bioplastic production: From monomer to the polymer. *European Polymer Journal, 193*, 112076. https://doi.org/10.1016/j.eurpolymj.2023.112076 (accessed on 15th November 2025).

Yu, J., Zeng, A., Yuan, X., Zhang, X., & Ju, J. (2015). Optimizing and scale-up strategy of molecular distillation for the purification of lactic acid from fermentation broth. *Separation Science and Technology, 50*, 2518–2524. https://doi.org/10.1080/01496395.2015.1058850 (accessed on 15th November 2025).

Yu, M. C., Wang, R. C., Wang, C. Y., Duan, K. J., & Sheu, D. C. (2007). Enhanced production of L(+)-lactic acid by floc-form culture of *Rhizopus oryzae. Journal of the Chinese Institute of Chemical Engineers, 38*, 223–228. https://doi.org/10.1016/j.jcice.2007.03.005 (accessed on 15th November 2025).

Yukawa, H., Omumasaba, C. A., Nonaka, H., Kos, P., Okai, N., Suzuki, N., Suda, M., Tsuge, Y., Watanabe, J., & Ikeda, Y. (2007). Comparative analysis of the *Corynebacterium glutamicum* group and complete genome sequence of strain R. *Microbiology, 153*, 1042–1058. https://doi.org/10.1099/mic.0.2006/003657-0 (accessed on 15th November 2025).

Zhang, B., He, P. J., Ye, N. F., & Shao, L. M. (2008). Enhanced isomer purity of lactic acid from the non-sterile fermentation of kitchen wastes. *Bioresource Technology, 99*(4), 855–862. https://doi.org/10.1016/j.biortech.2007.01.012 (accessed on 15th November 2025).

Zhang, B., Li, J., Liu, X., & Bao, J. (2022). Continuous simultaneous saccharification and co-fermentation (SSCF) for cellulosic L-lactic acid production. *Industrial Crops and*

Products, 187, 115527. https://doi.org/10.1016/j.indcrop.2022.115527 (accessed on 15th November 2025).

Zhang, M., Chen, L., Jiang, Z., & Ma, J. (2015). Effects of dehydration rate on the yield of ethyl lactate in a pervaporation-assisted esterification process. *Industrial & Engineering Chemistry Research, 54*, 6669–6676. https://doi.org/10.1021/acs.iecr.5b01140 (accessed on 15th November 2025).

Zhang, Y., & Vadlani, P. V. (2015). Lactic acid production from biomass-derived sugars via co-fermentation of *Lactobacillus brevis* and *Lactobacillus plantarum*. *Journal of Bioscience and Bioengineering, 119*, 694–699. https://doi.org/10.1016/j.jbiosc.2014.10.020 (accessed on 15th November 2025).

Zhang, Z. Y., Jin, B., & Kelly, J. M. (2007). Production of lactic acid from renewable materials by *Rhizopus* fungi. *Biochemical Engineering Journal, 35*, 251–263. https://doi.org/10.1016/j.bej.2007.01.024 (accessed on 15th November 2025).

Zhao, J., Xu, L., Wang, Y., Zhao, X., Wang, J., Garza, E., Manow, R., & Zhou, S. (2013). Homofermentative production of optically pure L-lactic acid from xylose by genetically engineered *Escherichia coli* B. *Microbial Cell Factories, 12*(1), Article 1. https://doi.org/10.1186/1475-2859-12-1 (accessed on 15th November 2025).

Zhao, W., Sun, X., Wang, Q., Ma, H., & Teng, Y. (2009). Lactic acid recovery from fermentation broth of kitchen garbage by esterification and hydrolysis method. *Biomass and Bioenergy, 33*, 21–25. https://doi.org/10.1016/j.biombioe.2008.04.005 (accessed on 15th November 2025).

Zheng, J., Wittouck, S., Salvetti, E., Franz, C. M. A. P., Harris, H. M. B., Mattarelli, P., O'Toole, P. W., Pot, B., Vandamme, P., Walter, J., Watanabe, K., Wuyts, S., Felis, G. E., Gänzle, M. G., & Lebeer, S. (2020). A taxonomic note on the genus *Lactobacillus*: Description of 23 novel genera, emended description of the genus *Lactobacillus* Beijerinck 1901, and union of *Lactobacillaceae* and *Leuconostocaceae*. *International Journal of Systematic and Evolutionary Microbiology, 70*, 2782–2858. https://doi.org/10.1099/ijsem.0.004107 (accessed on 15th November 2025).

Zhou, J., Ouyang, J., Xu, Q., & Zheng, Z. (2016). Cost-effective simultaneous saccharification and fermentation of L-lactic acid from bagasse sulfite pulp by *Bacillus coagulans* CC17. *Bioresource Technology, 222*, 431–438. https://doi.org/10.1016/j.biortech.2016.10.016 (accessed on 15th November 2025).

Zhou, S., Causey, T. B., Hasona, A., Shanmugam, K. T., & Ingram, L. O. (2003). Production of optically pure D-lactic acid in mineral salts medium by metabolically engineered *Escherichia coli* W3110. *Applied and Environmental Microbiology, 69*(1), 399–407. https://doi.org/10.1128/AEM.69.1.399-407.2003 (accessed on 15th November 2025).

CHAPTER 12

Microbial Butanol Production: A Sustainable Pathway to Biofuel Innovation

VISHWAJIT KUMAR,[1] MAHARUDRA PRATAP SINGH,[2] ANAND PRAKASH,[2] AKHILESH KUMAR SINGH,[2] and PIYUSH BEHARI LAL[1,*]

[1]*Department of Microbiology, Kasturba Medical College, Manipal Academy of Higher Education, Manipal, India*

[2]*Department of Biotechnology, School of Life Sciences, Mahatma Gandhi Central University, Motihari, Bihar, India*

*Corresponding author

ABSTRACT

Microbial production of butanol represents a promising and sustainable path for biofuel innovation, addressing the global need for alternative energy sources and reducing reliance on fossil fuels. With superior fuel properties and potential as a chemical feedstock, butanol offers higher energy content, lower vapor pressure, and a better octane rating compared to ethanol. Its production is achievable from various renewable sources, including food waste and lignocellulosic biomass. The primary microorganism used for butanol production is *Clostridium*, mainly through the acetone-butanol-ethanol (ABE) fermentation process. Utilizing lignocellulosic biomass waste as a renewable feedstock for butanol production holds great promise. Nevertheless, challenges exist in the pre-treatment and hydrolysis of hemicellulose and cellulose. Additionally, exploring

Industrial Microbiology: Current Progress and Novel Technologies.
Prakash Kumar Sarangi, Akhilesh Kumar Singh, Rajesh Kumar Srivastava, and Brijesh Pandey (Eds.)

DOI: 10.1201/9781779642219-12

new microbial strains and metabolic pathways is essential to enhance butanol production. This chapter highlights the importance of addressing current challenges and exploring new methods for microbial butanol production to establish it as a sustainable and economically viable biofuel pathway. Emphasizing advancements in genetic modification, metabolic engineering, and biomass utilization, the chapter advocates for ongoing research to refine and optimize microbial butanol production processes, ultimately contributing to a more sustainable and efficient biofuel landscape and reducing our dependence on fossil fuels.

12.1 INTRODUCTION

The rising global energy demand and the environmental impact of fossil fuel consumption have heightened the pursuit of sustainable energy sources. Biofuels, produced from renewable materials, offer a promising category of clean fuels capable of reducing the ecological footprint of traditional energy sources. With applications in transportation, power generation, and heating, biofuels present viable alternatives. However, the current literature lacks a thorough analysis of their potential to replace conventional fuels in power generation systems. This gap is addressed by evaluating biofuels in microgrid applications to achieve carbon neutrality. Overcoming technical and economic challenges could position biofuels as a key player in global power generation, aiding in the achievement of Net Zero Emissions targets by 2050 (Kiehbadroudinezhad et al., 2023). Butanol, in particular, stands out due to its higher energy content compared to ethanol, making it a more efficient fuel. Its reduced corrosiveness decreases engine and fuel system wear, and its lower vapor pressure lessens the risk of vapor lock, improving engine performance (Obergruber et al., 2021a). Butanol's compatibility with gasoline at higher concentrations requires minimal infrastructure modifications. Additionally, its lower water absorption compared to ethanol reduces the risk of phase separation, enhancing fuel stability. Ongoing research, especially into acetone-butanol-ethanol (ABE) fermentation, shows promise for increasing butanol production, solidifying its potential as an exceptional biofuel. Butanol's benefits include higher energy content, reduced corrosiveness, lower vapor pressure, easy integration into existing infrastructure, and decreased water absorption (Kolesinska et al., 2019a). The European Union's directive to derive 10%

of all fuels from bio-based sources by 2020 has driven the search for alternative biofuels, with biobutanol emerging as a strong candidate. Despite the inherent biobutanol production capabilities of native *Clostridial* strains, their industrial use presents challenges (Figure 12.1) (Bankar et al., 2013). To address these issues, expressing more favorable pathways in robust microorganisms like *Escherichia coli* has been suggested as a viable strategy. *In silico*-driven design approaches for *E. coli* strains enable the identification of new pathways for butanol production. Enumeration algorithms further expand the potential pathways, uncovering non-obvious enzyme combinations. Exploring microbial pathways in butanol production holds significant promise, potentially leading to the development of innovative strains capable of efficient butanol synthesis (Abdelaal and Yazdani, 2022). *Clostridial* strains have long been recognized for their role in biobutanol production. While traditional pathways remain important, introducing alternative routes in robust microorganisms like *E. coli* can strategically enhance butanol synthesis. A recent study developed a new

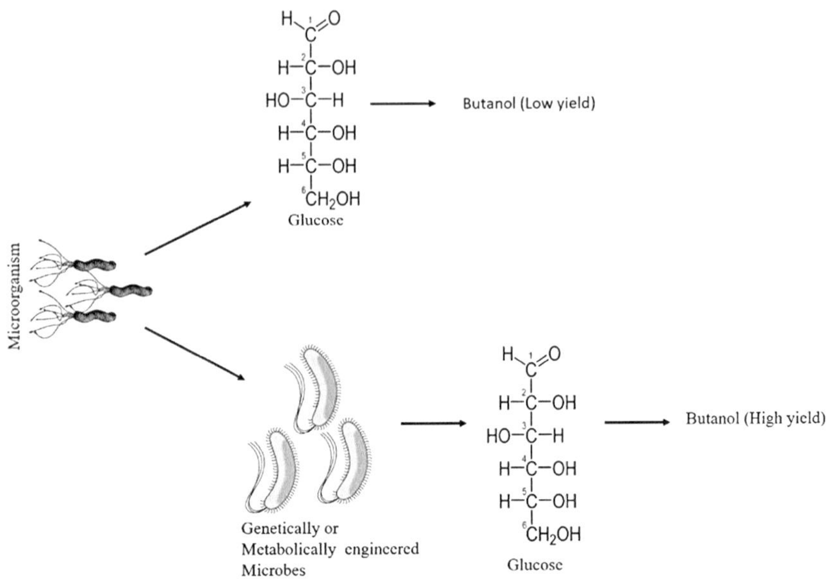

FIGURE 12.1 In the wild-type microorganism, the natural genetic makeup limits the efficiency of butanol production, resulting in lower yields. However, through genetic engineering, key genes involved in the butanol biosynthesis pathway can be manipulated or overexpressed, leading to increased production of butanol in the modified microorganism.

butanol production pathway using an *in silico* approach, with 2-oxoglutarate as a precursor. This pathway involves seven sequential catalytic processes and incorporates nine genes from various origins into *E. coli*. The study's findings suggest that expressing butanol-related genes in *E. coli* K12 and cultivating these microbes in a High-Density Medium can lead to substantial butanol accumulation via the 2-oxoglutarate pathway. Using enumeration algorithms, scientists can identify unconventional enzyme combinations, expand the range of potential butanol production pathways and paving the way for improvements in microbial production systems (Ferreira et al., 2019).

12.2 MICROBIAL STRAINS IN BUTANOL PRODUCTION

Clostridium species are highly valued in biofuel synthesis due to their exceptional ability to produce butanol. These bacteria are frequently employed in industrial butanol production techniques because of their natural capacity to generate this solvent (Sreekumar et al., 2015). Additionally, certain *Lactobacillus* species, particularly *Lactobacillus diolivorans,* exhibit notable solvent resistance and can naturally convert meso-2,3-butanediol into 2-butanol during anaerobic sugar fermentations. *L. diolivorans* is an efficient producer of 2-butanol because it can synthesize a variety of bulk compounds via a straightforward two-step metabolic route. The combined abilities of these microbial strains (Table 12.1), including *Lactobacillus* and *Clostridium*, present significant potential for enhancing microbial butanol production systems (Russmayer et al., 2019).

1. ***Clostridium* Species:** *Clostridium* species, such as *Clostridium acetobutylicum* and *Clostridium beijerinckii*, are renowned for their ability to produce butanol through the ABE fermentation process (Biebl, 1999). These microorganisms can convert a variety of carbon sources into butanol. *Clostridium acetobutylicum* is a prominent subject of research and is widely used in large-scale industrial fermentation processes for butanol production (Sreekumar et al., 2015). *Clostridium beijerinckii* has also shown promising results in terms of butanol yield and productivity. The use of *Clostridium* species in butanol fermentation offers several advantages, including their capacity to metabolize diverse carbon

sources, withstand high butanol concentrations, and naturally produce butanol as the main fermentation product (Tracy, 2012).

i. ***Clostridium acetobutylicum:*** *Clostridium acetobutylicum* is well-known for producing industrial solvents such as acetone, butanol, and ethanol. Recent research has revealed that this anaerobic bacterium can also produce polyketides, challenging the traditional belief that anaerobic organisms lack this capability. These polyketides are crucial for regulating biological processes within *C. acetobutylicum*, such as sporulation and granulose accumulation. This discovery expands our understanding of the metabolic capabilities of anaerobic microorganisms and suggests the potential for targeted manipulation of secondary metabolism to enhance industrial applications (Herman et al., 2017). The ABE pathway in *Clostridium acetobutylicum* involves the enzymatic conversion of glucose into butanol through a two-stage process. Initially, glucose is converted to butyrate, which is subsequently reduced to butanol. This bacterium's ability to convert sugars into butanol makes it pivotal in the development of butanol production methods (Buehler and Mesbah, 2016). Optimizing fermentation conditions to enhance the efficiency of the ABE pathway, along with genetic engineering efforts to increase butanol yields, underscores the importance of *C. acetobutylicum* in refining and improving butanol production processes. This contributes significantly to the advancement of sustainable biofuel technologies (Herman et al., 2017). The ABE pathway involves a series of enzymatic reactions where glucose is converted into butanol. It starts with glucose uptake, followed by glycolysis, which converts glucose to pyruvate. Pyruvate is then processed to yield acetyl-CoA, leading to the production of acetate. Acetyl-CoA is converted to butyrate, which undergoes enzymatic conversion to acetone, an intermediate that is reduced to butanol (Quintero-Díaz et al., 2023). Concurrently, ethanol is produced, highlighting the complex enzymatic network in *C. acetobutylicum* that enables glucose conversion into butanol through well-coordinated biochemical reactions (Buehler and Mesbah, 2016).

ii. ***Clostridium beijerinckii***: A newly identified strain, *Clostridium beijerinckii* XH 0906, exemplifies how Clostridium species produce solvents via the ABE fermentation process. This strain uses glucose as the carbon source during fermentation, producing butanol and isopropanol (BI) as the main byproducts. It demonstrates oxygen-tolerant BI fermentation and performs well under aerobic conditions. A mutant strain, *Clostridium beijerinckii* XH 29, developed through UV mutagenesis, shows increased BI generation capacity and adaptability in using glucose and maize stover hydrolysate for BI synthesis. This mutant strain can convert a glucose-acetone combination into isopropanol, highlighting its industrial potential as an efficient BI producer (Yao et al., 2022). *Clostridium beijerinckii* ATCC 55025 can produce ABE from wheat bran hydrolysate. This substrate, a by-product of the wheat milling industry, contains glucose, xylose, and arabinose, which *C. beijerinckii* ATCC 55025 can simultaneously utilize for ABE production. After a 72-hour fermentation period, the total ABE concentration was 11.8 g/l, with acetone, butanol, and ethanol concentrations at 2.2, 8.8, and 0.8 g/l, respectively. The productivity was 0.16 g/l/h, and the ABE yield was 0.32. This study demonstrates the potential of using wheat bran as a renewable material for ABE fermentation, offering an environmentally sustainable alternative amid challenges related to fossil fuel supply and petroleum price volatility (Liu et al., 2010). The butanol production process using *Clostridium beijerinckii* involves several methodical stages. Initially, a pure culture is inoculated and cultivated in an appropriate growth medium. A carbon source like glucose or molasses is prepared for the fermentation substrate. Fermentation occurs in an anaerobic environment with optimal pH and temperature levels (Daengbussadee et al., 2022). During the acidogenesis phase, *C. beijerinckii* converts the substrate into organic acids like butyric acid and acetic acid. In the subsequent solventogenesis phase, the bacterium produces solvents such as ethanol, acetone, and butanol (Figure 12.2). Finally, butanol is separated from the fermentation broth through techniques like distillation or extraction, completing the production process (Humphreys et al., 2023).

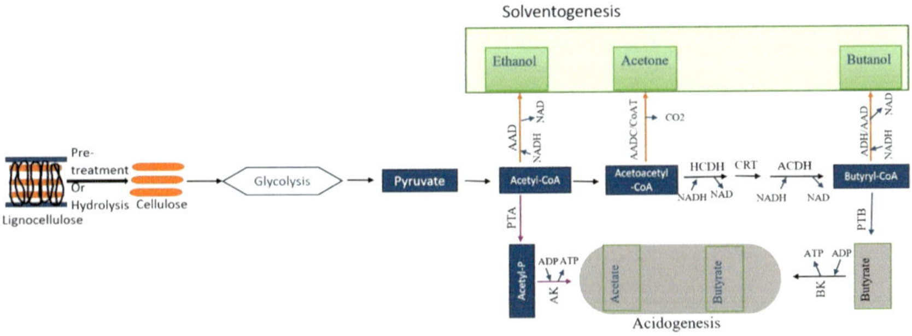

FIGURE 12.2 Lignocellulosic biomass, composed of cellulose, hemicellulose, and lignin, is a plentiful, sustainable, and economical resource for butanol production. Cellulose is a linear polysaccharide made of d-glucopyranosyl units, while hemicellulose comprises branched heteropolysaccharides with hexose and pentose sugar residues. Lignin, composed of phenylpropane units, provides structural support to cells, making lignocellulosic materials resistant to microbial degradation (Zoghlami and Paës, 2019). To facilitate successful lignocellulosic fermentation, lignin removal before cellulose and hemicellulose hydrolysis is crucial. Following pre-treatment or hydrolysis, lignocellulose is converted into its monomer forms, which then undergoes glycolysis to produce pyruvate. Pyruvate is subsequently converted to acetyl-CoA, which further transforms into acetoacetyl-CoA (Martín et al., 2022). The conversion of acetoacetyl-CoA to butyryl-CoA involves several enzymatic steps mediated by genes such as *THAL, HCDH, CRT, ACDH, ADH, AAD, PTB, BK, PTA, AK, AADC, and AAD*. Ultimately, butanol or other products like acetone or ethanol are synthesized, depending on the specific enzymatic pathways activated (Bennett and Rudolph, 1995).

2. **Other Microbial Candidates for Butanol Production:** *E. coli* is a commonly known bacterium that has been genetically engineered to produce butanol by incorporating vital genes from *Clostridium* species. However, its effectiveness in producing butanol is limited by its low tolerance to the compound. *Saccharomyces cerevisiae,* yeast traditionally used for ethanol production, can also be genetically modified to produce butanol by introducing the necessary genes for the butanol biosynthesis pathway (Humphreys et al., 2023). *Bacillus subtilis,* another bacterium, has been engineered to produce butanol by incorporating essential genes from *Clostridium* species and exhibits a higher tolerance to butanol compared to *E. coli* (Li et al., 2020). These approaches in microbial engineering demonstrate the versatility of different microorganisms in biofuel production and highlight the importance of selecting appropriate

host organisms based on their inherent characteristics and tolerance levels for optimal biofuel yields (Rao et al., 2016).

i. ***Bacillus* Species:** *Bacillus subtilis* GRSW2-B1 is a bacterium identified for its butanol tolerance. Its ability to withstand a wide range of organic solvents, including butanol, makes it a promising candidate for efficient butanol production (Singh et al., 2006). Researchers have enhanced its butanol tolerance through acclimatization processes. This bacterium can maintain high tolerance even at high cell densities and under growth conditions (Kataoka et al., 2011). By optimizing genetic transformation procedures, efficient gene expression has been achieved using various promoters, with the xylose promoter showing the highest expression levels (Feng et al., 2024). Importantly, the constructed vector carrying the desired genes remains stable in the transformants, even under butanol-induced stress. Overall, *Bacillus subtilis* GRSW2-B1's unique characteristics make it a promising host for butanol bioproduction, indicating its potential in biofuel development (Kataoka et al., 2011).

ii. **Engineered *Escherichia coli* Strains:** The efficiency of glycerol metabolism in *E. coli* is enhanced by rewiring its central metabolism, which is crucial for n-butanol production (Saini et al., 2017). This metabolic engineering redirects the gluconeogenic flux into the oxidative pentose phosphate pathway and forces the glycolytic flux through the oxidation pathway of pyruvate (Luo et al., 2023a). Additionally, the anaerobic catabolism of glycerol is intensified, and the tricarboxylic acid cycle is moderately suppressed. The engineered *E. coli* strain can produce n-butanol from crude glycerol under microaerobic conditions, achieving a conversion yield and productivity of 87% of the theoretical yield and 0.18 g/L/h, respectively. This tailored metabolic approach demonstrates the potential of *E. coli* as an efficient platform for biofuel production, particularly using glycerol as a renewable and economically viable feedstock for n-butanol synthesis (Saini et al., 2017).

iii. ***Lacto Bacillus*****:** *Lactobacillus diolivorans,* a lactic acid bacterium, can produce 2-butanol through a two-step enzymatic process during anaerobic sugar fermentation (Syu, 2001). This process involves the reduction of meso-2,3-butanediol (2,3-BTD) via two consecutive enzymatic reactions. The bacterium's high tolerance to stress conditions, including elevated concentrations of acids or alcohols, makes it a promising host for 2-butanol production. *L. diolivorans* can produce up to 35 g/L of 3-hydroxy propionaldehyde (3-HPA), an intermediate product in the metabolic pathway for 2-butanol production. Given the pathway's efficiency and metabolic balance in *L. diolivorans,* it is a viable option for 2-butanol production from 2,3-BTD and shows great potential in biofuel production processes (Russmayer et al., 2019).

iv. ***Thermoanaerobacterium* Species:** *Thermoanaerobacterium* species can produce butanol via the clostridial ABE pathway (Syu, 2001). The first step involves converting pyruvate to acetyl-CoA using the enzyme PFR. Acetyl-CoA is then transformed into acetate, which is further converted to butyrate through a series of enzymatic reactions involving acetyl-CoA acetyltransferase, acetoacetyl-CoA, butyryl-CoA, acetate CoA-transferase, butyrate kinase, and acetate/butyrate-CoA transferase. Ultimately, butyryl-CoA dehydrogenase, coronas, and butyraldehyde dehydrogenase convert butyrate into butanol. *Thermoanaerobacterium* spp. can produce butanol anaerobically from various carbon sources according to this metabolic pathway, highlighting its potential as a biofuel-producing microbe (Bhandiwad et al., 2014).

v. ***Ruminococcus* spp.:** *Ruminococcus* species, anaerobic bacteria, can generate butanol through ABE fermentation (Lin et al., 2023). This process uses various carbon sources, such as cellulose or carbohydrates, as substrates. The process involves multiple enzymatic reactions that convert sugars into intermediates like acetyl-CoA and butyryl-CoA. These intermediates undergo further metabolic processes through specific enzyme-catalyzed steps, ultimately yielding butanol as one of the fermentation products. Along with butanol,

Ruminococcus spp. can produce byproducts like acetone and ethanol during ABE fermentation. This microbial pathway has significant potential for biofuel production from renewable biomass sources like lignocellulosic materials (Gaffney et al., 2021).

vi. ***Synechococcus:*** Cyanobacteria of the genus *Synechococcus* are potential sources of butanol (Lau et al., 2015). By genetically engineering this cyanobacterium, researchers can modify its metabolic pathways to produce butanol (Lan et al., 2011). Typically, genes from other microorganisms capable of producing butanol, such as *Clostridium* species, are introduced into *Synechococcus.* These genes encode enzymes like hydroxybutyryl-CoA dehydrogenase, coronas, butyryl-CoA dehydrogenase, and butanol dehydrogenase. *Synechococcus* can utilize carbon dioxide and sunlight to produce butanol through metabolic engineering, offering a sustainable and renewable method for butanol production (Kehail and Brigham, 2015). Genetically engineered *Synechococcus elongatus* PCC7942 has been successfully used for phototrophic 2-methyl-1-butanol (2MB) synthesis from CO_2 (Shen and Liao, 2012).

12.3 METABOLIC PATHWAYS AND GENETIC ENGINEERING

1. **Exploring Metabolic Pathways for Butanol Synthesis:** Butanol synthesis involves various metabolic pathways depending on the organism and the engineering strategies implemented. Typically, the process starts with converting glucose to butyryl-CoA, a critical intermediate (Ndaba et al., 2015). This is followed by the reduction of butyryl-CoA, resulting in butanol production (Jang et al., 2012). Key enzymes, such as pyruvate decarboxylase, ADH, and butyraldehyde dehydrogenase, play significant roles in these biochemical reactions (Singh et al., 2024). The diversity of pathways and enzymatic reactions underscores the adaptability of metabolic engineering methods, allowing for customization based on the specific organism or microbial host used.

TABLE 12.1 Comparison of Different Microbial Strains, Substrates, Yield in Butanol Production and Their By-Products

Microorganism	Substrate	Strain Modification	Butanol Yield	By Products	References
Escherichia coli	Glucose, Xylose	Genes encoding the enzymes involved in the butanol biosynthesis pathway are inserted into the genome of *E. coli*. These genes may include thiolase *(thl)*, 3-hydroxybutyryl-CoA dehydrogenase (*hbd*), crotonase (*crt*), butyryl-CoA dehydrogenase (*bcd*), and alcohol dehydrogenase (*adhE2*)	14–16 g/L	Acetone, ethanol	Atsumi et al. (2008); Zverlov et al. (2006)
Clostridium acetobutylicum	Glucose	Modified TYA medium and P2 medium to grow *C. acetobutylicum*. The modified TYA medium contains glucose, ammonium acetate, yeast extract, tryptone, KH_2PO_4, K_2HPO_4, $MgSO_4 \cdot 7H_2O$, and $FeSO_4 \cdot 7H_2O$. The P2 medium contains glucose, K_2HPO_4, KH_2PO_4, CH_3COONH_4, $MgSO_4 \cdot 7H_2O$, $MnSO_4 \cdot H_2O$, NaCl, $FeSO_4 \cdot 7H_2O$, p-aminobenzoic acid, thiamine, and biotin	10.03 g/L	Acetone Ethanol Acetic acid Butyric acid	Khamaiseh et al. (2014); Zhou et al. (2023)
		Reinforcing the NADH *de novo* synthesis and VB1 biosynthesis			
Clostridium beijerinckii	glucose and xylose	*Clostridium beijerinckii* NCIMB 8052 was done by integrating an additional copy of a short-chain dehydrogenase/reductase (*SDR*) gene (*Cbei3904*) into the chromosome of the wildtype strain, controlled by the constitutive *thiolase* promoter This engineered strain, called *C. beijerinckii* SDR, showed improved fermentation of lignocellulosic biomass (LB)-derived hydrolysates to acetone-butanol-ethanol (ABE) compared to the wildtype strain	11.21 g/L	Acetone Ethanol	Olorunsogbon et al. (2022)
Clostridium saccharoperbutylacetonicum	Glucose		3.59 g/L		
C. pasteurianum	Glycerol	Immobilized cells/Bath fermentation	8.8 g/L		Khanna and Goyal (2013)

TABLE 12.1 *(Continued)*

Microorganism	Substrate	Strain Modification	Butanol Yield	By Products	References
Zymomonas mobilis	Glucose, Xylose	The important enzymes *Thiolase*, Hydroxybutyryl-CoA, dehydrogenase, Crotonase, Butyryl-CoA, dehydrogenase Butanol dehydrogenase	4.0 g/L	ethanol	Qiu et al. (2020)
		By adding these genes and their corresponding enzymes, we can reprogram Z. mobilis to produce butanol			
		Integration of the heterologous gene encoding 2-ketoisovalerate decarboxylase (*kdcA*) from Lactococcus lactis			
		Overexpression of a heterologous also gene and two native genes (*ilvC* and *ilvD*) involved in valine metabolism in a recombinant Z. mobilis strain expressing kdcA			
Bacillus subtilis	Glucose, xylose, sucrose,	Thiolase, hydroxybutyryl-CoA dehydrogenase, coronas, butyryl-CoA dehydrogenase, and butanol dehydrogenase. By introducing these genes into *Bacillus* subtilis and optimizing fermentation conditions	8.28 g/L	Acetone, Ethanol	Oliva-Rodríguez et al. (2019)
Thermoanaerobacterium spp.	Xylose	Expression of key genes responsible for n-butanol production, including *thl*, *hbd*, *crt*, *bcd*, *etfA*, *etfAB*, *adhE2*, and *adhE1*, which encode enzymes involved in the conversion of acetyl CoA and butyryl CoA to ethanol and n-butanol, respectively	0.85 g/L	Acetic acid	Bhandiwad et al. (2014)
Synechococcus	glucose, sucrose, xylose	genes encoding enzymes like alcohol dehydrogenase (*adh*), butanol dehydrogenase (*bdh*), and CoA transferase (*ctf*), to enhance their activity and expression level			Shen and Liao (2012)

i. **Glycolysis and Acetone-Butanol-Ethanol (ABE) Pathway:** Glycolysis, a fundamental metabolic pathway, is crucial for butanol production as it breaks down glucose into pyruvate. Pyruvate is then converted into butyrate with enzymes such as pyruvate: ferredoxin oxidoreductase and butyrate kinase, and subsequently, butyrate are reduced to butanol (Luo et al., 2023b). This pathway involves multiple reduction steps, including the transformation of butyrate to butyraldehyde and then to butanol. *Clostridium* species, particularly *Clostridium acetobutylicum*, utilize this pathway in the ABE fermentation process, producing butanol as one of the main products (Patakova et al., 2019). The ABE fermentation pathway not only converts carbohydrates into acids and solvents but also highlights the enzyme adhE, which is essential for butanol and ethanol synthesis. Acetoacetyl-CoA is a crucial precursor in this process, and enhancing its formation rate can significantly boost butanol and acetone production (Cho et al., 2019). Increasing the rate of acetate formation, regulated by factors such as K7, can further enhance butanol yields (Buehler and Mesbah, 2016).

ii. **Pathway Engineering for Improved Butanol Yield:** Various pathway engineering strategies can enhance butanol production in microorganisms. One effective approach is the overexpression of key enzymes in the ABE pathway, including crotonase (crt), butyryl-CoA dehydrogenase (bcd), 3-hydroxybutyryl-CoA dehydrogenase (hbd), ketoacid decarboxylase (KDC), and alcohol dehydrogenase (ADH) (Tian et al., 2019). Optimizing gene expression associated with butanol synthesis ensures the production of these enzymes at appropriate levels. Introducing heterologous pathways from organisms like *Clostridium acetobutylicum* into hosts such as *E. coli* or *Saccharomyces cerevisiae* can expand the metabolic capacity for butanol production (Schadeweg and Boles, 2016). Engineering the reductive tricarboxylic acid (TCA) cycle to convert carbon dioxide to butanol has shown promise in bacteria like *Clostridium ljungdahlii* (Lauer et al., 2022). Additionally, syngas fermentation, where synthesis gas is converted to butanol, is being explored in bacteria such as *Clostridium autoethanogenum*. The choice of strategy depends on the host organism's characteristics and the industrial application goals (Benito-Vaquerizo et al., 2020).

2. **Genetic Modifications to Enhance Microbial Butanol Production**

   i. **Overexpression of Key Enzymes:** Enhancing microbial butanol production can be significantly achieved by overexpressing essential enzymes within the butanol biosynthesis pathway. This genetic modification strategy targets enzymes such as acetyl-CoA acetyltransferase (thiolase), acetoacetyl-CoA reductase, and butanol dehydrogenase, aiming to boost butanol production levels (Nawab et al., 2020). Techniques to increase the expression of these enzymes include gene amplification and promoter engineering (Shi et al., 2016). For instance, studies have demonstrated that overexpressing the *thiolase* gene in *Clostridium acetobutylicum* leads to substantial improvements in butanol yield. Enhancing enzymes in precursor pathways, such as glycolysis and the pentose phosphate pathway, also increases precursor availability, further boosting butanol production (Jang et al., 2012). Methods such as plasmid-based gene expression, gene knockout, and CRISPR-Cas9-mediated genome editing are utilized to achieve these genetic modifications and enhance microbial butanol production (Rao et al., 2016).

   ii. **Synthetic Biology Approaches:** Synthetic biology provides a comprehensive approach to increasing butanol production through the genetic manipulation of microbial strains. This includes metabolic engineering techniques where key enzymes in the butanol biosynthesis pathway are overexpressed to enhance production (Wen et al., 2020). Pathway optimization involves fine-tuning gene expression and redirecting metabolic fluxes to improve yields, and advanced genome editing tools like CRISPR-Cas9 allow precise genetic modifications, enabling targeted gene knockouts or insertions to increase pathway efficiency (Zhao et al., 2020). The creation of modular biosynthetic pathways involves assembling functional modules from various microorganisms to optimize butanol synthesis (Bhan et al., 2013). Additionally, directed evolution and adaptive laboratory evolution (ALE) techniques enable microbial strains to evolve under selective pressure, leading to enhanced butanol production capabilities. These methods represent promising strategies for sustainable and efficient biofuel production (Singh et al., 2024).

12.4 ENVIRONMENTAL SUSTAINABILITY OF MICROBIAL BUTANOL PRODUCTION

Microbial butanol synthesis, a fourth-generation biofuel, holds significant potential for advancing environmental sustainability. Compared to traditional fossil fuels, microbial butanol boasts several benefits, such as reduced greenhouse gas emissions and decreased dependency on non-renewable resources (Jeswani et al., 2020). The integration of genetically engineered feedstocks and microorganisms in production processes further reduces environmental impacts. Nonetheless, it is vital to adopt a comprehensive strategy to recognize potential risks and develop mitigation measures to ensure the ongoing environmental benefits of microbial butanol production (Adegboye et al., 2021). Effective waste management, including the proper treatment of byproducts like wastewater and spent fermentation media, is crucial to prevent environmental contamination. Continuous research and development are necessary to improve the efficiency and sustainability of microbial butanol production, focusing on renewable feedstocks and more efficient microbial strains. Producing microbial butanol from lignocellulosic materials presents a sustainable and eco-friendly alternative to traditional biofuel methods (Nandhini et al., 2023). This approach utilizes renewable and sustainable feedstocks, such as rice straw, thereby reducing environmental impacts compared to non-renewable resources that contribute to deforestation and greenhouse gas emissions. The biorefinery approach optimizes resource use and minimizes waste, while modifying yeast and bacterial strains enhances butanol yield (Cai et al., 2016). These practices make microbial butanol production from lignocellulosic materials an attractive option for environmentally conscious individuals seeking sustainable biofuel solutions. Key environmental indicators, such as greenhouse gas emissions and carbon footprint, measure the impact of human activities on climate change (Qureshi and Ezeji, 2008). The carbon footprint quantifies the total greenhouse gases, particularly carbon dioxide (CO_2), emitted directly or indirectly by human activities, including energy production, transportation, and industrial processes. Greenhouse gas emissions, which include CO_2, methane (CH_4), nitrous oxide (N_2O), and fluorinated gases, contribute to climate change and global warming by trapping heat in the atmosphere. Reducing these emissions and the carbon footprint is essential for mitigating climate change and transitioning to a low-carbon, sustainable economy (Sun, 2023). Utilizing waste substrates and by-products is a key aspect of sustainable resource management and circular economy principles. Agricultural residues, food waste, and industrial by-products can be converted into valuable products through fermentation, anaerobic digestion, and biochemical conversion (Sharma et al., 2022). These processes produce biofuels, bio-based chemicals, animal feed, fertilizers, and other high-value products, reducing the environmental impact of waste disposal and enhancing resource efficiency. Repurposing waste substrates and by-products helps lower greenhouse gas emissions, conserve natural resources, and create economic opportunities in waste management and bio-based industries (Tiwari et al., 2021).

12.5 ECONOMIC VIABILITY AND INDUSTRIAL APPLICATIONS

The feasibility and scalability of microbial butanol production largely depend on its economic viability and industrial applications. Microbial butanol production offers several benefits, including higher energy density than ethanol, compatibility with current infrastructure, and potential uses as a biofuel, solvent, and chemical feedstock (Birgen et al., 2019). Nevertheless, challenges such as high production costs, low yields, and competition with petrochemical products must be resolved for widespread industrial use. Strategies to enhance economic viability include optimizing processes, genetically engineering microbial strains, and developing efficient fermentation methods (Zheng et al., 2009). Exploring niche markets and additional applications beyond biofuel can further improve the economic appeal of microbial butanol as a sustainable fossil fuel alternative. Collaboration among industry, academia, and government agencies is essential to advance research, scale up production, and reduce costs, making microbial butanol commercially competitive (Birgen et al., 2019). Economic production techniques are critical for the success and extensive use of microbial butanol synthesis. These techniques encompass various aspects of production, including waste management, downstream processing, feedstock selection, and fermentation optimization. Using inexpensive feedstocks like lignocellulosic biomass, agricultural waste, and other waste substrates can significantly reduce raw material costs (Guo et al., 2022). Optimizing fermentation conditions—such as temperature, pH, and nutrient supplementation—can enhance microbial growth and butanol yield while minimizing operational costs. Downstream processing methods, including *in situ* product recovery, solvent extraction, and membrane separation, can improve recovery efficiency and reduce energy use (Outram et al., 2017). Moreover, integrating waste utilization strategies allows for the valorization of by-products and co-products, thereby maximizing resource efficiency and reducing overall production costs. A comprehensive approach that incorporates cost-effective strategies throughout the production chain is crucial for achieving economic viability and competitiveness in microbial butanol production (Bankar et al., 2013). The economic challenges in microbial butanol production stem from various factors, including high production costs, competition with petroleum-based fuels, and market volatility. One significant challenge is the cost of feedstock, particularly lignocellulosic biomass or agricultural residues, which require extensive pre-processing and enzymatic hydrolysis (Patil et al., 2019). The fermentation process is also costly due to the need for specialized equipment, energy-intensive operations, and the production of co-products with limited market value. Market competition from cheaper petroleum-based fuels further complicates the economic viability of butanol production. To address these challenges, potential solutions include technological innovations to reduce production costs, such as developing more efficient enzymes and fermentation strains, and process intensification techniques to improve productivity and yield (Tracy, 2012). Integration with biorefinery concepts can enhance overall process economics by valorizing co-products and maximizing resource utilization. Additionally, policy support in the form of incentives, subsidies, and mandates for biofuel production can create a more favorable market environment for microbial butanol and stimulate investment in the industry. Overall, a combination of technological advancements, process optimization, and supportive policies is essential to overcome economic challenges and realize the full potential of microbial butanol as a sustainable alternative to petroleum-based fuels (Balan, 2014). The industrial-scale application of microbial butanol production holds significant promise for meeting the demand for renewable fuels and chemicals. Commercialization prospects are driven by butanol's versatility as a biofuel and chemical feedstock, offering various applications in the transportation, chemicals, and pharmaceuticals industries (Jiang et al., 2015). Butanol's compatibility with existing infrastructure, such as blending with gasoline without engine modifications, enhances its appeal as a drop-in biofuel. Additionally, butanol's superior properties, such as higher energy content and lower volatility compared to ethanol, make it an attractive option for blending in gasoline and diesel fuels, potentially reducing greenhouse gas emissions and improving fuel efficiency (Obergruber et al., 2021b). In the chemicals sector, butanol serves as a valuable intermediate for producing a variety of products, including solvents, plastics, and synthetic rubber, offering diverse revenue streams and market opportunities. The commercialization of microbial butanol production is further bolstered by advancements in biotechnology, process engineering, and downstream processing, which have led to improved yields, reduced production costs, and enhanced scalability (Kolesinska et al., 2019b). However, challenges remain in achieving cost competitiveness with conventional fossil fuels and securing investment for large-scale production facilities. Overcoming these challenges requires continued innovation, collaboration across the value chain, supportive policies, and market demand for sustainable alternatives. Despite these hurdles, the growing emphasis on decarbonization, renewable energy sources, and circular economy principles bodes well for the future commercialization and widespread adoption of microbial butanol as a key player in the bioeconomy.

12.6 ADVANCEMENTS AND CHALLENGES

Recent advancements in microbial butanol production have been achieved through significant progress in strain engineering, fermentation techniques, and downstream processing methods. Notably, a breakthrough in consolidated bioprocessing utilizing a metabolically engineered twin-clostridial consortium has demonstrated remarkable levels of butanol production, highlighting the potential of synergistic microbial interactions (Guo et al., 2022). Furthermore, metabolic and evolutionary engineering strategies have led to substantial increases in butanol production from

Clostridium cellulovorans, underscoring the effectiveness of genetic manipulation and selective pressure in enhancing strain performance (Wen et al., 2019). Despite these advancements, challenges persist in transferring the butanol pathway to non-clostridial strains, mainly due to pathway complexity and host compatibility issues. To address this, integrating *in situ* recovery processes has emerged as a promising strategy to alleviate product inhibition and enhance productivity (Patil et al., 2019). However, the high toxicity of butanol to fermentation microorganisms remains a significant obstacle, highlighting the need for further research to optimize strain tolerance and overall process efficiency. Addressing these challenges is imperative for unlocking the full economic and environmental potential of microbial butanol production (Veza et al., 2021). Technological hurdles and research gaps in butanol production must be addressed to optimize the efficiency and viability of this biofuel production process. One major challenge is achieving high final concentrations and yields of butanol, compounded by its toxicity to fermentation microorganisms. Efficient separation techniques are crucial for recovering butanol and mitigating its toxic effects, with methods like adsorption and pervaporation showing promise for separating butanol from dilute solutions. The complex composition of the solvent product further complicates commercialization, necessitating energy-efficient and cost-effective separation techniques, such as liquid-liquid extraction with nonfluorinated task-specific ionic liquids (TSILs). Overcoming product inhibition is another critical research gap that requires innovative strategies to enhance butanol production. Addressing these technological challenges and research gaps is crucial for advancing butanol production and realizing its potential as a sustainable biofuel alternative (Abdehagh et al., 2014). Challenges include low product yield, productivity, and high production costs. Developing specific catalysts with high selectivity and conversion titer is crucial, especially considering ethanol's challenging substrate nature. Ensuring the recovery of catalysts, particularly rare elements like ruthenium, is essential for sustainability. While n-butanol holds promise as an advanced biofuel due to its limited solubility and resemblance to gasoline, its current production heavily relies on petrochemical synthesis, making it vulnerable to fluctuations in crude oil prices. To overcome these challenges, research efforts should concentrate on optimizing fermentation conditions, enhancing catalyst efficiency, and exploring alternative feedstocks to improve the overall efficiency and cost-effectiveness of microbial butanol production processes (Branduardi and Porro, 2016).

12.7 FUTURE PROSPECTS AND POTENTIAL INNOVATIONS

Advancements in metabolic engineering techniques, such as synthetic biology and genome editing, present opportunities to customize microbial strains for enhanced butanol production through improvements in metabolic pathways and increased tolerance to butanol. Incorporating innovative fermentation methods like continuous fermentation and high-cell-density cultivation can augment productivity and slash production expenses. Broadening the range of resources utilized for butanol production involves exploring alternative feedstocks like lignocellulosic biomass and waste substrates. This strategy not only fosters sustainability but also bolsters economic feasibility. Intertwining butanol production with the creation of value-added products in a biorefinery context can additionally optimize process efficiency and overall profitability, exemplifying a commitment to sustainable production practices while maximizing economic returns. Collectively, forthcoming innovations in microbial butanol production are poised to propel progress toward scalable, environmentally friendly, and economically competitive biofuel manufacturing processes.

12.8 CONCLUSION

Producing butanol through microbial processes offers a sustainable alternative to traditional fossil fuels, catering to the increasing need for renewable energy sources. Butanol, with its superior fuel characteristics compared to ethanol, presents an efficient solution for energy demands. Notably, *Clostridium* species are adept at synthesizing butanol, while *Lactobacillus diolivorans* is recognized for its capability to produce 2-butanol. Utilizing lignocellulosic feedstocks promotes environmental sustainability by reducing waste and optimizing resource use. Innovations in genetic engineering and metabolic pathway optimization have significantly enhanced microbial strains and improved butanol production yields. Ongoing research is essential for refining these processes and enabling the large-scale implementation of microbial butanol production, ultimately contributing to a more sustainable and efficient biofuel industry.

KEYWORDS

- **butanol**
- ***Clostridial* strains**
- **genetic modification**
- **industrial fermentation**
- **lignocellulosic biomass**
- **sustainable energy**

REFERENCES

Abdehagh, N., Tezel, F., & Thibault, J. (2014). Separation techniques in butanol production: Challenges and developments. *Biomass and Bioenergy, 60*, 222–246. https://doi.org/10.1016/j.biombioe.2013.12.027 (accessed on 15th November 2025).

Abdelaal, A. S., & Yazdani, S. S. (2022). Engineering *E. coli* to synthesize butanol. *Biochemical Society Transactions, 50*(2), 867–876. https://doi.org/10.1042/BST20210870 (accessed on 15th November 2025).

Adegboye, M. F., Ojuederie, O. B., Talia, P. M., & Babalola, O. O. (2021). Bioprospecting of microbial strains for biofuel production: Metabolic engineering, applications, and challenges. *Biotechnology for Biofuels, 14*(1), 5. https://doi.org/10.1186/s13068-020-01852-1 (accessed on 15th November 2025).

Atsumi, S., Hanai, T., & Liao, J. C. (2008). Non-fermentative pathways for synthesis of branched-chain higher alcohols as biofuels. *Nature, 451*(7174), 86–89. https://doi.org/10.1038/nature06450 (accessed on 15th November 2025).

Balan, V. (2014). Current challenges in commercially producing biofuels from lignocellulosic biomass. *ISRN Biotechnology, 2014*, 463074. https://doi.org/10.1155/2014/463074 (accessed on 15th November 2025).

Bankar, S. B., Survase, S. A., Ojamo, H., & Granström, T. (2013). Biobutanol: The outlook of an academic and industrialist. *RSC Advances, 3*(47), 24734–24757. https://doi.org/10.1039/C3RA41501J (accessed on 15th November 2025).

Benito-Vaquerizo, S., Diender, M., Parera Olm, I., Martins Dos Santos, V. A. P., Schaap, P. J., Sousa, D. Z., & Suarez-Diez, M. (2020). Modeling a co-culture of *Clostridium autoethanogenum* and *Clostridium kluyveri* to increase syngas conversion to medium-chain fatty acids. *Computational and Structural Biotechnology Journal, 18*, 3255–3266. https://doi.org/10.1016/j.csbj.2020.10.004 (accessed on 15th November 2025).

Bennett, G. N. (1995). The central metabolic pathway from acetyl-CoA to butyryl-CoA in *Clostridium acetobutylicum. FEMS Microbiology Reviews, 17*(3), 241–249. https://doi.org/10.1111/j.1574-6976.1995.tb00209.x (accessed on 15th November 2025).

Bhan, N., Xu, P., & Koffas, M. A. (2013). Pathway and protein engineering approaches to produce novel and commodity small molecules. *Current Opinion in Biotechnology,*

24(6), 1137–1143. https://doi.org/10.1016/j.copbio.2013.04.001 (accessed on 15th November 2025).

Bhandiwad, A., Shaw, A. J., Guss, A. M., Guseva, A., Bahl, H., & Lynd, L. R. (2014). Metabolic engineering of *Thermoanaerobacterium saccharolyticum* for n-butanol production. *Metabolic Engineering, 21*, 17–25. https://doi.org/10.1016/j.ymben.2013.10.001 (accessed on 15th November 2025).

Biebl, H. (1999). *Clostridium | Clostridium acetobutylicum*. In *Encyclopedia of Food Microbiology* (pp. 445–451). Academic Press. https://doi.org/10.1016/B978-012227070-3.00066-2 (accessed on 15th November 2025).

Birgen, C., Dürre, P., Preisig, H. A., & Wentzel, A. (2019). Butanol production from lignocellulosic biomass: Revisiting fermentation performance indicators with exploratory data analysis. *Biotechnology for Biofuels, 12*, 167. https://doi.org/10.1186/s13068-019-1498-7 (accessed on 15th November 2025).

Branduardi, P., & Porro, D. (2016). n-Butanol: Challenges and solutions for shifting natural metabolic pathways into a viable microbial production. *FEMS Microbiology Letters, 363*(8), fnw070. https://doi.org/10.1093/femsle/fnw070 (accessed on 15th November 2025).

Buehler, E. A., & Mesbah, A. (2016). Kinetic study of acetone–butanol–ethanol fermentation in continuous culture. *PLOS ONE, 11*(8), e0158243. https://doi.org/10.1371/journal.pone.0158243 (accessed on 15th November 2025).

Cai, D., Dong, Z., Wang, Y., Chen, C., Li, P., Qin, P., Wang, Z., & Tan, T. (2016). Co-generation of microbial lipid and bio-butanol from corn cob bagasse in an environmentally friendly biorefinery process. *Bioresource Technology, 216*, 345 351. https://doi.org/10.1016/j.biortech.2016.05.093 (accessed on 15th November 2025).

Cho, C., Hong, S., Moon, H. G., Jang, Y. S., Kim, D., & Lee, S. Y. (2019). Engineering clostridial aldehyde/alcohol dehydrogenase for selective butanol production. *mBio, 10*(1), e02683–18. https://doi.org/10.1128/mBio.02683-18 (accessed on 15th November 2025).

Daengbussadee, C., Laopaiboon, L., & Laopaiboon, P. (2022). Butanol production by a novel efficient method using mixed cultures of *Clostridium beijerinckii* and *Arthrobacter* sp. in stirred-tank and gas-lift bioreactors. *Fermentation, 8*(4), 160. https://doi.org/10.3390/fermentation8040160 (accessed on 15th November 2025).

Dixit, S. K., Nidhi, T., Shukla, A., Upadhyay, S. K., & Verma, P. C. (2021). Molecular characterization of *N*-methyl-*d*-aspartate receptor from *Bemisia tabaci*. *Insect Molecular Biology, 30*(2), 231–240. https://doi.org/10.1111/imb.12702 (accessed on 15th November 2025).

Ferreira, S., Pereira, R., Liu, F., Vilaça, P., & Rocha, I. (2019). Discovery and implementation of a novel pathway for n-butanol production via 2-oxoglutarate. *Biotechnology for Biofuels, 12*, 230. https://doi.org/10.1186/s13068-019-1576-x (accessed on 15th November 2025).

Gaffney, J., Embree, J., Gilmore, S., & Embree, M. (2021). *Ruminococcus bovis* sp. nov., a novel species of amylolytic *Ruminococcus* isolated from the rumen of a dairy cow. *International Journal of Systematic and Evolutionary Microbiology, 71*(8), 004924. https://doi.org/10.1099/ijsem.0.004924.

Guo, Y., Liu, Y., Guan, M., Tang, H., Wang, Z., Lin, L., & Pang, H. (2022). Production of butanol from lignocellulosic biomass: Recent advances, challenges, and prospects. *RSC Advances, 12*(29), 18848–18872. https://doi.org/10.1039/D2RA01955K (accessed on 15th November 2025).

Herman, N. A., Kim, S. J., Li, J. S., Cai, W., Koshino, H., & Zhang, W. (2017). The industrial anaerobe *Clostridium acetobutylicum* uses polyketides to regulate cellular differentiation. *Nature Communications, 8*(1), 1514. https://doi.org/10.1038/s41467-017-01764-7 (accessed on 15th November 2025).

Humphreys, J. R., Debebe, B. J., Diggle, S. P., & Winzer, K. (2023). *Clostridium beijerinckii* strain degeneration is driven by the loss of Spo0A activity. *Frontiers in Microbiology, 13*, 1075609. https://doi.org/10.3389/fmicb.2022.1075609 (accessed on 15th November 2025).

Jang, Y. S., Lee, J. Y., Lee, J., Park, J. H., Im, J. A., Eom, M. H., Lee, J., Lee, S. H., Song, H., Cho, J. H., Lee, S. Y. (2012). Enhanced butanol production obtained by reinforcing the direct butanol-forming route in *Clostridium acetobutylicum. mBio, 3*(5), e00314-12. https://doi.org/10.1128/mBio.00314-12 (accessed on 15th November 2025).

Jeswani, H. K., Chilvers, A., & Azapagic, A. (2020). Environmental sustainability of biofuels: A review. *Proceedings of the Royal Society A: Mathematical, Physical and Engineering Sciences, 476*(2243), 20200351. https://doi.org/10.1098/rspa.2020.0351 (accessed on 15th November 2025).

Jiang, Y., Liu, J., Jiang, W., Yang, Y., & Yang, S. (2015). Current status and prospects of industrial bio-production of n-butanol in China. *Biotechnology Advances, 33*(7), 1493–1501. https://doi.org/10.1016/j.biotechadv.2015.07.004 (accessed on 15th November 2025).

Kataoka, N., Tajima, T., Kato, J., Rachadech, W., & Vangnai, A. S. (2011). Development of butanol-tolerant *Bacillus subtilis* strain GRSW2-B1 as a potential bioproduction host. *AMB Express, 1*, 10. https://doi.org/10.1186/2191-0855-1-10 (accessed on 15th November 2025).

Kehail, A., & Brigham, C. J. (2015). Biofuels production process: Microorganisms utilizing carbon to produce butanol and ethanol. *International Journal of Thermal & Environmental Engineering, 10*(1), 19–24. https://doi.org/10.5383/ijtee.10.01.004 (accessed on 15th November 2025).

Khamaiseh, E. I., Abdul Hamid, A., Abdeshahian, P., Wan Yusoff, W. M., & Kalil, M. S. (2014). Enhanced butanol production by *Clostridium acetobutylicum* NCIMB 13357 grown on date fruit as carbon source in P2 medium. *The Scientific World Journal, 2014*, 395754. https://doi.org/10.1155/2014/395754 (accessed on 15th November 2025).

Khanna, S., & Goyal, A. (2013). Production of n-butanol from biodiesel-derived crude glycerol using *Clostridium pasteurianum* immobilized on amberlite. *Fuel, 112*, 557–561. https://doi.org/10.1016/j.fuel.2013.05.031 (accessed on 15th November 2025).

Kiehbadroudinezhad, M., Merabet, A., Ghenaï, C., Abo-Khalil, A. G., & Salameh, T. (2023). The role of biofuels for sustainable microgrids: A path towards carbon neutrality and the green economy. *Heliyon, 9*(2), e13407. https://doi.org/10.1016/j.heliyon.2023.e13407 (accessed on 15th November 2025).

Kolesinska, B., Fraczyk, J., Binczarski, M., Modelska, M., Berlowska, J., Dziugan, P., Antolak, H., Kaminski, Z. J., Witonska, I. A., & Kregiel, D. (2019). Butanol synthesis routes for biofuel production: Trends and perspectives. *Materials, 12*(3), 350. https://doi.org/10.3390/ma12030350 (accessed on 15th November 2025).

Lan, E. I., & Liao, J. C. (2011). Metabolic engineering of cyanobacteria for 1-butanol production from carbon dioxide. *Metabolic Engineering, 13*(4), 353–363. https://doi.org/10.1016/j.ymben.2011.04.004 (accessed on 15th November 2025).

Lau, N. S., Matsui, M., & Abdullah, A. A. (2015). Cyanobacteria: Photoautotrophic microbial factories for the sustainable synthesis of industrial products. *BioMed Research International, 2015*, 754934. https://doi.org/10.1155/2015/754934 (accessed on 15th November 2025).

Lenzen, M., Sun, Y., Faturay, F., Ting, Y. P., Geschke, A., & Malik, A. (2018). The carbon footprint of global tourism. *Nature Climate Change, 8*(6), 522–528. https://doi.org/10.1038/s41558-018-0141-x (accessed on 15th November 2025).

Li, S., Huang, L., Ke, C., Pang, Z., & Liu, L. (2020). Pathway dissection, regulation, engineering and application: Lessons learned from biobutanol production by solventogenic clostridia. *Biotechnology for Biofuels, 13*(1), 84. https://doi.org/10.1186/s13068-020-01730-w (accessed on 15th November 2025).

Lin, Z., Cong, W., & Zhang, J. (2023). Biobutanol production from acetone–butanol–ethanol fermentation: Developments and prospects. *Fermentation, 9*(9), 847.

Liu, Z., Ying, Y., Li, F., Ma, C., & Xu, P. (2010). Butanol production by *Clostridium beijerinckii* ATCC 55025 from wheat bran. *Journal of Industrial Microbiology and Biotechnology, 37*(5), 495–501.

Luo, Q., Ding, N., Liu, Y., Zhang, H., Fang, Y., & Yin, L. (2023). Metabolic engineering of microorganisms to produce pyruvate and derived compounds. *Molecules, 28*(3), 1418.

Martín, C., Dixit, P., Momayez, F., & Jönsson, L. J. (2022). Hydrothermal pretreatment of lignocellulosic feedstocks to facilitate biochemical conversion. *Frontiers in Bioengineering and Biotechnology, 10*, 846592.

Nandhini, R., Rameshwar, S. S., Sivaprakash, B., Rajamohan, N., & Monisha, R. S. (2023). Carbon neutrality in biobutanol production through microbial fermentation technique from lignocellulosic materials – A biorefinery approach. *Journal of Cleaner Production, 413*, 137470.

Ndaba, B., Chiyanzu, I., & Marx, S. (2015). *N*-butanol derived from biochemical and chemical routes: A review. *Biotechnology, 8*, 1–9.

Obergruber, M., Hönig, V., Procházka, P., Kučerová, V., Kotek, M., Bouček, J., & Mařík, J. (2021). Physicochemical properties of biobutanol as an advanced biofuel. *Materials, 14*(4), 914.

Oliva-Rodríguez, A. G., Quintero, J., Medina-Morales, M. A., Morales-Martínez, T. K., Rodríguez-De la Garza, J. A., Moreno-Dávila, M., Aroca, G., & Rios González, L. J. (2019). *Clostridium* strain selection for co-culture with *Bacillus subtilis* for butanol production from agave hydrolysates. *Bioresource Technology, 275*, 410–415.

Olorunsogbon, T., Adesanya, Y., Atiyeh, H. K., Okonkwo, C. C., Ujor, V. C., & Ezeji, T. C. (2022). Effects of *Clostridium beijerinckii* and medium modifications on acetone-butanol-ethanol production from switchgrass. *Frontiers in Bioengineering and Biotechnology, 10*, 942701.

Outram, V., Lalander, C. A., Lee, J. G. M., Davies, E. T., & Harvey, A. P. (2017). Applied in situ product recovery in ABE fermentation. *Biotechnology Journal, 33*(3), 563–579.

Patakova, P., Branska, B., Sedlar, K., Vasylkivska, M., Jureckova, K., Kolek, J., Koscova, P., & Provaznik, I. (2019). Acidogenesis, solventogenesis, metabolic stress response and life cycle changes in *Clostridium beijerinckii* NRRL B-598 at the transcriptomic level. *Scientific Reports, 9*(1), 1371.

Qiu, M., Shen, W., Yan, X., He, Q., Cai, D., Chen, S., Wei, H., Knoshaug, E. P., Zhang, M., Himmel, M. E., & Yang, S. (2020). Metabolic engineering of *Zymomonas mobilis* for anaerobic isobutanol production. *Biotechnology for Biofuels, 13*, 15.

Quintero-Díaz, J. C., & Mendoza, D. F., & Avignone-Rossa, C. (2023). NADH-based kinetic model for acetone-butanol-ethanol production by *Clostridium*. *Frontiers in Bioengineering and Biotechnology, 11*, 1294355.

Qureshi, N., & Ezeji, T. C. (2008). Butanol, 'a superior biofuel' production from agricultural residues (renewable biomass): Recent progress in technology. *Biofuels, Bioproducts and Biorefining, 2*(4), 319–330.

Rao, A. G., Sathiavelu, A., & Sathiavelu, M. (2016). Genetic engineering in biobutanol production and tolerance. *Brazilian Archives of Biology and Technology, 59*(0).

Russmayer, H., Marx, H., & Sauer, M. (2019). Microbial 2-butanol production with *Lactobacillus diolivorans*. *Biotechnology for Biofuels, 12*, 262.

Saini, M., Wang, Z. W., Chiang, C. J., & Chao, Y. P. (2017). Metabolic engineering of *Escherichia coli* for production of *n*-butanol from crude glycerol. *Biotechnology for Biofuels, 10*, 173.

Schadeweg, V., & Boles, E. (2016). *n*-Butanol production in *Saccharomyces cerevisiae* is limited by the availability of coenzyme A and cytosolic acetyl-CoA. *Biotechnology for Biofuels, 9*, 44.

Sharma, V., Tsai, M. L., Nargotra, P., Chen, C. W., Kuo, C. H., Sun, P. P., & Dong, C. D. (2022). Agro-industrial food waste as a low-cost substrate for sustainable production of industrial enzymes: A critical review. *Catalysts, 12*(11), 1373.

Shen, C. R., & Liao, J. C. (2012). Photosynthetic production of 2-methyl-1-butanol from CO_2 in cyanobacterium *Synechococcus elongatus* PCC7942 and characterization of the native acetohydroxyacid synthase. *Energy & Environmental Science, 5*(11), 9574.

Shi, S., Si, T., Liu, Z., Zhang, H., Ang, E. L., & Zhao, H. (2016). Metabolic engineering of a synergistic pathway for *n*-butanol production in *Saccharomyces cerevisiae*. *Scientific Reports, 6*, 25675.

Singh, V. K., Jha, S., Rana, P., Soni, R., Lalnunpuii, R., Singh, P. K., Sinha, R. P., & Singh, G. (2024). Cyanobacteria as a biocatalyst for sustainable production of biofuels and chemicals. *Energies, 17*(2), 408.

Sreekumar, S., Baer, Z. C., Anbarasan, P., Günbaş, G., Grippo, A., Blanch, H. W., Clark, D. S., & Toste, F. D. (2015). Production of an acetone-butanol-ethanol mixture from *Clostridium acetobutylicum* and its conversion to high-value biofuels. *Nature Protocols, 10*(3), 528–537.

Sun, X. H., Zhang, S., Yang, Z., Chen, Z. L., Yue, S. J., Zhang, S., & Tang, Y. P. (2022). Efficacy and safety of Lianhua Qingwen for patients with COVID-19: A systematic review and meta-analysis. *Chinese Journal of Integrative Medicine, 28*(7), 650–660.

Syu, M. J. (2001). Biological production of 2,3-butanediol. *Applied Microbiology and Biotechnology, 55*(1), 10–18.

Tian, L., Conway, P. M., Cervenka, N. D., Cui, J., Maloney, M., Olson, D. G., & Lynd, L. R. (2019). Metabolic engineering of *Clostridium thermocellum* for *n*-butanol production from cellulose. *Biotechnology for Biofuels, 12*, 186.

Tiwari, A., & Khawas, R. (2021). Food waste and agro by-products: A step towards food sustainability. In *Innovation in the Food Sector Through the Valorization of Food and Agro-Food By-Products*. Elsevier.

Tracy, B. P. (2012). Improving butanol fermentation to enter the advanced biofuel market. *mBio, 3*(6), e00518-12.
Veza, I., Said, M. F. M., & Latiff, Z. A. (2021). Recent advances in butanol production by acetone-butanol-ethanol (ABE) fermentation. *Biomass and Bioenergy, 144*, 105919.
Wen, Z., Ledesma-Amaro, R., Lin, J., Jiang, Y., & Yang, S. (2019). Improved *n*-butanol production from *Clostridium cellulovorans* by integrated metabolic and evolutionary engineering. *Applied and Environmental Microbiology, 85*(7), e02560-18.
Wen, Z., Ledesma-Amaro, R., Lu, M., Jin, M., & Yang, S. (2020). Metabolic engineering of *Clostridium cellulovorans* to improve butanol production by consolidated bioprocessing. *ACS Synthetic Biology, 9*(2), 304–315.
Yao, X., Zhang, Q., Fan, Y., Xu, X., & Liu, Z. (2022). Butanol–isopropanol fermentation with oxygen-tolerant *Clostridium beijerinckii* XH29. *AMB Express, 12*(1), 57.
Zhao, J., Fang, H., & Zhang, D. (2020). Expanding application of CRISPR-Cas9 system in microorganisms. *Synthetic and Systems Biotechnology, 5*(4), 269–276.
Zheng, Y. N., Li, L. Z., Xian, M., Ma, Y. J., Yang, J. M., Xu, X., & He, D. Z. (2009). Problems with the microbial production of butanol. *Journal of Industrial Microbiology and Biotechnology, 36*(9), 1127–1138.
Zhou, Z., Jing, Y., Wei, S., Zhang, Q., Peng, S., An, X., & Li, H. (2023). Enhancement of butanol production in *Clostridium acetobutylicum* SE25 through oxidation–reduction potential regulation and analysis of its metabolic mechanisms. *Fuel, 331*, 125708.
Zoghlami, A., & Paës, G. (2019). Lignocellulosic biomass: Understanding recalcitrance and predicting hydrolysis. *Frontiers in Chemistry, 7*, 874.
Zverlov, V. V., Berezina, O., Velikodvorskaya, G. A., & Schwarz, W. H. (2006). Bacterial acetone and butanol production by industrial fermentation in the Soviet Union: Use of hydrolyzed agricultural waste for biorefinery. *Applied Microbiology and Biotechnology, 71*(5), 587–597.

Index

B

D

E

G

M

N

For Product Safety Concerns and Information please contact our EU representative GPSR@taylorandfrancis.com Taylor & Francis Verlag GmbH, Kaufingerstraße 24, 80331 München, Germany